D1195671

The IMA Volumes in Mathematics and its Applications

Volume 64

Series Editors
Avner Friedman   Willard Miller, Jr.

Institute for Mathematics and its Applications IMA

The **Institute for Mathematics and its Applications** was established by a grant from the National Science Foundation to the University of Minnesota in 1982. The IMA seeks to encourage the development and study of fresh mathematical concepts and questions of concern to the other sciences by bringing together mathematicians and scientists from diverse fields in an atmosphere that will stimulate discussion and collaboration.

The IMA Volumes are intended to involve the broader scientific community in this process.

Avner Friedman, Director
Willard Miller, Jr., Associate Director

* * * * * * * * * *

IMA ANNUAL PROGRAMS

1982–1983	**Statistical and Continuum Approaches to Phase Transition**
1983–1984	**Mathematical Models for the Economics of Decentralized Resource Allocation**
1984–1985	**Continuum Physics and Partial Differential Equations**
1985–1986	**Stochastic Differential Equations and Their Applications**
1986–1987	**Scientific Computation**
1987–1988	**Applied Combinatorics**
1988–1989	**Nonlinear Waves**
1989–1990	**Dynamical Systems and Their Applications**
1990–1991	**Phase Transitions and Free Boundaries**
1991–1992	**Applied Linear Algebra**
1992–1993	**Control Theory and its Applications**
1993–1994	**Emerging Applications of Probability**
1994–1995	**Waves and Scattering**
1995–1996	**Mathematical Methods in Material Science**

IMA SUMMER PROGRAMS

1987	**Robotics**
1988	**Signal Processing**
1989	**Robustness, Diagnostics, Computing and Graphics in Statistics**
1990	**Radar and Sonar (June 18 - June 29)**
	New Directions in Time Series Analysis (July 2 - July 27)
1991	**Semiconductors**
1992	**Environmental Studies: Mathematical, Computational, and Statistical Analysis**
1993	**Modeling, Mesh Generation, and Adaptive Numerical Methods for Partial Differential Equations**
1994	**Molecular Biology**

* * * * * * * * * *

SPRINGER LECTURE NOTES FROM THE IMA:

The Mathematics and Physics of Disordered Media

Editors: Barry Hughes and Barry Ninham
(Lecture Notes in Math., Volume 1035, 1983)

Orienting Polymers

Editor: J.L. Ericksen
(Lecture Notes in Math., Volume 1063, 1984)

New Perspectives in Thermodynamics

Editor: James Serrin
(Springer-Verlag, 1986)

Models of Economic Dynamics

Editor: Hugo Sonnenschein
(Lecture Notes in Econ., Volume 264, 1986)

Joe H. Chow Petar V. Kokotovic
Robert J. Thomas
Editors

Systems and Control Theory For Power Systems

With 140 Illustrations

Springer-Verlag
New York Berlin Heidelberg London Paris
Tokyo Hong Kong Barcelona Budapest

Joe H. Chow
Electrical, Computer and
Systems Engineering Department
Electric Power Engineering Department
Rensselaer Polytechnic Institute
Troy, NY 12180-3590 USA

Petar V. Kokotovic
Department of Electrical and
Computer Engineering
University of California-
Santa Barbara
Santa Barbara, CA 93106 USA

Robert J. Thomas
Electrical Engineering Department
Cornell University
396 Engineering & Theory Center
Building
Ithaca, NY 14853 USA

Series Editors:
Avner Friedman
Willard Miller, Jr.
Institute for Mathematics and its
Applications
University of Minnesota
Minneapolis, MN 55455 USA

Mathematics Subject Classifications (1991): 34A26, 34A47, 34C30, 34D20, 34E15, 93A15, 93B12, 93C10, 93D09

Library of Congress Cataloging-in-Publication Data
Systems and control theory for power systems / Joe H. Chow, Petar V.
Kokotovic, Robert J. Thomas, editors.
p. cm. – (The IMA volumes in mathematics and its
applications ; v. 64)
Based on the proceedings of the Control and Systems Theory in
Power Systems Workshop held at IMA in March, 1993.
Includes bibliographical references
ISBN 0-387-94438-9
1. Electric power systems–Control–Congresses. 2. Control
theory–Congresses. I. Chow, J.H. (Joe H.), 1951- .
II. Kokotovic, Petar V. III. Thomas, Robert J., 1942- .
IV. Series.
TK1005.S927 1995
621.31–dc20 94-44183

Printed on acid-free paper.

Production managed by Hal Henglein; manufacturing supervised by Jacqui Ashri.
Camera-ready copy prepared by the IMA.
Printed and bound by Braun-Brumfield, Ann Arbor, MI.
Printed in the United States of America.

9 8 7 6 5 4 3 2 1

ISBN 0-387-94438-9 Springer-Verlag New York Berlin Heidelberg

The IMA Volumes
in Mathematics and its Applications

Current Volumes:

Volume 1: Homogenization and Effective Moduli of Materials and Media
Editors: Jerry Ericksen, David Kinderlehrer, Robert Kohn, and J.-L. Lions

Volume 2: Oscillation Theory, Computation, and Methods of Compensated Compactness
Editors: Constantine Dafermos, Jerry Ericksen, David Kinderlehrer, and Marshall Slemrod

Volume 3: Metastability and Incompletely Posed Problems
Editors: Stuart Antman, Jerry Ericksen, David Kinderlehrer, and Ingo Muller

Volume 4: Dynamical Problems in Continuum Physics
Editors: Jerry Bona, Constantine Dafermos, Jerry Ericksen, and David Kinderlehrer

Volume 5: Theory and Applications of Liquid Crystals
Editors: Jerry Ericksen and David Kinderlehrer

Volume 6: Amorphous Polymers and Non-Newtonian Fluids
Editors: Constantine Dafermos, Jerry Ericksen, and David Kinderlehrer

Volume 7: Random Media
Editor: George Papanicolaou

Volume 8: Percolation Theory and Ergodic Theory of Infinite Particle Systems
Editor: Harry Kesten

Volume 9: Hydrodynamic Behavior and Interacting Particle Systems
Editor: George Papanicolaou

Volume 10: Stochastic Differential Systems, Stochastic Control Theory, and Applications
Editors: Wendell Fleming and Pierre-Louis Lions

Volume 11: Numerical Simulation in Oil Recovery
Editor: Mary Fanett Wheeler

Volume 12: Computational Fluid Dynamics and Reacting Gas Flows
Editors: Bjorn Engquist, M. Luskin, and Andrew Majda

Volume 13: Numerical Algorithms for Parallel Computer Architectures
Editor: Martin H. Schultz

Volume 14: Mathematical Aspects of Scientific Software
Editor: J.R. Rice

Volume 15: Mathematical Frontiers in Computational Chemical Physics
Editor: D. Truhlar

Volume 16: Mathematics in Industrial Problems
by Avner Friedman

Volume 17: Applications of Combinatorics and Graph Theory to the Biological and Social Sciences
Editor: Fred Roberts

Volume 18: q-Series and Partitions
Editor: Dennis Stanton

Volume 19: Invariant Theory and Tableaux
Editor: Dennis Stanton

Volume 20: Coding Theory and Design Theory Part I: Coding Theory
Editor: Dijen Ray-Chaudhuri

Volume 21: Coding Theory and Design Theory Part II: Design Theory
Editor: Dijen Ray-Chaudhuri

Volume 22: Signal Processing: Part I - Signal Processing Theory
Editors: L. Auslander, F.A. Grünbaum, J.W. Helton, T. Kailath, P. Khargonekar, and S. Mitter

Volume 23: Signal Processing: Part II - Control Theory and Applications of Signal Processing
Editors: L. Auslander, F.A. Grünbaum, J.W. Helton, T. Kailath, P. Khargonekar, and S. Mitter

Volume 24: Mathematics in Industrial Problems, Part 2
by Avner Friedman

Volume 25: Solitons in Physics, Mathematics, and Nonlinear Optics
Editors: Peter J. Olver and David H. Sattinger

Volume 41: Chaotic Processes in the Geological Sciences
Editor: David A. Yuen

Volume 42: Partial Differential Equations with Minimal Smoothness and Applications
Editors: B. Dahlberg, E. Fabes, R. Fefferman, D. Jerison, C. Kenig, and J. Pipher

Volume 43: On the Evolution of Phase Boundaries
Editors: Morton E. Gurtin and Geoffrey B. McFadden

Volume 44: Twist Mappings and Their Applications
Editors: Richard McGehee and Kenneth R. Meyer

Volume 45: New Directions in Time Series Analysis, Part I
Editors: David Brillinger, Peter Caines, John Geweke, Emanuel Parzen, Murray Rosenblatt, and Murad S. Taqqu

Volume 46: New Directions in Time Series Analysis, Part II
Editors: David Brillinger, Peter Caines, John Geweke, Emanuel Parzen, Murray Rosenblatt, and Murad S. Taqqu

Volume 47: Degenerate Diffusions
Editors: Wei-Ming Ni, L.A. Peletier, and J.-L. Vazquez

Volume 48: Linear Algebra, Markov Chains, and Queueing Models
Editors: Carl D. Meyer and Robert J. Plemmons

Volume 49: Mathematics in Industrial Problems, Part 5
by Avner Friedman

Volume 50: Combinatorial and Graph-Theoretic Problems in Linear Algebra
Editors: Richard A. Brualdi, Shmuel Friedland, and Victor Klee

Volume 51: Statistical Thermodynamics and Differential Geometry of Microstructured Materials
Editors: H. Ted Davis and Johannes C.C. Nitsche

Volume 52: Shock Induced Transitions and Phase Structures in General Media
Editors: J.E. Dunn, Roger Fosdick, and Marshall Slemrod

Volume 53: Variational and Free Boundary Problems
Editors: Avner Friedman and Joel Spruck

Volume 54: Microstructure and Phase Transitions
Editors: David Kinderlehrer, Richard James, Mitchell Luskin, and Jerry L. Ericksen

Volume 55: Turbulence in Fluid Flows: A Dynamical Systems Approach
Editors: George R. Sell, Ciprian Foias, and Roger Temam

Volume 56: Graph Theory and Sparse Matrix Computation
Editors: Alan George, John R. Gilbert, and Joseph W.H. Liu

Volume 57: Mathematics in Industrial Problems, Part 6
by Avner Friedman

Volume 58: Semiconductors, Part I
Editors: W.M. Coughran, Jr., Julian Cole, Peter Lloyd, and Jacob White

Volume 59: Semiconductors, Part II
Editors: W.M. Coughran, Jr., Julian Cole, Peter Lloyd, and Jacob White

Volume 60: Recent Advances in Iterative Methods
Editors: Gene Golub, Anne Greenbaum, and Mitchell Luskin

Volume 61: Free Boundaries in Viscous Flows
Editors: Robert A. Brown and Stephen H. Davis

Volume 62: Linear Algebra for Control Theory
Editors: Paul Van Dooren and Bostwick Wyman

Volume 63: Hamiltonian Dynamical Systems: History, Theory, and Applications
Editors: H.S. Dumas, K.R. Meyer, and D.S. Schmidt

Volume 64: Systems and Control Theory for Power Systems
Editors: Joe H. Chow, Petar V. Kokotovic, and Robert J. Thomas

Forthcoming Volumes:

Mathematics in Industrial Problems Part 7

1991-1992: *Applied Linear Algebra*

Linear Algebra for Signal Processing

1992 Summer Program: *Environmental Studies*

1992–1993: *Control Theory*

Robust Control Theory

Control and Optimal Design of Distributed Parameter Systems

Flow Control

Robotics

Nonsmooth Analysis & Geometric Methods in Deterministic Optimal Control

Adaptive Control, Filtering and Signal Processing

Discrete Event Systems, Manufacturing, Systems, and Communication Networks

Mathematical Finance

1993 Summer Program: *Modeling, Mesh Generation, and Adaptive Numerical Methods for Partial Differential Equations*

1993-1994: *Emerging Applications of Probability*

Discrete Probability and Algorithms

Random Discrete Structures

Mathematical Population Genetics

Stochastic Networks

Stochastic Problems for Nonlinear Partial Differential Equations

Image Models (and their Speech Model Cousins)

Stochastic Models in Geosystems

Classical and Modern Branching Processes

FOREWORD

This IMA Volume in Mathematics and its Applications

SYSTEMS AND CONTROL THEORY FOR POWER SYSTEMS

is based on the proceedings of a workshop that was an integral part of the 1992–93 IMA program on "Control Theory." We thank Joe H. Chow, Petar V. Kokotovic, and Robert J. Thomas for organizing the workshop and editing the proceedings. We also take this opportunity to thank the National Science Foundation and the Army Research Office, whose financial support made the workshop possible.

Avner Friedman

Willard Miller, Jr.

PREFACE

Power systems are rich in control and mathematical problems. The presentations given at the Control and Systems Theory in Power Systems Workshop held at IMA in March, 1993, clearly supported that claim. In this volume, we have collected 17 papers from the workshop. For papers with co-authors, the first author was the presenter. These papers deal with several topics of high current interest in power systems: modeling, stability, control, robustness, and computing.

Power system modeling is contained in several papers. Sauer's paper presents a time-scale analysis of load models using transient algebraic circuits. Ahmed-Zaid applies the same time-scale method to obtain reduced models of synchronous and induction machines. Chow's paper contains recent algorithms for identifying slow coherent groups of machines and aggregating the coherent machines. Vittal's paper develops an uncertainty model for analyzing system stability with respect to variations in loads and power transfer.

Most of the papers deal with either transient or voltage stability. Chiang's paper is a comprehensive overview of a direct transient stability method based on the stability region boundary and the controlling unstable equilibrium point. Several of the voltage stability papers are related. Venkatasubramanian introduces the stability regions and the surfaces that separate these regions for differential-algebraic systems. Hill's paper treats a differential-algebraic power system model with static and dynamic loads and analyzes voltage stability using Lyapunov functions. Sauer's paper also contains an analysis of the effect of load models on dynamic voltage stability. Liu discusses the mechanisms of tap-changers, load models, and generator excitation limits on the voltage collapse. Abed's paper contains a summary of several types of bifurcations observed in voltage stability. Pai's paper concerns the structural stability of power systems and whether small changes in parameter values would alter the nature of the trajectory behavior, and illustrates the results on dynamic and voltage stability. Dobson shows that a circuit Jacobian can be used to study damping and resonance effects in nonlinear high power switching circuits.

The control and robustness papers include the following. Kwatny presents results on the use of variable structure control for feedback linearizable systems and applies the design to obtain a nonlinear drum level control. Abed's paper also contains some ideas for controlling nonlinear systems operating close to bifurcation. Thorp reviews the applications of phasor measurement for real-time power system control, estimation, and direct stability prediction. DeMarco examines the determination of the stability of power system models in an operating region as a matrix polytope stability problem. Vittal's paper also proposes to use an L^∞ norm to provide an accurate assessment of the stability robustness of power system controls.

The final group of papers focuses on computing. Siljak presents results on applying overlapping decomposition in a parallel computing environment to large power systems. Huang's paper establishes the theoretical foundation of the convergence of a parallel textured algorithm for the active power economic dispatch problem and provides simulation results in parallel computing. DeMarco's paper also provides strong evidence that power system robust stability is unlikely to admit polynomial time solution algorithms. Talukdar proposes the super-agent organization aiming at designing parallel and distributed computer algorithms for solving large, difficult optimization problems.

The workshop was funded by IMA and the National Science Foundation. Despite the "Blizzard of 93" which delayed the travel plans of several workshop participants and required the shuffling of the program, the workshop achieved its objective of a fruitful interchange of ideas with lively discussions after every presentation. We thank the authors for their excellent contributions, and also thank James Alexander, Kevin Clements, Mark Damborg, Robert Fischl, Peter Hirsch, and Chika Nwankpa for active participation in the workshop program. We also thank Hector Sussmann for his help in the early stages of organizing the workshop and for his workshop participation. We also thank Avner Friedman and Willard Miller, Jr., for their support and their hospitality during the workshop period. The IMA staff deserves a lot of credit for their excellent local arrangement. Finally, we thank Patricia V. Brick, Stephan J. Skogerboe, and Kaye Smith for their assistance in the production of this volume.

Joe H. Chow
Petar V. Kokotovic
Robert J. Thomas

CONTENTS

BIFURCATION-THEORETIC ISSUES IN THE CONTROL OF VOLTAGE COLLAPSE

EYAD H. ABED*

Abstract. Power system instabilities, including voltage collapse phenomena, can involve an array of nonlinear dynamic phenomena. Bifurcations and chaos have recently been observed and studied in mathematical models of the dynamics of voltage collapse. The purpose of this paper is to consider some unconventional control problems that are motivated by these studies. Among these are the stabilization of bifurcated solutions and the introduction of certain degenerate bifurcations near existing saddle node bifurcations. Other ideas involving consideration of chaotic solutions are also presented. The use of bifurcation theory in the control of systems bordering on divergence instability ('collapse') is discussed.

1. Introduction. Voltage collapse was identified and analyzed for simple power systems in the first half of this century (see, for instance, the texts [16],[39],[75]). However, only recently has voltage collapse become a serious operating concern [24]. This is a result of the increasing stress being placed on today's complex power systems. Voltage instabilities leading to power disruptions have occurred in systems throughout the world. The power systems research community has responded by focusing significant effort toward the study of voltage instabilities and their control [22],[23],[35],[36],[52],[44], [66],[74]. It is interesting to observe that due to the importance and seeming permanence of the problems being considered, the impact of the research is being felt not only in the power systems engineering community, but also in the engineering community at-large (e.g., [37],[19]), in the popular scientific literature (e.g., [38]), and in university curricula and professional reference books (e.g., [29],[45],[64]).

One important outcome of the studies undertaken thus far is the general consensus that bifurcations in underlying mathematical models of a power system are closely linked with voltage collapse. This is especially true for voltage collapse precipitated by the slow variation in a system parameter such as load. Indeed, bifurcations may even be used to predict conditions under which collapse occurs. (A *bifurcation* is a qualitative change in the phase portrait of a dynamical system that occurs as a system *bifurcation parameter* is quasistatically varied.) On the technological side, sophisticated new devices for the control of voltage and reactive power have been developed and some are being field-tested. These components are an example of the overall effort in Flexible AC Transmission Systems (FACTS)

* Department of Electrical Engineering and the Institute for Systems Research, University of Maryland, College Park, Maryland 20742 USA. This research has been supported in part by the Electric Power Research Institute under contract RP8050-5, in part by the AFOSR under Grant F49620-93-1-0186, and in part by the NSF Engineering Research Centers Program: NSFD CDR-88-03012. The author is grateful to R.A. Adomaitis, J.C. Alexander, M. Golubitsky, A.M.A. Hamdan, D.-C. Liaw, A.N. Shoshitaishvili, M. Varghese and H.O. Wang for useful discussions.

(e.g., [37],[32],[21],[41]). Among the devices being developed, we mention static VAR compensators (SVCs) [33],[55],[29]; static compensators capable of combined active and reactive power control [26]; under-load tap changer (ULTC) blocking devices; and superconducting magnetic energy storage devices. Coordinated use of these devices along with the more traditional voltage control devices to minimize voltage fluctuations and increase line loadability in a complex power system presents a major research challenge [41].

In this paper, we discuss certain generic, albeit unconventional, feedback control problems which may prove to be of relevance in the context of voltage collapse. These problems are formulated for systems assumed to display bifurcation behavior. In particular, the control of local bifurcations from equilibrium points is addressed. The control problems discussed include some which have been considered previously and others given here for the first time. Only rough descriptions of the newly stated problems are given, details being left to future investigations. The discussion addresses control problems in a general setting: no specific power system model is assumed. However, specific models for simple systems, control device dynamics, etc., are given in some of the references.

Suppose a dynamical system depends on several parameters. Select one of these parameters as the distinguished bifurcation parameter. Suppose that for a nominal parameter range, the system operates at a stable equilibrium point. As the bifurcation parameter is varied outside this range, the fixed point can lose stability. This results in a bifurcation, in which new limit sets (i.e., steady state motions) arise. These new limit sets are known as bifurcated solutions. Typical bifurcated solutions include fixed points and periodic solutions in the vicinity of the nominal fixed point. The bifurcation can entail the disappearance of the nominal fixed point, through a *saddle node bifurcation* (also known as a *fold bifurcation*).

The saddle node bifurcation was hypothesized by some authors as the 'cause' of voltage collapse [62],[46],[18],[7]. Later work demonstrated the possible role of other bifurcations in power system voltage instability. For instance, the occurrence and implications of Andronov-Hopf bifurcations and/or routes to chaotic behavior in power system models exhibiting voltage collapse were considered in [3],[1],[72],[7], [12],[63],[69]. (The Andronov-Hopf bifurcation will be described below.) A new bifurcation mechanism, the *singularity-induced bifurcation*, was introduced and applied to power system voltage stability analysis in [69],[68]. This latter development illustrates that power system dynamics is a rich subject which can motivate the development of new tools in applied mathematics.

Control of bifurcations and chaos is a subject which is currently being researched actively by scientists and engineers from many disciplines. Applications are a driving force behind this research. In particular, a main motivation for the study of control of bifurcation and chaos relates to a (nonlinear) performance vs. stability trade-off that appears in a variety of forms in various applications. It is often the case that significant improvement in performance is achieved by operation near a system's stability boundary. As noted above, this operation can, in the presence of small disturbances, lead to bifurcations. Achieving increases in performance for such systems while maintaining an acceptable safety margin is an important current engineering challenge. An essential aspect of this challenge is the design of controllers which facilitate operation of systems in regimes characterized by a negligible margin of stability. It is important to note that linearized models are not adequate for prediction or control of a system's response near the stability boundary.

Power systems provide a rich source of problems in the control of systems exhibiting bifurcation and chaos. A performance vs. stability tradeoff occurring in power system operation results from the attempt to maintain system operation at high loading levels, i.e., under stressed conditions. Voltage collapse tends to occur near the theoretical maximum loading level. Note, however, that this maximum loading level does not necessarily correspond to a bifurcation point. This fact is sometimes overlooked, although it was pointed out as early as 1938 that the power limit of a system does not necessarily coincide with its stability limit (Dahl [16], p. 264). In more modern language, it is observed [50] that power-voltage curves are not necessarily bifurcation diagrams.

The remainder of the paper proceeds as follows. In Section 2, we recall some basic bifurcation terminology and theorems. The discussion focuses on one-parameter families of nonlinear systems. Discussed are local generic bifurcations, local degenerate bifurcations, and some global bifurcations. In Section 3, we discuss problems in the control of nonlinear systems exhibiting bifurcation. The problems, some of which are open and presented here for the first time, are chosen for their potential relevance to the control of voltage stability. Conclusions are collected in Section 4.

2. Bifurcations. In this section we briefly discuss several local and global bifurcations that have been observed in analytical and numerical studies of voltage collapse. Since an electric power system normally functions at a stable operating *point*, it is this condition (an equilibrium) which, upon slow variation of a system parameter, yields, after one or many bifurcations, the decisive bifurcation at which collapse occurs. This bifurcation is distinguished by the 'local' vanishing of any stable limit set of the system model (2.1).

The discussion below is in the context of general nonlinear systems rather than for a specific power system model or class of models. For several of the bifurcation phenomena, we provide explicit theorem statements. The bifurcations are discussed either in the setting of one-parameter families of autonomous systems of ordinary differential equations

$$\dot{x}(t) = f_\mu(x(t)), \tag{2.1}$$

or one-parameter families of autonomous systems of ordinary difference equations

$$x_{k+1} = F_\mu(x_k). \tag{2.2}$$

In either description (2.1) or (2.2), the state $x \in \mathbb{R}^n$. Time is continuous in (2.1) and integer-valued in (2.2). In both descriptions, $\mu \in \mathbb{R}$ is the bifurcation parameter. The mappings f_μ and F_μ are sufficiently smooth in x and μ. The difference equation setting (2.2) is especially useful in the study of period doubling bifurcation.

Let us begin by focusing on (2.1). Suppose that (2.1) possesses an equilibrium point $x_0(\mu)$ for a range of parameter values of interest. We assume that this is an asymptotically stable equilibrium for a large portion of this range. Thus, the equilibrium can be viewed as a possible operating condition for the physical system (say, a power system) modeled by (2.1). When the power system operates in a highly stressed environment, it is possible for the equilibrium $x_0(\mu)$ to lose stability for some parameter value μ_c. At such a loss of stability, the nonlinear system (2.1) typically undergoes a local bifurcation. *Local bifurcations* for Eq. (2.1) (resp. (2.2)) are bifurcations from an equilibrium point of (2.1) (resp. (2.2)). Among the possible local bifurcations for (2.1), only two are *generic*, i.e., preserved under small perturbations of the right side of (2.1). These are the saddle node bifurcation and the Andronov-Hopf bifurcation. The former bifurcation involves the coalescence and disappearance of two neighboring equilibrium points as a parameter is varied, while the latter involves the emergence of a small-amplitude periodic orbit from an equilibrium point as a parameter is varied.

Before proceeding to state the basic theorems on saddle node bifurcation and Andronov-Hopf bifurcation, we note that limit sets emerging from these bifurcations often undergo further bifurcations. The first bifurcation, which occurs along the nominal solution branch, is termed a primary bifurcation. A bifurcation from the solution emanating from the primary bifurcation is termed a secondary bifurcation. Successive bifurcations can become rather complex, and can lead to chaotic (or turbulent) behavior. Chaos is an irregular, seemingly random dynamic behavior displaying extreme sensitivity to initial conditions. Nearby initial conditions result, at least initially, in trajectories that diverge exponentially fast.

2.1. Saddle node bifurcation. Figure 2.1 depicts the *bifurcation diagram* for a saddle node (or fold) bifurcation of Eq. (2.1), as well as the dynamics associated with this bifurcation. The amplitudes of two equilibrium solutions are plotted in the diagram against the bifurcation parameter μ. A dark curve indicates an asymptotically stable equilibrium point (node), while a thin curve indicates an unstable equilibrium point (saddle). The arrows indicate the dynamics of the system for various parameter values.

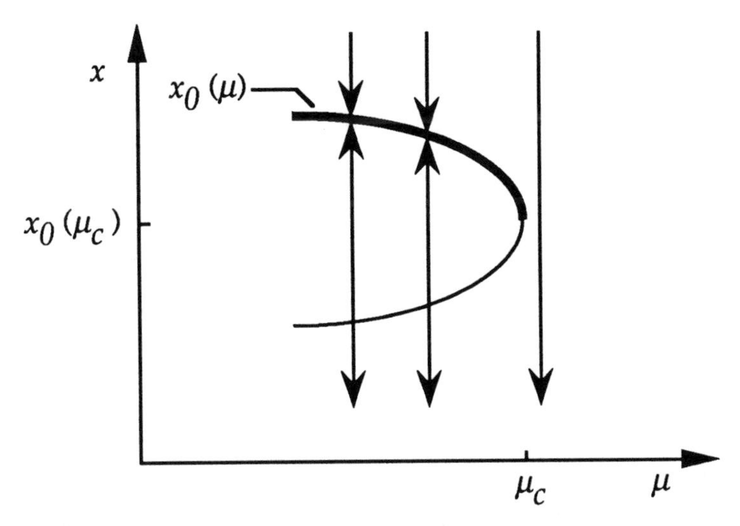

FIG. 2.1. *Saddle node bifurcation*

To state a theorem on saddle node bifurcation, we consider the system (2.1) where f is sufficiently smooth and $f_0(0) = 0$. Express $f_\mu(x)$ in a Taylor series about $x = 0, \mu = 0$ as follows:

$$f_\mu(x) = Ax + g\mu + Q(x,x) + \cdots \tag{2.3}$$

Here, $A = D_x f_0(0)$ is simply the Jacobian matrix of f_μ at the origin for $\mu = 0$, $g = D_\mu f_0(0)$, $Q(x,x)$ includes all quadratic terms in x in the Taylor expansion of $f_0(x)$ near $x = 0$, and the dots denote terms of still higher order. The quadratic form $Q(x,x)$ is written such that it is generated by a symmetric bilinear form $Q(x,y)$. The following two hypotheses will be invoked.

(SN1) The Jacobian A is of rank $n-1$. Thus, A possesses a zero eigenvalue to which there correspond unique right and left eigenvectors.

With this hypothesis in force, denote by r (resp. l) the right column (resp. left row) eigenvector of the critical Jacobian A corresponding to $\lambda(0) = 0$. Normalize r and l by setting the first component of r to 1 and then chosing l so that $lr = 1$. (This may require interchanging the first state variable with another state variable.)

(SN2) $lg \neq 0$ and $lQ(r,r) \neq 0$.

Under these hypotheses, Eq. (2.1) undergoes a saddle node bifurcation from the origin at $\mu = 0$. The precise statement is given in the following theorem. Note that usually this result is stated for a one-dimensional reduced system model, whereas we give a statement which applies directly to a general n-dimensional system [43].

THEOREM 2.1. *(Saddle Node Bifurcation Theorem) With the notation above, if (SN1) and (SN2) hold, then there is an* $\epsilon_0 > 0$ *and a function*

$$\mu(\epsilon) = \mu_2\epsilon^2 + O(\epsilon^3) \tag{2.4}$$

such that $\mu_2 \neq 0$ *and for each* $\epsilon \in [-\epsilon_0, \epsilon_0]$, *(2.1) possesses an equilibrium* $x_0(\epsilon)$ *near* 0 *for* $\mu = \mu(\epsilon)$. *These are the only equilibrium points of (2.1) near the origin for* μ *near* 0. *Thus, (2.1) undergoes a saddle node (or fold) bifurcation from the origin at* $\mu = 0$. *Specifically, for each* μ *sufficiently small and positive (if* $\mu_2 > 0$*), or negative (if* $\mu_2 < 0$*), there are exactly two equilibrium points near the origin, one occurring for some* $\epsilon > 0$ *and another occurring for some* $\epsilon < 0$. *Moreover, the origin is unstable at the critical parameter value* $\mu = 0$.

The fold bifurcation has been linked to voltage collapse in [62],[46],[18], [7]. An important feature of the saddle node bifurcation is the *disappearance, locally, of any stable bounded solution of the system* (2.1). In the next subsection, we discuss the Andronov-Hopf bifurcation, which can give rise to this same feature.

2.2. Andronov-Hopf bifurcation. Suppose that the instability of the nominal equilibrium $x_0(\mu)$ is the result of a pair of eigenvalues of the system linearization crossing the imaginary axis transversely in the complex plane. Then, as is well known, generically it follows that a small amplitude periodic orbit of (2.1) emerges from the equilibrium $x_0(\mu)$. More accurately, the following hypotheses are invoked in the Andronov-Hopf Bifurcation Theorem, detailed statements of which can be found in many references (e.g., [4],[8],[14],[30],[31],[34], [40],[42],[53]). The essence of this bifurcation theorem was originally stated by Poincaré and used in his study of lunar orbital dynamics ([58], Secs. 51-52).

(AH1) f of system (2.1) is sufficiently smooth in x, μ, and $f_0(0) = 0$. The Jacobian $D_x f_\mu(0)$ possesses a complex-conjugate pair of (algebraically) simple eigenvalues $\lambda(\mu) = \alpha(\mu) + i\omega(\mu)$, $\overline{\lambda(\mu)}$, such that $\alpha(0) = 0$, $\alpha'(0) \neq 0$ and $\omega_c := \omega(0) > 0$.

(AH2) $\pm i\omega_c$ are the only pure imaginary eigenvalues of the critical Jacobian $D_x f_0(0)$.

The Andronov-Hopf Bifurcation Theorem asserts that, under (AH1) and (AH2), a small-amplitude nonconstant periodic orbit of Eq. (2.1) emerges from the origin at $\mu = \mu_c$. Moreover, locally it is generically true that the bifurcated periodic orbit is either orbitally asymptotically stable or is unstable. In the former case the bifurcation is said to be supercritical, while in the latter case it is termed subcritical.

Figure 2.2 depicts bifurcation diagrams for both situations. In this figure, a solid circle indicates a stable periodic orbit, while an open circle indicates an unstable periodic orbit. As in Figure 2.1, a dark line represents a stable equilibrium branch, and a thin line represents an unstable equilibrium branch. Note the general property that supercritical solutions are stable while subcritical solutions are unstable.

The Andronov-Hopf Bifurcation Theorem also involves a stability computation. In this computation, one calculates a coefficient of a Taylor expansion of the Floquet exponent determining stability of the periodic orbits. This 'bifurcation stability coefficient' is sometimes denoted β_2 [34],[40]. If $\beta_2 < 0$ then the bifurcation is supercritical, while if $\beta_2 > 0$ the bifurcation is subcritical. Stability formula derivations can be found in several references (e.g., [34],[53],[40],[25]).

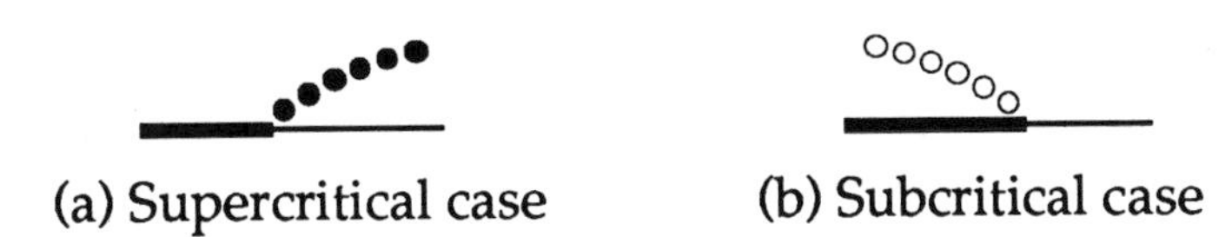

FIG. 2.2. *Andronov-Hopf bifurcation*

2.3. Degenerate Andronov-Hopf bifurcation. The hypotheses (AH1)-(AH2) above under which Andronov-Hopf bifurcation occurs are robust to small perturbations in the model (2.1). If one or more ingredients of these hypotheses fails, or if the stability coefficient β_2 vanishes, one says that a degenerate situation prevails. In our discussion of new control problems in Section 4 we shall employ degenerate Andronov-Hopf bifurcations. We therefore discuss them briefly at this juncture.

Golubitsky and Scaeffer [30] discuss the degenerate Andronov-Hopf bifurcations of low 'codimension' (i.e., the simplest ones) which arise upon the failure of either the eigenvalue crossing condition $\alpha'(0) \neq 0$ or the condition $\beta_2 \neq 0$. Here our interest is mainly in the degenerate bifurcations which result from failure of the eigenvalue crossing (transversality) condition $\alpha'(0) \neq 0$. The simplest of these bifurcations are illustrated (using bifurcation diagrams) in Figure 2.3(a) and Figure 2.3(b). Note that in Figure 2.3(a) and (b) the nominal equilibrium is stable for all parameter values in a neighborhood of the critical value. In Figure 2.3(a), no periodic orbits bifurcate from the equilibrium at criticality. However, in Figure 2.3(b), two branches of unstable periodic orbits bifurcate from the equilibrium, one supercritical and one subcritical. The bifurcation diagrams in Figure 2.3(c),(d) depict two unfoldings of the degenerate bifurcation in Figure 2.3(b). That is, the bifurcation diagram in Figure 2.3(b) is not robust to small perturbations in the model (2.1), and this bifurcation diagram is replaced by those in Figure 2.3(c) or (d) for small perturbations in the model. We could similarly display unfoldings of the bifurcation diagram in Figure 2.3(a). However these will not be employed in this paper and are hence not given here. See [30] for details.

2.4. Period doubling bifurcation. Next we give a basic period doubling bifurcation result for systems (2.2) described by an n-dimensional map. Note that several authors [7],[1],[12] have detected period doubling

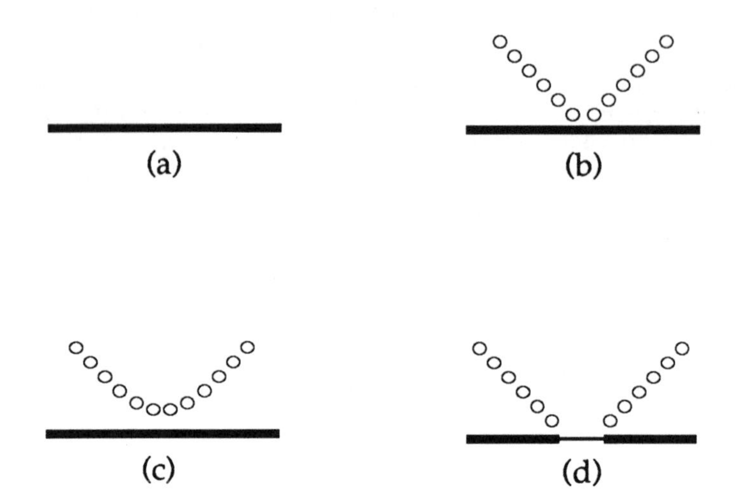

FIG. 2.3. *(a)-(b) Two degenerate Andronov-Hopf bifurcations. (c)-(d) Two unfoldings of the bifurcation diagram of (a)*

bifurcation in an example power system model of Dobson and Chiang [18].

Period doubling bifurcations are most readily analyzed in a discrete-time setting. In discretetime, the nominal periodic orbit (fixed point) is given whereas in the continuous-time setting it is a waveform that must usually be approximated. Of course, to obtain a discrete-time model from a continuous-time model, a device such as the Poincaré return map must be used, and this also involves approximation.

The next hypothesis is invoked in the result below on period doubling bifurcation.

(PD) The map F_μ of Eq. (2.2) is sufficiently smooth in x, μ and has a fixed point at $x = 0$ for $\mu = 0$. The linearization of (6) along the fixed point which is the continuous extension of the origin possesses an eigenvalue $\lambda_1(\mu)$ with $\lambda_1(0) = -1$ and $\lambda_1'(0) \neq 0$. All remaining eigenvalues of the linearization at $\mu = 0$ have magnitude less than unity.

For brevity we do not include the stability computation determining supercriticality or subcriticality for the period doubling bifurcation predicted in the theorem below. A bifurcation stability coefficient β_2 may, however, be evaluated in analogy to the case for Andronov-Hopf bifurcation [6]. The period doubled orbit is supercritical if $\beta_2 < 0$ but is subcritical if $\beta_2 > 0$. It is reassuring to note that specialization of the formula for β_2 in [6] to the case in which F_μ is a scalar map agrees with Theorem 3.5.1 in Guckenheimer and Holmes [31]. In fact, for scalar maps $\beta_2 = -2a$ where a is as given in ([31], p. 158).

THEOREM 2.2. *(Period Doubling Bifurcation Theorem) If (PD) holds, then a period doubled orbit bifurcates from the origin of (2.2) at* $\mu = 0$. *The period doubled orbit is supercritical and stable if* $\beta_2 < 0$ *but is subcritical and unstable if* $\beta_2 > 0$, *where* β_2 *is as noted in the preceding paragraph.*

2.5. Global bifurcation. The bifurcations considered above can all be viewed as local bifurcations. That is, they are bifurcations of new limit sets from nominal equilibrium points or fixed points. When viewed in a continuous-time setting, the period doubling bifurcation is a global, rather than local, bifurcation since it then involves considerations over a large domain in state space (a neighborhood of the nominal periodic solution). Other global bifurcations also arise in the analysis of power system voltage collapse. Periodic solutions can collide with saddle equilibria in a *saddle-connection bifurcation*. Two periodic solutions can merge and disappear in a saddle node bifurcation of periodic orbits, or *cyclic fold bifurcation*. A strange attractor can collide with a saddle equilibrium or periodic orbit and disappear in a *boundary crisis*. Examples of these various global bifurcations in electric power system models along with implications for voltage collapse are given in [72],[69],[63]. References on global bifurcations and chaos include Thompson and Stewart [65], Guckenheimer and Holmes [31], Ott [56] and Wiggins [73].

3. Bifurcation control problems. Bifurcation control deals with design of control inputs to modify the bifurcation characteristics of a parametrized system. The control signal can take many forms, including open-loop control or static or dynamic feedback. The objective of control can be stabilization and/or delay of a given bifurcation, reduction of the amplitude of bifurcated solutions, introduction of bifurcations at desired parameter values, optimization of a performance index near bifurcation, re-shaping of a bifurcation diagram, or a combination of these. A closely related subject is the control of qualitative behavior, as developed by Shoshitaishvili [60].

As noted above, the role of bifurcation phenomena in voltage collapse has been established in several references. It is to be expected that bifurcation-theoretic concepts will have some bearing on the control of voltage collapse. In this section, we discuss problems in the control of nonlinear systems exhibiting bifurcation. We begin by briefly describing results of [2]. Some of the problems discussed below are open and presented here for the first time.

3.1. Stabilization of the Andronov-Hopf bifurcation. Consider a one-parameter family of nonlinear control systems

$$\dot{x} = f_\mu(x, u), \tag{3.1}$$

where $x \in \mathbb{R}^n$ is the state vector, u is the scalar control, $\mu \in \mathbb{R}$ is the bifurcation parameter, and the vector function f is smooth in x, u and μ. Suppose that Eq. (3.1) with the control set to 0 undergoes either a subcritical Andronov-Hopf bifurcation from a nominal equilibrium point

$x_0(\mu)$ at the critical parameter value $\mu = \mu_c$. For definiteness, let the conjugate pair of eigenvalues crossing the imaginary axis at $\mu = \mu_c$ traverse the axis from the left half of the complex plane to the right half as μ *increases*. Then the subcritical bifurcation gives rise to an unstable periodic solution of small amplitude in the vicinity of the equilibrium point for values of the parameter μ less than μ_c. This means that for μ slightly greater than μ_c, system trajectories beginning near the (unstable) nominal equilibrium point will tend to diverge away from this point. If the control u can be designed so as to render the bifurcation supercritical, then this situation will be remedied to a certain extent. That is, the same system trajectories will now converge to a small amplitude periodic solution. Since this periodic solution is *near the nominal equilibrium point*, divergence of the system to a distant (and unacceptable) operating mode is thus prevented.

Feedback control laws rendering supercritical an initially subcritical Andronov-Hopf bifurcation are derived in [2]. Such control laws are said to 'stabilize' the bifurcation. These control laws were taken to be smooth and of the general form $u = u(x)$, i.e., in the form of *static state feedbacks*. In fact, in [2] the feedback $u(x)$ was not allowed to include constant or linear terms in its Taylor series expansion. A constant term would physically represent a continuous expenditure of control energy. The absence of a linear term in the control reduces the complexity of the calculations, and facilitates treatment of the bifurcated solution stabilization problem separately from that of delaying the occurrence of the bifurcation to higher parameter values. Thus, this choice of structure of the control law reflects a two-stage control design philosophy in which linear terms in the control are used to *relocate* a bifurcation in parameter space and nonlinear terms are used to modify its stability.

In [2], stability of the bifurcated solutions was measured using the bifurcation stability coefficient β_2 discussed earlier. The projection method of bifurcation analysis [42] was employed.

Application of local bifurcation control results to voltage dynamics in electric power systems was reported in [47] and [71]. References to applications in other areas are given in [5].

3.2. Dynamic feedback in bifurcation control. Use of a static state feedback control law $u = u(x)$ has potential disadvantages in nonlinear control, especially for systems with multiple equilibria. In general, a static state feedback

$$u = u(x - x_0(\mu)), \tag{3.2}$$

designed with reference to the nominal equilibrium path $x_0(\mu)$ of (3.1) will affect not only the stability of this equilibrium but also the location and stability of other equilibria. Indeed, even a linear feedback control law can introduce extraneous equilibrium points in the system [11]. Now suppose that (3.1) is only an approximate model for the physical system of interest.

Then the nominal equilibrium branch will also be altered by the feedback. A main disadvantage of such an effect is the wasted control energy that is associated with the forced alteration of the system equilibrium structure. Another disadvantage is that system performance is often degraded by operating at an equilibrium which differs from the one at which the system is designed to operate.

For these reasons, we have developed [49] bifurcation control laws for systems (3.1) which are *dynamic state feedback* control laws of a special form. Specifically, we have incorporated high pass filters known as washout filters into the structure of the allowed controllers. In this way, we guarantee preservation of all system equilibria even under model uncertainty. Note that washout filters are common in power system stabilizers and aircraft control systems.

A washout filter is a stable high pass filter with transfer function

$$G(s) = \frac{y(s)}{x(s)} = \frac{s}{(s+d)}. \tag{3.3}$$

In the following, washout filters are incorporated into bifurcation control laws for (3.1). Specifically, in (3.1), for each system state variable x_i, $i = 1, \ldots, n$, introduce a washout filter governed by the dynamic equation

$$\dot{z}_i = x_i - d_i z_i \tag{3.4}$$

along with output equation

$$y_i = x_i - d_i z_i. \tag{3.5}$$

Here, the d_i are positive parameters (this corresponds to using stable washout filters). Finally, we require that the control u depend only on the measured variables y, and that $u(y)$ satisfy $u(0) = 0$.

In this formulation, n washout filters, one for each system state, are present. In fact, the actual number of washout filters needed, and hence also the resulting increase in system order, can usually be taken less than n.

The advantages of using washout filters stem from the resulting properties of equilibrium preservation and automatic equilibrium (operating point) following. Indeed, since $u(0) = 0$, it is clear that y vanishes at steady state. Hence the x subvector of a closed loop equilibrium point (x, z) agrees exactly with the open loop equilibrium value of x. Also, since

$$y_i = x_i - d_i z_i = (x_i - x_{0_i}(\mu)) - d_i(z_i - z_{0_i}(\mu)), \tag{3.6}$$

the control function $u = u(y)$ is guaranteed to center at the correct operating point.

3.3. Stabilization of the period doubling bifurcation. In this section, we summarize results in [6] on stabilization of period doubling bifurcations for discrete-time nonlinear control systems

$$x_{k+1} = f_\mu(x_k, u_k). \tag{3.7}$$

Here k is an integer, $x_k \in \mathbb{R}^n$ is the state, u_k is a scalar control input, $\mu \in \mathbb{R}$ is the bifurcation parameter, and the f is sufficiently smooth in x, u and μ. The continuous-time setting is considered in [27]. As stated in Theorem 2, a period doubling bifurcation may be subcritical or supercritical. The problem of rendering supercritical an existing subcritical period doubling bifurcation is analogous to the problem of stabilizing an Andronov-Hopf bifurcation. The motivation for this problem is also analogous to that in the case of Andronov-Hopf bifurcation.

Suppose the zero-input version of (3.7) satisfies hypothesis (PD). This leads to a period doubling bifurcation for the zero-input version of (3.7). As was the case for Andronov-Hopf bifurcation, take the control u to be of the form $u = u(x_k)$ (a smooth static state feedback), with the constant and linear terms in the Taylor expansion of u set to zero.

The following theorem summarizes stabilization results for period doubling bifurcations which are given in detail in [6]. These results have also been extended to incorporate washout filters in the control laws. Discrete-time versions of washout filters and washout filter-aided bifurcation control laws are also discussed in [6] and in other work in preparation.

THEOREM 3.1. *Suppose that hypothesis (PD) holds for the zero-input version of (3.7). If the critical eigenvalue -1 is controllable for the associated linearized system, then there is a feedback $u_k(x_k)$, containing only third order terms in the components of x_k, that results in a locally stable bifurcated period-2 orbit for μ near 0. This feedback also stabilizes the origin for $\mu = 0$. If, on the other hand, the critical eigenvalue -1 is uncontrollable for the associated linearized system, then generically there is a feedback $u_k(x_k)$, containing only second order terms in the components of x_k, that results in a locally stable bifurcated period-2 orbit for μ near 0. This feedback also stabilizes the origin of (3.7) for $\mu = 0$.*

3.4. Control of routes to chaos. The bifurcation control techniques discussed in the foregoing have direct relevance for issues of control of chaotic behavior of dynamical systems.

There are many scenarios by which bifurcations can result in a chaotic invariant set. The current state of understanding differs considerably among the various known routes to chaos. These include the period doubling route, the Ruelle-Takens route, homoclinic bifurcation, the period adding route and intermittency [56]. What is important about these scenarios from a control of chaos perspective is that chaos may be suppressed by controlling a bifurcation in a given route to chaos. The bifurcation control approach to this problem entails designing feedback control laws which ensure a sufficient degree of stability for a primary bifurcation in such a scenario. We have successfully addressed the homoclinic and period doubling routes to chaos using this approach [6],[70].

3.5. Control of distance to bifurcation. The prevailing approach to real-time control of voltage collapse (as opposed to system planning) consists of security monitoring coupled with algorithms for commitment of resources for minimizing risk of collapse (e.g., [9],[48],[67]). A related bifurcation-theoretic problem is that of maintaining a maximum distance (in parameter space) to the set of parameter values at which bifurcation from the nominal operating point occurs. The development of voltage stability indices is essential to studies of this kind (e.g., [17],[13],[20],[51]).

Most voltage stability indices measure the distance in parameter space to conditions in which the system linearization becomes singular. This is connected to an assumption that voltage instability arises through static bifurcation, the main example of which is the saddle node bifurcation.

However, it may happen that the nominal equilibrium actually loses stability *before* the fold bifurcation [3],[1],[72], [7], [12],[63],[69]. This can be made precise as follows. Consider a nonlinear control system (3.1) depending on a scalar parameter μ. Recall that x denotes the finite-dimensional state vector of the system, and u denotes the control input. Suppose that, for a range of parameter values, say $\mu \leq \mu_c$, the open loop system has a nominal equilibrium point $x = x_0(\mu)$ which represents the desired operating point. At $\mu = \mu_c$, the equilibrium $x_0(\mu)$ merges with another equilibrium in a saddle node bifurcation, and no longer appears for $\mu > \mu_c$. Now suppose that the operating point $x_0(\mu)$ loses stability at $\mu = \mu_1 < \mu_c$. The situation is sketched in Figure 3.1. This figure is reminiscent of a power-voltage curve, with x being the voltage magnitude and μ the power (real or reactive) delivered to a load. It is clear that a control which maximizes the distance to instability (i.e., maintains μ far from μ_1) is preferable to one which simply maintains μ far from μ_c.

It is often also possible to use the available control means to move bifurcations in parameter space. This is especially useful for dynamic (rather than static) bifurcations, such as Andronov-Hopf and period doubling bifurcations. Controls can be designed which relocate these bifurcations without modifying the system's equilibrium structure, as briefly discussed in Section 3.1. In terms of the situation depicted in Figure 3.1, this could correspond to using feedback to increase the critical parameter value μ_1. Having achieved this delay in the occurrence of the Andronov-Hopf bifurcation, the closed loop system will be able to deliver a greater load without a reduction in the stability margin. An example of a controller which delays Andronov-Hopf bifurcations in a model of thermal convection is given in [70].

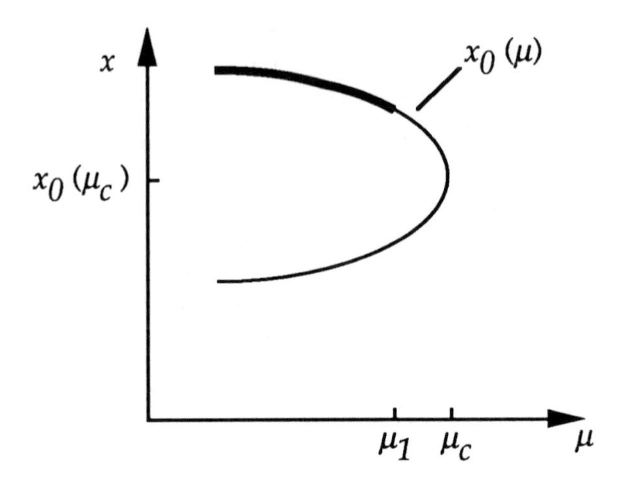

FIG. 3.1. *Loss of stability before fold bifurcation*

3.6. Control of the saddle node bifurcation. Finally, we briefly discuss control problems for systems exhibiting saddle node (or fold) bifurcations. Fold bifurcations are severe since they entail the *disappearance*—as opposed to loss of stability—of the nominal equilibrium, the desired operating condition. There is no obvious 'natural' feedback control problem for fold bifurcations. None the less, we present some reflections on possible viewpoints for this problem, and briefly discuss associated needs in analysis and application. Our considerations generalize to the multiparameter setting, but for simplicity we focus on the single parameter case. None of the control design strategies considered below involves modifying the equilibria of the open loop system. This is in accord with the motivation for using washout filter-aided feedback as discussed in Sec. 3.2. An alternative approach was presented in [15].

Suppose that the nominal equilibrium point loses stability before the fold bifurcation, as depicted in Figure 3.1. Then we might wish to stabilize this equilibrium over a larger parameter range. This would employ a bifurcation-delaying controller, of the type discussed in the preceding subsection. Indeed it may be possible to stabilize the nominal equilibrium up to the parameter value μ_c at which the fold bifurcation occurs. That is, we might construct a control in state feedback form ($u = u(x)$) which stabilizes the equilibrium $x_0(\mu)$ for all $\mu < \mu_c$. Alternately, we could allow the control to depend on μ in addition to x, or to depend on x only through specified measured output variables. Achieving this large range of stability would seem to be highly desirable, but it comes with a hidden risk. For an uncertain dynamic model, it is not possible to ascertain the value μ_c. If the nominal equilibrium is stable up to μ_c, and if the parameter μ drifts slowly toward μ_c, then it is likely that collapse will occur *without warning* for μ close to μ_c.

An alternative involves stabilizing the nominal equilibrium up to a value of μ close to, but less than, μ_c. Then we obtain Figure 3.1 with $\mu_c - \mu_1$ small. Generically, it will then be the case that an Andronov-Hopf bifurcation will occur at the nominal equilibrium point for $\mu = \mu_1$. There are generically two possibilities for this bifurcation: it can be subcritical or supercritical. These possibilities are illustrated in Figure 3.2 and Figure 3.3, respectively.

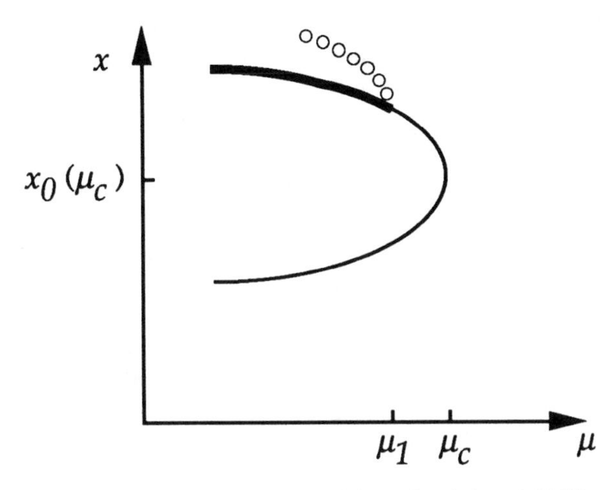

FIG. 3.2. *Subcritical Andronov-Hopf bifurcation before fold bifurcation*

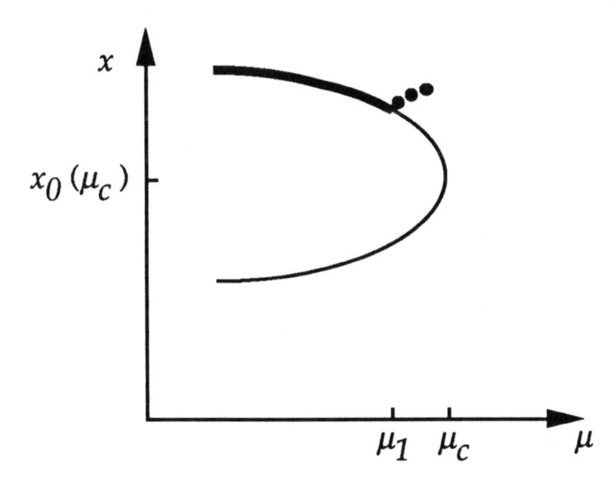

FIG. 3.3. *Supercritical Andronov-Hopf bifurcation before fold bifurcation*

It is now natural to ask if a design of this type can facilitate automatic recognition of the proximity to the fold even in the absence of an accurate model. If the Andronov-Hopf bifurcation occurring at $\mu = \mu_1$ is subcritical, then trajectories starting at the nominal equilibrium for μ slightly greater than μ_1 will diverge. However, if the Andronov-Hopf bifurcation is supercritical, then these same trajectories will converge to small-amplitude periodic motions near the nominal (now unstable) equilibrium. An automatic monitoring system could conceivably then detect proximity to the fold bifurcation by detecting the periodic solutions. Thus periodic solutions, which usually are considered undesirable, might provide a warning signal for impending voltage collapse. Note that if the open loop system does not exhibit Andronov-Hopf bifurcation from the nominal equilibrium, by this discussion it appears that in some situations it may be advantageous to introduce such bifurcations.

Let us continue this thought experiment. Suppose that a feedback control law has been implemented which achieves the foregoing objectives. That is, the closed loop system has the following bifurcation behavior. The equilibrium branches of the open loop system coincide with those of the closed loop system. The nominal equilibrium is stable for parameter values up to $\mu = \mu_1$, where μ_1 is close to μ_c. A supercritical Andronov-Hopf bifurcation occurs at the nominal equilibrium for the parameter value $\mu = \mu_1$. Thus, Figure 3.3 applies to the closed loop system. If the state x is two-dimensional, then it is guaranteed that the stable periodic solution emerging at $\mu = \mu_1$ cannot exist past the parameter value $\mu = \mu_c$. (It would need to encircle an equilibrium point, which would not exist for $\mu > \mu_c$.) However, it would be desirable to maintain operation in the vicinity of the nominal equilibrium point even for parameter values $\mu > \mu_c$. This would allow operation even past the fold bifurcation.

To see whether or not this is achievable, note that before the introduction of a stable periodic solution for $\mu > \mu_c$, system trajectories beginning in the vicinity of $x_0(\mu_c)$ would tend to diverge. By continuity, then, introduction of the stable periodic solution for $\mu > \mu_c$ must also entail the (unintentional) introduction of some other unstable invariant set(s). Skimming the literature on applications of bifurcation theory, we find many examples of systems exhibiting Andronov-Hopf bifurcation near a fold bifurcation (e.g., [28],[54],[57],[72]). However, we have found no published examples in which the periodic solution generated by Andronov-Hopf bifurcation remains stable for all parameter values *and* penetrates the fold bifurcation. However, recent work of Song and Liaw [61] may provide an example in the context of control of high angle-of-attack flight dynamics.

As noted above, achieving the existence of a small-amplitude stable periodic orbit past a fold bifurcation would entail introduction of an unstable invariant set as well. Thus we consider the use of feedback to introduce a degenerate Andronov-Hopf bifurcation of a special type, as depicted in Figure 3.4. This figure shows the simultaneous occurrence of a fold bifurcation and a degenerate Andronov-Hopf bifurcation of the type given in Figure 2.3(b)—with the added condition that the subcritical branch of unstable periodic orbits experiences a cyclic fold bifurcation. The stable periodic orbit resulting from the cyclic fold bifurcation, and the unstable periodic orbit which bifurcates supercritically at $\mu = \mu_1$, are required to persist for $\mu > \mu_c$, as shown in Figure 3.4.

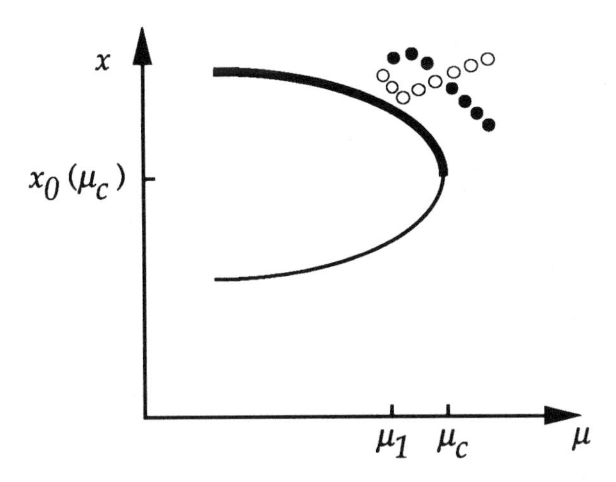

FIG. 3.4. *Special degenerate bifurcation before fold bifurcation*

Other possibilities exist for control of systems past fold bifurcations. One involves designing control laws which introduce new equilibria (rather than periodic solutions) past the fold, one of which is stable. This would preferably be achieved without altering the open loop equilibria. This idea is actually closely linked with the foregoing scheme of using degenerate Andronov-Hopf bifurcations, and involves introduction of certain degenerate static bifurcations. Indeed, projecting the periodic solutions onto a lower dimensional space can sometimes result in the introduction of new equilibria as we have just described. Another possibility involves introducing a chaotic attractor near $x_0(\mu_c)$ for $\mu > \mu_c$. This could possibly be achieved by synthesizing a feedback control which results in the occurrence of an intermittent route to chaos [59],[65]. Again, to respect the original divergent nature of the system's dynamics near the fold, an unstable limit set would need to be introduced into the dynamics simultaneously with the chaotic attractor. As yet a third possibility, the use of differential inclusion descriptions of nonlinear systems may prove useful in determining the feasibility of various control objectives for uncertain nonlinear systems operating with negligible margin of stability [10].

4. Concluding remarks. Mechanisms of loss of stability are inherently connected to bifurcations. Voltage collapse is a power system instability which may arise in different ways for different systems. A variety of bifurcation phenomena, including routes to chaos, have been detected for power system models exhibiting voltage collapse. We have briefly summarized some of the bifurcation-theoretic problems associated with control of voltage collapse.

REFERENCES

[1] E.H.ABED, H.O.WANG, J.C.ALEXANDER, A.M.A.HAMDAN, H.-C.LEE, *Dynamic bifurcations in a power system model exhibiting voltage collapse*, Internat. J. Bifurcation and Chaos, **3** (5) (1993), 1169–1176.

[2] E.H.ABED, J.-H.FU, *Local feedback stabilization and bifurcation control, I. Hopf bifurcation*, Systems and Control Letters, **7** (1) Feb (1986), 11–17.

[3] E.H.ABED, A.M.A.HAMDAN, H.-C.LEE, A.G.PARLOS, *On bifurcations in power system models and voltage collapse*, Proceedings of the 29th IEEE Conference on Decision and Control, Honolulu 1990, 3014–3015.

[4] E.H.ABED, P.P.VARAIYA, *Nonlinear oscillations in power systems*, Int. J. Elec. Power and Energy Syst., **6** (1) Jan (1984), 37–43.

[5] E.H.ABED, H.O. WANG, *Feedback control of bifurcation and chaos in dynamical systems*, Recent Developments in Stochastic and Nonlinear Dynamics: Applications to Mechanical Systems, (W.KLIEMANN, N.SRI NAMACHCHIVAYA, eds.) CRC Press (in press).

[6] E.H.ABED, H.O.WANG, R.C.CHEN, *Stabilization of period doubling bifurcations and implications for control of chaos*, Physica D, **70** (1-2) (1994), 154–164.

[7] V.AJJARAPU, B.LEE, *Bifurcation theory and its application to nonlinear dynamical phenomena in an electrical power system*, IEEE Trans. on Power Systems, **7** (1992), 424–431.

[8] A.A.ANDRONOV, A.A.VITT, S.E.KHAIKIN, *Theory of oscillators*, (English translation of the 2nd Russian edition) Pergamon Press, Oxford, UK 1966 (especially Section VI.5.4).

[9] M.M.BEGOVIĆ, A.G.PHADKE, *Control of voltage stability using sensitivity analysis*, IEEE Trans. on Power Systems, **7** (1992), 114–123.

[10] A.G.BUTKOVSKIY, *Phase portraits of control dynamical systems*, Kluwer, Dordrecht, The Netherlands 1991.

[11] L.-H.CHEN, H.-C.CHANG, *Global effects of controller saturation on closed-loop dynamics*, Chemical Engineering Science **40** (1985), 2191–2205.

[12] H.-D.CHIANG, C.-W.LIU, P.P.VARAIYA, F.F.WU, M.G.LAUBY, *Chaos in a simple power system*, IEEE Trans. on Power Systems, **8** (4) Nov (1993), 1407–1417.

[13] J.C.CHOW, R.FISCHL, H.YAN, *On the evaluation of voltage collapse criteria*, IEEE Trans. on Power Systems, **5** (1990), 612–620.

[14] S.N.CHOW, J.K.HALE, *Methods of bifurcation theory*, Springer-Verlag, New York 1982.

[15] M.CIBRARIO, J. LÉVINE, *Saddle-node bifurcation control with application to thermal runaway of continuous stirred tank reactors*, Proc. 30th IEEE Conference on Decision and Control, Brighton, United Kingdom, December 1991, 1551–1552.

[16] O.G.C.DAHL, *Electric power circuits, Vol. II: Power system stability*, McGraw-Hill, New York 1938.

[17] I.DOBSON, F.ALVARADO, C.L.DEMARCO, *Sensitivity of Hopf bifurcations to power system parameters*, Proc. 31st IEEE Conference on Decision and Control, Tucson, AZ, December 1992, 2928–2933.

[18] I.DOBSON, H.-D.CHIANG, *Towards a theory of voltage collapse in electric power systems*, Systems and Control Letters, **13** (1989), 253–262.

[19] I.DOBSON, H.GLAVITSCH, C.-C.LIU, Y.TAMURA, K.VU, *Voltage collapse in power systems*, IEEE Circuits and Devices Magazine, **8** (3) May (1992), 40–45.

[20] I.DOBSON, L.LU, *New methods for computing a closest saddle node bifurcation and worst case load power margin for voltage collapse*, IEEE Trans. on Power Systems, **8** (1993), 905–913.

[21] J.DOUGLAS, *FACTS: The delivery system of the future*, EPRI Journal, **17** (7) Oct (1992), 4–11.

[22] L.H.FINK, (ed.) *Proceedings: Bulk power system voltage phenomena—voltage stability and security*, Potosi, Missouri, Jan. 1989, Report EL-6183, Electric Power Research Institute 1989.

[23] L.H.FINK, (ed.) *Proceedings: Bulk power system voltage phenomena II—voltage stability and security*, An ECC/NSF Workshop, Deep Creek Lake, MD, Aug 1991, ECC, Inc., 4400 Fair Lakes Court, Fairfax, VA 22033-3899, 1991.

[24] L.H.FINK, *Introductory overview of voltage problems*, (in [23], 3–7).

[25] J.-H.FU, E.H.ABED, FAMILIES OF LYAPUNOV FUNCTIONS FOR NONLINEAR SYSTEMS IN CRITICAL CASES, IEEE Trans. Automatic Control, **33** (1993), 3–16.

[26] G.D.GALANOS, C.I.HATZIADONIU, X.-J.CHENG, D.MARATUKULAM, *Advanced static compensator for flexible AC transmission*, IEEE Trans. on Power Systems, **8** (1993), 231–238.

[27] R.GENESIO, A.TESI, H.O.WANG, E.H.ABED, *Control of period doubling bifurcations using harmonic balance*, Proc. 32nd IEEE Conf. on Decision and Control, San Antonio, TX, Dec 1993, 492–497.

[28] R.GILMORE, *Catastrophe theory for scientists and engineers*, Wiley, New York 1981.

[29] J.D.GLOVER, M.SARMA, *Power system analysis and design*, second edition, PWS Publishing, Boston 1994.

[30] M.GOLUBITSKY, D.G.SCHAEFFER, *Singularities and groups in bifurcation theory*, Volume I, Springer-Verlag, New York 1985.

[31] J. Guckenheimer and P. Holmes, *Nonlinear oscillations, dynamical systems, and bifurcations of vector fields*, Springer-Verlag, New York 1983.

[32] L.GYUGYI, *Unified power-flow control concept for flexible AC transmission systems*, IEE Proceedings–C: Generation, Transmission and Distribution, **139** (4) July (1992), 323–331.

[33] A.E.HAMMAD, M.Z. EL-SADEK, *Prevention of transient voltage instabilities due to induction motor loads by static VAR compensators*, IEEE Trans. on Power Systems, **4** (3) Aug (1989), 1182–1190.

[34] B.D.HASSARD, N.D.KAZARINOFF, Y.-H.WAN, *Theory and applications of Hopf bifurcation*, Cambridge University Press, Cambridge, UK 1981.

[35] J.F.HAUER, *Robustness issues in stability control of large electric power systems*, Proc. 32nd IEEE Conf. on Decision and Control, San Antonio, TX, Dec 1993, 2329–2334.

[36] D.J.HILL, I.A.HISKENS, Y.WANG, *Robust, adaptive or nonlinear control for modern power systems*, Proc. 32nd IEEE Conf. on Decision and Control, San Antonio, TX, Dec 1993, 2335–2340.

[37] N.G.HINGORANI, *Flexible ac transmission*, IEEE Spectrum, **30** (4) Apr (1993), 40–45.

[38] N.G.HINGORANI, K.E.STAHLKOPF, *High-power electronics*, Scientific American, **269** (5) Nov (1993), 78–85.

[39] R.A.HORE, *Advanced studies in power system design*, Chapman and Hall, London 1966.

[40] L.N.HOWARD, *Nonlinear oscillations*, Nonlinear Oscillations in Biology, F.C.HOPPENSTEADT, (ed.) American Mathematical Society, Providence, RI 1979, 1–67.

[41] M.ILIC, *A survey of the present state-of-the-art voltage control with emphasis on potential role of FACTS devices*, (in [23], 379–393).

[42] G.IOOSS, D.D.JOSEPH, *Elementary stability and bifurcation theory*, second edition, Springer-Verlag, New York 1990.

[43] N.KOPELL, L.N.HOWARD, *Bifurcations and trajectories joining critical points*, Advances in Mathematics, **18** (1975), 306–358.

[44] A.B.RANJIT KUMAR, A.IPAKCHI, F.ALVARADO, I.DOBSON, S.MUKHERJEE, W.H.ESSELMAN, (eds.) *Proceedings: EPRI/NSF workshop on application of*

advanced mathematics to power systems, Redwood City, CA, Sept 1991, Report TR-101795, Electric Power Research Institute 1993.

[45] P.Kundur, *Power system stability and control*, **Vol. 1** of the EPRI Power System Engineering Series, McGraw-Hill, New York 1994.

[46] H.G.Kwatny, A.K.Pasrija, L.Y.Bahar, *Static bifurcation in electric power networks: Loss of steady-state stability and voltage collapse*, IEEE Trans. Circuits Syst., **CAS-33** (10) Oct (1986), 981–991.

[47] H.G.Kwatny, X.M.Yu, *Voltage regulation and stabilization in power networks*, Proc. American Control Conference, Chicago, 1992, 2079–2083.

[48] B.H.Lee, K.Y.Lee, *Dynamic and static voltage stability enhancement of power systems*, IEEE Trans. on Power Systems, **8** (1993), 231–238.

[49] H.-C.Lee, E.H.Abed, *Washout filters in the bifurcation control of high alpha flight dynamics*, Proc. American Control Conference, Boston 1991, 206–211.

[50] B.C.Lesieutre, P.W.Sauer, M.A.Pai, *Why power/voltage curves are not necessarily bifurcation diagrams*, Proc. 1993 North American Power Symposium, Washington, DC, Oct 1993, 30–37.

[51] P.-A.Löf, G.Andersson, D.J.Hill, *Voltage stability indices for stressed power systems*, IEEE Trans. on Power Systems, **8** (1993), 326–335.

[52] Y.Mansour, (ed.) *Voltage stability of power systems: concepts, analytical tools, and industry experience*, IEEE Press, Publ. 90TH0358-2-PWR, New York 1990.

[53] J.E.Marsden, M.McCracken, *The Hopf bifurcation and its applications*, Springer-Verlag, New York 1976.

[54] J.L.Moiola, G.Chen, *Frequency domain approach to computation and analysis of bifurcations and limit cycles: a tutorial*, Internat. J. Bifurcation and Chaos, **3** (1993), 843–867.

[55] S.Mori, K.Matsuno, T.Hasegawa, S.Ohnishi, M.Takeda, M.Seto, S.Murakami, F.Ishiguro, *Development of a large static VAR generator using self-commutated inverters for improving power system stability*, IEEE Trans. on Power Systems, **8** (1993), 371–377.

[56] E.Ott, *Chaos in dynamical systems*, Cambridge University Press, Cambridge, UK 1993.

[57] J.B.Planeaux, J.A.Beck, D.D.Baumann, *Bifurcation analysis of a model fighter aircraft with control augmentation*, AIAA Atmospheric Flight Mechanics Conference, Paper No. AIAA-90-2836, Portland, OR, Aug 1990.

[58] H.Poincaré, *New methods of celestial mechanics*, Parts 1,2 and 3 (edited and introduced by D.L.Goroff) **Vol. 13** of the History of Modern Physics and Astronomy Series, American Institute of Physics, USA 1993. (English translation of) *Les méthodes nouvelles de la mécanique céleste* (originally published in) 1892–1899.

[59] Y.Pomeau, P.Manneville, *Intermittent transition to turbulence in dissipative dynamical systems*, Commun. Math. Phys., **74** 1980, 189–197.

[60] A.N.Shoshitaishvili, *Singularities for projections of integral manifolds with applications to control and observation problems*, Advances in Soviet Mathematics, **1** (1990), 295–333.

[61] C.-C.Song, D.-C.Liaw, *Bifurcation analysis and control of high angle-of-attack flight dynamics*, (preprint) Dept. of Control Engineering, National Chiao Tung University, Hsinchu, Taiwan 1994.

[62] Y.Tamura, H.Mori, S.Iwamoto, *Relationship between voltage instability and multiple load flow solutions in electric power systems*, IEEE Trans. Power Apparatus and Systems, **PAS-102** (5) May (1983), 1115–1125.

[63] C.-W.Tan, M.Varghese, P.Varaiya, F.Wu, *Bifurcation and chaos in power systems*, SADHANA: Proceedings in Engineering Sciences of the Indian Academy of Sciences, Bangalore, India, **18** (Part 5) Sept (1993), 761–786.

[64] C.W. Taylor, *Power system voltage stability*, **Vol. 2** of the EPRI Power System Engineering Series, McGraw-Hill, New York 1994.

[65] J.M.T.THOMPSON, H.B.STEWART, *Nonlinear dynamics and chaos*, Wiley, Chichester, United Kingdom 1986.

[66] P.VARAIYA, F.WU, H.-D.CHIANG, *Bifurcation and chaos in power systems: A survey*, Report TR-100834, Electric Power Research Institute, Palo Alto, CA, August 1992.

[67] V.VEERA RAJU, A.KUPPURAJULU, *A closed loop controller for voltage stability*, Int. J. Elec. Power and Energy Syst., **15** (1993), 283–292.

[68] V.VENKATASUBRAMANIAN, H.SCHÄTTLER, J.ZABORSZKY, *A taxonomy of the dynamics of the large electric power system with emphasis on its voltage stability*, (in [23], 9–52).

[69] V.VENKATASUBRAMANIAN, H.SCHÄTTLER, J.ZABORSZKY, *Voltage dynamics: study of a generator with voltage control, transmission and matched MW load*, IEEE Trans. Automatic Control, **37** (1992), 1717–1733.

[70] H.O.WANG, E.H.ABED, *Bifurcation control of a chaotic system*, (preprint) Institute for Systems Research, University of Maryland, Jan 1994. (See also *Bifurcation control of chaotic dynamical systems*, Proc. of the Second NOLCOS (Nonlinear Control System Design) Conference, June 1992, Bordeaux, France, (published by the International Federation of Automatic Control) 57–62.)

[71] H.O.WANG, E.H.ABED, *Control of nonlinear phenomena at the inception of voltage collapse*, Proc. 1993 American Control Conference, San Francisco, June 1993, 2071–2075.

[72] H.O.WANG, E.H.ABED, A.M.A.HAMDAN, *Bifurcations, chaos and crises in voltage collapse of a model power system*, IEEE Transactions on Circuits and Systems—I: Fundamental Theory and Applications, **41** (4) Apr (1994), 294–302.

[73] S.WIGGINS, *Introduction to applied nonlinear dynamical systems and chaos*, Springer-Verlag, New York 1990.

[74] J.ZABORSZKY, *Some basic issues in voltage stability and viability* (Sec. 1.3 in [22]).

[75] J.ZABORSZKY, J.W.RITTENHOUSE, *Electric power transmission*, Ronald Press, New York 1954.

REDUCED-ORDER MODELING OF ELECTRIC MACHINES USING INTEGRAL MANIFOLDS

SAID AHMED-ZAID*

Abstract. Recent results are reviewed on the application of integral manifolds in the simplification of power system models. The existence of such integral manifolds in electric machine models, cast as singularly-perturbed dynamical systems, leads to a rigorous and systematic generation of improved reduced-order models for use in power system studies. This mathematical tool for order reduction is applied to the modeling of small and large three-phase synchronous and induction machines. The dominant behavior in each type of machine is extracted in the form of a first- or second-order dynamical model, depending on the size of the machine. Small synchronous machines are characterized by a second-order swing model, and large synchronous machines by a first-order voltage model. In contrast, the dominant behavior in small induction machines is characterized by a well-known first-order speed model, and by a first-order voltage model in large induction machines.

1. Introduction. The existence of integral (or invariant) manifolds in systems of ordinary differential equations has been known to mathematicians for a long time, along with geometric proofs dating back at least to 1901 [1]. Integral manifolds are useful in qualitative investigations near critical points and periodic orbits, and in bifurcation theory [2,3]. Using singular perturbation theory [4] or center-manifold theory [5], similar results can be derived on the existence of such integral manifolds in certain physical systems which are characterized by dynamical phenomena of widely-different speeds and can be cast in singular perturbation form.

Recently, integral manifolds have emerged as a powerful tool for the modeling of such physical systems including, for example, electrical circuits [6] and electric machines [7]–[10]. The use of integral manifolds in these systems leads to a systematic generation of reduced-order models which can be used to simplify the order and complexity of the system at hand. Through the use of asymptotic expansions, these reduced models can be improved to any desired level of accuracy and yield more insight into the effect of critical parameters on the dynamic performance of the system.

The objective of this paper is to introduce the method of integral manifolds in the reduced-order modeling of synchronous machines and induction machines in power systems. This exercise in the application of integral manifolds leads to a rigorous justification of an assumption commonly referred to as the "neglect of stator transients." Once these fast dynamics are eliminated from the system through the use of stator manifolds, different rotor manifolds can be extracted depending both on the size as well the type of machine under consideration, as will be shown in the following sections.

* Clarkson University, Department of Electrical and Computer Engineering, Potsdam, NY 13699. This research was supported in part by the National Science Foundation under Grant No. ECS-9058174.

2. Integral manifold theory. Consider a set of nonlinear ordinary differential equations

$$\dot{x} = f(x), \qquad x(t_o) = x_o, \qquad x \in \mathbf{R}^n \tag{2.1}$$

where x is an n-dimensional vector and the dot means differentiation with respect to time t. A manifold or hypersurface S of dimension m less than n is called an invariant manifold for (2.1) if $x(t)$ belongs to S for all t whenever $x(t_o)$ is in S. If S exists for a finite time only, it is said to be a local invariant manifold. In essence, an invariant manifold is the extension of the notion of invariant subspace in linear systems. However, existence theorems of such manifolds are much more involved and have only been derived for special classes of nonlinear systems, e.g., for multi-time-scale nonlinear systems [4,5].

In the following, consider singularly-perturbed systems of the form

$$\dot{x} = \epsilon f(x, y, \epsilon), \qquad x(t_o) = x_o, \qquad x \in \mathbf{R}^n \tag{2.2}$$

$$\dot{y} = g(x, y, \epsilon), \qquad y(t_o) = y_o, \qquad y \in \mathbf{R}^m \tag{2.3}$$

where f and g are n- and m-dimensional vector functions, respectively, and ϵ is a small positive parameter. Under a certain hypothesis to be specified below, the x-variables are slow-rate variables whereas the y-variables can be considered fast-rate variables.

Let us formally look for an invariant "slow" manifold of the form [8]

$$y = \overline{y}(x, \epsilon) \tag{2.4}$$

If $y(t_o) = \overline{y}(x_o, \epsilon)$, then we should have

$$\frac{dy}{dt} = \frac{\partial \overline{y}}{\partial x}\frac{dx}{dt} \tag{2.5}$$

or

$$g(x, \overline{y}, \epsilon) = \epsilon \frac{\partial \overline{y}}{\partial x} f(x, \overline{y}, \epsilon) \tag{2.6}$$

Thus, the invariant manifold $y = \overline{y}(x, \epsilon)$ is seen to be the solution of the above partial differential equation. Assuming that $\overline{y}$, f and g have smooth expansions in ϵ of the form

$$\overline{y}(x, \epsilon) = \overline{y}_o(x) + \epsilon \overline{y}_1(x) + \epsilon^2 \overline{y}_2(x) + \dots \tag{2.7}$$

$$f(x, \overline{y}, \epsilon) = f_o(x, \overline{y}_o) + \epsilon f_1(x, \overline{y}_o, \overline{y}_1) + \epsilon^2 f_2(x, \overline{y}_o, \overline{y}_1, \overline{y}_2) + \dots \tag{2.8}$$

$$g(x, \overline{y}, \epsilon) = g_o(x, \overline{y}_o) + \epsilon g_1(x, \overline{y}_o, \overline{y}_1) + \epsilon^2 g_2(x, \overline{y}_o, \overline{y}_1, \overline{y}_2) + \dots \tag{2.9}$$

where

$$f_r(x, \overline{y}_o, \dots, \overline{y}_r) = \frac{1}{r!}\frac{\partial^r f}{\partial \epsilon^r}\bigg|_{\epsilon=0} \tag{2.10}$$

and

$$g_r(x, \overline{y}_o, ..., \overline{y}_r) = \frac{1}{r!} \frac{\partial^r g}{\partial \epsilon^r}\bigg|_{\epsilon=0} \tag{2.11}$$

Substituting these equations into (2.6) and matching equal powers of ϵ, the series terms of $\overline{y}$ can be solved for recursively from the following algebraic subproblems:

$$g_o(x, \overline{y}_o) = 0 \tag{2.12}$$

$$g_1(x, \overline{y}_o, \overline{y}_1) = \frac{\partial \overline{y}_o}{\partial x} f_o(x, \overline{y}_o) \tag{2.13}$$

$$g_2(x, \overline{y}_o, \overline{y}_1, \overline{y}_2) = \frac{\partial \overline{y}_o}{\partial x} f_1(x, \overline{y}_o, \overline{y}_1) + \frac{\partial \overline{y}_1}{\partial x} f_o(x, \overline{y}_o) \tag{2.14}$$

An important condition which is at the heart of the existence of a manifold $\overline{y}$ is

$$\det\left(\frac{\partial \overline{y}_o}{\partial x}\right) \neq 0 \tag{2.15}$$

that is, the Jacobian submatrix corresponding to the fast-rate variables y must not be singular. From the implicit function theorem, this is not surprising if we have to solve for $\overline{y}_o$ as a function of x from (2.12).

Once the slow-manifold is determined to sufficient accuracy [5], we define off-manifold variables

$$\tilde{y} = y - \overline{y}(x, \epsilon) \tag{2.16}$$

Differentiating (2.16) with respect to time,

$$\dot{\tilde{y}} = g(x, \tilde{y} + \overline{y}(x, \epsilon), \epsilon) - \epsilon \frac{\partial \overline{y}}{\partial x} f(x, \tilde{y} + \overline{y}(x, \epsilon), \epsilon) \tag{2.17}$$

with

$$\tilde{y}(t_o) = y(t_o) - \overline{y}(x(t_o), \epsilon) \tag{2.18}$$

If $\tilde{y}(t_o) = 0$, then $\tilde{y}(t) = 0$ for all time by definition of the slow manifold. If $\tilde{y}(t_o) \neq 0$ and lies in the region of attraction of the (locally) asymptotically stable equilibrium point $\tilde{y} = 0$ of (2.17), then $\tilde{y}(t)$ quickly decays to zero. Once on the slow-manifold, the y-variables follow the variations of the x-variables according to (2.4) which is referred to as its quasi-steady state. The reduced equation governing the x-variables on the slow-manifold is

$$\dot{x} = \epsilon f(x, \overline{y}(x, \epsilon), \epsilon) \tag{2.19}$$

3. Synchronous machine modeling

3.1. A fifth-order synchronous machine model. The synchronous machine model used in this study is represented by a fifth-order model in direct and quadrature-axes stator flux linkages (ψ_d,ψ_q), a field flux variable E'_q, and electromechanical variables (δ,ω). The machine is connected to a Thevenin representation of an infinite bus through an equivalent transmission line with resistance R_e and reactance X_e as shown in Figure 3.1. Using a standard notation [7], this fifth-order model is given by

$$\frac{1}{\omega_s}\frac{d\psi_d}{dt} = -\frac{R_s}{X'_d}\psi_d + \frac{\omega}{\omega_s}\psi_q + \frac{R_s}{X'_d}E'_q + V\sin\delta \tag{3.1}$$

$$\frac{1}{\omega_s}\frac{d\psi_q}{dt} = -\frac{\omega}{\omega_s}\psi_d - \frac{R_s}{X_q}\psi_q + V\cos\delta \tag{3.2}$$

$$T'_{do}\frac{dE'_q}{dt} = -\frac{X_d}{X'_d}E'_q + (\frac{X_d - X'_d}{X'_d})\psi_d + E_{fd} \tag{3.3}$$

$$\frac{d\delta}{dt} = \omega - \omega_s \tag{3.4}$$

$$M\frac{d\omega}{dt} = T_m + (\frac{1}{X_q} - \frac{1}{X'_d})\psi_d\psi_q + \frac{1}{X'_d}E'_q\psi_q \tag{3.5}$$

where $\omega_s = 2\pi 60$ (rad/s) is the synchronous electrical speed and M is an inertia constant. The stator resistance R_s and the machine reactances X_d, X_q, X'_d and X'_q include the effects of the external transmission line resistance and reactance. The field voltage E_{fd} and the infinite bus voltage V are assumed constant in this analysis.

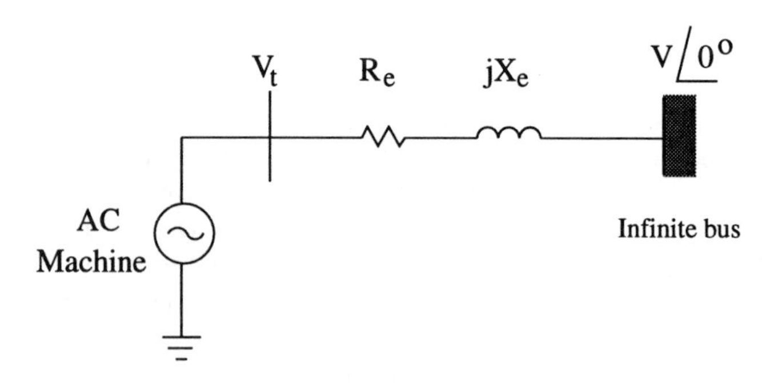

FIG. 3.1. *Study System*

3.2. Exact stator integral manifolds with zero stator resistance. For the special case of zero stator resistance, $R_s = 0$, a third-order reduced model can be found by searching for a manifold which gives the relationship between the stator transients (ψ_d, ψ_q) and the field and rotor electromechanical variables (E'_q, δ, ω). It can be verified by direct substitution that the following expressions of ψ_d and ψ_q

$$\psi_d = V \cos\delta \tag{3.6}$$

$$\psi_q = -V \sin\delta \tag{3.7}$$

satisfy the definition of an integral manifold. That is, if the initial conditions $\psi_d(0)$, $\psi_q(0)$ and $\delta(0)$ satisfy (3.6) and (3.7), then $\psi_d(t)$ and $\psi_q(t)$ will satisfy (3.1) and (3.2) for all time and may be eliminated from the remaining differential equations. That Equations (3.6) and (3.7) satisfy (3.1) and (3.2) with $R_s = 0$ can be checked by direct substitution into these differential equations. It is also interesting to note that, while (3.6) and (3.7) just happen to be the steady-state relationships, they are also exact relationships when speed ω is not equal to ω_s and when δ and, therefore, ψ_d and ψ_q are not constant. The existence of exact integral manifolds such as these are rare in nonlinear dynamic models. Mathematically speaking, the integral manifold is a type of particular solution of a subset of differential equations. It is not the complete solution since the integral manifold does not depend on initial conditions. If the initial conditions do not satisfy (3.6) and (3.7), then ψ_d and ψ_q will not satisfy (3.1) and (3.2). In this case, we introduce off-manifold variables

$$\eta_d = \psi_d - V \cos\delta \tag{3.8}$$

$$\eta_q = \psi_q + V \sin\delta \tag{3.9}$$

which are the deviations of ψ_d and ψ_q from the manifolds (3.6) and (3.7). Substitution of (3.8)–(3.9) into (3.1)–(3.2) with $R_s = 0$ gives the following equations for these new variables

$$\frac{1}{\omega_s}\frac{d\eta_d}{dt} = \frac{\omega}{\omega_s}\eta_q, \qquad \eta_d(0) = \psi_d(0) - V\cos\delta(0) \tag{3.10}$$

$$\frac{1}{\omega_s}\frac{d\eta_q}{dt} = -\frac{\omega}{\omega_s}\eta_d, \qquad \eta_q(0) = \psi_q(0) + V\sin\delta(0) \tag{3.11}$$

where $\omega = d\delta/dt + \omega_s$ appears as a time-varying coefficient. It is easy to verify that these equations have the following explicit solutions

$$\eta_d(t) = \sqrt{\eta_d^2(0) + \eta_q^2(0)} \cos\left[\omega_s t + \delta(t) - \delta(0) - \tan^{-1}\left(\tfrac{\eta_q(0)}{\eta_d(0)}\right)\right] \tag{3.12}$$

$$\eta_q(t) = -\sqrt{\eta_d^2(0) + \eta_q^2(0)} \sin\left[\omega_s t + \delta(t) - \delta(0) - \tan^{-1}\left(\tfrac{\eta_q(0)}{\eta_d(0)}\right)\right] \tag{3.13}$$

These are the exact deviations of ψ_d and ψ_q from the manifold. With any initial conditions, these expressions give the exact solutions of the stator transients ψ_d and ψ_q as a function of δ, t and the initial conditions. When substituted into the remaining differential equations, η_d and η_q yield the following exact reduced-order model for the case $R_s = 0$,

$$
\begin{aligned}
(3.14)\quad T'_{do}\frac{dE'_q}{dt} &= -\frac{X_d}{X'_d}E'_q + (\frac{X_d - X'_d}{X'_d})(V\cos\delta + \eta_d) + E_{fd} \\
(3.15)\quad \frac{d\delta}{dt} &= \omega - \omega_s \\
M\frac{d\omega}{dt} &= T_m - \frac{1}{2}\left(\frac{1}{X_q} - \frac{1}{X'_d}\right)V^2\sin 2\delta - \frac{VE'_q}{X'_d}\sin\delta \\
(3.16)\quad &+ \left(\frac{1}{X_q} - \frac{1}{X'_d}\right)(\eta_d\eta_q - \eta_d V\sin\delta + \eta_q V\cos\delta) + \frac{E'_q\eta_q}{X'_d}
\end{aligned}
$$

The off-manifold stator solutions η_d and η_q are known explicitly as functions of δ and t as given by (3.12) and (3.13). In this way, the additional torques due to the initial conditions of the stator transients have been identified and appear in the last line of (3.16). This model includes the effects of stator transients without including their differential equations. It is usually simplified by assuming that $\eta_d(0) \cong 0$ and $\eta_q(0) \cong 0$, that is, the initial conditions $\psi_d(0)$ and $\psi_q(0)$ are assumed "close" to the manifold (3.6)–(3.7). In this case, the third-order model simplifies to the familiar one-axis model [7]

$$
\begin{aligned}
(3.17)\quad T'_{do}\frac{dE'_q}{dt} &= -\frac{X_d}{X'_d}E'_q + \left(\frac{X_d - X'_d}{X'_d}\right)V\cos\delta + E_{fd} \\
(3.18)\quad \frac{d\delta}{dt} &= \omega - \omega_s \\
(3.19)\quad M\frac{d\omega}{dt} &= T_m + \frac{1}{2}\left(\frac{1}{X'_d} - \frac{1}{X_q}\right)V^2\sin 2\delta - \frac{VE'_q}{X'_d}\sin\delta
\end{aligned}
$$

3.3. Approximate rotor manifolds with zero stator resistance

3.3.1. Second-order speed model for small synchronous machines. Small synchronous machines differ from large synchronous machines by certain design parameters which endow these machines with different dynamical structures. If we define $T'_d = T'_{do}X'_d/X_d$, then small machines are characterized here by the following relationship

$$(3.20)\qquad MX'_d \;<\; T'_d \;<\; \sqrt{MX'_d}$$

Rescaling the rotor speeds ω and ω_s by $\sqrt{MX'_d}$ as

$$(3.21)\qquad \Omega - \Omega_s \;=\; \sqrt{MX'_d}(\omega - \omega_s)$$

the synchronous machine equations are rewritten as

$$T'_d \frac{dE'_q}{dt} = -E'_q + \left(\frac{X_d - X'_d}{X_d}\right) V\cos\delta + \frac{X'_d}{X_d} E_{fd} \tag{3.22}$$

$$\sqrt{MX'_d}\frac{d\delta}{dt} = \Omega - \Omega_s \tag{3.23}$$

$$\sqrt{MX'_d}\frac{d\Omega}{dt} = X'_d T_m + \frac{1}{2}\left(1 - \frac{X'_d}{X_q}\right) V^2 \sin 2\delta - VE'_q \sin\delta \tag{3.24}$$

Treating $T'_d/\sqrt{MX'_d}$ as a small parameter ϵ, the above third-order model can be considered as a two-time-scale singularly-perturbed system. A slow model in E'_q can be obtained by searching for an integral manifold of the form

$$E'_q = h(\delta, \Omega - \Omega_s, \frac{T'_d}{\sqrt{MX'_d}}) \tag{3.25}$$

After differentiating both sides of this equation and solving subproblems similar to (2.12)–(2.14), the zero- and first-order approximations of the integral manifold h can be solved for as

$$h_o = \left(\frac{X_d - X'_d}{X_d}\right) V\cos\delta + \frac{X'_d}{X_d} E_{fd} \tag{3.26}$$

$$h_1 = (\Omega - \Omega_s)\left(\frac{X_d - X'_d}{X_d}\right) V\sin\delta \tag{3.27}$$

Substituting these expressions into (3.23)–(3.24), a second-order model in the original electromechanical variables (δ,ω) is obtained as

$$\frac{d\delta}{dt} = \omega - \omega_s \tag{3.28}$$

$$\begin{aligned} M\frac{d\omega}{dt} &= T_m + \frac{1}{2}\left(\frac{1}{X_d} - \frac{1}{X_q}\right) V^2 \sin 2\delta - \frac{VE_{fd}}{X_d}\sin\delta \\ &- T'_{do}\left(\frac{X_d - X'_d}{X_d^2}\right)(\omega - \omega_s)V^2 \sin^2\delta \end{aligned} \tag{3.29}$$

3.4. First-order voltage model for large synchronous machines Using the same definition of T'_d, large synchronous machines are characterized here as those satisfying the following relationship

$$MX'_d < \sqrt{MX'_d} < T'_d \tag{3.30}$$

To extract a reduced-order voltage model, we search for a two-dimensional integral manifold of the form

$$\delta = h_1\left(E'_q, \frac{\sqrt{MX'_d}}{T'_d}\right) \tag{3.31}$$

$$\Omega = h_2\left(E'_q, \frac{\sqrt{MX'_d}}{T'_d}\right) \tag{3.32}$$

Differentiating both sides of these equations and solving subproblems similar to (2.12)–(2.14), the zero-order algebraic subproblems yield

$$h_2^o = 0 \tag{3.33}$$

$$T_m + \frac{1}{2}\left(\frac{1}{X'_d} - \frac{1}{X_q}\right)V^2 \sin 2h_1^o - \frac{VE'_q}{X'_d}\sin h_1^o = 0 \tag{3.34}$$

If transient saliency is neglected, that is, if we assume that $X'_d = X_q$, then (3.34) yields

$$h_1^o = \sin^{-1}\left(\frac{X'_d T_m}{VE'_q}\right) \tag{3.35}$$

Substituting (3.35) into (3.17), a first-order voltage model for large synchronous machines is

$$T'_{do}\frac{dE'_q}{dt} = -\frac{X_d}{X'_d}E'_q + \left(\frac{X_d - X'_d}{X'_d}\right)V\sqrt{1 - \left(\frac{X'_d T_m}{VE'_q}\right)^2} + E_{fd} \tag{3.36}$$

4. Induction machine modeling

4.1. Fifth-order induction machine model. A single-cage induction motor is generally represented by a fifth-order dynamic model of the stator flux linkages (ψ_{ds}, ψ_{qs}), rotor flux linkages (ψ_{dr}, ψ_{qr}) and shaft speed ω [10,11]

$$\frac{1}{\omega_s}\frac{d\psi_{ds}}{dt} = -\frac{R_s}{X'}\psi_{ds} + \psi_{qs} + \frac{R_s}{X'}E'_q + v_{ds} \tag{4.1}$$

$$\frac{1}{\omega_s}\frac{d\psi_{qs}}{dt} = -\psi_{ds} - \frac{R_s}{X'}\psi_{qs} - \frac{R_s}{X'}E'_d + v_{qs} \tag{4.2}$$

$$T'_o\frac{dE'_q}{dt} = -\frac{X}{X'}E'_q + \left(\frac{X - X'}{X'}\right)\psi_{ds} - (\omega_s - \omega)T'_o E'_d \tag{4.3}$$

$$T'_o\frac{dE'_d}{dt} = -\frac{X}{X'}E'_d - \left(\frac{X - X'}{X'}\right)\psi_{qs} + (\omega_s - \omega)T'_o E'_q \tag{4.4}$$

$$M\frac{d\omega}{dt} = \frac{1}{X'}(\psi_{qs}E'_q + \psi_{ds}E'_d) - T_m \tag{4.5}$$

where ω_s is the synchronous electrical speed in electrical radians per second. X and X' are machine reactances and M is an inertia constant. Here, the mechanical load torque T_m is assumed to be constant. The parameter R_s, X and X' include any external series resistance or reactance between the terminal of the induction machine and the infinite bus as shown in Figure 3.1. The dq stator voltages (v_{ds}, v_{qs}) are found from the abc infinite-bus voltages.

4.2. Exact stator manifolds with zero stator resistance. A third-order induction machine model is obtained by neglecting stator transients, a process which can be rigorously justified using integral manifold theory. Exact expressions for stator integral manifolds are possible under the simplifying assumption of zero stator resistance. In the case of a machine connected to an infinite bus with constant voltages $v_{ds}(t) = 0$ and $v_{qs}(t) = V$, the stator manifolds are

$$\psi_{ds} = V \tag{4.6}$$
$$\psi_{qs} = 0 \tag{4.7}$$

Thus, if the initial conditions of the stator flux linkages satisfy $\psi_{ds}(0) = V$ and $\psi_{qs}(0) = 0$, then (4.6)–(4.7) are the exact time solutions for $\psi_{ds}(t)$ and $\psi_{qs}(t)$, and can easily be checked by direct substitution into (4.1)–(4.2) with $R_s = 0$. In mathematical terms, these solutions correspond to the particular solutions when the stator transients are not excited. If the initial conditions of ψ_{ds} and ψ_{qs} do not lie on the manifold (4.6)–(4.7), then off-manifold variables

$$\eta_{ds} = \psi_{ds} - V \tag{4.8}$$
$$\eta_{qs} = \psi_{qs} \tag{4.9}$$

are defined as the deviations of ψ_{ds} and ψ_{qs} from the manifold (4.6)–(4.7). Substitution of (4.8)–(4.9) into (4.1)–(4.2) with $R_s = 0$ yields the decoupled set of differential equations

$$\frac{1}{\omega_s}\frac{d\eta_{ds}}{dt} = \eta_{qs}, \qquad \eta_{ds}(0) = \psi_{ds}(0) - V \tag{4.10}$$
$$\frac{1}{\omega_s}\frac{d\eta_{qs}}{dt} = -\eta_{ds}, \qquad \eta_{qs}(0) = \psi_{qs}(0) \tag{4.11}$$

The exact solutions of these differential equations are

$$\eta_{ds}(t) = A\sin(\omega_s t + \phi) \tag{4.12}$$
$$\eta_{qs}(t) = A\cos(\omega_s t + \phi) \tag{4.13}$$

where $A = \sqrt{\eta_{ds}^2(0) + \eta_{qs}^2(0)}$ and $\tan\phi = \eta_{ds}(0)/\eta_{qs}(0)$. Then, the complete solutions for the stator transients are

$$\psi_{ds}(t) = \eta_{ds}(t) + V \tag{4.14}$$
$$\psi_{qs}(t) = \eta_{qs}(t) \tag{4.15}$$

These two equations can be substituted into (4.3)–(4.5) to reduce the number of differential equations from five to three. The resulting third-order model is exact and captures all of the fifth-order dynamics of the induction machine.

Other exact manifolds can be derived in the case of zero stator resistance and ramping input voltages $v_{ds}(t) = 0$ and $v_{qs}(t) = Vt$. This type of input voltage is actually used to start induction machines to reduce inrush currents and to diminish the effects of the dc braking torque. The exact manifolds in this case are found as

$$\psi_{ds}(t) = Vt \tag{4.16}$$

$$\psi_{qs}(t) = \frac{V}{\omega_s} \tag{4.17}$$

In this case, the off-manifold variables satisfy the differential equations

$$\frac{1}{\omega_s}\frac{d\eta_{ds}}{dt} = \eta_{qs}, \qquad \eta_{ds}(0) = \psi_{ds}(0) \tag{4.18}$$

$$\frac{1}{\omega_s}\frac{d\eta_{qs}}{dt} = -\eta_{ds}, \qquad \eta_{qs}(0) = \psi_{qs}(0) - \frac{V}{\omega_s} \tag{4.19}$$

The exact solutions for these variables are identical to (4.12)–(4.13) with the appropriate initial conditions shown in (4.18)–(4.19). The stator flux linkages now have the form

$$\psi_{ds}(t) = Vt + \eta_{ds}(t) \tag{4.20}$$

$$\psi_{qs}(t) = \frac{V}{\omega_s} + \eta_{qs}(t) \tag{4.21}$$

The previous two results can be combined by assuming the following input voltages

$$v_{ds}(t) = 0 \text{ for all time} \tag{4.22}$$

$$v_{qs}(t) = \begin{cases} Vt & \text{for } 0 \le t \le 1 \\ V & \text{for } t \ge 1 \end{cases} \tag{4.23}$$

In this case, the stator flux linkages are given by (4.20)–(4.21) for $0 \le t \le 1$ and by (4.14)–(4.15) for $t \ge 1$. At $t = 1$ second, the final values $\psi_{ds}(1)$ and $\psi_{qs}(1)$ satisfy (4.6)–(4.7) and, thus, $\eta_{ds}(t)$ and $\eta_{qs}(t)$ are zero for $t \ge 1$. For $0 \le t \le 1$, $\eta_{ds}(t)$ and $\eta_{qs}(t)$ are not zero but their magnitude is reduced by ω_s if $\psi_{ds}(0) = \psi_{qs}(0) = 0$. In summary, the complete solution for the stator flux linkages is

$$\psi_{ds}(t) = \begin{cases} A\sin(\omega_s t + \phi) + Vt & \text{for } t \le 1 \\ V & \text{for } t \ge 1 \end{cases} \tag{4.24}$$

$$\psi_{qs}(t) = \begin{cases} A\cos(\omega_s t + \phi) + V/\omega_s & \text{for } t \le 1 \\ 0 & \text{for } t \ge 1 \end{cases} \tag{4.25}$$

When the stator resistance is small but not zero, the corresponding manifolds exist but do not have an exact closed-form expression. Higher-order corrections can be used to find these manifolds to any desired degree of accuracy. Usually, the first-order approximation yields a sufficiently accurate model for engineering purposes. Approximate integral manifolds for the rotor flux linkages can also be found using the same procedure. The interesting feature about these manifolds is that, just as in the synchronous machine case, they also depend on the size of the machine considered as shown in the next sections.

4.3. Approximate rotor manifolds. To simplify the exposition of the derivation of approximate rotor integral manifolds, we assume zero stator resistance and a constant positive load torque T_m in the following analysis. Neglecting stator resistance, a simplified third-order model of an induction motor connected to an infinite bus is

$$T_o' \frac{dE_q'}{dt} = -\frac{X}{X'}E_q' + (\frac{X - X'}{X'})V + T_o'(\omega - \omega_s)E_d' \tag{4.26}$$

$$T_o' \frac{dE_d'}{dt} = -\frac{X}{X'}E_d' - T_o'(\omega - \omega_s)E_q' \tag{4.27}$$

$$M\frac{d\omega}{dt} = \frac{VE_d'}{X'} - T_m \tag{4.28}$$

4.3.1. Polar form of third-order model. Rather than pursue the decomposition of (4.26)–(4.28) in rectangular variables E_d' and E_q', it is advantageous to convert them to polar variables E' and δ by means of the transformation [10]

$$E' = \sqrt{E_d'^2 + E_q'^2} \tag{4.29}$$

$$\delta = \tan^{-1}(-\frac{E_d'}{E_q'}) \tag{4.30}$$

The new model in polar form is

$$T_o' \frac{dE'}{dt} = -\frac{X}{X'}E' + (\frac{X - X'}{X'})V\cos\delta \tag{4.31}$$

$$\frac{d\delta}{dt} = \omega - \omega_s - (\frac{X - X'}{X'})\frac{V\sin\delta}{T_o'E'} \tag{4.32}$$

$$M\frac{d\omega}{dt} = -\frac{VE'\sin\delta}{X'} - T_m \tag{4.33}$$

This model presents some striking similarities with a one-axis model of a synchronous machine with the following differences:

- Even though E' is a voltage behind a transient reactance, it does not arise from an external excitation current.
- The angle δ is not a stroboscopic angle of rotation of the shaft but the angle of the rotor flux magnitude (equivalently, of an induced rotor voltage magnitude) with respect to the synchronously-rotating reference frame.

- There is no excitation voltage in (4.31) as the rotor circuits are short-circuited.
- There is an extra term in (4.32) when comparing this equation with the angle equation of a synchronous machine due to the variable speed operation of the induction machine.
- The maximum steady-state torque occurs at an angle $\delta = -45^o$ as opposed to -90° for a synchronous motor with constant field excitation. Stable operation ranges between 0° and -45° and unstable operation between -45° and an angle less than -90° at standstill [10].

The differences in the interactions between various state variables in small and large induction machines are attributed to certain design parameters which endow these machines with different dynamical structures. If $T' = T_o' X'/X$, then a small induction machine is characterized as satisfying the parameter relationship $MX' < T' < \sqrt{MX'}$. On the other hand, a large induction machine is assumed to satisfy the relationship $MX' < \sqrt{MX'} < T'$. Since these parameters can be made to appear on the left-hand side of (4.31)–(4.33), a natural approach is to express this latter model as a two-time-scale singularly-perturbed system for both types of machines. Depending on the order of magnitude of these key parameters, a different decomposition of slow and fast variables will result.

4.3.2. First-order speed model for small induction machines. In the case of small machines, a scaling of the speeds ω and ω_s by the factor T' yields a singularly-perturbed system where the small parameter is T'^2/MX'. The slow variable is the modified speed variable $\Omega - \Omega_s = T'(\omega - \omega_s)$ and E' and δ are the fast variables. To find a slow model we search for time-independent rotor manifolds of the form [10]

$$E' = h_1(\Omega, \frac{T'^2}{MX'}) \quad (4.34)$$

$$\delta = h_2(\Omega, \frac{T'^2}{MX'}) \quad (4.35)$$

The zero-order approximations of these manifolds are obtained as

$$E' \cong \frac{(X - X')V}{X\sqrt{1 + (\Omega - \Omega_s)^2}} \quad (4.36)$$

$$\delta \cong \tan^{-1}(\Omega - \Omega_s) \quad (4.37)$$

These approximations can be viewed as the solutions to (4.31)–(4.32) taken as algebraic equations although this may not be true in general. A first-order speed model governing the dominant dynamics of small induction machines on these manifolds is, in terms of the original shaft speed ω,

$$M\frac{d\omega}{dt} = -V^2(\frac{X - X'}{XX'})\frac{T'(\omega - \omega_s)}{1 + T'^2(\omega - \omega_s)^2} - T_m \quad (4.38)$$

An alternate approach using (4.26)–(4.28) yields the same speed model and the following equivalent approximations of the rotor manifolds

$$E'_q \cong \left(\frac{X-X'}{X}\right)\frac{V}{1+T'^2(\omega-\omega_s)^2} \tag{4.39}$$

$$E'_d \cong -\left(\frac{X-X'}{X}\right)\frac{VT'(\omega-\omega_s)}{1+T'^2(\omega-\omega_s)^2} \tag{4.40}$$

We can also derive a second-order model in the fast deviations of E' and δ from the manifold (4.36)–(4.37). However, this model does not present any interest as the corresponding off-manifold variables decay exponentially fast towards the rotor manifolds. Hence, after a short time following an initial transient, these variables reach the manifolds (4.36)–(4.37) or (4.39)–(4.40) and remain there. The remainder of the dynamics of the induction machine is governed by the speed model (4.38) which becomes the dominant behavior.

4.3.3. First-order voltage model for large induction machines. Due to the fact that the parameter T' is greater than $\sqrt{MX'}$ for large machines, a different decomposition of slow and fast variables results. Whereas this decomposition is not apparent in (4.26)–(4.28), it is readily available from (4.31)–(4.33). Indeed, it can be checked that a rescaling of the speed variables ω and ω_s by the factor $\sqrt{MX'}$, i.e., $\Omega = \sqrt{MX'}\omega$ and $\Omega_s = \sqrt{MX'}\omega_s$, produces a singularly-perturbed system in standard form. The slow variable is the voltage E', the fast variables are δ and $\Omega - \Omega_s$, and the small parameter is $\sqrt{MX'}/T'$.

In this case, we search for rotor manifolds of the form [10]

$$\delta = h_1\left(E', \frac{\sqrt{MX'}}{T'}\right) \tag{4.41}$$

$$\Omega - \Omega_s = h_2\left(E', \frac{\sqrt{MX'}}{T'}\right) \tag{4.42}$$

The zero-order approximations of these manifolds are

$$\delta = \sin^{-1}\left(-\frac{X'T_m}{VE'}\right) \tag{4.43}$$

$$\Omega - \Omega_s = 0 \tag{4.44}$$

Equation (4.44) does not mean that $\omega - \omega_s = 0$ which is clearly incorrect. It should be remembered that we are performing an asymptotic analysis where $\sqrt{MX'}/T'$ is an infinitesimal parameter that is capable of assuming arbitrary small values in the limiting process. By virtue of (4.44), the Ω-manifold has a nontrivial first-order approximation in the parameter $\sqrt{MX'}/T'$ given by [10]

$$\Omega - \Omega_s \cong -\frac{\sqrt{MX'}}{T'}\frac{X'T_m}{\sqrt{(VE')^2-(X'T_m)^2}} \tag{4.45}$$

A corresponding approximate ω-manifold is deduced as

$$\omega - \omega_s \cong -\frac{X'T_m}{T'\sqrt{(VE')^2 - (X'T_m)^2}} \tag{4.46}$$

and is different from the solution obtained by taking (4.32)–(4.33) as algebraic equations. Substitution of (4.43) into (4.31) results in the following first-order voltage model

$$T_o'\frac{dE'}{dt} = -\frac{X}{X'}E' + (\frac{X - X'}{X'})V\sqrt{1 - (\frac{X'T_m}{VE'})^2} \tag{4.47}$$

which describes the dominant behavior of large induction machines. This new result has an important bearing on the corresponding dynamic behavior of large and small induction motors. Clearly, the dominant behaviors are quite different and are governed by the reduced models (4.38) and (4.47), respectively. Further analytical work is needed to understand the mechanisms of voltage instabilities associated with large induction machines.

4.4. A Lyapunov function for third-order induction machine models. Unlike for the synchronous machine, relatively little work has been devoted to the application of Lyapunov stability theory to the induction machine. Consider the following scalar function in the variables E', δ and ω:

$$\begin{aligned} V(E', \delta, \omega) &= \frac{1}{2}M(\omega - \omega_o)^2 \\ &+ \frac{1}{2}\frac{X}{X'}\frac{1}{X - X'}(E' - (\frac{X - X'}{X})V\cos\delta)^2 \\ &+ \int_{\delta_o}^{\delta}((\frac{1}{X'} - \frac{1}{X})\frac{V^2\sin 2\delta}{2} + T_m)d\delta \end{aligned} \tag{4.48}$$

where $(E_o', \delta_o, \omega_o)$ is the stable equilibrium point among two possible equilibrium points of (4.31)–(4.33). (See details in [10].) The function inside the integral sign satisfies a sector condition around the stable equilibrium point. Thus, this function satisfies the following conditions:

- $V(E_o', \delta_o, \omega_o) = 0$.
- $V(E', \delta, \omega) > 0$ locally around the stable equilibrium point.

After some algebraic manipulations, the time derivative of the function V(E', δ, ω) along trajectory solutions can be written in the form

$$\frac{dV}{dt} = -\frac{(X - X')V^2}{T_o'X'^2}\phi(E', \delta) \tag{4.49}$$

where

$$\begin{aligned} \phi(E', \delta) &= \frac{E'^2}{E_o'^2}\cos^2\delta_o + 1 - \frac{2E'}{E_o'}\cos\delta\cos\delta_o + \sin^2\delta_o \\ &- \sin\delta\sin\delta_o(\frac{E'}{E_o'} + \frac{E_o'}{E'}) \end{aligned} \tag{4.50}$$

The Hessian matrix of ϕ evaluated around the stable equilibrium point yields the following 2×2 matrix

$$\Phi = \begin{bmatrix} \frac{2\cos 2\delta_o}{E_o'^2} & \frac{\sin 2\delta_o}{E_o'} \\ \frac{\sin 2\delta_o}{E_o'} & 2 \end{bmatrix} \tag{4.51}$$

The conditions for local positive definiteness of ϕ are:

$$\cos 2\delta_o > 0 \tag{4.52}$$

$$\cos^2 2\delta_o + 4\cos 2\delta_o - 1 > 0 \tag{4.53}$$

The first condition yields $\delta_o < -45^o$ which is to be expected. However, the second one yields $\cos\delta_o > -2 + \sqrt{5}$ or $\delta_o < -38.17^o$.

Thus, the proposed scalar function $V(E', \delta, \omega)$ is a (local) Lyapunov function provided the stable equilibrium point $\delta_o < -38.17^o$. Open areas for research in this area include estimating the domain of attraction of a stable equilibrium point using the proposed Lyapunov function; comparing the regions of attraction for both small and large induction machines; and, finally, pursuing the search for "better" Lyapunov functions.

5. Conclusion. The method of integral manifolds has been used in this paper to illustrate the systematic generation of improved reduced-order models of three-phase synchronous and induction machines. Exact stator manifolds for both types of machines exist under the assumption of zero stator resistance. The traditional assumption of neglecting stator transients can be justified when the initial conditions of the stator variables are close enough to these stator manifolds. After eliminating stator transients, various approximate rotor manifolds can be derived depending on the size of the synchronous or induction machine considered. The dominant dynamical behavior in synchronous machines is characterized by a second-order electromechanical model for small machines and by a first-order voltage model in large machines. In contrast, the dominant behavior in induction machines is characterized by a first-order speed model for small machines and by a first-order voltage model for large machines. Finally, a local Lyapunov function for third-order induction motor models was found around stable equilibrium points characterized by torque angles less than about -38.17^o.

REFERENCES

[1] N. Fenichel, *Persistence and Smoothness of Invariant Manifolds for Flows*, Indiana University Mathematics Journal, vol. 21, no. 3, (1971), pp. 193–226.

[2] A. Kelley, *The Stable, Center-Stable, Center, Center-Unstable, Unstable Manifolds*, J. Differential Equations, vol. 3, (1967), pp. 546-570.

[3] P. Holmes and J. Marsden, *Dynamical Systems and Invariant Manifolds*, New Approaches to Nonlinear Problems in Dynamics, (edited by) P. Holmes, SIAM, Philadephia, 1980.

[4] N. Fenichel, *Geometric Singular Perturbation Theory of Ordinary Differential Equations*, J. Differential Equations, vol. 31, (1979), pp. 53-98.

[5] J. Carr, *Application of Centre Manifold Theory*, New York: Springer-Verlag, 1981.

[6] M. Odyniec and L. O. Chua, *Integral Manifolds for Non-Linear Circuits*, Circuit Theory and Applications, vol. 12, (1984), pp. 293-328.

[7] P. W. Sauer, S. Ahmed-Zaid, and P. V. Kokotovic, *An Integral Manifold Approach to the Reduced Order Modeling of Synchronous Machines*, IEEE Transactions on Power Systems, vol. PWRS-3, no. 1, February (1988), pp. 17-23.

[8] P. V. Kokotovic and P. W. Sauer, *Integral Manifold as a Tool for Reduced-Order Modeling of Nonlinear Systems: A Synchronous Machine Case Study*, IEEE Transactions on Circuits and Systems, vol. 36, no. 3, March (1989), pp. 403–410.

[9] E. Drennan, S. Ahmed-Zaid and P. W. Sauer, *Invariant Manifolds and Start-up Dynamics of Induction Machines*, Proceedings of the 21st Annual North-American Power Symposium, University of Missouri-Rolla, Rolla, Missouri, Oct. 9-10, 1989.

[10] S. Ahmed-Zaid and M. Taleb, *Structural Modeling of Small and Large Induction Machines Using Integral Manifolds*, IEEE Transactions on Energy Conversion, vol. 6, no. 3, Sept (1991), pp. 529-535.

[11] P. C. Krause, *Analysis of Electric Machinery*, New York: McGraw-Hill Book Co., 1986.

THE BCU METHOD FOR DIRECT STABILITY ANALYSIS OF ELECTRIC POWER SYSTEMS: THEORY AND APPLICATIONS

HSIAO-DONG CHIANG*

Abstract. The controlling unstable equilibrium point (u.e.p.) method is considered to be the most viable for direct stability analysis of practical power systems. The success of the controlling u.e.p. method, however, hinges upon its ability to find the (correct) controlling u.e.p. Recently, a systematic method, called boundary of stability region based controlling unstable equilibrium point method (BCU method), to find the controlling u.e.p. was developed. The BCU method has been evaluated in a large-scale power system and it compared favorably with other methods in terms of its reliability and the required computational efforts. This paper presents an overview of the BCU method for the network-reduction classical machine stability model in conjunction with its theoretical basis. In addition to an overview, some new material is offered. The application of the BCU method to large-scale power systems is also presented. Furthermore, this paper develops a method which can find the (exact) controlling u.e.p. of general power system models relative to any given fault as long as such a controlling u.e.p. exists. This method is based on the time-domain simulation approach and therefore it is slow in nature. This method therefore is not intended to be used as a 'direct' method for transient stability analysis, rather the role of this method is to serve as a bench-mark to check the correctness of the controlling u.e.p. obtained by any of the direct methods.

1. Introduction. A power system is continually experiencing disturbances. These may be classified as: *event disturbances* and *load disturbances*. Event disturbances include generator outages, short-circuits caused by lightning, sudden large load changes, or a combination of such events. Event disturbance usually lead to a change in the configuration of the power system. Load disturbances, on the other hand, are the small random fluctuations in load demands. The system configuration usually remains unchanged after load disturbances. Stability analysis is concerned with a power system's ability to reach an acceptable steady-state (operating condition) following an event disturbance. The power system under this circumstance can be considered as going through changes in configuration in three stages: from pre-fault, fault-on to post-fault systems. The pre-fault system is usually in a stable steady state. The fault occurs (e.g., a short circuit), and the system is then in the fault-on condition before it is cleared by the protective system operation. Stability analysis is the study of whether the post-fault trajectory will converge (tend) to an acceptable steady-state as time passes.

---

* School of Electrical Engineering, Cornell University, Ithaca, NY 14850. The author gratefully acknowledges support in part from National Science Foundation under grant number ECS-8957878 and in part from Electric Power Research Institute under grant number RP2473-61. The author wishes to express appreciation to Prof. F. F. Wu and Prof. P. P. Varaiya of University of California, Berkeley, and to Mr. C. C. Chu, Prof. J. S. Thorp, Dr. J. Tong, Mr. T. P. Conneen and Ms. K. N. Miu of Cornell University for their useful discussions.

Stability analysis programs are used by power system planning and operating engineers to predict the response of the system to various disturbances. In these simulations, the behavior of a present or proposed power system is evaluated to determine its operating limits or to determine the need for additional facilities. Important conclusions and decisions are made based on the results of stability studies. It is therefore important to ensure that the results of stability studies are as prompt and accurate as possible.

Until recently, stability analysis has been performed in power companies exclusively by means of numerical integrations to calculate the behavior of the generators for a given fault (i.e. disturbance). By examining the behavior of the generators, one determines whether stability has been maintained or lost. The chief disadvantage with this practice is that it requires time-consuming numerical integration. In contrast, direct methods determine system stability directly based on energy function. These methods determine whether or not the system will remain stable once the fault is cleared by comparing the system energy (when the fault is cleared) to a calculated threshold value. Direct methods not only avoid the time-consuming solutions of step-by-step time-domain stability analysis of the post-fault system, but also provide a quantitative measure of the degree of system stability. This additional information makes direct methods very attractive when the relative stability of different plans must be compared or when stability limits must be calculated quickly. Furthermore, information regarding the degree of stability can be used to quickly determine critical clearing times for given disturbances. In this paper, direct methods refer to methods based on energy function that can directly determine system stability without the explicit numerical integrations of post-fault system models.

Another important advantage of the direct method over the conventional time-domain approach is its ability to quickly determine the stability limit due to changes in critical parameters such as tie-line flows, real power generations, etc. Very often operators in power companies are faced with the following situation which they must evaluate under time pressure. With a given power system and a fault contingency, called a base case, parameters (such as inter-tie line flows, real power generations) in the power system will change (due to, say, some emergency, or generation rescheduling, and load shedding) and form a new system. The effect on system stability must be determined as quickly as possible. Again, the direct method can meet this need.

The direct method requires an energy function but not every power system model has it. Hence, the direct method suffer from limited modeling capability. At present, the direct method can deal with the following models: the classical generator model or a two-axis generator model, an one gain/one time constant exciter model, two-terminal HVDC, and certain nonlinear static loads. It is fair to conclude that the direct method superiors over the traditional time-domain approach in 2.1) its speed in determining stability and stability loading limits, 2.2 its ability to derive preventive control, and 2.3 it ability to provide information on degree of stability. Direct methods are intended to complement traditional time-domain stability analysis in these regards.

Among the existing direct methods, the closest unstable equilibrium point (u.e.p.) method gives very conservative stability assessments, mainly because this method is independent of the fault location [1]. The potential energy boundary surface (PEBS) method proposed by Kakimoto et al. [2] gives fairly fast and accurate stability assessments but may give inaccurate results (both over-estimates and under-estimates) [3,4]. A fault-dependent method using the concept of controlling u.e.p. makes the direct method more practical [5]. From an analytical viewpoint, transient stability analysis is essentially the problem of determining whether or not the fault-on trajectory at clearing time is lying inside the stability region of a desired stable equilibrium point (s.e.p.) of its post-fault system. Hence, the main point in transient stability analysis is not to estimate the whole stability boundary of the post-fault system. Instead, only the relevant part of stability boundary toward which the fault-on trajectory is heading is of main concern. When the closest u.e.p. method is applied to power system transient stability analysis, this method has been found to yield conservative results. In fact, in the context of transient stability analysis, the closest u.e.p. method provides an approximated stability boundary for the post-fault system, and is independent of the fault-on trajectory. Thus, the closest u.e.p. method gives conservative results for transient stability analysis.

A desirable method for transient stability analysis would be the one which can provide the most accurate approximation of the part of the stability boundary toward which the fault-on trajectory is heading, even though it might provide a very poor estimate of the other part of stability boundary. To this end, the controlling u.e.p. method uses the (connected) constant energy surface passing through the controlling u.e.p. to approximate the part of stability boundary of the post-fault system toward which the fault-on trajectory is heading. If, when the fault is cleared, the trajectory lies inside the (connected) energy surface passing through the controlling u.e.p., then the post-fault system will be stable (i.e. the post-fault trajectory will settle down to a stable operating point); otherwise, the post-fault system will be unstable. This is the essence of the controlling u.e.p. method.

To appreciate the difficulty in finding the controlling u.e.p. for large-scale power systems, recall that the controlling u.e.p. of a fault-on trajectory is the u.e.p. whose stable manifold contains the exit point of the trajectory. Thus, the difficulty derives in part from the following complexities:

- the controlling u.e.p. is a particular u.e.p. imbedded in a large-degree state-space

- the controlling u.e.p. is the first u.e.p. whose stable manifold is hit by the fault-on trajectory (at the exit point).
- the exit point is very involved to compute.

A consensus seems to have emerged that among several direct methods, the controlling u.e.p. method is the most viable for direct stability analysis of practical power systems [6,7,8,9]. The success of the controlling u.e.p. method, however, hinges upon its ability to find the (correct) controlling u.e.p. Thus, there is a strong need for a reliable and systematic method to find the controlling u.e.p.

Recently, a systematic method, called boundary of stability region based controlling unstable equilibrium point method (BCU method), to find the controlling u.e.p. was developed [10,11]. The method was also given other names such as the exit point method [12,13,7] and the hybrid method [6]. The BCU method has been evaluated in a large-scale power system and it compared favorably with other methods in terms of its reliability and the required computational efforts [12,13]. The BCU method has been applied to the fast derivation of power transfer limits [14] and applied to real power rescheduling to increase dynamic security [16].

The fundamental ideas behind the BCU method can be explained in the following. Given a power system stability model (which admits an energy function), the BCU method first explores special properties of the underlying model with the aim to define an artificial, dimension-reduced system such that the following conditions are met:

Static properties :

- the locations of equilibrium points of the dimension-reduced system correspond to the locations of equilibrium points of the original system. For example, $\hat{x}$ is an e.p. of the dimension-reduced system if and only if $(\hat{x}, 0)$ is an e.p. of the original system, where $0 \in R^m$, m is an appropriate positive integer.
- the types of equilibrium points of the dimension-reduced system are the same as those of the original system. For example, x_s is a stable equilibrium point of the dimension-reduced system if and only if $(x_s, 0)$ is a stable equilibrium point of the original system.

Dynamical properties :

- there exists an energy function for the artificial, dimension-reduced system.
- an equilibrium point, say x_i, is on the stability boundary $\partial A(x_s)$ of the dimension-reduced system if and only if the equilibrium point $(x_i, 0)$ is on the stability boundary $\partial A(x_s, 0)$ of the original system.
- it is much easier to identify the stability boundary $\partial A(x_s)$ of the dimension-reduced system than to identify the stability boundary $\partial A(x_s, 0)$ of the original system.

The BCU method then finds the controlling u.e.p. of the dimension-reduced system by exploring the special structure of the stability boundary and the energy function of the dimension-reduced system. Third, it relates the controlling u.e.p. of the artificial system to the controlling u.e.p. of the original system. In summary, given a power system stability model, the fundamental ideas behind the BCU method can be roughly described as: (i) it explores the special structure of the underlying model so as to define an artificial, dimension-reduced system which captures all the equilibrium points on the stability boundary of the original system, and (ii) it finds the controlling u.e.p. of the original system via the controlling u.e.p. of the dimension-reduced system, which is much easier to find than that of the original system. Therefore, given a power system stability model with certain properties, there exists a corresponding version of the BCU method.

The main purpose of this paper is to present an overview of the BCU method for the network-reduction classical machine stability model in conjunction with its theoretical basis. In addition to an overview, some new material is offered. The application of the BCU method to large-scale power systems is also presented. Furthermore, this paper develops a method which can find the (exact) controlling u.e.p. of general power system models relative to any given fault as long as such a controlling u.e.p. exists. This method is based on the time-domain simulation approach and therefore it is slow in nature. Consequently, it is not intended to be used as a 'direct' method for transient stability analysis, rather the role of this method is to serve as a bench-mark to check the correctness of the controlling u.e.p. obtained by any of the direct methods.

2. Preliminaries. Mathematically the power system stability problem can be phrased as follows. In the pre-fault regime the system is at a known stable equilibrium point, say x_i. At some time t_f, the system undergoes a fault, which results in a structural change in the system. Suppose the fault duration is confined to the time interval $[t_f, t_p]$. During this interval, the system is governed by fault-on dynamics described by

$$\dot{x}(t) = f_F(x(t)), \qquad t_f \leq t < t_p \tag{2.1}$$

where x(t) is the vector of state variables of the system at time t. The fault is cleared at time t_p and the system is henceforth governed by post-fault dynamics described by

$$\dot{x}(t) = f(x(t)) \qquad t_p \leq t < \infty \tag{2.2}$$

Next, assume that the post-fault system (2.2) has a (asymptotically) stable equilibrium point x_s which satisfies operational constraints. The fundamental problem of transient stability is the following: starting from the post-fault initial state x(t_p), will the post-fault system settle down to the steady state condition x_s? In other words, transient stability analysis is to determine whether the initial point of post-fault trajectory (i.e. the final point of fault-on trajectory) is located inside the stability region (domain of attraction) of an acceptable stable equilibrium point (acceptable steady state). The problem of direct stability analysis can be simply translated into the following: given a set of nonlinear equations with an initial condition, determine whether or not the ensuing trajectories will settle down to a desired steady-state without resorting to explicit numerical integrations.

It is further assumed in the area of direct methods that the following conditions are satisfied:

[A1] the pre-fault stable equilibrium point, x_s^{pre}, and the post-fault equilibrium point, x_s, are sufficiently close to each other (so a nonlinear algebraic solver, such as the Newton method, with x_s^{pre} as the initial guess will find x_s) and

[A2] the pre-fault stable equilibrium point, x_s^{pre}, lies inside the stability region of the post-fault stable equilibrium point x_s.

Next, we state two theorems and physical arguments to support the two assumptions.

THEOREM 2.1. *(robustness of a stable equilibrium point) Let* $f: R^n \rightarrow R^n$ *be a* C^1 *vector field and* x_s *be a stable equilibrium point of* $\dot{x} = f(x)$. *Then, there exist a neighborhood* $U \in R^n$ *of* x_s *and a* C^1 *neighborhood* ϑ *of f such that for any* $g \subset \vartheta$ *there is a unique stable equilibrium point* $\hat{x}$ *of* $\dot{x} = f(x)$. *Moreover, for any* $\epsilon > 0$ *we can choose* ϑ *so that* $||\hat{x} - x_s|| \leq \epsilon$.

THEOREM 2.2. *[15]: (Property of stability regions) The stability region of general nonlinear dynamical systems is an open, invariant set which is diffeomorphic to* R^n, *where n is the dimension of the system.*

We employ these two theorems to argue that the above two assumptions are generally true as follows: the energy imbalance due to a fault, large or small, is relatively small compared with the total energy level of the system. The effect of the fault can be mathematically viewed as a small perturbation to the algebraic equations describing the steady-state of the pre-fault system. If the pre-fault stable equilibrium point is hyperbolic, which is a generic property, then the post-fault stable equilibrium point can be shown to be close to the pre-fault equilibrium point, provided the perturbation is sufficiently small as stated in Theorem 2.1. The second assumption can be explained as follows: since the stability region is an open set (see Theorem 2.2) with noticable size (unless the system is close to a bifurcation which affects the stability property of the stable equilibrium point), it is very plausible that assumption (ii) is also satisfied.

We next review some notions from nonlinear dynamical systems theory. To unify our notation, let

$$\dot{x}(t) = f(x(t)) \tag{2.3}$$

be the power system model under study, where the state vector x(t) belongs to the Euclidean space R^n, and the function $f : R^n \rightarrow R^n$ satisfies the sufficient condition for the existence and uniqueness of solutions. The solution curve of (2.3) starting from x at t=0 is called a (system) trajectory, denoted by $\Phi(x,\cdot) : R \rightarrow R^n$. Note that $\Phi(x,0) = x$.

The concepts of *equilibrium point* (e.p.), *invariant set, limit set* including *α-limit set* and *ω-limit set, stable* and *unstable manifolds* are important in dynamical system theory. Each of these concepts is defined next. A detailed discussion of these concepts and their implications may be found in [17]–[21].

A state vector $\hat{x}$ is called an equilibrium point of system (2.3) if $f(\hat{x}) = 0$. We denote E to be the set of equilibrium points of the system. A state vector x is called a *regular point* if it is not an equilibrium point. We say that an equilibrium point of (2.3) is *hyperbolic* if the Jacobian of $f(\cdot)$ at $\hat{x}$, denoted $J_f(x)$, has no eigenvalues with a zero real part. For a hyperbolic equilibrium point, it is a *(asymptotically) stable equilibrium point* if all the eigenvalues of its corresponding Jacobian have negative real parts; otherwise it is an *unstable equilibrium point*. If the Jacobian of the equilibrium point $\hat{x}$ has exactly one eigenvalue with positive real part, we call it a *type-one equilibrium point*. Likewise, $\hat{x}$ is called a *type-k equilibrium point* if its corresponding Jacobian has exactly k eigenvalues with positive real part. Throughout this section, it will be assumed that all the equilibrium points of system (2.3) are hyperbolic.

A set $M \in R^n$ is called an *invariant set* of (2.3) if every trajectory of (2.3) starting in M remains in M for all t. A point p is said to be in the ω-limit set (respectively, α-limit set) of x if corresponding to each $\epsilon > 0$ and each $T > 0$ (respectively, $T < 0$) there is a $t > T$ (respectively, $t < T$) with the property that $||\Phi(x,t) - p|| < \epsilon$. This is equivalent to saying that there is a sequence t_i in R, $t_i \rightarrow \infty$ (or $t_i \rightarrow -\infty$), with the property that $p = lim_{i\rightarrow\infty}\Phi(x,t_i)$. Thus, the ω-limit set of a trajectory captures the asymptotic behavior of the trajectory in positive time while the α-limit set of a trajectory captures the asymptotic behavior of the trajectory in negative time. The limit set of a trajectory includes its α-limit set and its ω-limit set. Usually, determining the limit set of a nonlinear dynamical system is not an easy task.

Let $\hat{x}$ be a hyperbolic equilibrium point. Its stable and unstable manifolds, $W^s(\hat{x})$ and $W^u(\hat{x})$, are defined as follows:

$$W^s(\hat{x}) := \{ x \in R^n : \Phi(x,t) \rightarrow \hat{x} \ as \ t \rightarrow \infty \}$$

$$W^u(\hat{x}) := \{ x \in R^n : \Phi(x,t) \rightarrow \hat{x} \ as \ t \rightarrow -\infty \}$$

These two sets are invariant sets (see Fig. 2.1). Clearly, $\hat{x}$ is the ω-limit set of every point in $W^s(\hat{x})$ and $\hat{x}$ is the α-limit set of every point in $W^u(\hat{x})$.

The idea of *transversality* is basic in the study of dynamical systems. If A and B are two manifolds, we say that they satisfy the *transversality*

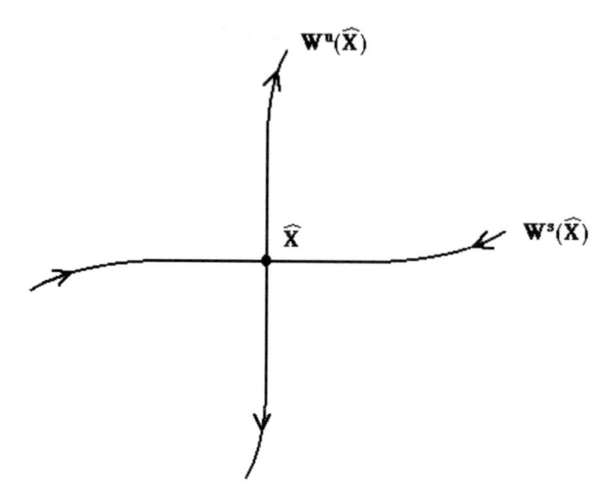

FIG. 2.1. *The stable and unstable manifolds of an equilibrium point. They are invariant sets.*

condition if either (i) at every point of intersection $x \in A \cap B$, the tangent spaces of A and B span the tangent spaces of M at x, i.e.

$$T_x(A) \bigoplus T_x(B) = T_x(M) \quad for\ x \in A \cap B$$

or (ii) they do not intersect at all. The transversal intersection persists under perturbation of the vector field.

For a stable equilibrium point, it can be shown that there exists a number $\delta > 0$ such that $||x_0 - \hat{x}|| < \delta$ implies $\Phi(x_0, t) \to \hat{x}$ as $t \to \infty$. If δ is arbitrarily large, then $\hat{x}$ is called a *global stable equilibrium point*. There are many physical systems containing stable equilibrium points but not global stable equilibrium points. A useful concept for these kinds of systems is that of *stability region* (or *region of attraction*). The stability region of a stable equilibrium point x_s is defined as

$$A(x_s) \ := \ \{x \in R^n \ : \ \lim_{t \to \infty} \Phi(x, t) = x_s\}$$

From a topological point of view, the stability region $A(x_s)$ is an open, invariant and connected set. The boundary of stability region $A(x_s)$ is called the *stability boundary* (or *separatrix*) of x_s and will be denoted by $\partial A(x_s)$ (see Fig.2.2). The stability boundary is topologically an (n-1)-dimensional closed, invariant set.

We say a function $V : R^n \mapsto R$ is an energy function for system (2.3) if the following three conditions are satisfied:

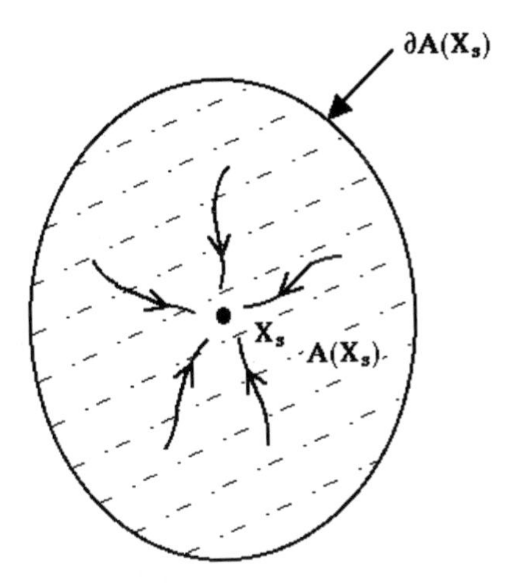

FIG. 2.2. *The stability region of a stable equilibrium point is an open, invariant and connected set. The boundary of stability region $A(x_s)$ is called the stability boundary which is topologically an (n-1)-dimensional closed, invariant set.*

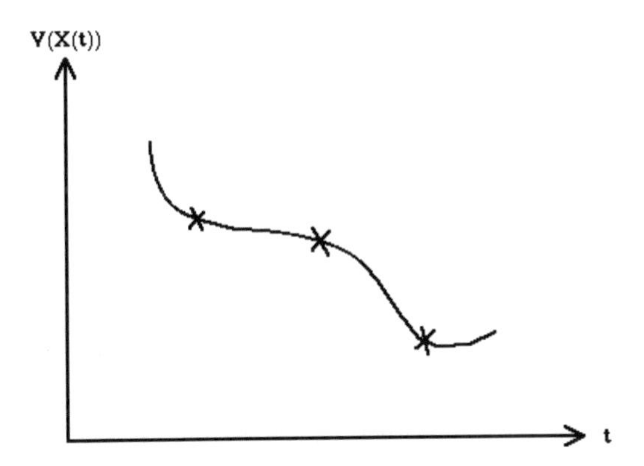

FIG. 2.3. *Along the trajectory $x(t)$, there are countable points at which the derivative of energy functions is zero. The energy function value strictly decreases along any system trajectory. Furthermore, along any system trajectory, the energy function is a proper map.*

(i) the derivative of the energy function $V(x)$ along any system trajectory $x(t)$ is non-positive, i.e.

$$\dot{V}(x(t)) \leq 0$$

(ii) If x(t) is a non-trivial trajectory (i.e. x(t) is not an equilibrium point (e.p.)), then there does not exist a time interval, say $[t_1, t_2]$, $t_2 > t_1$, such that $\dot{V}(x(t)) = 0$ for $t \in [t_1, t_2]$. Mathematically, this can be expressed as follows: along any non-trivial trajectory x(t) the set

$$\{t \in R \,:\, \dot{V}(x(t)) = 0\}$$

has measure zero in R.

(iii) If a trajectory $x(t)$ has a bounded value of $V(x(t))$ for $t \in R^+$, then the trajectory x(t) is also bounded. Stating this in brief:
if $V(x(t))$ is bounded, then x(t) is also bounded.

Property (i) indicates that the energy is non-increasing along its trajectory, but does not imply that the energy is strictly decreasing along its trajectory. There may exist a time interval $[t_1, t_2]$ such that $\dot{V}(x(t)) = 0$ for $t \in [t_1, t_2]$. Properties (i) and (ii) imply that the energy is strictly decreasing along any system trajectory. Property (iii) states that, along any system trajectory, the energy function is a proper map but its energy need not be a proper map for the entire state space (see Fig.2.3). Obviously, an energy function is not a Lyapunov function.

3. Energy function theory. This section reviews some analytical results on energy function theory that are needed for subsequent analysis. A more comprehensive development of energy function theory can be found in [9]. This section also shows that all equilibrium points generically have different values of energy.

In general, the behaviors of trajectories of general nonlinear dynamical systems could be very complicated, unless the underlying dynamical system has some special properties. For instance, every trajectory of system (2.3) having an energy function has only two modes of behaviors: it either converges to an equilibrium point or goes to infinity (becomes unbounded) as time increases or decreases.

THEOREM 3.1. *[11,9] (Global behavior of trajectories)*
If there exists a function, satisfying the condition (i) and (ii) of the energy function conditions, for system (2.3), then every trajectory of system (2.3) either converges to one of the equilibrium points or diverges in both forward and backward time.

Theorem 3.1 indicates that there does not exist any limit cycle (oscillation behavior) or bounded complicated behavior such as almost periodic trajectory, chaotic motion, etc. in the system. Applying this result to power system models, it indicates that for a power system model with an energy function, there is no complicated behavior such as chaotic motion and closed orbit (limit cycle). Note that the chaotic motion of a simple power system shown by Kopell and Washburn [22] is based on a classical model with zero damping, and the function they employed to analyze the simple system is not an energy function.

In Theorem 3.1 we have shown that the trajectory of system (2.3) either converges to one of the equilibrium points or goes to infinity. However, in the following, we will show that every trajectory on the stability boundary must converge to one of the equilibrium points on the stability boundary as time increases.

THEOREM 3.2. *[8] (The behavior of trajectories on the stability boundary)*

If there exists an energy function for system (2.3), then every trajectory on the stability boundary $\partial A(x_s)$ converges to one of the equilibrium points on the stability boundary $\partial A(x_s)$.

The significance of this theorem is that it offers a way to characterize the stability boundary. In fact, Theorem 3.2 implies that the stability boundary $\partial A(x_s)$ is composed of several stable manifolds of the u.e.p.'s on the stability boundary.

COROLLARY 3.3. *(Energy function and stability boundary)*

If there exists an energy function for the system (2.3) which has an asymptotically stable equilibrium point x_s (but not globally asymptotically stable), then the stability boundary $\partial A(x_s)$ is contained in the set which is the union of the stable manifolds of the u.e.p.'s on the stability boundary $\partial A(x_s)$, i.e.

$$\partial A(x_s) \subseteq \cup_{x_i \in \{E \cap \partial A(x_s)\}} W^s(x_i)$$

Next, we consider the following question: is it possible that two equilibrium points of (2.3) have the same energy function value $V(\cdot)$ (i.e. $V(x_1) = V(x_2)$ for $x_1, x_2 \in E$)? This question is related to the issue of the uniqueness of the closest u.e.p. and the controlling u.e.p. Note, all the potential energy $V_p(x)$ in the (total) energy functions $V(\cdot)$ for different power system models have the property that $\nabla V_p(x) = 0$ at the equilibrium point. To answer this question, first notice that for two equilibrium points x_1, x_2 to have the relationship that $V_p(x_1) = V_p(x_2)$ and $\nabla V_p(x_1) = \nabla V_p(x_2)$. Note that this is a set of (2n+1) equations with the 2n unknowns, (x_1, x_2), it is unlikely that there exists a solution. A rigorous justification of this observation is presented in Theorem 3.4 below. This theorem states that generically, all the equilibrium points of system (2.3) have distinct energy function values. Note that let X be a complete metric space and P(x) a statement about points x in X. We say that P(x) is a *generic property* if the set of points where it holds true contains a countable intersection of open dense sets.

THEOREM 3.4. *(Energy functions and equilibrium points)*
Let $V(\cdot)$ be an energy function for system (3) and $\nabla\ V(x) = 0$ at the equilibrium points of system (2.3). Then, the property that all the equilibrium points of system (2.3) have distinct values of energy function $V(\cdot)$ is generic.

Proof. Let $f : R^n \mapsto R^m$ be a C^r function, $r \geq 1$, and M $\in R^m$ a C^r submanifold of co-dimension q. Let p be a point in R^n. We say that f is *transversal* to M at p if either (i) f(p) does not belong to M, or (ii) f(p) belongs to M, and for some local chart ψ of (M, R^m) around f(p), the corresponding matrix has rank q. We say that the function f($\cdot$) is *transversal* to M if it is transversal to M at every point p $\in R^n$.

Recall Thom's famous transversality theorem. Let Ω be an open set of R^n, $f : V \times \Omega \mapsto R^p$ be a C^r function, $r \geq 1$, and V is a parameter space. Also let M be a C^∞ submanifold of R^p with codimension q.

Consider the following property of the parameter value u

$$P(u) = \{ f(u, \cdot) \text{ is transversal to } M \}$$

THEOREM 3.5. *: (Thom's transversality theorem)*
If r $\geq$ max (1, n-q+1) and f : $V \times \Omega \mapsto R^p$ is transversal to M, then P(u) is a generic property on V.

COROLLARY 3.6. *If r >= max (1, n-q+1), then the property P(f) = f : $\Omega \mapsto R^p$ is transversal to M is generic in $C^r(\Omega; R^p)$.*

This means that if any particular function we are dealing with is not transversal to M, then any function which is an arbitrarily small C^r perturbation of $f(\cdot)$ will be transversal to M.

We are now in a position to prove Theorem 3.4. We shall prove that the property $Q(V) = \{$all the values of energy function V($\cdot$) at the equilibrium points are distinct$\}$ is generic in $C^r(\Omega,\ R)$. The following proof closely follows [23]. Let

Ω be an open set of R^n,

$C^r_\infty(\Omega; R^p)$ be the space of all C^r functions f : $\Omega \mapsto R^p$ such that f($\cdot$) and all its derivatives up to order r are uniformly bounded over Ω,

U $= C^r_\infty(\Omega; R^p)$ and

$\hat{\Omega} := \{(x_1, x_2) \in \Omega \times \Omega : x_1 \neq x_2\}$ an open subset of R^{2n}.

Define a map $Q : U \times \hat{\Omega} \mapsto R^2 \times R^{2n}$ by $Q(f, x_1, x_2) := (f(x_1), f(x_2), \nabla f(x_1), \nabla f(x_2))$. This map $Q(\cdot, \cdot, \cdot)$ is linear in f and C^r in (x_1, x_2).

Let us choose M = (s,s,0,0) : $s \in R$, which is a one dimensional submanifold. According to Thom's transversality theorem, the property P(f) $= \{Q(f, \cdot, \cdot)$ is transversal to M$\}$ is generic. Since the map $Q(f, \cdot, \cdot)$ is from an open subset of R^{2n} into $R^{2(n+1)}$, to say $Q(f, \cdot, \cdot)$ is transversal to M means that whenever $Q(\bar{f}, \bar{x}_1, \bar{x}_2) \in M$, any vector in $R^{2(n+1)}$ can be expressed as the sum of a vector in the tangent space of M (which is of one dimension) and a vector in the tangent space of Q (which is at most a 2n-dimensional manifold). Their vector sum can not be of dimension 2(n+1). Thus, $Q(f, x_1, x_2)$ can't belong to M, and consequently, Q(V) and P(f) have the same property. This completes the proof. □

4. The controlling U.E.P. method. This section presents an overview and analysis of the controlling u.e.p. method for direct stability analysis. Section 4.1 uses a simple example to illustrate the concept of the controlling u.e.p. Section 4.2 studies the controlling u.e.p. method which is justified in Section 4.3.

4.1. Concept. In order to illustrate the concept of the controlling u.e.p., we use the following simple numerical example, which closely represents a three-machine system, with machine number 3 as the reference machine.

$$\begin{aligned}
\dot{\delta}_1 &= \omega_1 \\
\dot{\omega}_1 &= -\,sin\delta_1 - 0.5sin(\delta_1 - \delta_2) - 0.4\delta_1 \\
\dot{\delta}_2 &= \omega_2 \\
\dot{\omega}_2 &= -\,0.5sin\delta_2 - 0.5sin(\delta_2 - \delta_1) - 0.5\delta_2 + 0.05
\end{aligned}$$

It is easy to show that the following function is an energy function for this system.

$$\begin{aligned}
V(\delta_1, \delta_2, \omega_1, \omega_2) = {} & \omega_1{}^2 + \omega_2{}^2 - 2cos\delta_1 - cos\delta_2 - cos(\delta_1 - \delta_2) \\
& - 0.1\delta_2
\end{aligned} \tag{4.1}$$

The point $x^s = (\delta_1{}^s, \omega_1{}^s, \delta_2{}^s, \omega_2{}^s) = (0.02001,\ 0,\ 0.06003,\ 0)$, is a stable equilibrium point of the above post-fault system. There are six type-one equilibrium points and four type-two equilibrium points on the stability boundary of x^s because the unstable manifold of each of these equilibrium points converges to the s.e.p., x^s.

The stability boundary, $\partial A(x^s)$, is contained in the set which is the union of the stable manifolds of these six type-one e.p.'s and four type-two e.p.'s, as shown in Corollary 3.3. Fig. 4.1 shows numerically the intersection between the stability boundary $\partial A(x^s)$ and the angle space, (δ_1, δ_2).

For illustrational purposes, we assume the fault-on trajectory, $x_f(t)$, shown in Fig. 4.2 is due to a fault (of course, the fault-on trajectory $x_f(t)$ in practice has non-zero components of ω.) The fault-on trajectory, $x_f(t)$, leaves the stability boundary, $\partial A(x^s)$, by passing through the stable manifold of the type-one e.p. $x_{co} = (0.03333,\ 0,\ 3.10823,\ 0)$ at the exit point, x_e. x_{co} is termed the controlling u.e.p. relative to the fault-on trajectory $x_f(t)$. The time duration between p and x_e along the fault-on trajectory $x_f(t)$ is the so-called *critical clearing time*. If the fault is cleared before the

TABLE 4.1
Coordinates of type-one and type-two equilibrium points

type-one e.p.	δ_1	ω_1	δ_2	ω_2
1	0.03333	0	-3.17496	0
2	3.24512	0	0.31170	0
3	3.04037	0	3.24307	0
4	0.03333	0	3.10823	0
5	-2.69489	0	1.58620	0
6	-3.03807	0	1.58620	0
7	-3.24282	0	3.24387	0
8	3.04037	0	-3.03931	0
type-two e.p.	δ_1	ω_1	δ_2	ω_2
1	-2.67489	0	1.58620	0
2	-3.66392	0	-2.02684	0
3	2.61926	0	-2.02684	0
4	3.60829	0	1.58620	0

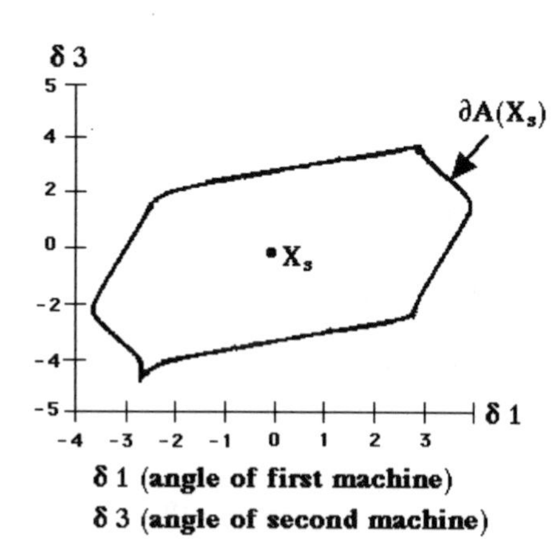

FIG. 4.1. *The intersection between the stability boundary $\partial A(x^s)$ and the angle space (δ_1, δ_2)*

fault-on trajectory reaches x_e, then the initial condition of the post-fault trajectory lies inside the stability region $A(x_s)$; and hence, the corresponding post-fault trajectory will converge to x_s. Otherwise, the corresponding post-fault trajectory will, as stated in Theorem 3.1, will either converge to another stable equilibrium point (physically, this implies a pole-slip phenomenon) or diverge (it becomes unbounded; of course power system protective devices will be activated and the underlying model becomes invalid).

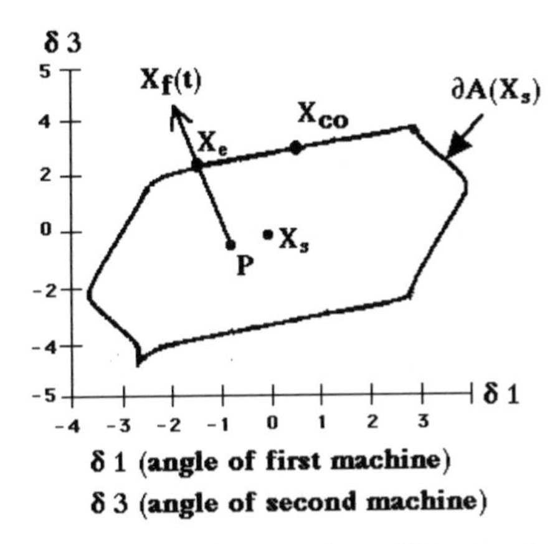

FIG. 4.2. *The fault-on trajectory $x_f(t)$ leaves the stability boundary $\partial A(x^s)$ via passing through the stable manifold of the type-one e.p. x_{co} = (0.03333, 0, 3.10823, 0) at the exit point x_e.*

In general, it is computationally very involved to obtain the exit point, x_e. The only available method seems to be the 'bruce-force' approach; i.e., the traditional time-domain approach. Next, we explain how the controlling u.e.p. method approximates the exit point, x_e. Suppose one can somehow obtain the controlling u.e.p. x_{co}. There are at least two ways to approximate the exit point, x_e: (i) derive the stable manifold of x_{co} (called the stable manifold approach), (ii) use the (connected) constant energy surface of $V(\cdot)$ passing through x_{co} (called the energy function approach). The former one offers little promise because presently, it is not an easy task to derive an explicit expression for stable manifolds. The latter approach is the spirit of the controlling u.e.p. method. Let us go back to the simple example. Fig. 4.3 shows the numerical relationship between the stable manifold $W^s(x_{co})$, the stability boundary and the constant energy surface passing through x_{co}. It is clear from this figure (and it will be rigorously proven in Thereom 4.1) that the fault-on trajectory $x_f(t)$ must pass through the constant energy surface $\partial S(V(x_{co}))$, say at $\hat{x}$, before it passes through the stable manifold of x_{co}. The controlling u.e.p. method uses $\hat{x}$ to approximate x_e.

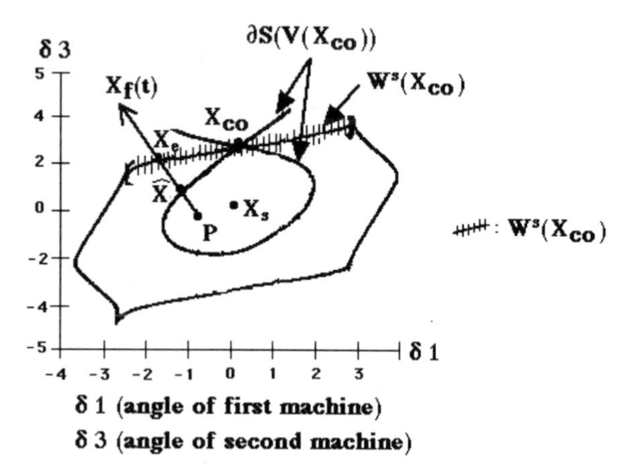

FIG. 4.3. *Numerical relationship between the stable manifold $W^s(x_{co})$, the stability boundary and the constant energy surface passing through x_{co}*

In the operation or operational planning environment, calculating the critical clearing time relative to a fault is not practical because the reclosure times of protective relays in the system are already set. The key concern under this circumstance is whether the system after the fault is cleared will remain stable or not. This concern also favors the energy function approach. This is explained as follows: Given a fault-on trajectory $x_f(t)$ with a preset fault clearing time. Let x_I be the corresponding initial point of the post-fault trajectory and $W^s(x_{co})$ be the stable manifold of the corresponding controlling u.e.p. x_{co}. The task of determining whether the segment of the fault-on trajectory, $x_f(t)$, from p to x_I lies inside the stability region $A(x_s)$ based on the stable manifold $W^s(x_{co})$ is numerically very difficult, if not impossible. The main reason is because of the lack of an explicit expression for stable manifolds. However, the task is relatively easy if an energy function instead of the stable manifold $W^s(x_{co})$, is given. To elaborate on this point, let us go back to the previous numerical example and examine Fig. 4.3 again. The relationship between the stable manifold $W^s(x_{co})$ and the constant energy surface passing through x_{co} is shown. In order for the fault-on trajectory $x_f(t)$ to pass through the constant energy surface $\partial S(V(x_{co}))$, the point x_I must have an energy value greater than the energy value at the controlling u.e.p. x_{co}; i.e., $V(x_I) > V(x_{co})$. Hence, the task of determining whether the segment of the fault-on trajectory $x_f(t)$ from p to x_I lies inside the stability region $A(x_s)$ is boiled down to the task of comparing two scalars: $V(x_I)$ and $V(x_{co})$. If $V(x_I) \leq$

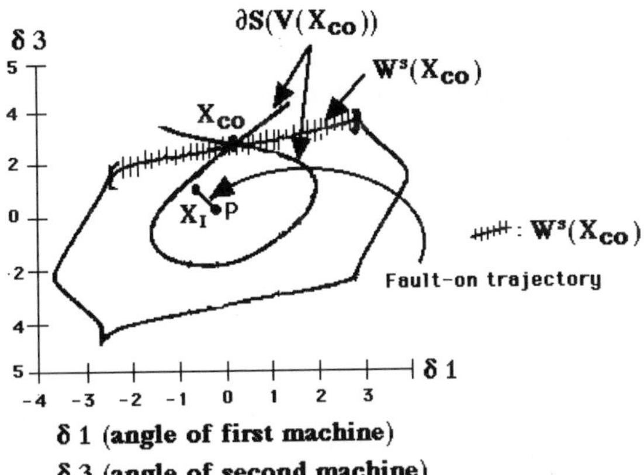

FIG. 4.4. *The intersection between the stability boundary $\partial A(x^s)$ and the angle space (δ_1, δ_2)*

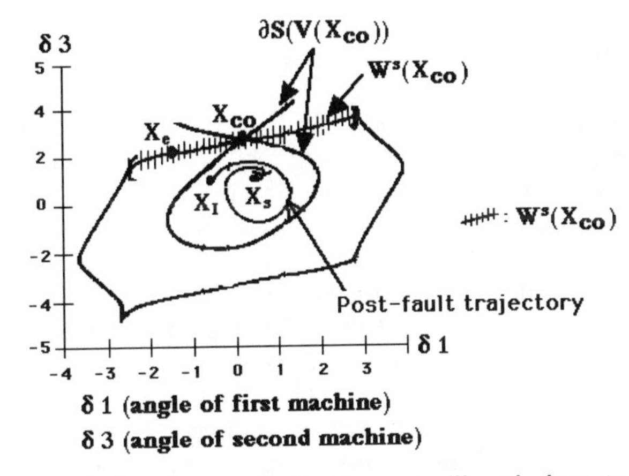

FIG. 4.5. *The post-fault trajectory starting from x_I will settle down to x_s because x_I lies inside the stability region $A(x_s)$*

V(x_{co}), then the controlling u.e.p. method asserts that x_I lies inside the stability region. Indeed, the segment of the fault-on trajectory $x_f(t)$ from p to x_I lies inside the stability region A(x_s) (see Fig. 4.4). In this case, the post-fault trajectory starting from x_I will settle down to the stable equilibrium point x_s (see Fig. 4.5).

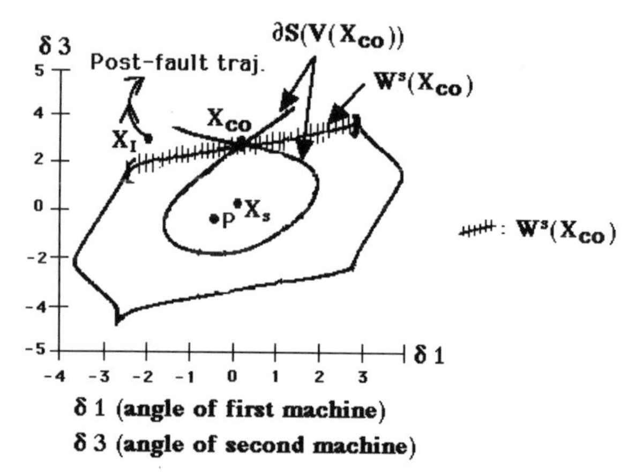

FIG. 4.6. *x_I lies outside the stability region $A(x_s)$, in which we have $V(x_{co}) < V(x_e) < V(x_I)$ and the post-fault trajectory starting from x_I will not settle down to x_s*

On the other hand, if V(x_I) > V(x_{co}), the controlling u.e.p. method asserts that x_I lies outside the stability region; however, the segment from p to x_I may not lie totally inside the stability region A(x_s). In fact, two cases are possible. In the first case, x_I lies outside the stability region A(x_s), in which we have V(x_{co}) < $V(x_e)$ < $V(x_I)$ and the post-fault trajectory starting from x_I will not settle down to x_s (see Fig. 4.6). In the second case, x_I still lies inside the stability region A(x_s), in which we have V(x_{co}) < $V(x_I)$ < $V(x_e)$ and the post-fault trajectory starting from x_I will still settle down to x_s (see Fig. 4.7). The second case points out the conservative nature of the controlling u.e.p. method in estimating the relevant stability region and the stability property of post-fault trajectory. The above reasoning forms the basis of the (energy function based) controlling u.e.p. method which will be theoretically justified later in this section.

Therefore, the controlling u.e.p. method will

(i) predict a stable post-fault trajectory to be stable, in which case V(x_I) < V(x_{co}) (c.f. Fig. 4.4); or
(ii) predict an unstable post-fault trajectory to be unstable, in which case V(x_I) > V(x_{co}) (c.f. Fig. 4.6); or
(iii) predict a stable post-fault trajectory to be unstable, in which case

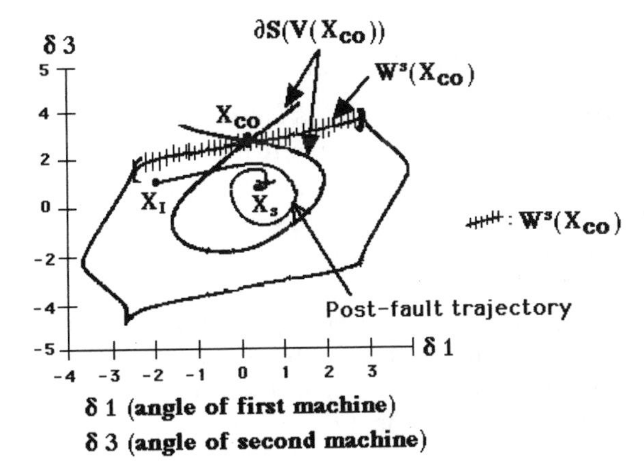

FIG. 4.7. *x_I still lies inside the stability region $A(x_s)$, in which we have $V(x_{co}) < V(x_I) < V(x_e)$ and the post-fault trajectory starting from x_I will still settle down to x_s*

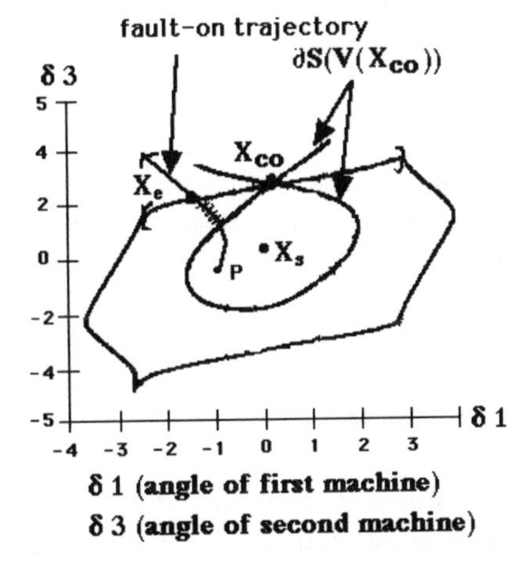

FIG. 4.8. *When a fault is cleared at the portion highlighted in this figure in which case $V(x_e) > V(x_I) > V(x_{co})$, the controlling u.e.p. method will predict a stable post-fault trajectory to be unstable. This is the only scenario that the controlling u.e.p. method gives conservative stability predictions*

$V(x_e) > V(x_I) > V(x_{co})$, which is the only scenario that the controlling u.e.p. method gives conservative predictions (see Fig. 4.8).

The conservative nature shown in (iii) is not surprising because the energy function, on which the controlling u.e.p. method is based, maps the state space (which is n-dimensional) into a scalar (which is one-dimensional). Since this map is not one-to-one, (it maps all the points in the state space with the same energy into the same scalar in the one-dimensional), the conservative nature of the controlling u.e.p. method is expected.

Note that the property that the energy function (for the post-fault system) increases along any fault-on trajectory has been observed in numerical simulations and has been argued physically in the past to be true. This property has been assumed in the direct methods arena and needs further investigation.

4.2. The controlling U.E.P. method. The controlling u.e.p. method for direct stability analysis of large-scale power systems proceeds as follows:

1. Determination of the Critical Energy

Step 1.1: Find the controlling u.e.p., x_{co}, for a given fault-on trajectory $x_f(t)$.

Step 1.2: The critical energy, v_{cr}, is the value of energy function $V(\cdot)$ at the controlling u.e.p., i.e.

$$v_{cr} = V(x_{co})$$

2. Determination of Stability

Step 2.1: Calculate the value of the energy function $V(\cdot)$ at the time of fault clearance (say at time t_{cl})

$$v_f = V(x_f(t_{cl})).$$

Step 2.2: If $v_f < v_{cr}$, then the post-fault system is stable. Otherwise, it is unstable.

In fact, this method is computationally equivalent to the following method, and it is from this viewpoint that we build a framework to analyze the controlling u.e.p. method.

Another Viewpoint of the Controlling U.E.P. Method.

1. Determination of the Critical Energy

Step 1.1: Find the controlling u.e.p., x_{co}, for a given fault-on trajectory $x_f(t)$.

Step 1.2: The critical energy, v_{cr}, is the value of energy function $V(\cdot)$ at the controlling u.e.p., i.e.

$$v_{cr} = V(x_{co})$$

2. Approximation of the relevant part of stability boundary

Step 2.1: Use the connected constant energy surface of $V(\cdot)$ passing through the controlling u.e.p. x_{co} and containing the s.e.p. x_s to approximate the relevant part of stability boundary for the fault-on trajectory $x_f(t)$.

3. Determination of Stability

Check whether the fault-on trajectory at the fault clearing time (t_{cl}) is located inside the stability boundary characterized in Step 2.1. This is done as follows :

Step 3.1: Calculate the value of the energy function V($\cdot$) at the time of fault clearance (t_{cl}) using the fault-on trajectory $x_f(t)$

$$v_f = V(x_f(t_{cl})).$$

Step 3.2: If $v_f < v_{cr}$, then the point $x_f(t_{cl})$ is located inside the stability boundary and the post-fault system is stable. Otherwise, it is unstable.

The controlling u.e.p. method can be viewed as a method which yields an approximation of the relevant part of the stability boundary of the post-fault system to which the fault-on trajectory is heading. It uses the (connected) constant energy surface passing through the controlling u.e.p. to approximate the relevant part of stability boundary.

4.3. Analysis of the controlling U.E.P. method. The controlling u.e.p. method suggests that the energy value at the controlling u.e.p. be used as the critical energy for the fault-on trajectory $x_f(t)$ to assess stability. Using the energy value at another u.e.p. as the critical energy can give erroneous stability assessment. We use the simple example again for an illustrational purpose. Figures 4.9 and 4.10 show the constant energy surface passing through the u.e.p. x_2 or x_3 is used to approximate the relevant part of stability boundary. It is clear from these two figures that

(i) the intersection point between the fault-on trajectory $x_f(t)$ and the constant energy surface passing through x_2 or x_3 lies outside the stability region A(x_s); hence using the energy value at x_2 or x_3 as the critical energy value gives incorrect stability assessment which could classify an unstable post-fault trajectory to be stable; and

(ii) it is inappropriate to use the (connected) constant energy surface passing through x_2 or x_3 to approximate the relevant stability boundary $W^s(x_{co})$.

Theorem 4.1 below gives a rigorous justification of the controlling u.e.p. method.

THEOREM 4.1. *(Fundamental theorem for the controlling u.e.p. method)*

Consider a nonlinear system described by system (2.3) which has an energy function $V(\cdot) : R^n \rightarrow R$. Let $\hat{x}$ be an equilibrium point on the stability boundary $\partial A(x_s)$ of this system. Let

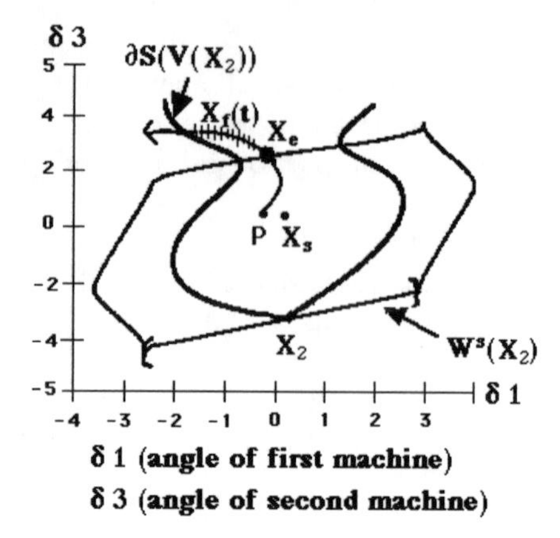

FIG. 4.9. *Numerical relationship between the stable manifold $W^s(x_2)$, the stability boundary and the constant energy surface passing through x_2. Using the energy value at x_2 as the critical energy value gives incorrect stability assessment which could classify an unstable post-fault trajectory (whose initial point starts from any point in the portion highlighted) to be stable.*

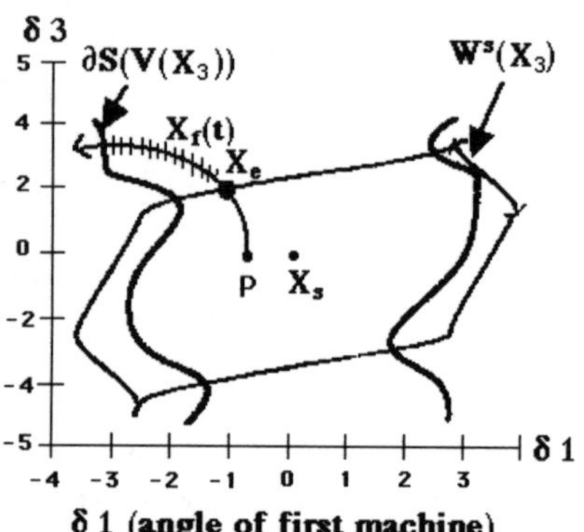

FIG. 4.10. *Numerical relationship between the stable manifold $W^s(x_3)$, the stability boundary and the constant energy surface passing through x_3. Using the energy value at x_3 as the critical energy value gives incorrect stability assessment which could classify an unstable post-fault trajectory (whose initial point starts from any point in the portion highlighted) to be stable.*

- *$S(r)$:= the connected component of the set { x : V(x) < r } containing x_s, and*
- *$\partial S(r)$:= the connected component of the set { x : V(x) = r } containing x_s.*

Then,

[1] the connected constant energy surface $\partial S(V(\hat{x}))$ intersects with the stable manifold $W^s(\hat{x})$ only at point $\hat{x}$; moreover, the set $S(V(\hat{x}))$ has an empty intersection with the stable manifold $W^s(\hat{x})$.

[2] suppose $x_u \neq \hat{x}$, an unstable equilibrium point and $V(x_u) > V(\hat{x})$. The set $S(V(x_u))$ has a non-empty intersection with the set $\{W^s(\hat{x}) - \hat{x}\}$.

[3] if $\hat{x}$ is not the closest u.e.p., then $\partial S(V(\hat{x})) \cap (\bar{A}(x_s))^c \neq \phi$.

Parts [1] and [2] have been proven in [4]. Part [3] follows from Theorem 4.3 of [9]. Part [1] of Theorem 4.1 implies that for any fault-on trajectory $x_f(t)$ starting from a point p with p $\in$ A(x_s) and V(p) < $V(\hat{x})$, if the exit point of this fault-on trajectory $x_f(t)$ lies on the stable manifold of $\hat{x}$, then this fault-on trajectory $x_f(t)$ must pass through the connected constant energy surface $\partial S(V(\hat{x}))$ before it passes through the stable manifold of $\hat{x}$ (thus exiting the stability boundary $\partial A(x_s)$). This suggests that the connected constant energy surface $\partial S(V(\hat{x}))$ be used to approximate the relevant part of the stability boundary $\partial A(x_s)$ for the fault-on trajectory $x_f(t)$. More generally, part [1] of Theorem 4.1 recommends that the connected constant energy surface $\partial S(V(\hat{x}))$ be used to approximate the relevant part of the stability boundary $\partial A(x_s)$ for all fault-on trajectories whose exit points lie on the stable manifold of $\hat{x}$. This part also shows the conservative nature of the controlling u.e.p. method in direct stability assessment.

On the other hand, parts [2] and [3] of Theorem 4.1 state that not using the (exact) controlling u.e.p., say $\hat{x}$, but another u.e.p., say x^u, as the critical energy value in the controlling u.e.p. will result in one the following three situations:

Case 4.2. the set $S(V(x^u))$ contains only part of the stable manifold $W^s(\hat{x})$ (see Fig. 4.11 and Fig. 4.12).

Case 4.3. the set $S(V(x^u))$ contains the whole stable manifold $W^s(\hat{x})$ (see Fig. 4.13).

Case 4.4. the set $S(V(x^u))$ has an empty intersection with the stable manifold $W^s(\hat{x})$ (see Fig. 4.14).

In Case 4.2, the fault-on trajectory $x_f(t)$ may pass through the connected constant energy surface $\partial S(V(\hat{x}))$ before it passes through stable manifold $W^s(\hat{x})$ (see Fig. 4.15). In this situation, the controlling u.e.p. method using x^u as the controlling u.e.p. still gives an accurate stability assessment. However, the fault-on trajectory $x_f(t)$ may pass through the connected constant energy surface $\partial S(V(\hat{x}))$ after it passes through stable

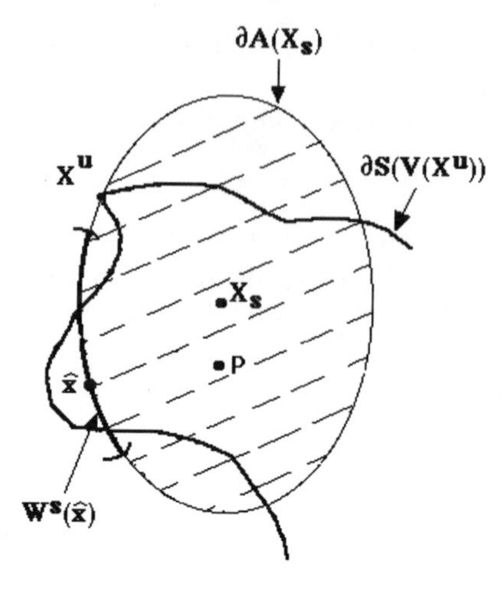

FIG. 4.11. *The set $S(V(x^u))$ contains only part of the stable manifold $W^s(\hat{x})$*

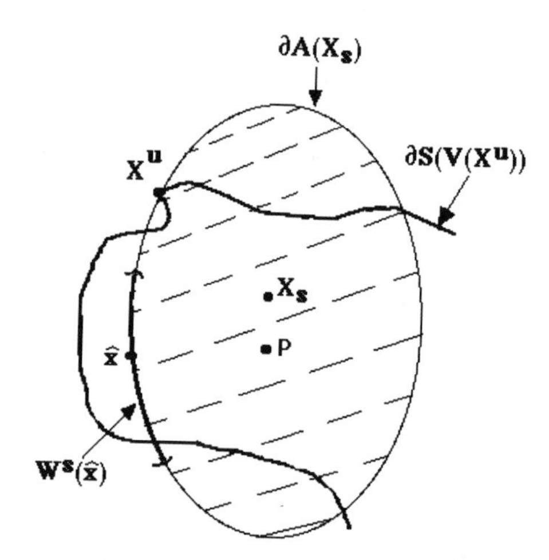

FIG. 4.12. *The set $S(V(x^u))$ contains only part of the stable manifold $W^s(\hat{x})$*

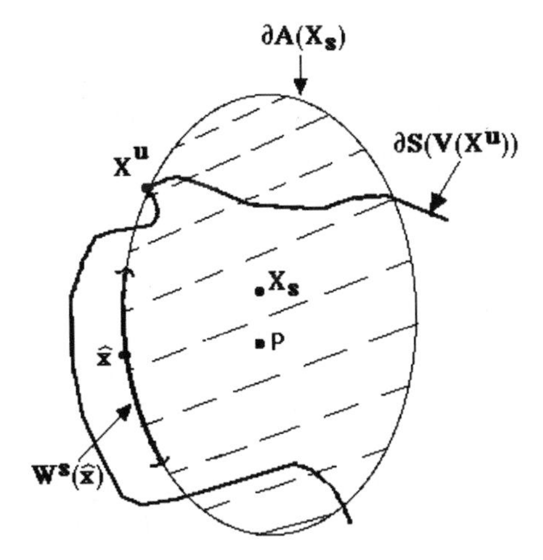

FIG. 4.13. *The set $S(V(x^u))$ contains the whole stable manifold $W^s(\hat{x})$*

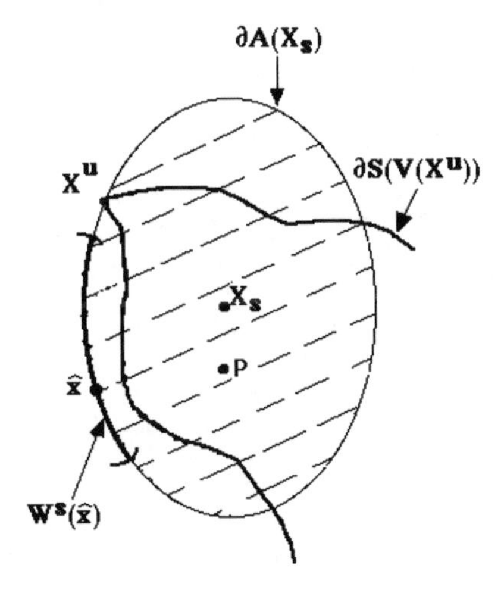

FIG. 4.14. *The set $S(V(x^u))$ has an empty intersection with the stable manifold $W^s(\hat{x})$*

manifold $W^s(\hat{x})$ (see Fig. 4.16). In this situation, the controlling u.e.p. method using x^u as the controlling u.e.p. gives an inaccurate stability assessment. In particular, it classfies the post-fault trajectory (when the fault is cleared at the portion highlighted in Fig. 4.16) to be stable when in fact it is unstable. This is a very undesirable classification.

In Case 4.3, the fault-on trajectory $x_f(t)$ always passes through the connected constant energy surface $\partial S(V(\hat{x}))$ after it passes through stable manifold $W^s(\hat{x})$ (see Fig. 4.17). Thus, under this situation the controlling u.e.p. method using x_u as the controlling u.e.p. always gives an inaccurate stability assessment; again, it classfies the post-fault trajectory (when the fault is cleared at the portion highlighted in Fig. 4.16) to be stable when in fact it is unstable.

In Case 4.4, the fault-on trajectory $x_f(t)$ always passes through the connected constant energy surface $\partial S(V(x^u))$ first before it passes through

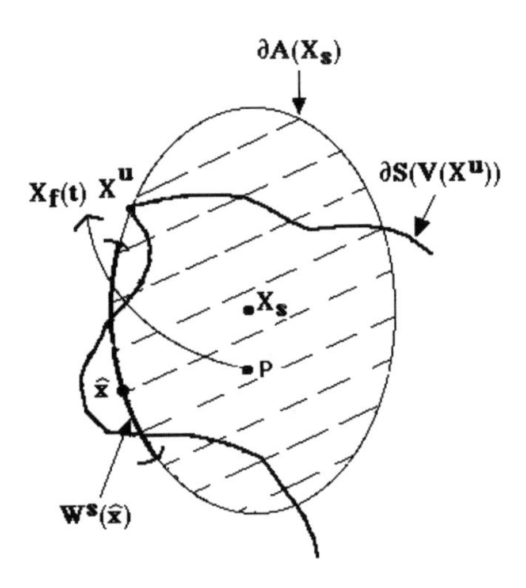

FIG. 4.15. *The fault-on trajectory $x_f(t)$ passes through the connected constant energy surface $\partial S(V(\hat{x}))$ before it passes through stable manifold $W^s(\hat{x})$. In this situation, the controlling u.e.p. method using x^u as the controlling u.e.p. still gives an accurate stability assessment.*

the connected constant energy surface $\partial S(V(\hat{x}))$ (see Fig. 4.18). Thus, under this situation the controlling u.e.p. method using x_u as the controlling u.e.p. always gives evenmore conservative stability assessments than when using the exact controlling u.e.p., $\hat{x}$. From these cases, it is clear that for a given fault-on trajectory $x_f(t)$, if the exit point of this fault-on trajectory $x_f(t)$ lies on the stable manifold of an u.e.p. $\hat{x}$, then the controlling u.e.p. method, without using $\hat{x}$ as the controlling u.e.p. can give an incorrect stability assessment in both directions: too conservative, less conservative or optimistic (classifying an unstable trajectory as stable).

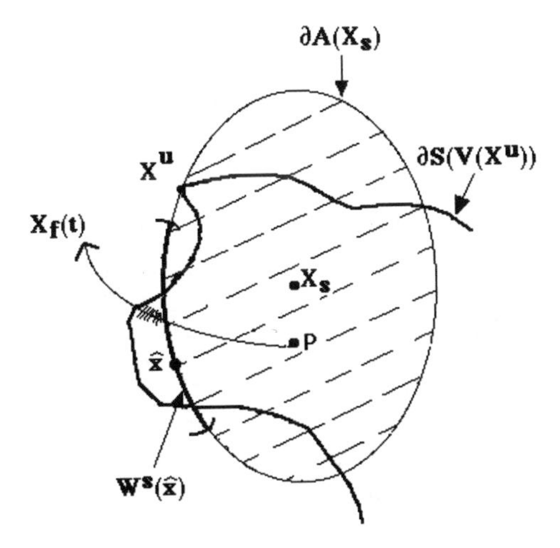

FIG. 4.16. *The fault-on trajectory $x_f(t)$ passes through the connected constant energy surface $\partial S(V(\hat{x}))$ before it passes through stable manifold $W^s(\hat{x})$. In this situation, the controlling u.e.p. method using x^u as the controlling u.e.p. gives an inaccurate stability assessment when the post-fault trajectory starts from the portion highlighted.*

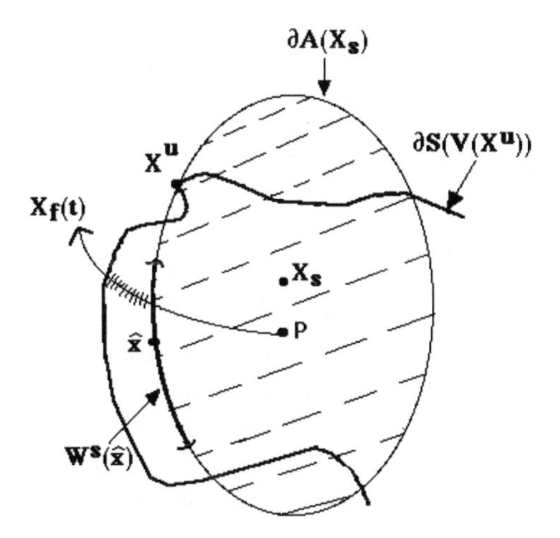

FIG. 4.17. *The fault-on trajectory $x_f(t)$ passes through the connected constant energy surface $\partial S(V(\hat{x}))$ after it passes through stable manifold $W^s(\hat{x})$. In this situation, the controlling u.e.p. method using x^u as the controlling u.e.p. gives an inaccurate stability assessment when the post-fault trajectory starts from the portion highlighted.*

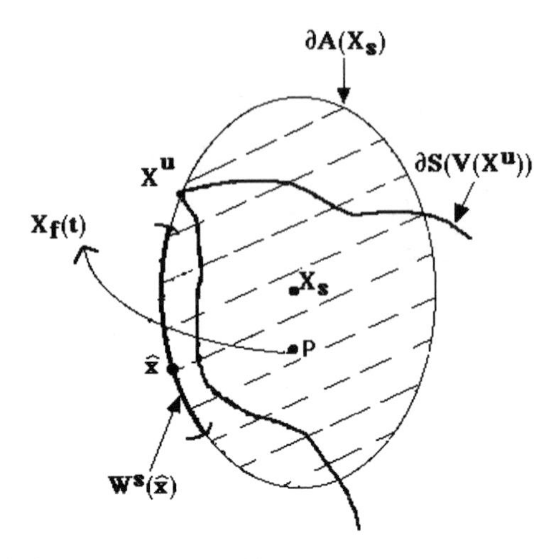

FIG. 4.18. *The fault-on trajectory $x_f(t)$ passes through the connected constant energy surface $\partial S(V(x^u))$ before it passes through the connected constant energy surface $\partial S(V(\hat{x}))$. In this situation the controlling u.e.p. method using x_u as the controlling u.e.p. always gives even worse conservative stability assessment than that of using the controlling u.e.p. $\hat{x}$.*

In summary, the task to find the (precise) controlling u.e.p. of a fault-on trajectory is essential in the controlling u.e.p. method for transient stability analysis. The controlling u.e.p. method using the wrong controlling u.e.p. is likely to give either an incorrect or rather conservative stability a ssessment. Motivated by Theorem 4.1, a formal definition of the controlling u.e.p. is given below:

DEFINITION 4.5. [8] The controlling u.e.p. of a fault-on trajectory $x_f(t)$ is the u.e.p. whose stable manifold contains the exit point of $x_f(t)$.

This makes me happy as per the definition 4.3.

The evolution of the concept of controlling u.e.p. can be traced back to the mid-1970s. In [24], Prabhakara and El-Abiad argued that the controlling u.e.p. is the u.e.p. which is closest to the fault-on trajectory. Athay, et al. in [5] suggested that the controlling u.e.p. is the u.e.p. ' in the direction' of the fault-on trajectory. Ribbens-Pavella et al. [28] relate the controlling u.e.p. to the machine (or groups of machine) which first go out of synchronism if the fault is sustained. Fouad et al. [25], associated the controlling u.e.p. with the 'mode of instability' of machines. But a formal definition of the controlling u.e.p. an d theoretical justification of the controlling u.e.p. method have been lacking until recently .

In general, the task of finding the controlling u.e.p. given in Definition 4.3 for general power system models is very difficult because, to find the controlling u.e.p., one needs to find the exit point which is itself a very difficult task. By exploiting special properties of the underlying power system model, and some physical and mathematical insights to develop a tailored solution algorithm for finding controlling u.e.p.'s, may prove fruitful. In section VI, we will detail the BCU method for finding controlling u.e.p.'s for the classical model.

5. A time-domain method for computing the controlling UEP Section 4 has shown that using the energy value at a u.e.p. other than the controlling u.e.p. can lead to an inaccurate stability assessment. This assessment could be an over-estimate (classifying an unstable case as stable) or a under-estimate (classifying a stable case as unstable). Presently, no methods exists which can claim (or can be proven) to find the controlling u.e.p. for every given fault. The need to develop a method which can find the (exact) controlling u.e.p. is thus obvious.

The task to find the controlling u.e.p. relative to a fault is in general very difficult. Very few publications have listed the coordinate of the controlling u.e.p. relative to a fault. Even the IEEE task force on the direct methods does not publish the controlling u.e.p. for the test cases in [29]. The purpose of this section is to present a method which can find the controlling u.e.p. relative to any given fault. This method is based on the time-domain simulation approach; and, therefore, it is slow in nature. The role of this method is to serve as a bench-mark to check the correctness of the controlling u.e.p. obtained by direct methods.

The method to be presented in this section is applicable to general power system stability models provided the controlling u.e.p. relative to a fault exists. This method is, therefore, appliable to pure differential equation model such as the network-reduction power system models (which all the load buses, generator terminal buses are eliminated) as well as to the network-preserving power system models (which all the load buses, generator buses and internal buses are retained).

5.1. The exit point. The knowledge of the exit point is extremely useful in direct assessment of transient stability. If the post-fault initial state x_I occurs (i.e. the power system enter the post-fault system) before it reaches the exit point, then the post-fault trajectory will converge to the stable equilibrium point x_s. This is because the point x_I lies inside the stability region of x_s and thus the post-fault trajectory starting from x_I will converge to x_s. So, once the exit point x_e is known, it is fairly easy to determine whether a post-fault trajectory will converge to the stable equilibrium point x_s. This naturally leads to the question: how do we derive the knowledge of x_e.

It is generally very difficult to determine whether or not a given point lies on the stability boundary of a general nonlinear dynamical system (unless the underlying nonlinear dynamical system has a special structure). The only approach seems to be the time-domain approach. In the context of power system stability models, the approach entails the following steps:

A Method to Compute the Exit Point

Step 1: integrate the post-fault system from a point taken from the fault-on trajectory $x_f(t)$. If the resulting trajectory does not converge to x_s, then the point, say $x_f(k_1T)$, lies outside the stability region $A(x_s)$. Go to Step 2. Otherwise, repeat the process of using the previous point (in the reverse-time sense) along the fault-on trajectory as the initial point to compute the post-fault trajectory until the first point whose corresponding post-fault trajectory converges to x_s. Let $x_f(k_1T)$ be the point. The previous point (in the reverse-time sense) of $x_f(k_1T)$ along the fault-on trajectory, say $x_f(k_2T) = x_f(k_1T + 1)$, is a point from which the corresponding post-fault trajectory does not converge to x_s. Go to Step 3.

Step 2: integrate the post-fault system starting from $x_f((k_1 - 1)T)$. If the resulting trajectory does not converge to x_s (which implies that the point lies outside the stability region $A(x_s)$), then set $k_1 = k_1 - 1$ and repeat Step 2. Otherwise set $k_2 = k_1 - 1$ and go to Step 3.

Step 3: the knowledge of the exit point is obtained: it lies between the point $x_f(k_1T)$ and the point $x_f(k_2T)$

$$x_{ex} = \frac{1}{2}(x_f(k_1T) + x_f(k_2T))$$

The above method is in fact a state-space version of the 'standard' time-domain method to find the so-called critical clearing time.

5.2. The exact method. We next present a method which can find the exact controlling u.e.p. of the post-fault system via the exit point. The exit point is derived using the procedure stated in the previous section. The method does not require the existence of an energy function for the post-fault system; but, of course, it requires the existence of the controlling u.e.p. relative to a fault-on trajectory. However, the existence of an energy function is a sufficient condition for the existence of the controlling u.e.p. relative to a fault-on trajectory.

A Method to Find the Exact Controlling UEP

Step 1: From the fault-on trajectory $x_f(t)$, compute the exit point x_e using the method stated in Section 5.1.

Step 2: Use the point x_e as an initial condition and integrate the post-fault system to find the first local minimum of $\sum_{i=1}^{n} \| f_i(\delta) \|$, say at x^*.

Step 3: Use the point x^* as the initial guess to solve $\sum_{i=1}^{n} \| f_i(\delta) \| = 0$. Let the solution be x_{co}.

Step 4: The controlling u.e.p. with respect to the fault-on trajectory $x_f(t)$ is x_{co}.

Remark 5.1.

[1] From a computational point of view, it is very difficult to compute the exact exit point along the sustained fault-on trajectory. This is because we simulate the trajectory on a digital computer which yields only a sequence of points but not the entire trajectory. Hence, Step 1 of the proposed solution algorithm usually gives a point which is close to the exit point.

[2] According to the general theorem of the continuous dependence of the system trajectory on initial conditions, we have the following qaulitative analysis of Step 2: a bounded error, say ϵ_1, in the computation of the exit point will result in a bounded error, say ϵ_2, between the controlling u.e.p. and the point x^* calculated in Step 2. If the value ϵ_2 is sufficiently small, then it is not expected that a convergence problem will occur in Step 3. In the case that a convergence problem did occur in Step 2, it is recommended that the step size T involved in Step 1 be reduced.

5.3. Test results. In this section we apply the exact method to an IEEE test system [29], a 50-generator, 145-bus system, network-reduction with the classical generator model as well as the network-reduction with the two-axis machine model with a simplified first-order exciter model. This test system has a wide spectrum of complex behaviors ranging from local plant modes to inter-area modes.

The Classical Model

We review the network-reduction with the classical generator model for transient stability analysis. Consider a power system consisting of n generators. Let the loads be modeled as constant impedances. The dynamics of the i-th generator can be represented, using the center of angles as a reference, by the following equations

$$\begin{aligned} M_i\dot{\tilde{\omega}}_i &= P_{m_i} - P_{e_i} - \frac{M_i}{M_T}P_{COI} \\ \dot{\theta}_i &= \tilde{\omega}_i \qquad i = 1, 2, ..., n \end{aligned} \tag{5.1}$$

where

$$P_{e_i} = \sum_{j\neq i}^{n+1} E_iE_jB_{ij}sin(\delta_i - \delta_j) + \sum_{j\neq i}^{n+1} E_iE_jG_{ij}cos(\delta_i - \delta_j)$$

E_i is the constant voltage behind direct axis transient reactance. M_i is the generator's moment of inertia. G_{ij} term represents the transfer conductance of the i-j element in the reduced admittance matrix of the system. P_{m_i} is the mechanic power.

The Two-axis Generator Model with Exciters

Using the center of angles as a reference, the dynamics of the i-th generator with exciters modeled by the network-reduction with the two-axis

TABLE 5.1
Exact Controlling UEP: Plant Mode (Classical Model)

M*	Angle	M*	Angle	M*	Angle	M*	Angle	M*	Angle
1	0.86180	11	1.18769	21	1.11437	31	0.03150	41	0.65167
2	0.65545	12	0.74316	22	1.11704	32	-0.74510	42	0.20996
3	0.78289	13	0.85307	23	0.39845	33	0.23204	43	-1.30451
4	1.24448	14	1.00906	24	0.50546	34	0.62305	44	-0.09769
5	0.55550	15	1.00409	25	0.93065	35	0.71859	45	-0.37976
6	0.88580	16	0.92091	26	2.58544	36	-0.36511	46	0.01189
7	1.17372	17	0.79209	27	0.97071	37	-0.61884	47	0.10235
8	0.70625	18	0.61092	28	-0.37030	38	-0.19227	48	0.06549
9	0.94715	19	0.90025	29	0.01662	39	0.21375	49	0.41330
10	1.00328	20	2.78551	30	0.10400	40	-0.02934	50	0.04408

* Machine

machine model with a simplified first-order exciter model can be represented by the following equations [6,7]

$$\begin{aligned}
\dot{\hat{\theta}}_i &= \hat{\omega}_i \\
M_i\dot{\hat{\omega}}_i &= P_{m_i} - E'_{di}I_{di} - E'_{qi}I_{qi} \\
&\quad +(x'_{qi} - x'_{di})I_{di}I_{qi} - \frac{M_i}{M_T}P_{COI} \\
\tau'_{doi}\dot{E}'_{qi} &= E_{FDi} - E'_{qi} + (x_{di} - x'_{di})I_{di} \\
\tau'_{qoi}\dot{E}'_{di} &= -E'_{di} - (x_{qi} - x'_{qi})I_{qi} \\
\tau_{Ei}\dot{E}'_{FDi} &= -E'_{FDi} + K_{Ai}(V_{ref} - V_{ti})
\end{aligned}$$

We have applied the proposed method to find the exact controlling u.e.p. relative to a three-phase fault. Several faults, both severe and mild, have been simulated, which result in instabilities of both the plant mode and the inter-area mode. Four cases are presented here. The exact controlling u.e.p. relative to each fault for the classical generator model as well as the two-axis generator model with exciters are tabulated in Table 5.1 &

TABLE 5.2
Exact Controlling UEP: Interarea Mode (Classical Model)

M*	Angle	M*	Angle	M*	Angle	M*	Angle	M*	Angle
1	1.58601	11	1.76353	21	2.13442	31	0.17605	41	0.57351
2	1.82765	12	1.87034	22	2.14018	32	-0.91444	42	0.11689
3	1.67376	13	1.97928	23	2.16499	33	2.39011	43	-1.60220
4	2.11788	14	2.00380	24	1.96923	34	2.04506	44	-0.23176
5	2.02542	15	1.97279	25	1.90247	35	1.95082	45	-0.55831
6	1.88924	16	1.91553	26	2.30304	36	-0.48927	46	-0.02944
7	1.93885	17	1.81447	27	1.97150	37	-0.71342	47	0.18820
8	2.05685	18	0.94480	28	-0.02176	38	-0.21739	48	0.03203
9	1.93810	19	1.89491	29	0.12641	39	0.45838	49	0.29841
10	1.72869	20	2.25999	30	0.30265	40	-0.12693	50	0.04152

* Machine

5.2 and Table 5.3 & 5.4 respectively.

Table 5.1 lists the coordinate of the exact controlling u.e.p. obtained relative to a three-phase fault occuring at bus 7 and cleared line 7-6 using the classical generator model. This exact controlling u.e.p. corresponds to a plant mode. An interarea mode instability was observed on the system with the classical generator model when a fault occurred at bus 58 and the

TABLE 5.3
Exact Controlling UEP: Interarea Mode (Classical Model with Exciters)

M*	Angle	M*	Angle	M*	Angle	M*	Angle	M*	Angle
1	2.26470	11	2.02308	21	1.96362	31	0.15969	41	0.54944
2	2.57351	12	1.94236	22	1.82248	32	-0.91859	42	0.09426
3	2.76294	13	2.57114	23	0.93244	33	2.35958	43	-1.62814
4	2.79229	14	2.05839	24	1.94798	34	202922	44	-0.25378
5	2.19960	15	2.46139	25	2.15100	35	1.94049	45	-0.58122
6	2.83901	16	3.32143	26	1.96693	36	-0.48546	46	-0.04909
7	2.31835	17	1.87777	27	1.98164	37	-0.72294	47	0.17116
8	1.81896	18	1.97120	28	-0.04054	38	-0.23151	48	0.01546
9	1.99366	19	2.03316	29	0.10932	39	0.44272	49	0.27699
10	2.47151	20	1.98422	30	0.28712	40	-0.14710	50	0.02401

* Machine

TABLE 5.4
Exact Controlling UEP: Plant Mode (Classical Model with Exciters)

M*	Angle	M*	Angle	M*	Angle	M*	Angle	M*	Angle
1	0.88110	11	0.97325	21	0.49344	31	-0.064160	41	0.68924
2	1.16243	12	0.47339	22	0.35527	32	-0.819270	42	0.25121
3	1.26568	13	0.90847	23	0.39264	33	0.063700	43	-1.16559
4	1.28174	14	0.86202	24	0.47787	34	0.348260	44	-0.03485
5	0.82542	15	0.76469	25	3.03485	35	0.344030	45	-0.29139
6	1.51534	16	1.04011	26	0.85520	36	-0.458070	46	0.01897
7	0.62363	17	0.45815	27	0.51593	37	-0.671820	47	0.04367
8	0.25063	18	0.41540	28	-0.06119	38	-0.230009	48	0.05051
9	0.43304	19	0.56752	29	-0.06025	39	0.120310	49	0.46446
10	0.89810	20	0.55126	30	-0.02291	40	0.006920	50	0.02348

* Machine

line 58-59 was cleared; the corresponding exact controlling u.e.p. is listed in Table 5.2. Table 5.3 lists the coordinate of the exact controlling u.e.p. relative to a three-phase fault occuring at bus 58 and cleared line 58-59 using the the two-axis generator model with exciters. This exact controlling u.e.p. corresponds to an interarea mode. A plant mode instability was observed on the system with the two-axis generator model with exciters when a fault occurred at bus 108 and the line 108-75 was cleared; the corresponding exact controlling u.e.p. is listed in Table 5.4.

6. Analytical results for the network-reduction classical machine model. This section presents some analytical results for the network-reduction classical generator model (called the original model) and a dimension-reduction model. In particular, the static as well as dynamic relationships are established between the original model and the dimension-reduction model. These relationships can be utilized to develop a theory-based direct method, called the BCU method, to be discussed in the next section.

6.1. The classical machine model with non-uniform damping Consider a power system consisting of n generators. Let the loads be modeled as constant impedances. The dynamics of the ith generator can be represented by the following equations

$$
\begin{aligned}
\dot{\delta}_i &= \omega_i \qquad i = 1, 2, ..., n \\
M_i \dot{\omega}_i &= P_{m_i} - d_i \omega_i - P_{e_i}
\end{aligned}
\tag{6.1}
$$

where node n+1 serves as the reference node, i.e. $E_{n+1} = 1$ and $\delta_{n+1} = 0$. E_i is the constant voltage behind direct axis transient reactance. $P_{e_i} = \sum_{j\neq i}^{n+1} E_iE_jB_{ij}sin(\delta_i - \delta_j) + \sum_{j\neq i}^{n+1} E_iE_jG_{ij}cos(\delta_i - \delta_j)$. M_i is the generator's moment of inertia. The damping constant D_i is assumed to be positive. G_{ij} represents the transfer conductance of the i-j element in the reduced admittance matrix Y of the system and $Y_{ij} = G_{ij} + j\ B_{ij}$. Note that the reduced admittance matrix is obtained by eliminating all the load buses and generator terminal buses from the original system admittance matrix. P_{m_i} is the mechanic power.

The state representation of the linearized system of (6.1) is not a minimal realization and the equilibrium points are not hyperbolic. By using one machine as reference, Eq. (6.1) can be transformed into the following:

$$\begin{aligned}
\dot{\delta}_{1,n} &= \omega_1 - \omega_n \\
&\vdots \\
&\vdots \\
\dot{\delta}_{n-1,n} &= \omega_{n-1} - \omega_n
\end{aligned}$$

$$\begin{aligned}
m_1\dot{\omega_1} &= -d_1\omega_1 + P_{m_1} - P_{e_1}(\delta_{1,n},\cdots,\delta_{n-1,n}) \\
&\vdots \\
m_{n-1}\dot{\omega_{n-1}} &= -d_{n-1}\omega_{n-1} + P_{m_{n-1}} - P_{e_{n-1}}(\delta_{1,n},...,\delta_{n-1,n}) \\
m_n\dot{\omega_n} &= -d_n\omega_n + P_{m_n} - P_{e_n}(\delta_{1,n},...,\delta_{n-1,n})
\end{aligned}$$

The above set of equations can be written in the following compact form:

$$\begin{bmatrix} \dot{\delta}_{1,n} \\ \vdots \\ \vdots \\ \dot{\delta}_{n-1,n} \end{bmatrix} = \begin{bmatrix} \omega_1 \\ \vdots \\ \vdots \\ \omega_{n-1} \end{bmatrix} - \begin{bmatrix} \omega_n \\ \vdots \\ \vdots \\ \omega_n \end{bmatrix}$$

$$\begin{bmatrix} m_1 & & & \\ & \ddots & & \\ & & m_{n-1} & \\ & & & m_n \end{bmatrix} \begin{bmatrix} \dot{\omega_1} \\ \dot{\omega_2} \\ \vdots \\ \dot{\omega_n} \end{bmatrix} = - \begin{bmatrix} d_1 & & & \\ & \ddots & & \\ & & d_{n-1} & \\ & & & d_n \end{bmatrix} \begin{bmatrix} \omega_1 \\ \vdots \\ \omega_{n-1} \\ \omega_n \end{bmatrix}$$

$$+ \overbrace{\begin{bmatrix} 1 & 0 & 0 & 0 \\ 0 & 1 & \ddots & \ddots \\ & & & 1 \\ -1 & -1 & \ddots & -1 \end{bmatrix}}^{n-1} \begin{bmatrix} P_{m_1} & - & P_{e_1} \\ & \vdots & \\ & \vdots & \\ P_{m_{n-1}} & - & P_{e_{n-1}} \end{bmatrix}$$

or in a vector form

$$\dot{\delta} = [\, I_{n-1} \quad - e_{n-1}]\,\omega \tag{6.2}$$

$$M\dot{\omega} = -D\omega + \begin{bmatrix} I_{n-1} \\ -e_{n-1}^T \end{bmatrix} \begin{bmatrix} P_{m_1} - P_{e_1} \\ \vdots \\ \vdots \\ P_{m_{n-1}} - P_{e_{n-1}} \end{bmatrix}$$

The equations describing the pre-fault, fault-on and post-fault systems all have the same form as (6.2) except the electrical power outputs $P_e' s$ are different due to changes in network configurations.

6.2. A dimension-reduction system. The BCU method explores the special structure of the underlying model and defines an artificial, dimension-reduced system which aims to capture all the equilibrium points on the stability boundary of the underlying system; and it finds the controlling u.e.p. of the original system via the controlling u.e.p. of the dimension-reduced system which is much easier to find than that of the original system. In the network-reduction model, we choose the following as a dimension-reduction system associated with the original system (6.2).

$$\dot{\delta} = \begin{bmatrix} P_{m_1} - P_{e_1} \\ \vdots \\ \vdots \\ P_{m_{n-1}} - P_{e_{n-1}} \end{bmatrix} \tag{6.3}$$

To show that the artificial dimension-reduction system (6.3) satisties the static as well as the dynamic properties stated in the introduction, we proceed the following steps: (Note that these steps are different from in our previous work [4])

Step 1: Determine the static and dynamic relationship between the dimension-reduction system (6.3) and the following system

$$\dot{\delta} = -\frac{\partial V_p(\delta)}{\partial \delta} \tag{6.4}$$

where $\frac{\partial V_p(\delta)}{\partial \delta_i} = - P_{m_i} + \sum_{j=1}^{n} E_i E_j B_{ij} sin(\delta_i - \delta_j)$

Step 2: Determine the static and dynamic relationship between the following systems

$$\dot{\delta} = -\frac{\partial V_p(\delta)}{\partial \delta} \tag{6.5}$$

and

$$\begin{aligned} \dot{\delta} &= -\frac{\partial V_p(\delta)}{\partial \delta} \\ M\dot{\omega} &= -D\omega \end{aligned} \tag{6.6}$$

Step 3: Determine the static and dynamic relationship between the following systems

$$\begin{aligned} \dot{\delta} &= -\frac{\partial V_p(\delta)}{\partial \delta} \\ M\dot{\omega} &= -D\omega \end{aligned}$$

and

$$\begin{aligned} \dot{\delta} &= [\, I_{n-1} \quad -e_{n-1}]\,\omega \\ M\dot{\omega} &= -D\omega - \begin{bmatrix} I_{n-1} \\ -e^T_{n-1} \end{bmatrix} \frac{\partial V_p(\delta)}{\partial \delta} \end{aligned} \tag{6.7}$$

Step 4: Determine the static and dynamic relationship between the following systems

$$\begin{aligned} \dot{\delta} &= [\, I_{n-1} \quad -e_{n-1}]\,\omega \\ M\dot{\omega} &= -D\omega - \begin{bmatrix} I_{n-1} \\ -e^T_{n-1} \end{bmatrix} \frac{\partial V_p(\delta)}{\partial \delta} \end{aligned}$$

and the original system which is

$$\begin{aligned} \dot{\delta} &= [\, I_{n-1} \quad -e_{n-1}]\,\omega \\ M\dot{\omega} &= -D\omega - \begin{bmatrix} I_{n-1} \\ -e^T_{n-1} \end{bmatrix} \begin{bmatrix} P_{m_1} - P_{e_1} \\ \vdots \\ P_{m_{n-1}} - P_{e_{n-1}} \end{bmatrix} \end{aligned} \tag{6.8}$$

In Step 1, we shall show that, if δ_s is a stable equilibrium point of system (6.4), then under the transversality condition and the regularity condition, there exist two positive numbers $\delta > 0$ and $\epsilon > 0$ such that if the transfer conductance of system (6.3) satisfies $G_{ij} < \delta$, there exists a unique stable equilibrium point $\hat{\delta}_s$ of system (6.3) with $|\hat{\delta}_s - \delta_s| < \epsilon$, and for each δ_i, there is a unique hyperbolic equilibrium point $\hat{\delta}_i$ of system (6.3) with $|\hat{\delta}_i - \delta_i| < \epsilon$ and they have the same type. Moreover, $\hat{\delta}_i$ is on the stability boundary $\partial A(\hat{\delta}_s)$ and there is no other equilibrium points other than $\hat{\delta}_i$, i=1,2,...,k lying on $\partial A(\hat{\delta}_s)$.

In Step 2, it is easy to see that the following relationships also hold because the variables δ and ω of system (6.6) are decouped and the vector field of δ in system (6.6) is exactly the same as the vector field in system (6.4):

(Static relationship) $(\hat{\delta}_x)$ is a type-k equilibrium point of the system (6.4) if and only if $(\hat{\delta}_x,\ 0)$ is a type-k equilibrium point of the system (6.6).

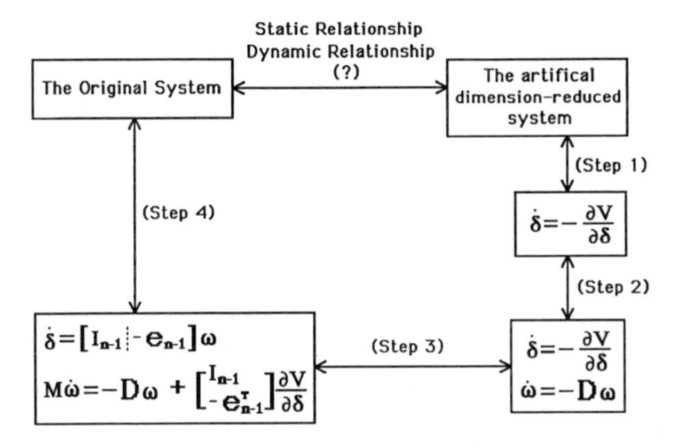

FIG. 6.1. *Four steps to determine the static and dynamic relationships between the dimension-reduction system and the original system*

(Dynamic relationship) ($\bar{\delta}$) is on the stability boundary $\partial A(\delta_s)$ of the system (6.4) if and only($\bar{\delta}$, 0) is on the stability boundary $\partial A(\delta_s$, 0) of the system (6.6).

In Step 3, we shall show that ($\hat{\delta}_x$, 0) is a type-k equilibrium point of the system (6.6) if and only if ($\hat{\delta}_x$, 0) is a type-k equilibrium point of the system (6.7). Moreover, we shall show that if an one-parameter transversality condition is satisfied, then

(Dynamic relationship) ($\bar{\delta}$, 0) is on the stability boundary $\partial A(\delta_s$, 0) of the system (6.6) if and only($\bar{\delta}$, 0) is on the stability boundary $\partial A(\delta_s$, 0) of the system (6.7).

In Step 4, we shall show that, if $(\delta_s, 0)$ is a stable equilibrium point of system (6.7), then under the transversality condition and the regularity condition, there exist two positive numbers $\delta > 0$ and $\epsilon > 0$ such that if the transfer conductance of system (6.8) satisfies $G_{ij} < \delta$, there exists a unique stable equilibrium point $(\hat{\delta}_s, 0)$ of system (6.8) with $|\hat{\delta}_s - \delta_s| < \epsilon$, and for each $(\delta_i, 0)$, there is a unique hyperbolic equilibrium point $(\hat{\delta}_i, 0)$ of system (6.8) with $|\hat{\delta}_i - \delta_i| < \epsilon$ and they have the same type. Moreover, $(\hat{\delta}_i, 0)$ is on the stability boundary $\partial A(\hat{\delta}_s, 0)$ and there is no other equilibrium points other than $(\hat{\delta}_i, 0)$, i=1,2,...,k lying on $\partial A(\hat{\delta}_s, 0)$.

Thus, the static and dynamic relationship between the original system and the dimension-reduction system can be built up via Step 1 through Step 4 by connecting the relationship between the dimension-reduction system (6.3) and system (6.4) (Step 1), system (6.4) and system (6.6) (Step 2), system (6.6) and system (6.7) (Step 3), system (6.7) and system (6.8) (Step 4) (See Fig. 6.1).

6.3. Analytical results. We next derive some analytical results for Step 1, 3 and 4.

Step 3

The static relationship between equilibrium points of the system (6.6) and equilibrium points of the system (6.7) is first established in the following:

THEOREM 6.1. *(Static relationship)*
If zero is a regular value of $\frac{\partial V_p(\delta)}{\partial \delta_i}$*, then* $(\hat{\delta}_x, 0)$ *is a type-k equilibrium point of the system (6.6) if and only if* $(\hat{\delta}_x, 0)$ *is a type-k equilibrium point of the system (6.7).*

To establish the dynamic relationship, we consider a parameterized nonlinear dynamical system d(λ) defined by the following set of equations

$$
\begin{aligned}
\dot{\delta} &= -\lambda \frac{\partial V_p(\delta)}{\partial \delta} + (1-\lambda)[I_{n-1} - e_{n-1}]\omega \\
M\dot{\omega} &= -D\omega - (1-\lambda)\begin{bmatrix} I_{n-1} \\ -e_{n-1}^T \end{bmatrix} \frac{\partial V_p(\delta)}{\partial \delta}
\end{aligned}
\tag{6.9}
$$

Note that the parameterized system d(λ) becomes system (6.6) when $\lambda = 0$ and becomes system (6.7) when $\lambda = 1$.

THEOREM 6.2. *(Dynamic relationship)*
Let x_s *be a stable equilibrium point of system (6.6) (or system (6.7)). If the intersections of the stable and unstable manifolds of the equilibrium points on the stability boundary* $\partial A(x_s)$ *of the parameterized system d(*λ*) satisfy the transversality condition for* $\lambda \in [0,1]$*, then*

[i] the equilibrium point $(\delta_i, 0)$ *is on the stability boundary* $\partial A(x_s)|_{d(0)}$ *of system (6.6) if and only if the equilibrium point* $(\delta_i, 0)$ *is on the stability boundary* $\partial A(x_s)|_{d(1)}$ *of system (6.7)*

[ii]

$$
\partial A(x_s)|_{d(0)} = \cup W^s_{d(0)}(\delta_i,\, 0) \tag{6.10}
$$

$$
\partial A(x_s)|_{d(1)} = \cup W^s_{d(1)}(\delta_i,\, 0) \tag{6.11}
$$

Theorem 6.2 asserts that if the one-parameter transversality condition is satisfied, then the equilibrium points on the stability boundary $\partial A(x_s)$ of system d(0) are exactly the same as the equilibrium points on the stability boundary $\partial A(x_s)$ of system d(1). Moreover, the stability boundary $\partial A(x_s)$ of system d(0) is the union of the stable manifolds (with respect to the dynamics of system d(0)) of the equilibrium points, say $(\delta_i,\, 0)$, i = 1,2,...,n, of system d(0) and the stability boundary $\partial A(x_s)$ of system d(1) is the union of the stable manifolds (with respect to the dynamics of system d(1)) of the equilibrium points $(\delta_i, 0)$, i = 1,2,...,n.

Step 1 and Step 4

Mathematically speaking, we study in these two steps the robustness of both the types equilibrium points and the equilibrium points on the stability boundary under a small perturbation of vector field. Here, we consider a c^1 perturbations of the vector field. Let f:M $\longrightarrow$ E a c^1 vector field, M be an open set in a vector space E. Let v(M) be the set of all c^1 vector fields on M. If E has a norm, the c^1-norm of a vector field h $\in$ v(M) is the least upper bound of all the numbers

$$|h(x)|, \; ||Dh(x)|| \; ; \quad x \in M$$

With this topology, a neighborhood of f $\in$ v(M) is a subset of v(M) that contains a set of the form

$$\{g \in v(M)| \; |g - f|_1 < \epsilon\}$$

for some $\epsilon > 0$ and some norm on M. We call g is a $\epsilon - c^1$ perturbation of f.

The property of persistence is a weak condition, much weaker than the notion of structural stability. In the following theorem it is shown that this property holds for "almost all" c^1 vector fields, while the stronger property does not.

THEOREM 6.3. *Let f:M $\longrightarrow$ E be a c^1 vector field, M is an open set in the vector space E. Let $\hat{x}$ be a hyperbolic equilibrium point on the stability boundary $\partial A(x_s)$ satisfying the transversality condition, then for any $\epsilon > 0$ there exists a δ-neighborhood h $\subset$ v(M) of f and a neighborhood u $\subset$ M of $\hat{x}$ such that for any g $\in$ h*

[i] There is a unique hyperbolic equilibrium point $\hat{x}^g$ of the vector field g such that $|\hat{x}^g - \hat{x}| < \epsilon$; moreover,

[ii] $\hat{x}^g$ is on the stability boundary $\partial A(x_s^g)$, where x_s^g is a stable equilibrium point of vector field g and $|x_s^g - x_s| < \epsilon$.

Proof. Since $\hat{x}$ is a hyperbolic equilibrium point, there exists a δ_1-neighborhood of f such that for any $|x_s^g - x_s| < \epsilon$ is true. For the second part, since $\hat{x}$ is on the stability boundary $\partial A(x_s^g)$, by Theorem 3.6 in [9] we have

$$W^u(\hat{x}) \cap A(x_s) \neq \phi$$

in other words, $W^u(\hat{x})$ intersects with $W^s(x_s)$ transversally. By the openess of transversality condition [15], there exists a δ_2-neighborhood of f such that $W^u(\hat{x}^g)$ intersects with $W^s(x_s^g)$ transversally for all vector field g in this neighborhood. It follows from Corollary 3.3 that $\hat{x}^g$ is on the stability boundary $\partial A(x_s^g)$. Taking $\delta = min\ \{\delta_1, \delta_2\}$ we complete the proof. □

We next generalize the above result to the robustness of all the equilibrium points on a stability boundary. Let the equilibrium point of the vector field f be hyperbolic. Theorem 6.3 implies that for an equilibrium point, say x_i, there exists a neighborhood h_i of f and a neighborhood u_i of x_i such that any vector field g $\in$ h will have a unique critical element $x_i(g) \in u_i$. It can be shown using the notion of filtration that the neighborhoods h_i and u_i can be shrunk so that $x_i(g)$ is the maximal invariant set of g entirely contained in u. Moreover, there exists a neighborhood h of f such there is no other non-wandering point outside the union of u_i. Next, suppose that the vector field f satisfies the transversality condition. By the open condition of transversality condition, it follows that for a pair of critical elements x_i and x_j with x_j stable and $W^u(x_i) \cap W^s(x_j) \neq \phi$, there exists a neighborhood ϵ of f such that, for any vector field g in this neighborhood, the intersection between $W^u(x_i(g))$ *and* $W^s(x_j(g))$ is non-empty and transverse. Based on the above arguments, we have the following:

THEOREM 6.4. *Let x_s be a stable equilibrium point of the vector field f and let x_i, i=1,2,...,k be the hyperbolic equilibrium points on $\partial A(x_s)$ and the intersection between $W^u(x_i)$ and $W^s(x_s)$ satisfies the transversality condition and $\partial A(x_s) = \cup W^s(x_i)$. There exists a positive number ϵ such that, for any vector field g which is a $\epsilon - C^1$ perturbation of f, the equilibrium points $h(x_i)$, $i = 1,2,...,k$ are also on the stability boundary of h $(x_s(g))$ and preserve their types, and $\partial A(x_s(g)) = \cup W^s(x_i(g))$, where $h(\cdot)$ is an one-to-one correspondence and $h(x_s(g))$ is a stable equilibrium point of the vector field g.*

Applying Theorem 6.4 to system (6.4) and system (6.7), we have the following results

PROPOSITION 6.5. *(Properties for Step 1) Consider the system (6.4). Let δ_s be a stable equilibrium point and δ_i, i=1,2,...,k be the equilibrium points on $\partial A(\delta_s)$ and the intersection between $W^u(\delta_i)$ and $A(\delta_s)$ satisfies the transversality condition. If zero is a regular value of $\frac{\partial V_p(\delta)}{\partial \delta_i}$, then there exist two positive numbers $\delta > 0$ and $\epsilon > 0$ such that if the transfer conductance of system (6.3) satisfies $G_{ij} < \delta$, there exists a unique stable equilibrium point $\hat{\delta}_s$ of system (6.3) with $|\hat{\delta}_s - \delta_s| < \epsilon$, and for each δ_i, there is a unique hyperbolic equilibrium point $\hat{\delta}_i$ of system (6.3) with $|\hat{\delta}_i - \delta_i| < \epsilon$ and they have the same type. Moreover, $\hat{\delta}_i$ is on the stability boundary $\partial A(\hat{\delta}_s)$ and there is no other equilibrium points other than $\hat{\delta}_i$, i=1,2,...,k lying on $\partial A(\hat{\delta}_s)$.*

PROPOSITION 6.6. *(Properties for Step 4) Consider the system (6.7). Let $(\delta_s, 0)$ be a stable equilibrium point and $(\delta_i, 0)$, i=1,2,...,k be the equilibrium points on $\partial A(\delta_s, 0)$ and the intersection between $W^u(\delta_i, 0)$*

and $A(\delta_s, 0)$ satisfies the transversality condition. If zero is a regular value of $\frac{\partial V_p(\delta)}{\partial \delta_i}$, then there exist two positive numbers $\hat{\delta} > 0$ and $\hat{\epsilon} > 0$ such that if the transfer conductance of system (6.6) satisfies $G_{ij} < \hat{\delta}$, there exists a unique stable equilibrium point $(\hat{\delta}_s, 0)$ of system (6.6) with $|(\hat{\delta}_s, 0) - (\delta_s, 0)| < \epsilon$, and for each $(\delta_i, 0)$, there is a unique hyperbolic equilibrium point $(\hat{\delta}_i, 0)$ of system (6.6) with $|(\hat{\delta}_i, 0) - (\delta_i, 0)| < \hat{\epsilon}$ and they have the same type. Moreover, $(\hat{\delta}_i, 0)$ is on the stability boundary $\partial A(\hat{\delta}_s, 0)$ and there is no other equilibrium points other than $(\hat{\delta}_i, 0)$, i=1,2,...,k lying on $\partial A(\hat{\delta}_s, 0)$.

We are now in a position to establish a static and a dynamic relationship between the original system (6.2) and the artificial dimension-reduction system (6.3)

THEOREM 6.7. : *(Relationships between the original system and the artificial system)*

Consider the original system (6.2) and the artificial dimension-reduction system (6.3). If zero is a regular value of $\frac{\partial V_p(\delta)}{\partial \delta_i}$, then there exists a positive number $\epsilon > 0$ such that if the transfer conductance of system (6.2) satisfies $G_{ij} < \epsilon$, we have

(R1) (δ) is a type-k equilibrium point of the dimension-reduction system (6.3) if and only if $(\delta, 0)$ is a type-k equilibrium point of the original system (6.2). In particular, (δ_s) is a stable equilibrium point of the dimension-reduction system (6.3) if and only if $(\delta_s, 0)$ is a stable equilibrium point of the original system (6.2).

Moreover, if the intersections of the stable and unstable manifolds of the equilibrium points on the stability boundary $\partial A(\delta_s, 0)$ of the parameterized system $d(\lambda)$ satisfy the transversality condition for $\lambda \in [0, 1]$, then

(R2) the equilibrium point $(\delta_i, 0)$ is on the stability boundary $\partial A(\delta_s, 0)$ of system (6.2) if and only if the equilibrium point (δ_i) is on the stability boundary $\partial A(\delta_s)$ of system (6.3).

(R3)

$$\partial A(\delta_s,\ 0) \ = \ \cup\, W^s(\delta_i,\ 0) \tag{6.12}$$

$$\partial A(\delta_s) \ = \ \cup\, W^s(\delta_i) \tag{6.13}$$

Result (R1) justifies the efforts of establishing the relationship between the stability region $A(\delta_s)$ of the dimension-reduction system (6.3) and the stability region $A(\delta_s,0)$ of the original system (6.2). Results (R2) and (R3) establish a dynamic relationship between the stability boundary $\partial A(\delta_s)$ and the stability boundary $\partial A(\delta_s,0)$ and suggests the plausibility of finding the controlling u.e.p. of the original system (6.2) via finding the controlling u.e.p. of the dimension-reduction system (6.3).

7. BCU method for the network-reduction classical machine model. Based on the analytical results derived in previous section, a theory-based method, called boundary of the stability region based controlling unstable equilibrium point method (BCU method) for direct analysis of power system transient stability will be presented in this section. One feature distinghishes the BCU method from the existing direct methods is that it consistently find the exact controlling u.e.p. relative to a fault-on trajectory and it has a sound theoretical basis. Moreover, the BCU method appears to be fast.

The BCU method : find the controlling u.e.p. relative to a fault-on trajectory (version 1)

Step 1: From the fault-on trajectory $(\delta(t), \omega(t))$, detect the exit point δ^* at which the projected trajectory $\delta(t)$ exits the stability boundary of the dimension-reduction system (6.3).

Step 2: Use the point δ^* as initial condition and integrate the post-fault dimension-reduction system (6.3) to find the first local minimum of $\sum_{i=1}^{n} \| f_i(\delta) \|$, say at δ_o^*.

Step 3: Use the point δ_o^* as the initial guess to solve $\sum_{i=1}^{n} \| f_i(\delta) \| = 0$, say at δ_{co}^*.

Step 4: The controlling u.e.p. with respect to the fault-on trajectory is $(\delta_{co}^*, 0)$.

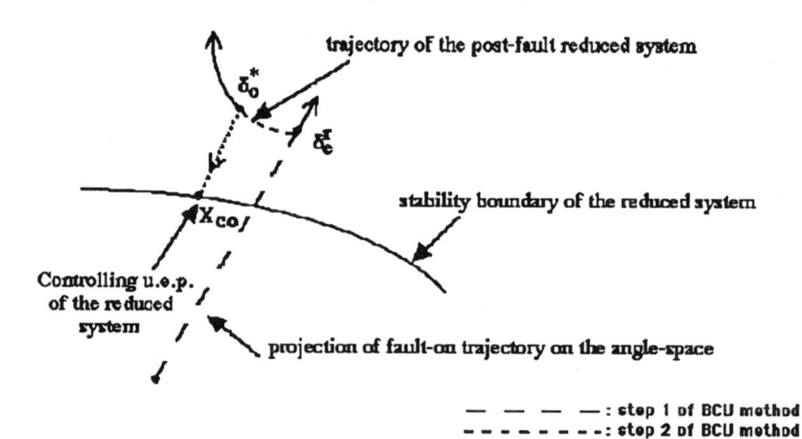

FIG. 7.1. *Three main steps of the BCU method for the network-reduction classical machine model*

Once the controlling u.e.p. is found, the BCU method uses the same procedure as that of the conceptual controlling u.e.p. method presented in Section 4 (i.e. Step 2 - Step 4) to perform stability assessment. The essence of the BCU method is that it finds the controlling u.e.p. via the controlling u.e.p. of the dimension-reduction system (6.3) which is only defined in the angle space and whose controlling u.e.p. is easier to compute. Steps 1–3 find the controlling u.e.p. of the dimension-reduction system and Step 4 relates the controlling u.e.p. of the dimension-reduction system to the controlling u.e.p. of the original system (see Fig. 7.1).

Remark 7.1.

[1] An effective computation scheme to implement Step 1 is: From the fault-on trajectory $(\delta(t), \omega(t))$, detect the exit point δ^* at which the projected trajectory $\delta(t)$ reaches the first local maximum of $V_p(\cdot)$.

[2] The dimension-reduction system (6.3) can be stiff. In such a case, a stiff differential equation solver is recommended to implement Step 2.

[3] Several existing direct methods can be viewed as finding the controlling u.e.p. in the angle space. For example, Fouad et al. found the controlling u.e.p. via the 'mode of disturbance' of machines. Pavella et al. located the controlling u.e.p. via the 'accelerating machines'. The BCU method finds the controlling u.e.p. of the dimension-reduction system (in the angle space) via two steps. The first step locates the exit point of the dimension-reduction system. Recall that the exit point must lie on the stable manifold of the controlling u.e.p. Hence, the second step integrates the dimension-reduction system starting from the exit point and the resulting trajectory will converge to the controlling u.e.p. In practical computation, the point calculated in Step 1 can not be exactly the exit point but lies in a neighborhood of the exit point. Consequently, the resulting trajectory will just pass by the controlling u.e.p.

We next present a sufficient condition under which the BCU method finds the correct controlling u.e.p. relative to a given fault: the exit point of the projected fault-on trajectory $(\delta(t))$ is on the stable manifold $W^s(\hat{\delta})$ if and only if $(\delta(t), \omega(t))$ is on the stable manifold $W^s(\hat{\delta}, 0)$. Theorem 6.7 shows that, under the one-parameter transversality condition, the equilibrium point $(\hat{\delta})$ is on the stability boundary $\partial A(\delta_s)$ of the dimension-reduction system (6.3) if and only if the equilibrium point $(\hat{\delta}, 0)$ is on the stability boundary $\partial A(\delta_s, 0)$ of the original system (6.2).

7.1. Computational considerations. There are three major computational tasks in the BCU method: (i) compute the point with the first local maximum of potential energy along the projected fault-on trajectory, (ii) compute the trajectory of the post-fault dimension-reduction system, and (iii) compute the controlling u.e.p. via solving the nonlinear algebraic equations related to the dimension-reduction system. It is very difficult to compute the exact point where the first local maximum of potential energy along the projected fault-on trajectory occurs because the simulation of the trajectory on a digital computer yields only a sequence of points and not the entire trajectory. Hence, Step 1 of the BCU method usually gives a point which is close to the exit point. In order to achieve a speed-up in computation without sacrificing accuracy, the following method is recommended when a large integration step size is used in integrating the fault-on system:

The BCU method (version 2)

Step 1: From the fault-on trajectory $(\delta(t), \omega(t))$, detect the point δ^* at which the projected trajectory $\delta(t)$ reaches the first local maximum of $V_p(\cdot)$. Also, compute the point δ^+ that is one step after δ^*.

Step 2: Use the point δ^* as the initial condition and integrate the post-fault dimension-reduction system to find the first local minimum of $\sum_{i=1}^{n} \| f_i(\delta) \|$, say at δ_o^*.

Step 3: Use δ^+ as an initial condition and repeat Step 2 to find the corresponding points, say δ_o^+.

Step 4: Compare the values of $\|f(\delta)\|$ at δ_o^* and δ_o^+. The one with the smallest value is used as the initial guess to solve $f_i(\delta) = 0$, say the solution is δ_{co}.

Step 5: The controlling u.e.p. with respect to the fault-on trajectory is $(\delta_{co}, 0)$.

Using a large integration step size in Step 1 may cause the first local minimum to go undetected in Step 2. Should this happen, go back to Step 1 and take a smaller step size to detect the new points δ^* and δ^+ and repeat Step 2 to Step 5.

7.2. Numerical studies. The BCU method has been tested on several power systems. This section presents several test results on a 50-generator, 145-bus system due to several three-phase faults with fault locations at both generator and load buses. Both severe and mild faults have been considered. A majority of the material presented in this section was taken from [11].

Table 7.1 displays the estimated critical clearing time of several faulted systems using three different methods: step-by-step numerical integration technique, the BCU method (version 1), the BCU method (version 2) with corrected kinetic energy [25] and the MOD method [7]. The results from the numerical integration technique is used as a benchmark. A few observations and comments on the simulation results follow.

The BCU method consistently finds the exact controlling u.e.p. There are a few cases where the MOD method did not converge to any u.e.p. And in some cases, the MOD method converges to a u.e.p. which is not the controlling u.e.p. In all but one case, the BCU method gives slightly conservative results in CCT (under-estimate). The only case that the BCU

TABLE 7.1
*The Critical Clearing Time. The sign *** means that the method does not converge to any U.E.P.*

Fault location at bus	Line clearing between buses	Critical Clearing Time estimated by different methods			
		Step-by-Step simulation Unit: sec	The BCU Method Unit: sec	The BCU Method + corrected energy Unit: sec	The MOD Method Unit: sec.
6	6 - 12	0.200	0.170	0.185	***
7	6 - 7	0.119	0.115	0.122	0.113
12	12 - 14	0.190	0.170	0.190	***
58	58 - 87	0.51	0.59	-	***
63	61 - 63	0.91	0.855	0.91	***
90	90 - 92	0.270	0.265	0.270	***
96	96 - 73	0.245	0.242	0.245	0.240
97	97 - 66	0.260	0.260	0.260	0.250
98	98 - 72	0.205	0.205	0.205	0.190
100	100 - 72	0.315	0.315	0.320	0.315
102	63 - 102	0.200	0.195	0.195	0.195
116	116 - 63	0.28	0.28	0.28	0.285

method gives an over-estimate is the case that a fault occurs at bus 58. In this case, the CCT from benchmark is 0.51 sec. while the estimated CCT by the BCU method is 0.59 sec. This case was then further studied and it is found that the BCU method still computed the correct controlling u.e.p. However, a whole group of machines (interarea mode) in this case tend to become unstable such that the assumption in deriving the energy function that the fault-on trajectory is a straight line does not hold. Thus, it is the accuracy of the energy function (i.e. the conditions required for energy functions are not satisfied) for this case that causes the over-estimate. It has been our experience that cases causing a whole group of machines to go unstable usually damage the accuracy of energy functions.

For those cases that the MOD method converges, the estimated CCT's are fairly good even though the u.e.p. found by the method may not be the controlling u.e.p. (c.f. the discussions following Theorem 4.1). The simulation results also reveal that the MOD method gives both slightly over-estimates and under-estimates. The BCU method with the corrected kinetic energy method gives on the average very good results except in some cases it gives over-estimate results such as the faults occurring at bus #7 and bus #100. In all the other cases, it reduces the conservativeness of the BCU method. This indicates that the corrected kinetic energy method is quite effective and deserves further (theoretical) investigation.

8. A sensitivity-based BCU method for the network-reduction classical machine model. Quite often, utility companies need to cope with the following situation: for a given power system with a given contingency, called the base case, if some of the parameters in the system change (due to an emergency, perhaps), called the new case, one must know how these changes affect the system stability as quickly as possible. Also, there is a great need of a fast, yet accurate method to determine the transient stability-constrained loading limits, power transfer limits and, furthermore, to decide how to make changes in the parameters, such as

Angle Space

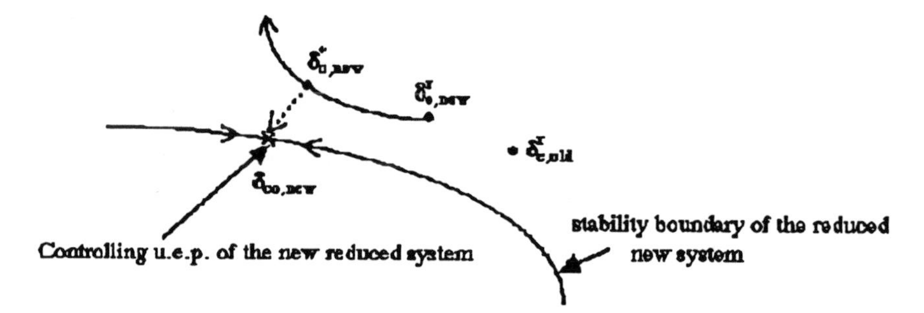

FIG. 8.1. *The sensitivity-based BCU method fast estimates the state variables at the fault clearing time X_I (i.e. the initial state of post-fault system) and at the exit point X_e^r (of the reduced system) when changes are made to the parameters of the base case to which the BCU method has already been applied.*

generation rescheduling, and load shedding in order to improve system transient behaviors. This section illustrates an application of the BCU method: a sensitivity-based BCU method for fast derivation of stability limits in electric power systems [14].

Numerous previous methods, such as those described in [31,32,33,34] have attempted to use the sensitivity of energy functions to changes in system parameters to determine the stability limits of systems. It is assumed in these methods that the mode of controlling u.e.p. does not change. In general, this assumption may not be true for every case. A counter-example to the assumption on a 50-gerenator, 145-bus system will be provided, A principal difference between the sensitivity-based BCU method and those which have proceeded it is its freedom from the assumption that the mode of controlling u.e.p. dose not change for any case.

The main idea behind the sensitivity-based BCU method is that the values of the state variables at the fault clearing time, X_I, (i.e. the initial state of post-fault system) and at the exit point, X_e^r, (of the reduced system) can be quickly estimated when changes are made to the parameters of the base case to which the BCU method has already been applied. (See

Fig. 8.1). The controlling u.e.p. of the new case is then found following Steps 2–4 of the BCU method. Let $X_I^{new} = X_f^{new}(t_{cl})$ and $X^r_{e,new}$ be the state variables at the fault clearing time and at the exit point of the new case and X_I^{old}, $X^r_{e,old}$ be the corresponding ones of the base case. And let

$$X_I^{new} = X_I^{old} + \Delta X_I \tag{8.1}$$

$$X^r_{e,new} = X^r_{e,old} + \Delta X^r_e \tag{8.2}$$

where ΔX_I, ΔX^r_e are the changes in X_I *and* X^r_e, respectively, caused by the parameter changes. They satisfy the following equations:

$$\Delta X_{I,i} = \sum_{k=1}^{m} \frac{\partial X_{I,i}}{\partial \alpha_k} \Delta \alpha_k \tag{8.3}$$

$$\Delta X^r_{e,i} = \sum_{k=1}^{m} \frac{\partial X^r_{e,i}}{\partial \alpha_k} \Delta \alpha_k \tag{8.4}$$

where α_k, k=1,2,...,m are the parameters in the system, and $\frac{\partial X_{I,i}}{\partial \alpha_k}$ *and* $\frac{\partial X^r_{e,i}}{\partial \alpha_k}$ are the first-order derivatives, satisfying the following equation:

$$\begin{aligned}
M_i \frac{\partial \dot{\tilde{\omega}}_i}{\partial \alpha_k} &= \frac{\partial P_{mi}}{\partial \alpha_k} - 2E_i \frac{\partial E_i}{\partial \alpha_k} G^f_{ii} - E_i^2 \frac{\partial G^f_{ii}}{\partial \alpha_k} \\
&- \sum_{j=1, j\neq i}^{n} \left[C^f_{ij} cos\theta_{ij} \left(\frac{\partial \theta_i}{\partial \alpha_k} - \frac{\partial \theta_j}{\partial \alpha_k} \right) \right. \\
&\qquad \left. - D^f_{ij} sin\theta_{ij} \left(\frac{\partial \theta_i}{\partial \alpha_k} - \frac{\partial \theta_j}{\partial \alpha_k} \right) \right] \\
&- \sum_{j=1, j\neq i}^{n} \left[\left(\frac{\partial E_i}{\partial \alpha_k} E_j + \frac{\partial E_j}{\partial \alpha_k} E_i \right) \right. \\
&\qquad \left. \cdot \left(B^f_{ij} sin\theta_{ij} + G^f_{ij} cos\theta_{ij} \right) \right] \\
&- \sum_{j=1, j\neq i}^{n} \left[E_i E_j sin\theta_{ij} \frac{\partial B^f_{ij}}{\partial \alpha_k} + E_i E_j cos\theta_{ij} \frac{\partial G^f_{ij}}{\partial \alpha_k} \right] \\
&- \frac{M_i}{M_T} \left[\sum_{j=1}^{n} \left(\frac{\partial P_{mj}}{\partial \alpha_k} - 2E_j \frac{\partial E_j}{\partial \alpha_k} G^f_{jj} - E_j^2 \frac{\partial G^f_{jj}}{\partial \alpha_k} \right) \right. \\
&\qquad + \sum_{i=1}^{n} \sum_{j=1, j\neq i}^{n} \left[D^f_{ij} sin\theta_{ij} \left(\frac{\partial \theta_i}{\partial \alpha_k} - \frac{\partial \theta_j}{\partial \alpha_k} \right) \right. \\
&\qquad - \left(\frac{\partial E_i}{\partial \alpha_k} E_j + \frac{\partial E_j}{\partial \alpha_k} E_i \right) G^f_{ij} cos\theta_{ij} \\
&\qquad \left. \left. - E_i E_j cos\theta_{ij} \frac{\partial G^f_{ij}}{\partial \alpha_k} \right] \right]
\end{aligned} \tag{8.5}$$

$$\frac{\partial \dot{\theta}_i}{\partial \alpha_k} = \frac{\partial \hat{\omega}_i}{\partial \alpha_k} \qquad i = 1, 2, \ldots, n-1$$

where $C_{ij}^f = E_i E_j B_{ij}^f$, $D_{ij}^f = E_i E_j G_{ij}^f$ and B_{ij}^f, G_{ij}^f represent the susceptance and transfer conductance of the i-j element of the reduced admittance matrix of the faulted system.

This is a set of linear differential equations which can be integrated if knowing the values of θ_i and $\hat{\omega}_i$, $i = 1, \ldots, n$ at each time step. The equations are solved to determine the conditions at the end of the disturbance and at the exit point (of the reduced system) in the base case. Thus,

$$\frac{\partial \theta_{I,i}}{\partial \alpha_k}, \frac{\partial \hat{\omega}_{I,i}}{\partial \alpha_k}, \frac{\partial \theta^r_{e,i}}{\partial \alpha_k}, \frac{\partial \hat{\omega}^r_{e,i}}{\partial \alpha_k} \qquad i = 1, 2, \ldots, n$$

can be obtained.

A sensitivity-based BCU Method for fast stability assessment is as follows:

Step 1: Calculate the point $X^r_{e,new}$ using equations (8.2) and (8.4). Let $X^r_{e,new} = (\delta^*_{e,new}, \omega^*_{e,new})$

Step 2: Use the point $\delta^*_{e,new}$ as an initial condition and integrate the post-fault reduced system (of the new case) to find the first local minimum point, say at $\delta^*_{0,new}$.

Step 3: Solve the post-fault reduced system, say $f(\delta) = 0$, using $\delta^*_{0,new}$, as the initial guess; call the solution $\delta_{co,new}$. The controlling u.e.p. with respect to the fault-on trajectory of the new case is ($\delta_{co,new}$, 0).

Step 4: Use the energy function value at the controlling u.e.p. ($\delta_{co,new}$, 0) as the critical energy $V_{cr} = V(\delta_{co,new}, 0)$.

Step 5: Calculate the point X_I^{new} using equations (8.1) and (8.3).

Step 6: Calculate the energy function value at the fault clearing point X_I^{new}, $V_{cl} = V(X_I^{new})$.

Step 7: Compare V_{cl} with V_{cr}; if $V_{cl} < V_{cr}$, the (new post-fault) system will be stable. Otherwise, it is unstable.

Remark 8.1.

[1] The above sensitivity-based BCU method computes the controlling u.e.p. of the new system via Step 1 through Step 3. Previous fast methods based on the energy margin's sensitivity to the changes in system parameters were dependent upon the invalid assumption that the mode of controlling u.e.p. of the new case does not change. The method proposed here performs a fast stability assessment without this assumption.

[2] An inherent problem with any sensitivity-based method is the size of the range in which the linear approximation holds. It is conceivable that the linear approximate is invalid for large changes in system

TABLE 8.1

Energy margin of the base case by the BCU method and energy margin of the new system by the sensitivity-based BCU method, and the stability assessment results of the new system using the step-by-step time domain simulations for stable base cases

Fault Bus	Line Clear-ing	Fault Clearing Time Unit:sec	Energy Margin by BCU Method Base System	Allowable Change in Generation Due to Stability Limit Unit: MW				Energy Margin by Proposed Method New System	Step-by-step Simu-lation Result
				Gen. 9	Gen. 20	Gen. 25	Gen. 25		
116	116-63	0.267	1.832/ Stable	100	70	100	50	0.0705/ Stable	Stable
67	67-66	0.170	1.491/ Stable	40	0	50	50	0.0195/ Stable	Stable
98	98-72	0.174	1.451/ Stable	100	50	50	70	0.0506/ Stable	Stable
89	89-59	0.246	1.046/ Stable	40	50	100	50	0.0532/ Stable	Stable
91	91-75	0.217	1.443/ Stable	50	120	50	100	0.0101/ Stable	Stable

TABLE 8.2

Energy margin of the base case by the BCU method and energy margin of the new system by the sensitivity-based BCU method, and the stability assessment results of the new system using the step-by-step time domain simulations for unstable base cases

Fault Bus	Line Clear-ing	Fault Clearing Time Unit: sec	Energy Margin by BCU Method Base System	Allowable Change in Generation Due to Stability Limit Unit: MW				Energy Margin by Proposed Method New System	Step-by-step Simu-lation Result
				Gen. 9	Gen. 20	Gen. 25	Gen. 26		
7	7-6	0.082	-1.132/ Unstable	-20	-20	-20	-20	0.0291/ Stable	Stable
116	116-63	0.279	-0.817/ Unstable	-60	-70	-70	-70	0.0154/ Stable	Stable
67	67-66	0.176	-2.550/ Unstable	-70	-70	-70	-80	0.0197/ Stable	Stable
98	98-72	0.196	-0.772/ Unstable	-60	-60	-60	-50	0.0011/ Stable	Stable
89	89-59	0.261	-0.816/ Unstable	-50	-50	-50	-60	0.0241/ Stable	Stable
90	90-92	0.245	-0.388/ Unstable	-60	-60	-60	-50	0.0011/ Stable	Stable
91	91-75	0.291	-0.597/ Unstable	-60	-60	-70	-70	0.0086/ Stable	Stable
100	100-72	0.315	-0.734/ Unstable	-60	-60	-50	-60	0.0007/ Stable	Stable

parameters. Second-order approximation may be required in equations (8.1) and (8.2) for some large changes in system parameters.

Numerical Studies

The sensitivity-based BCU method has been tested on a 50-generator, 145-bus power system with three-phase faults located at both load and generator buses. Both mild and severe faults were considered. During testing, the points $X^r_{e,old}$ at which the projected trajectory X(t) reaches the first local maximum of potential energy, and X_I^{old}, or X(t) at t=t_{ce}, wew computed for the base case.

Tables 8.1 and 8.2 list fault clearing times, changes in (real power) generation at several generators (the $\partial\alpha$ in equations (8.3) and (8.4)), the base case's energy margins, and energy margins of the new cases found

TABLE 8.3
Demonstration of a change in 'MOD' between the controlling u.e.p. of the base system and that of the new system

Controlling U.E.P. Base System			
Generator	Angle (δ)	Generator	Angle (δ)
1	64.45995	26	160.42375
2	55.20459	27	79.63201
3	65.25882	28	-1.92878
4	91.36189	29	3.13811
5	52.25932	30	9.97981
6	74.72792	31	4.77953
7	83.69713	32	-38.37576
8	62.01586	33	27.59951
9	89.48299	34	49.85472
10	72.60353	35	58.02808
11	78.58679	36	-14.73075
12	65.91161	37	-32.46071
13	66.86551	38	-9.08975
14	82.16170	39	16.74927
15	86.94065	40	-4.04800
16	76.85336	41	35.14288
17	69.39467	42	9.46479
18	41.91384	43	-82.45346
19	75.57879	44	-10.06099
20	172.34465	45	-31.59456
21	87.89543	46	-0.33017
22	88.04443	47	7.87006
23	40.56034	48	4.31619
24	49.43820	49	20.20789
25	89.23398	50	3.09340

Controlling U.E.P. New System			
Generator	Angle (δ)	Generator	Angle (δ)
1	128.85747	26	179.65063
2	71.79848	27	101.20805
3	106.25616	28	-1.84082
4	134.72824	29	4.38711
5	70.77628	30	12.34212
6	99.20209	31	6.54915
7	142.98723	32	-34.84964
8	81.47444	33	41.59850
9	111.41263	34	63.87070
10	137.09608	35	74.13261
11	199.50250	36	-9.40540
12	86.78194	37	-29.99523
13	83.81512	38	-7.45915
14	105.22060	39	20.74225
15	108.87260	40	-5.43824
16	101.08370	41	33.89392
17	90.77351	42	7.80966
18	46.72015	43	-88.39584
19	100.09409	44	-12.36244
20	-172.35338	45	-34.34732
21	109.65578	46	-0.93494
22	109.77702	47	9.32712
23	56.99943	48	4.90092
24	67.97100	49	18.29047
25	111.19093	50	3.62722

through the sensitivity-based BCU method and through the step-by-step time simulation with information altered to reflect the system changes under consideration. Table 8.1 lists data for cases where the base cases are stable. For small changes in generation, the sensitivity-based BCU method quickly found the new case's stability energy margin with a high degree of accuracy. The proposed method consistently gave a good estimate of the energy margin for both the stable cases of Table 8.1 and the unstable cases of Table 8.2.

The sensitivity-based BCU method also allows one to quickly determine the extent to which one may change real power generations in the system before jeopardizing its transient stability, or how much change is necessary in order to gain stability. The change in real power generations for each of the stable base cases listed in Table 8.1 represents a 'maximum' increase which can be tolerated at these generators before stability is lost. Similary, the change in real power generation for each of the unstable base cases listed in Table 8.2 represents a 'minimum' necessary change at these generators in order to re-establish stability. Since only one complete run of the BCU method is necessary for each case in the determination of these transient stability limits, the proposed method can fast derive stability limits.

The chief advantage of the proposed method is that it does not make the assumption that the new case will have roughly the same controlling u.e.p. as the base case. Other methods have employed sensitivity equations to estimate the new, slightly changed position of the controlling u.e.p. The proposed method estimates the position of the point where the potential energy reaches its first local maximum-the exit point-and integrates along the stability boundary of the reduced system associated with the new case to find the controlling u.e.p. Occasionally, the new exit point will fall within the stable manifold of a different u.e.p. The proposed method accounts for this possibility and allows the finding of a new controlling u.e.p. in such cases. Table 8.3 lists the coordinate of the controlling u.e.p. of a base case, followed by the coordinate of the new case found by the proposed method. The fault for this case was at bus 7, with line 7-6 cleared; the fault clearing time was 0.08 seconds, and generation changes of 450MW were made at generators No.20 and No.26. The results were checked using the step-by step method with appropriately altered load-flow information.

9. Discussion and conclusion. The controlling u.e.p. method is the most viable method for direct stability analysis based on energy functions of practical power systems. The success of the controlling u.e.p. method hinges upon the ability to find the (correct) controlling u.e.p. This paper has developed a method based on the time-domain simulation approach to compute the controlling u.e.p. The method is applicable to general power system stability models provided the controlling u.e.p. relative to a fault exists. The method is intended to serve as a bench-mark method to check the correctness of the controlling u.e.p. obtained by direct methods.

The BCU method is a systematic method of computing the controlling u.e.p. for large-scale power systems. It explores the special structure of the underlying model so as to define an artificial, dimension-reduced system which can capture all of the equilibrium points on the stability boundary of the underlying system. The BCU method then finds the controlling u.e.p. of the original system via the controlling u.e.p. of the dimension-reduced system which is much easier to find than that of the original system. Given a power system stability model with certain properties, there exists a corresponding version of the BCU method.

This paper has presented a general framework for the BCU method and developed a theoretical foundation for the BCU method for the network-reduction classical machine stability model. Sufficient conditions for the BCU method to find the correct controlling u.e.p. relative to a given fault have also been derived. It remains to be verified whether or not these sufficient conditions are always satisfied on practical large-scale power systems.

Further improvements in direct methods' limited modeling capability are desirable. It is well known that inaccurate load modeling can lead to a power system operating in modes that result in actual system collapse or separation. Accurate load models capturing load behaviors during dynamics are therefore necessary for more precise calculations of power system controls and stability limits. There is a strong need to use an accurate load model in direct methods. Traditionally, direct methods have been based on the network-reduction model where all the load representations are expressed in constant impedance and the network is reduced to the generator internal buses. As a result, the network-reduction model has two critical shortcomings for the analysis of modern power systems.

First, in the context of system modeling, it precludes consideration of dynamic load behaviors (i.e. voltage and frequency variations) at load buses. The so-called static load model, where loads are represented as an appropriate combination of constant impedance, constant current, and constant power, may be adequate to capture load behaviors during transient for some cases, but not adequate for other cases in which proper dynamic load models (a model that expresses the real and reactive powers at any instant of time as functions of voltage magnitude and frequency at past instants of time and, usually, including the present instant [30]) are required. A summary of the state-of-the-art load representations for power system dynamic performance analysis can be found in [30]. Second, in the context of physical explanation of results, reduction of the transmission network leads to loss of network topology and, hence, precludes study of transient energy shifts among different components of the entire power network. Considerable efforts have been made to improve the modeling capability of the network-reduction model. The work reported in [26] includes the effect of exciters represented by a first-order transfer function in the network-reduction model. An attempt to incorporate the static load model in the network-reduction model can be found in [27].

Network-preserving models (structure-preserving models) have been proposed in the last decade to overcome some of the shortcomings of the classical model and to improve the modeling of generators, exciters, automatic voltage regulators and load representations. The first network-preserving model was developed by Bergen and Hill [35], who assumed frequency dependent real power demands and constant reactive power demands. Narasimhamurthy and Musavi [37] moved a step further by considering constant real power and voltage dependent reactive power loads. Padiyar and Sastry [38] have included nonlinear voltage dependent loads for both real and reactive powers. Tsolas, Araposthasis and Varaiya [36] developed a network-preserving model with the consideration of flux decay and constant real and reactive power loads. An energy function for a network-preserving model accounting for static var compensators and their operating limits was developed by Hiskens and Hill [39]. Therefore, it seems a logical step is to develop the BCU method for network-preserving models with adequate load representations as well as its underlying theoretical foundations.

Another stability problem receiving great concerns in power industry is the voltage-dip problem [40]. This problem can be stated as follows: following a disturbance (event disturbance or load disturbance), the underlying power system will settle down to an acceptable steady-state (in the context of frequency and voltage) and during the transient, the system voltage will not dip to un-acceptable values. The voltage dip problem can also be stated as the following: given a set of nonlinear equations with an initial condition, determine whether or not the ensuing trajectories will settle down to a desired steady-state without resort to explicit numerical integrations of whole set of nonlinear equations and, furthermore, one wants to know whether during the transient part of the state variables (related to voltage magnitudes at load buses) drop below a certain value, say 0.9 p.u. Alternatively, from a nonlinear system viewpoint, the problem can be stated as: given a set of nonlinear equations with an initial condition, determine whether or not the initial condition lies inside the stability region (region of attraction) of a desired stable equilibrium point and, whether or not the projected system trajectory on the subspace spanned by the voltage magnitudes at load buses does not fall below a certain value. Obviously, the voltage-dip problem is very unique and challenging. Research efforts on extending the BCU method to cope with the voltage dip problem may prove fruitful.

REFERENCES

[1] P.P.Varaiya, F.F.Wu, R-LChen, *Direct methods for transient stability analysis of power systems: recent results*, Proceedings of the IEEE, December 1985, 1703–1715.

[2] N.Kakimoto, Y.Ohsawa, M.Hayashi, *Transient stability analysis of electric*

power system via lure-type Lyapunov function, part I and II, Trans. IEE of Japan **98** (1978), 516.

[3] TH.VAN GUTSEM, M.RIBBENS-PAVELLA, *Structure preserving direct methods for transient stability analysis of power systems*, Proc. 24th IEEE Conference on Decison and Control, Ft. Lauderdale, FL., Dec 1985, 70–76.

[4] H.D.CHIANG, F.F.WU, P.VARAIYA, *Foundation of PEBS method for power system transient stability analysis*, IEEE Trans. on circuits and systems **CAS-35** June (1988), 712–728.

[5] T.ATHAY, R.PODMORE, S.VIRMANI, *A practical method for direct analysis of transient stability*, IEEE Transaction on Power Apparatus and System **PAS-98** (1979), 573–584.

[6] M.A.PAI, Energy Function Analysis for Power System Stability, Kluwer Academic Publishers 1989.

[7] A.A.FOUAD, V.VITTAL, Power System Transient Stability Analysis: Using the Transient Energy Function Method, Prentice-Hall, New Jersey 1991.

[8] H.D.CHIANG, F.F.WU, P.P.VARAIYA, *Foundation of direct methods for power system transient stability analysis*, IEEE Trans. on Circuits and Systems **CAS-34** Feb (1987), 712–728.

[9] H.D.CHIANG, *Analytical results on direct methods for power system transient stability analysis*, Advances in Control and Dynamic Systems, **XL** Theme: Advances in Electric Power and Energy Conversion Systems Dynamics and Control, Academic Press, **43, Part 3** 1991, 275–334.

[10] H.D.CHIANG, *A theory-based controlling u.e.p. method for direct analysis of transient stability*, IEEE 1989 International Symposium of Circuits and Systems, May 1989.

[11] H.D.CHIANG, F.F.WU, P.P.VARAIYA, *A BCU method for direct analysis of power system transient stability*, IEEE Transactions on Power Systems (*to appear*).

[12] F.A. RAHIMI, *Evaluation of transient energy function method software for dynamic security analysis*, EPRI Project RP 4000-18, Final Report, Dec 1990.

[13] F.A.RAHIMI, M.G.LAUBY, J.N.WRUBEL, K.L.LEE *Evaluation of the transient energy function method for on-line dynamic security assessment*, IEEE Transaction on Power Systems, **8** (2) May (1993), 497–507.

[14] J.TONG, H.-D.CHIANG, T.P.CONNEEN, *A sensitivity-based BCU method for fast derivation of stability limits in electric power systems*, IEEE Transaction on Power Systems, **8** (4) Nov (1993), pp. 1418–1428.

[15] S.SMALE, *Differentiable dynamical systems*, Bull. Amer. Math. Soc. **73** (1967), 747.

[16] D.H.KUO, A.BOSE, *A generation rescheduling method to increase the dynamic security of power systems*, IEEE Transactions on Power Systems (*to appear*).

[17] M.W.HIRSCH, S.SMALE, Differential Equations, Dynamical Systems and Linear Algebra, Academic Press 1974.

[18] J.GUCKENHEIMER, P.HOLMES, Nonlinear Oscillations, Dynamical Systems, and Bifurcation Of Vector Fields, Springer-Verlag 1983.

[19] J.K.HALE, Ordinary Differential Equations, Robert E. Krieger Publishing Company, Huntington, New York 1980.

[20] P.HARTMAN, Ordinary Differential Equations, John Wiley, New York 1973.

[21] M.VIDYASAGAR, Nonlinear Systems Analysis, Prentice-Hall Inc., N.J. 1992.

[22] N.KOPELL, R.B.WASHBURN,JR., *Chaotic motion in two-degree-freedom swing equations*, IEEE Trans. on Circuit and Systems, **CAS-29** Sept (1982), 612–623.

[23] J.P.AUBIN, I.EKELAND, Applied Nonlinear Analysis, John Wiley & Sons 1984.

[24] F.S.PRABHAKARA, A.H.EL-ABIAD, *A simplified determination of stability regions for Lyapunov method*, IEEE Transaction on Power Apparatus and System, **PAS-94** (1975), 672–689.

[25] A.A.FOUAD, S.E.STANTON, *Transient stability of a multimachine power systems. Part I: investigation of system trajectories*, IEEE Transaction on Power Ap-

paratus and System, **PAS-100** (1981), 3408–3414.

[26] A.A.Fouad, V.Vittal, Y-X.Ni, A.R.Pota, K.Nodehi, H.M.Zein-Eldin, J.Kim, *Direct transient stability assessment with excitation control*, IEEE Trans. on Power Systems, **PWRS-4** Feb (1989), 75–82.

[27] V.Vittal, N.Bhatia, A.A.Fouad, G.A.Maria, H.M.Zein El-Din, *Incorporation of nonlinear load models in the transient energy function method*, IEEE Trans. on Power Systems, **4** (3) Aug (1989), 1031–1036.

[28] M.Ribbens-Pavella, P.G.Murthy, J.L.Horward, *The acceleration approach to practical transient stability domain estimation in power systems*, Proc. of the 20th IEEE Conference on Decision and Control, San Diego, CA, Dec 16-18, 1981, 471–477.

[29] IEEE Committee Report, *Transient stability test systems for direct stability methods*, IEEE Trans. on Power Systems, **7** Feb (1992), 37–43.

[30] IEEE Committee Report, *Load representation for dynamic performance analysis*, IEEE Trans. on Power Systems, **8** May (1993), 472–482.

[31] P.W.Sauer, K.D.Demaree, M.A.Pai, *Stability limited load supply and interchange capability*, IEEE Trans. on Power Apparatus and Systems, **PAS-102** Nov (1983), 3637–3643.

[32] M.A.El-kady, C.K.Tang, V.F.Carvalho, A.A.Fouad, V.Vittal, *Dynamic security assessment utilizing the transient energy function method*, IEEE Trans. on Power Systems, **PWRSl** (3) Aug (1986), 284–291.

[33] M.A.Pai, P.W.Sauer, K.D.Demaree, *Direct methods of stability analysis in dynamic security assessment*, Proc. 9th IFAC World Congress, Budapest, July 1984.

[34] Y.Xue, Th.Van Custem, M.Ribbens-Pavella, *Real-time analytic sensitivity method for transient security assessment and preventive control*, Proc. IEE **135 Part C** (2) March (1988), 107–117.

[35] A.R.Bergen, D.J.Hill, *A structure preserving model for power system stability analysis*, IEEE Trans on Power Apparatus and Systems, **PAS-100** January (1981), 25–35.

[36] N.Tsolas, A.Arapostathis, P.P.Varaiya, *A structure preserving energy function for power system transient stability analysis*, IEEE Trans. on Circuits and Systems **CAS-32** Oct (1985), 1041–1049.

[37] N.Narasimhamurthi, M.R.Musavi, *A general energy function for transient stability analysis of power systems*, IEEE Trans. on Circuits and Systems **CAS-31** (1984), 637–645.

[38] K.R.Padiyar, K.K.Ghosh, *Direct stability evaluation of power systems with detailed generator models using structure-preserving energy functions*, International Journal of Electrical Power and Systems, **11** (1) Jan (1989), 47–56.

[39] I.A.Hiskens, D.J.Hill, *Incorporation of SVCS into energy function methods*, IEEE Trans. on Power Systems **7** Feb (1992), 133–140.

[40] A.Debs, *Voltage dip at maximum swing in the context of direct stability analysis*, IEEE Trans. on Power Systems **5** Nov (1990), 1497–1502.

NEW ALGORITHMS FOR SLOW COHERENCY AGGREGATION OF LARGE POWER SYSTEMS

JOE H. CHOW*

Abstract. This paper presents new algorithms for the slow coherency aggregation of power system models for dynamic analysis and control design. In a typical large power system, groups of stiffly connected machines exhibit coherent low frequency oscillations. These groups of machines form slow coherent areas which are linked by sparse connections. The paper describes a tight coherency grouping algorithm for identifying the slow coherent areas, and aggregation algorithms for obtaining slow coherency and inertial aggregate models. The results are illustrated with a 48-machine system.

1. Introduction. With the introduction of the slow coherency concept in [1], slow coherency has become popular in large power system dynamic analysis [2,3,4] due to two main reasons. First for economic generation, utilities are transmitting increasingly larger amount of power over a few high voltage transmission corridors. Such operating strategies draw attention to the stability of interarea oscillations between groups of machines [5,6]. These oscillations are lower in frequencies than the oscillations between close by machines. Second, software packages [7,8] for computing these low frequencies in very large power systems allow the analysis of the participation of the machines in these oscillations. This information is critical in designing damping controllers for enhancing interarea stability.

Slow coherency is motivated by the observation that when a large power system is subject to disturbances, groups of stiffly connected machines exhibit coherent motions with respect to the low frequency interarea modes. For system analysis and control design that do not involve the fast dynamics outside a study region, the slow coherency phenomenon can be used to obtain reduced order models for the coherent machines outside the study region.

The slow coherency approach exploits the two-time-scale properties of power system models [1,9]. The small parameter to denote the separation of the time scales is the ratio of the sparseness of the connections between the coherent machines to the stiffness of the connections within the coherent machines. We will present here an analysis based on linearized electromechanical models. The theory is readily extended to nonlinear electromechanical models [10,11]. The model will include the voltages of the load buses as algebraic variables, that is, the model will contain algebraic as well as differential equations. This model is more desirable since it encompasses both the aggregation of coherent machines and the preservation

* On sabbatical leave at Power Systems Engineering Department, General Electric Company, Schenectady, New York 12345. Electrical, Computer, & Systems Engineering Department, Electric Power Engineering Department, Rensselaer Polytechnic Institute Troy, New York 12180-3590.

of critical load buses and major transmission lines. The slow coherency approach is contained in Section 2.

The application of the slow coherency results requires first identifying the slow coherent groups of machines. In [1], a slow coherency grouping algorithm that identifies coherent machines based on the slow eigenvector matrix has been proposed. The algorithm fixes the number of coherent areas, and does not consider the disparity within a coherent group. Section 3 proposes a new rule-based tight coherency grouping algorithm which starts from the same slow eigenvector matrix, but groups machines only if their disparity is within a specified tolerance. When applied to a 48-machine NPCC (Northeast Power Coordinated Council) system, the aggregate models obtained from the tight coherency algorithm provide better eigenvalue approximations than those obtained from the slow coherency grouping algorithm.

Model reduction is achieved when the slow coherent machines are combined into aggregates with conventional network and machine parameters. The aggregation can be performed on a per coherent area basis since the parameters of the aggregate of a coherent area do not depend, to a first order approximation, on the parameters of the other coherent areas [13]. In the Podmore algorithm [14], the generator terminal buses within a coherent area are linked with infinite admittances. This process stiffens the resulting reduced network [15]. A more accurate algorithm is the inertial aggregation where the coherent generator internal nodes are linked with infinite admittances. Further improvement is possible with the slow coherency aggregation which modifies the impedances to more properly represent that the coherent machines are connected with finite admittances. A transient simulation of a disturbance on the NPCC system is used to compare the three aggregation algorithms.

The simulation results on the NPCC system presented in this paper are obtained using the Power System Toolbox (PST) [16], a Matlab-based [17] power system simulation and control design package. The tight coherency grouping algorithm and the network and machine aggregation algorithms are available as functions in PST.

2. Two-time-scale power system models. In the slow coherency approach, singular perturbation techniques [18] are used to separate the slow and fast dynamics in large power systems. The relevant slow dynamics are the low frequency oscillations between coherent groups of stiffly connected machines and the less significant fast dynamics are the higher frequency oscillations between machines within the coherent groups. This two-time-scale behavior can be captured using the electromechanical models. The slow coherency analysis is valid for both linearized and nonlinear electromechanical models of power systems [1,10,11]. However, we will present only the linearized analysis, since the slow coherency aggregation algorithm can reconstruct the network from the linearized matrices.

We consider the nonlinear model of an n-machine, p-bus power system

$$\begin{aligned} M\ddot{\delta} &= f(\delta, V), \\ 0 &= g(\delta, V), \end{aligned} \tag{2.1}$$

where δ is an n-vector of the machine angles, V is a p-vector of the complex load bus voltages, M is the diagonal inertia matrix, f is a vector of the acceleration torques, and g is a vector of the network flow equations. Note that damping is neglected in the model (2.1) since it does not impact on the network aggregation procedure. We linearize (2.1) about a nominal load flow (δ_o, V_o) to obtain the linear model

$$\begin{aligned} M\Delta\ddot{\delta} &= \left.\frac{\partial f(\delta,V)}{\partial \delta}\right|_{\delta_o,V_o} \Delta\delta + \left.\frac{\partial f(\delta,V)}{\partial V}\right|_{\delta_o,V_o} \Delta V = K_1\Delta\delta + K_2\Delta V, \\ 0 &= \left.\frac{\partial g(\delta,V)}{\partial \delta}\right|_{\delta_o,V_o} \Delta\delta + \left.\frac{\partial g(\delta,V)}{\partial V}\right|_{\delta_o,V_o} \Delta V = K_3\Delta\delta + K_4\Delta V, \end{aligned} \tag{2.2}$$

where $\Delta\delta$ is an n-vector of the machine angle deviations from δ_o, and ΔV is a $2p$-vector of the real and imaginary parts of the load bus voltage deviations from V_o. The matrices K_1, K_2 and K_3 consist of partial derivatives of the power transfer between the machines and their terminal buses. In particular, K_1 is diagonal and K_4 is the network admittance matrix. The analytical expressions of the sensitivity matrices K_i can be found in [19]. Alternatively, the sensitivities can be computed numerically by introducing perturbations.

We define a slow coherent area to be composed of a slow coherent group of machines and the load buses that interconnect these machines. The slow coherency phenomenon is attributed primarily to the connections between the machines in the same coherent areas being stiffer than those between different areas. To model this characteristic, we separate the admittance matrix K_4 into

$$K_4 = K_4^I + \epsilon K_4^E, \tag{2.3}$$

where K_4^I is the matrix of internal connections between the buses in the same coherent areas, K_4^E is the matrix of external connections between different coherent areas, and ϵ is a parameter representing the ratio of the external connection strength to the internal connection strength. The parameter ϵ is small if the external connections are weak or sparse [20]. For large power systems, it is more realistic to model ϵ as representing the sparseness of the connections between the coherent areas. Thus the partition of an interconnected power system into slow coherent areas is often along utility company boundaries, since most utilities have more connections within the company service areas than those connecting to the neighboring utilities.

Assume that the system (2.2) has r slow coherent areas of machines and define

$\Delta\delta_i^\alpha$ = the rotor angle of the ith machine in area α,
m_i^α = the inertia of the ith machine in area α.

We also order the machines such that $\Delta\delta_i^\alpha$ from the same coherent areas appears consecutively in $\Delta\delta$. To describe the slow motion, we define for each area an inertia weighted *aggregate variable*

$$y_\alpha = \sum_{i=1}^{n_\alpha} m_i^\alpha \Delta\delta_i^\alpha / m_\alpha, \quad \alpha = 1, 2, ..., r, \tag{2.4}$$

where n_α is the number of machines in area α and

$$m_\alpha = \sum_{i=1}^{n_\alpha} m_i^\alpha, \quad \alpha = 1, 2, ..., r. \tag{2.5}$$

Denoting by y the r-vector whose α^{th} entry is y_α, the matrix form of (2.4) is

$$y = C\Delta\delta = M_a^{-1} U^T M \Delta\delta, \tag{2.6}$$

where

$$U = \text{block-diagonal}(u_1, u_2, \ldots, u_r) \tag{2.7}$$

is the grouping matrix with $n_\alpha \times 1$ vectors

$$u_\alpha = \begin{bmatrix} 1 & 1 & \ldots & 1 \end{bmatrix}^T, \quad \alpha = 1, 2, ..., r, \tag{2.8}$$

and

$$M_a = \text{diag}\,(m_1, m_2, \ldots, m_r) = U^T M U \tag{2.9}$$

is the $r \times r$ aggregate inertia matrix.

For the fast motion, we select the first machine in each area to be the reference machine and define the motions of the other machines in the same area relative to this reference machine by the *local variables*

$$z_{i-1}^\alpha = \Delta\delta_i^\alpha - \Delta\delta_1^\alpha, \quad i = 2, 3, \ldots, n_\alpha, \quad \alpha = 1, 2, ..., r. \tag{2.10}$$

Denoting by z^α the $(n_\alpha - 1)$-vector whose i^{th} entry is z_i^α and considering z^α as the αth subvector of the $(n - r)$-vector z, we rewrite (2.10) as

$$z = G\Delta\delta = \text{block-diagonal}(G_1, G_2, \ldots, G_r)\ \Delta\delta, \tag{2.11}$$

where G_α is the $(n_\alpha - 1) \times n_\alpha$ matrix

$$G_\alpha = \begin{bmatrix} -1 & 1 & 0 & . & 0 \\ -1 & 0 & 1 & . & 0 \\ . & . & . & . & . \\ -1 & 0 & 0 & . & 1 \end{bmatrix}. \tag{2.12}$$

Combining (2.6) and (2.11), the transformation of the original state $\Delta\delta$ into the aggregate and local variables is

$$\begin{bmatrix} y \\ z \end{bmatrix} = \begin{bmatrix} C \\ G \end{bmatrix} \Delta\delta, \tag{2.13}$$

whose inverse is

$$\Delta\delta = \begin{bmatrix} U & G^+ \end{bmatrix} \begin{bmatrix} y \\ z \end{bmatrix}, \tag{2.14}$$

where

$$G^+ = G^T (GG^T)^{-1} \tag{2.15}$$

is also block-diagonal.

Applying the transformation (2.13) to the model (2.2), we obtain

$$\begin{aligned} M_a \ddot{y} &= K_{11} y + K_{12} z + K_{13} \Delta V, \\ M_d \ddot{z} &= K_{21} y + K_{22} z + K_{23} \Delta V, \\ 0 &= K_{31} y + K_{32} z + K_4 \Delta V, \end{aligned} \tag{2.16}$$

where

$$\begin{aligned} K_{11} &= U^T K_1 U, \quad K_{12} = U^T K_1 G^+, \quad K_{13} = U^T K_2, \\ K_{21} &= (G^+)^T K_1 U, \quad K_{22} = (G^+)^T K_1 G^+, \quad K_{23} = (G^+)^T K_2, \\ K_{31} &= K_3 U, \quad K_{32} = K_3 G^+. \end{aligned} \tag{2.17}$$

In the appendix, it is shown that the elimination of the ΔV variables in (2.16) results in a singularly perturbed system with y being the slow variables and z the fast variables.

From the singular perturbation theory [18], the slow subsystem can be obtained as an asymptotic series expansion in the small parameter ϵ. We will consider only the zero and the first order terms in the series expansion. From the appendix, we observe that the slow variable y is coupled to the fast variable equation through ϵ, such that as a zero order approximation, z can be considered to be a constant equal to zero. Consequently, (2.16) reduces to

$$\begin{aligned} M_a \ddot{y} &= K_{11} y + K_{13} \Delta V, \\ 0 &= K_{31} y + K_4 \Delta V. \end{aligned} \tag{2.18}$$

This is the *inertial aggregate model* [1,15] which is equivalent to linking the internal nodes of the coherent machines by infinite admittances. The model (2.18) is different from the model obtained using the Podmore technique in [14], which aggregates the coherent machines at the terminal buses.

Thus the inertial aggregate model will be more accurate than the machine terminal bus aggregate model.

To improve on the inertial aggregate model, we consider z to vary slowly with y. As a first order approximation, the *quasi-steady state* of z in (2.16) is obtained as

$$z = -K_{22}^{-1}(K_{21}y + K_{23}\Delta V). \tag{2.19}$$

Eliminating z from (2.16) results in the *slow coherency aggregate model*

$$\begin{aligned} M_a\ddot{y} &= K_{11s}y + K_{13s}\Delta V, \\ 0 &= K_{31s}y + K_{4s}\Delta V, \end{aligned} \tag{2.20}$$

where

$$\begin{aligned} K_{11s} &= K_{11} - K_{12}K_{22}^{-1}K_{21}, \quad K_{13s} = K_{13} - K_{12}K_{22}^{-1}K_{23}, \\ K_{31s} &= K_{31} - K_{32}K_{22}^{-1}K_{21}, \quad K_{4s} = K_4 - K_{32}K_{22}^{-1}K_{23}. \end{aligned} \tag{2.21}$$

The physical interpretation of the singular perturbation correction is that the impedance terms in the connection matrices K_{11}, K_{13}, K_{31} and K_4 are modified such that the internal nodes of the coherent machines are now connected by finite admittances. The improvement in the accuracy of the slow coherency aggregate over the Podmore aggregate and the inertial aggregate are illustrated in Sections 4 and 5.

To complete the model reduction process, the voltage variables for the load buses that need not be retained can be eliminated from (2.18) or (2.20). The reduced order model would then consist of the aggregate machines and the critical load buses. The machine aggregation and load bus elimination can be performed in any order, that is, the two processes commute.

It is shown in [13] that the computation of the slow coherency aggregation (2.20) can be performed on a per coherent area basis, that is, the aggregation of a coherent area does not depend on the parameters of the other coherent areas. This property allows the aggregation of the coherent areas to be performed sequentially or in parallel. In addition, the aggregate of a coherent area remains unchanged if any other coherent areas, for some specific analysis needs, are retained in detail. The Podmore and inertial aggregations can also be performed on a per coherent area basis, as they only involve linking the generator terminal or internal nodes of the machines in the same coherent areas. However, if second or higher order terms are used to construct higher order aggregates from the singularly perturbed model (2.16), this per area aggregation property is lost since the corrections will depend on the parameters in other coherent areas.

3. Identification of slow coherent areas. For a group of machines to have slow coherent responses, their mode shapes with respect to the low frequency modes must be similar. In other words, if V is the matrix of the eigenvectors corresponding to the small eigenvalues of the state matrix $M^{-1}K$ (A.2) derived in the appendix with the voltage variables eliminated, then a slow coherent group of machines must have similar row vectors in V. This slow coherency condition is independent of the power system initial conditions which may vary according to the applied disturbance. Algorithms to find the slow coherent machines include the slow coherency grouping algorithm in [1,9], the signed coherency grouping algorithm in [21], the clustering algorithm in [22], the RMS coherency technique in [23] and the weak link algorithm in [24]. The slow coherency and signed coherency algorithms use the slow eigenvector matrix, while the clustering and weak link algorithms use only the linearized state matrix $M^{-1}K$. With the development of eigenvalue computation software packages [7], the slow coherency and signed coherency grouping algorithms are now applicable to large power systems.

In this section we develop a tight coherency grouping algorithm to obtain slow coherent areas for practical power systems. The algorithm is an extension to the slow coherency algorithm and is based on practical experience in developing reduced models for transient and small signal stability analysis and control design. In the slow coherency algorithm, for a system with r coherent groups, the row vectors of V form r clusters in an r-dimensional space. To identify the slow coherent groups, first, a set of the r most linearly independent vectors V_α from V is found and used as the reference vectors. Then a machine with the row vector V_i is grouped in the same area with the reference machine whose row vector V_α is closest to V_i. The algorithm fixes the number of coherent areas, r, to be equal to the number of slow eigenvalues, and does not take into account the disparity within a coherent group. The application of this technique can result in coherent areas having geographically distant machines. An example is a 9-area partition of the 48-machine NPCC system shown in Figure 3.1, where machine 33 is quite far away from the other machines in area 9. This partition is obtained using the 9 slowest modes (including the system mode) of the system.

The tight coherency grouping algorithm requires a measure of the slow coherency between the machines. Let the columns of V be normalized to unity, and define the slow coherency measure

$$d_{ij} = V_i V_j^T / (|V_i||V_j|) \tag{3.1}$$

as the cosine of the angle between V_i and V_j. If machines i and j are perfectly coherent with respect to the slow modes, then $V_i = V_j$. A tolerance, γ, typically in the range of 0.9 to 0.95, can be selected such that two machines are said to be coherent if $d_{ij} > \gamma$. We define a coherency matrix or map C_m to be a matrix whose (i,j) entry is given by

$$(C_m)_{ij} = d_{ij} - \gamma. \tag{3.2}$$

A 3-dimensional plot of C_m of the NPCC system using the eigenvector matrix of the 9 slowest modes and $\gamma = 0.95$ is shown in Figure 3.2. Note

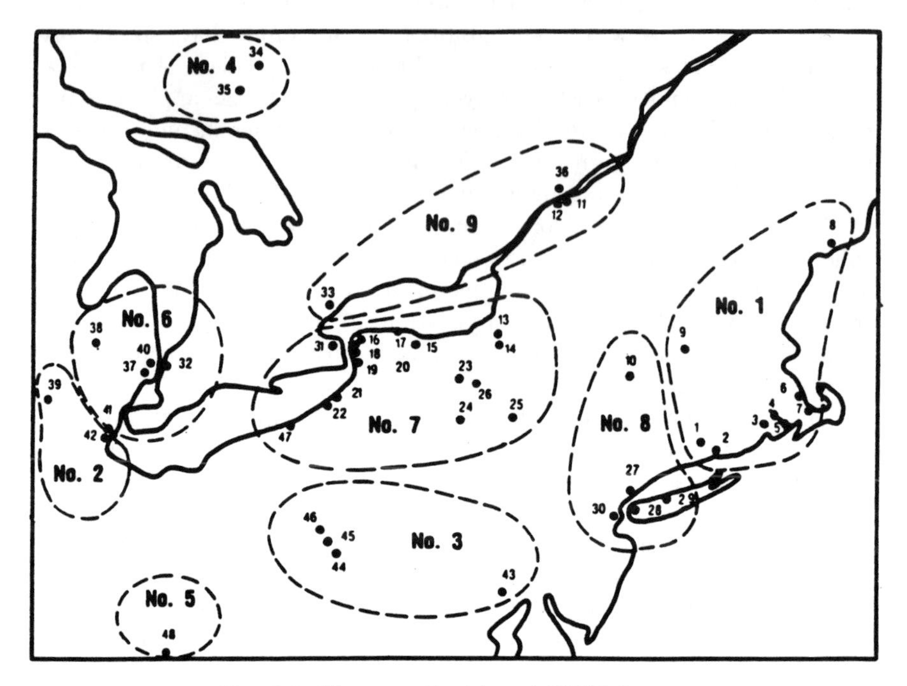

FIG. 3.1. *Nine-area Partition of NPCC System*

that in Figure 3.2, the negative values in C_m have been set to 0 to accentuate the coherency. The (1,1) entry of C_m is located at the farthest corner. If the (i, j) grid has a spike, then machines i and j are coherent. The C_m plot depicts clearly the coherency between machines 1-9 in New England and machines 13-26 in New York.

Based on the coherency measure, we formulate a set of coherency rules:

1. Machines i and j are coherent if $(C_m)_{ij} > 0$.
2. If machines i and j are coherent and machines j and k are coherent, then machines i and k are also coherent.
3. A *loose* coherent area J_α is formed by the machines that are coherent under Rules 1 and 2. Let $(C_m)_\alpha$ be a submatrix of C_m corresponding to J_α.
4. If the column sums of $(C_m)_\alpha$ excluding the diagonal entries are all positive, then J_α is a *tight* coherent area.
5. If any of the column sums of $(C_m)_\alpha$ excluding the diagonal entries is negative, then J_α should be decomposed into tight coherent areas.
6. The least coherent machine in J_α corresponds to the columns of $(C_m)_\alpha$ with the smallest sum.
7. The coherency of J_α may be improved by removing the least co-

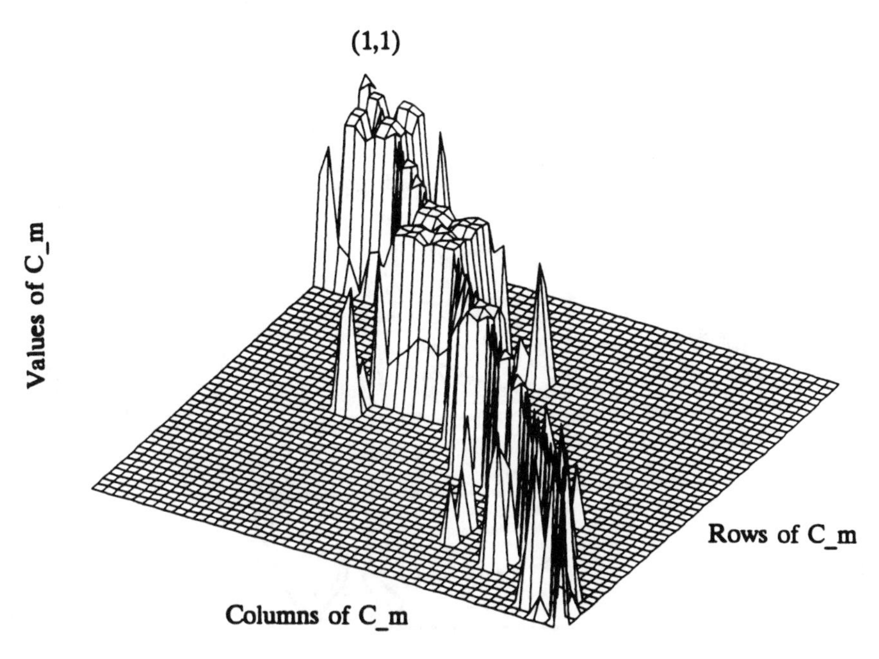

FIG. 3.2. *Coherency Map of NPCC System with 9 Slowest Modes and* $\gamma = 0.95$

herent machine and reassigning it to a different area.

8. Given two partitions I_1 and I_2 of J_α, I_1 is tighter than I_2 if the sum of the off-diagonal entries of $(C_m)_\alpha$ corresponding to I_1 is larger than that of I_2.

The coherency rules can be used to construct the following algorithm.

Tight Coherency Grouping Algorithm

1. Find the loose coherent areas using Rules 1 - 3.
2. For each coherent area J_α,
   1. Use Rule 4 to determine tight coherent area, which requires no further decomposition.
   2. If the area is not tight, decompose the area into tight coherent areas. Start the decomposition by identifying the least coherent machine using Rule 6 and reassigning it using Rule 8. Continue until the loose coherent area has been decomposed into tight coherent areas and no improvement is possible under Rule 8.

When the tight coherency grouping algorithm is applied to the NPCC system with L containing the eigenvectors of the 9 slowest modes and $\gamma = 0.95$, 17 slow coherent areas are found. The machine groups are listed in Table 3.1 and shown in Figure 3.3. For a comparison, the slow

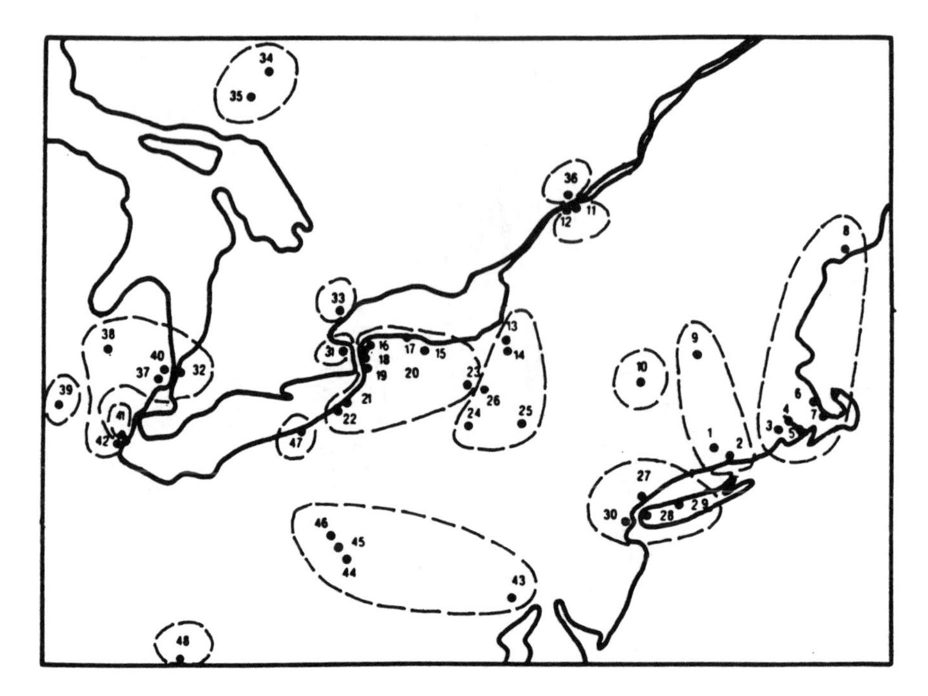

FIG. 3.3. *Seventeen-area Tight Coherency Partition of NPCC System*

coherency grouping algorithm with L containing the eigenvectors of the 17 slowest modes is used to find 17 areas for the NPCC system. The resulting areas are also listed in Table 3.1. Note that the first machine listed in each group is the reference machine for that particular group. These two 17-area partitions have a lot of similarities, but also some differences.

To assess the significance between the two different 17-area partitions, and to show the improvement of the 17-area partition over the 9-area partition, we perform the aggregation based on these partitions and compute the slow eigenvalues of the linearized reduced order models. For a comprehensive illustration, we use all three aggregation algorithms: the Podmore, inertial aggregation and slow coherency aggregations. In the Podmore and inertial aggregations, nonlinear reduced networks are first constructed using the algorithms described in the next section. Eigenvalues are then computed from the linearized models. For the slow coherency aggregation, eigenvalues are directly computed from (2.20). The frequencies of the 8 slowest modes (excluding the system mode) are shown in Tables 3.2–3.4.

From Tables 3.2–3.4, we conclude that in all the aggregation techniques, the 17-area partition from the slow coherency grouping algorithm does not necessarily provide a better eigenvalue approximation than the 9-area partition. However, the 17-area tight coherency partition shows a

TABLE 3.1
Seventeen-area Partitions using Tight Coherency and Slow Coherency

Area	Tight Coherency, 9 Slowest Modes	Slowest Coherency, 17 Slowest Modes
1	3 4 5 6 7 8	6 3 7 9
2	1 2 9	1 2
3	10	8
4	11 12	11 12
5	13 14 24 25 26	13 10 14 24 25 26
6	15 16 17 18 19 20 21 22 23	16 17 18 19 20 21 22
7	27 28 29 30	29 27 28 30
8	31	15 23
9	32 37 38 40 42	32 31 33 37 38
10	33	40
11	34 35	34 35
12	36	36
13	39	39 42
14	41	41
15	43 44 45 46	44 43 45 46 47
16	47	5 4
17	48	48

TABLE 3.2
8 Slowest Modes from Podmore Aggregate Model

Full Model	Slow Coherency 9 Areas		Slow Coherency 17 Areas		Tight Coherency 17 Areas	
Freq. (Hz)	Freq. (Hz)	error (%)	Freq. (Hz)	error (%)	Freq. (Hz)	error (%)
0.2697	0.3178	17.8	0.3175	17.7	0.2810	4.2
0.3815	0.4237	11.1	0.4510	18.2	0.4091	7.2
0.4873	0.5971	22.5	0.5883	20.7	0.5202	6.8
0.5328	0.6433	20.8	0.6525	22.5	0.5795	8.8
0.7069	0.8655	22.4	0.7798	10.3	0.7157	1.3
0.7405	0.9079	22.6	0.8469	14.4	0.7718	4.2
0.8040	0.9711	20.8	0.8679	7.9	0.8450	5.1
0.8411	1.0437	24.1	0.9087	8.0	0.8903	5.9

TABLE 3.3
8 Slowest Modes from Inertial Aggregation

Full Model	Slow Coherency 9 Areas		Slow Coherency 17 Areas		Tight Coherency 17 Areas	
Freq. (Hz)	Freq. (Hz)	error (%)	Freq. (Hz)	error (%)	Freq. (Hz)	error (%)
0.2697	0.2958	9.7	0.2958	9.7	0.2747	1.8
0.3815	0.4049	6.1	0.4277	12.1	0.3932	3.0
0.4873	0.5378	10.4	0.5352	9.8	0.5034	3.3
0.5328	0.5847	9.7	0.5822	9.3	0.5472	2.7
0.7069	0.8315	17.6	0.7582	7.3	0.7123	0.8
0.7405	0.8514	15.0	0.7995	8.0	0.7515	1.5
0.8040	0.8918	10.9	0.8397	4.4	0.8365	4.0
0.8411	0.9678	15.1	0.8592	2.1	0.8562	1.8

TABLE 3.4
8 Slowest Modes of Slow Coherency Aggregation

Full Model	Slow Coherency 9 Areas		Slow Coherency 17 Areas		Tight Coherency 17 Areas	
Freq. (Hz)	Freq. (Hz)	error (%)	Freq. (Hz)	error (%)	Freq. (Hz)	error (%)
0.2697	0.2772	2.8	0.2786	3.3	0.2709	0.4
0.3815	0.4087	7.1	0.4348	14.0	0.3862	1.2
0.4873	0.5554	14.0	0.5580	14.5	0.4904	0.6
0.5328	0.5745	7.8	0.5999	12.6	0.5297	0.6
0.7069	0.8532	20.7	0.7629	7.9	0.7119	0.7
0.7405	0.8545	15.4	0.7987	7.9	0.7463	0.8
0.8040	0.8827	9.8	0.8369	4.1	0.8304	3.3
0.8411	0.9708	15.4	0.8606	2.3	0.8485	0.9

substantial improvement over both the 9-area and 17-area partitions from the slow coherency grouping algorithm. Tables 3.2–3.4 also support the singular perturbation theory that the slow coherency and inertial aggregations are better than the Podmore aggregation.

4. Network aggregation. One of the applications of dynamic equivalencing is to use reduced order models for dynamic security assessment, in which many transient stability analyses are performed to screen critical cases for more detailed analysis. In this application, a dynamic equivalencing technique must generate a reduced order model having a power network representation. Although linearized models are derived for the inertial and slow coherency aggregations, aggregates with conventional network and machine models can be reconstructed from the linearized reduced models. For completeness, we will discuss all three aggregation techniques and show the incremental improvement between the techniques. The discussion on aggregation will be based on a coherent area with two machines. Extensions to more than two machines are straightforward.

The Podmore aggregation for two machines 'A' and 'B' is shown in Figure 4.1 where 'a' and 'b' are the generator terminal buses. In the Podmore technique, it is assumed that coherency occurred at the generator terminal buses 'a' and 'b' (Figure 4.1-i) . As a result these buses are tied together to a common bus 'q' with infinite admittances (Figure 4.1-ii). The voltage at bus 'q' can be set either to an average of the voltages at buses 'a' and 'b' or a weighted average with respect to the active and reactive power generation. To preserve the power flow, ideal transformers with complex turns ratios $\alpha_a \arg(\phi_a)$ and $\alpha_b \arg(\phi_b)$ and zero impedances link the buses 'a' and 'b' to bus 'q'. The phases ϕ_a and ϕ_b can be represented separately as phase shifters. The machines 'A' and 'B' are then aggregated into a single equivalent machine with an inertia m_{eq} and a transient reactance $(x'_d)_{eq}$ equal to

$$(4.1) \qquad m_{eq} = m_A + m_B, \quad (x'_d)_{eq} = (1/(x'_d)_A + 1/(x'_d)_B)^{-1}$$

where m_A and m_B are the inertias and $(x'_d)_A$ and $(x'_d)_B$ the transient reactances of machines 'A' and 'B'.

The inertial aggregation for two machines 'A' and 'B' is shown in Figure 4.2 where 'a' and 'b' are the generator terminal buses. In the inertial

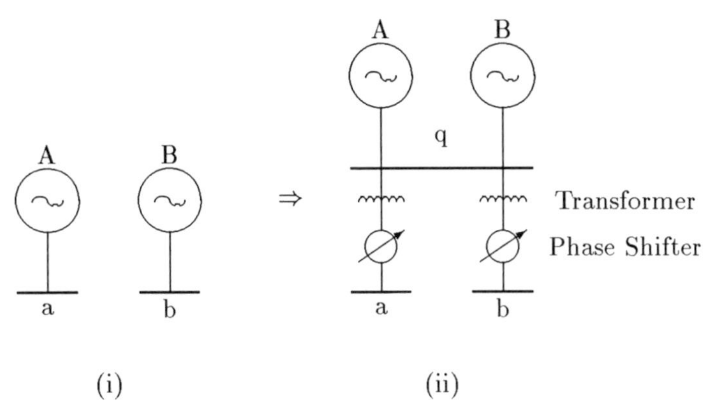

FIG. 4.1. *Podmore Aggregation*

aggregation technique, the machine internal node voltages E'_A and E'_B are computed (Figure 4.2-i). From (2.18), the generator internal nodes are tied together to a common bus 'p' with infinite admittances (Figure 4.2-ii). The voltage at bus 'p', E'_{eq}, can be set either to the average of E'_A and E'_B or a weighted average with respect to the active and reactive power generation. To preserve the power flow, ideal transformers with complex turns ratios $\alpha_a \arg(\phi_a)$ and $\alpha_b \arg(\phi_b)$ and zero impedances link the buses 'a' and 'b' to bus 'p'.

The linking of the internal nodes creates an equivalent generator with multiple terminal buses, which is not a conventional power system network representation. Taking $(x'_d)_{eq}$ from (4.1) as an equivalent transient reactance, the network is extended beyond node 'p' by two buses with reactances of $-(x'_d)_{eq}$ and $(x'_d)_{eq}$ (Figure 4.2-iii). The node 'r' serves as the internal node of the equivalent machine, and the node 'q' the generator terminal bus. The node 'p' can be eliminated if desired. Finally, the inertia of the equivalent machine, m_{eq}, is computed from (4.1).

The slow coherency aggregation for two machines 'A' and 'B' is shown in Figure 4.3. In slow coherency aggregation, the machine internal node voltages E'_A and E'_B are computed to obtain the linearized model (2.20) (Figure 4.3-i). In the construction of (2.20), only the fast variable z is eliminated, while all the bus voltage variables are retained. This allows the reconstruction from the connection matrices K_{11s}, K_{13s}, K_{31s} and K_{4s} of a power network consisting of impedances and phase shifters (Figure 4.3-ii). Although branch parameters can be reconstructed from the connection matrices, the recovered network, in general, would not have a balanced load flow. For tightly connected areas, the load flow mismatch would be small and loads can be added to the generation terminal buses to balance the load flow. The elimination of the fast variable z results in K_{4s} being a dense matrix. Thus in the reconstruction, all the generator terminal buses in the same area will be interconnected. This interconnection is dependent only

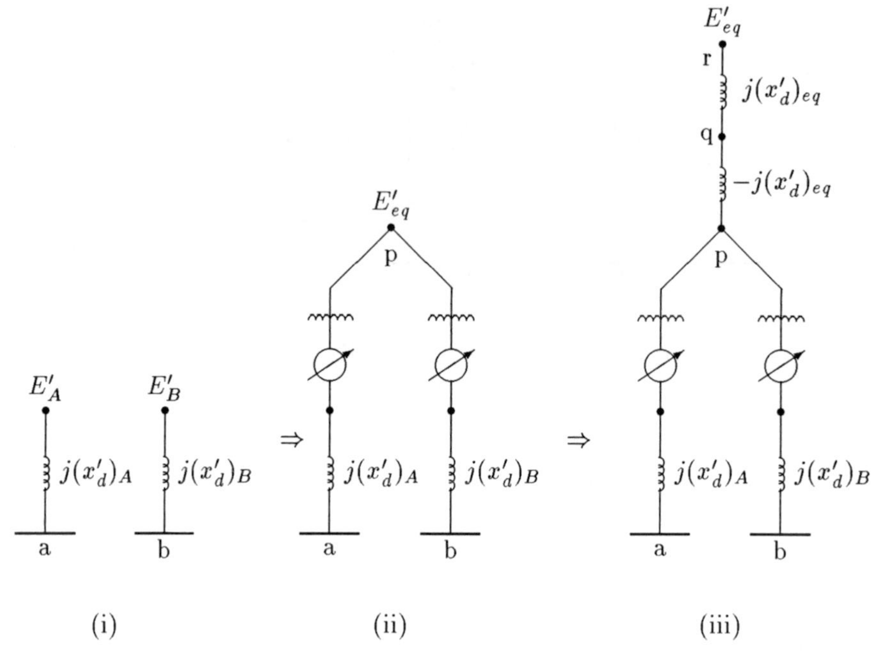

FIG. 4.2. *Inertial Aggregation*

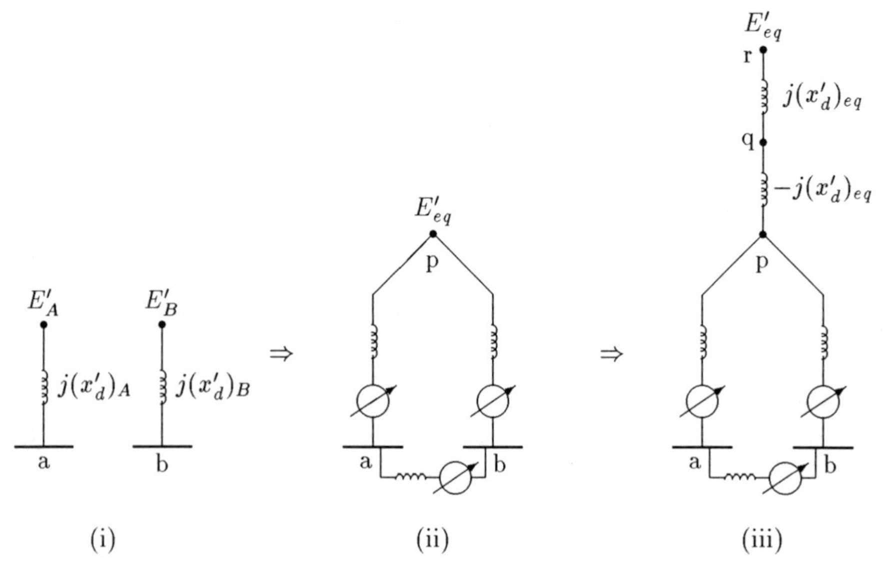

FIG. 4.3. *Slow Coherency Aggregation*

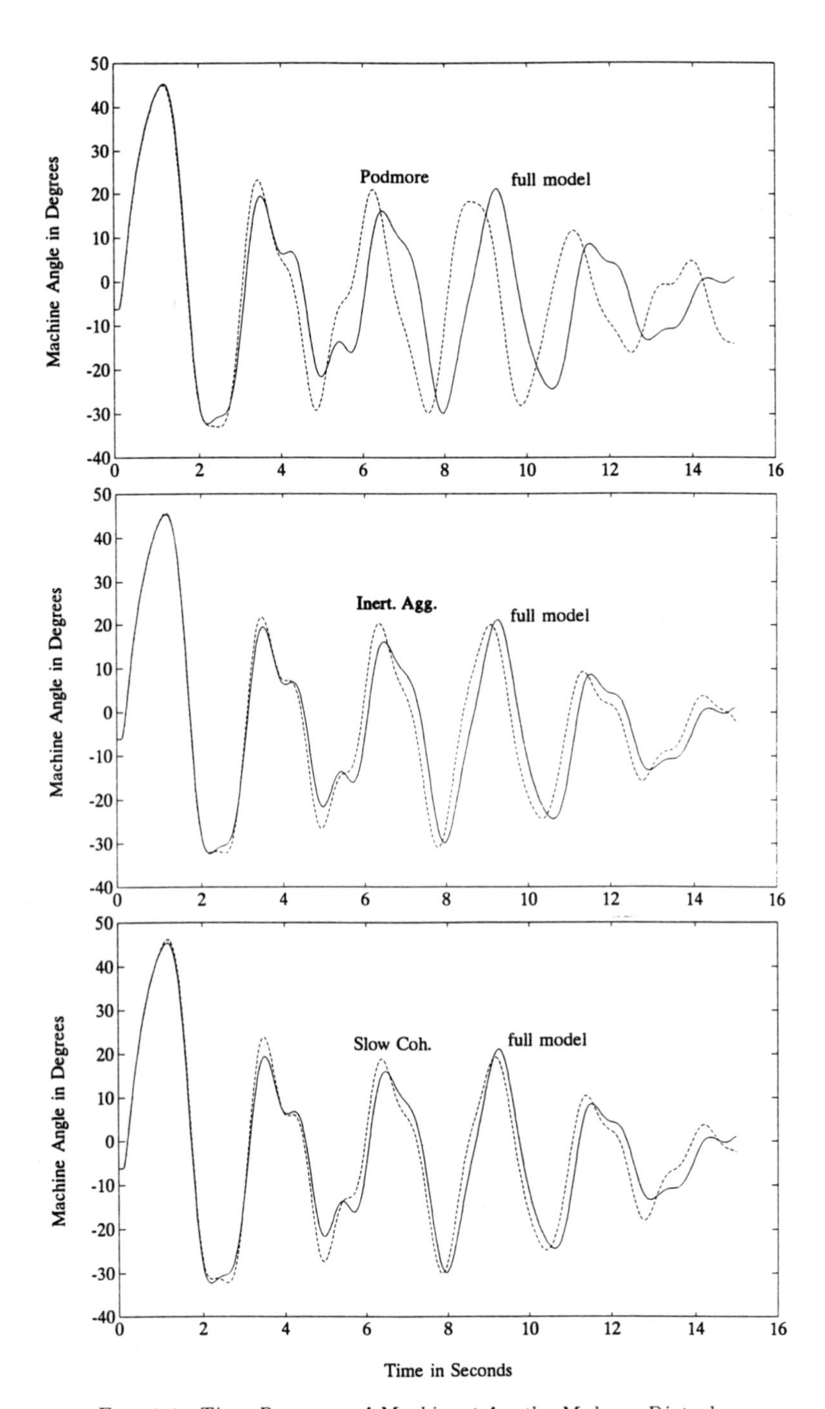

FIG. 4.4. *Time Response of Machine 1 for the Medway Disturbance*

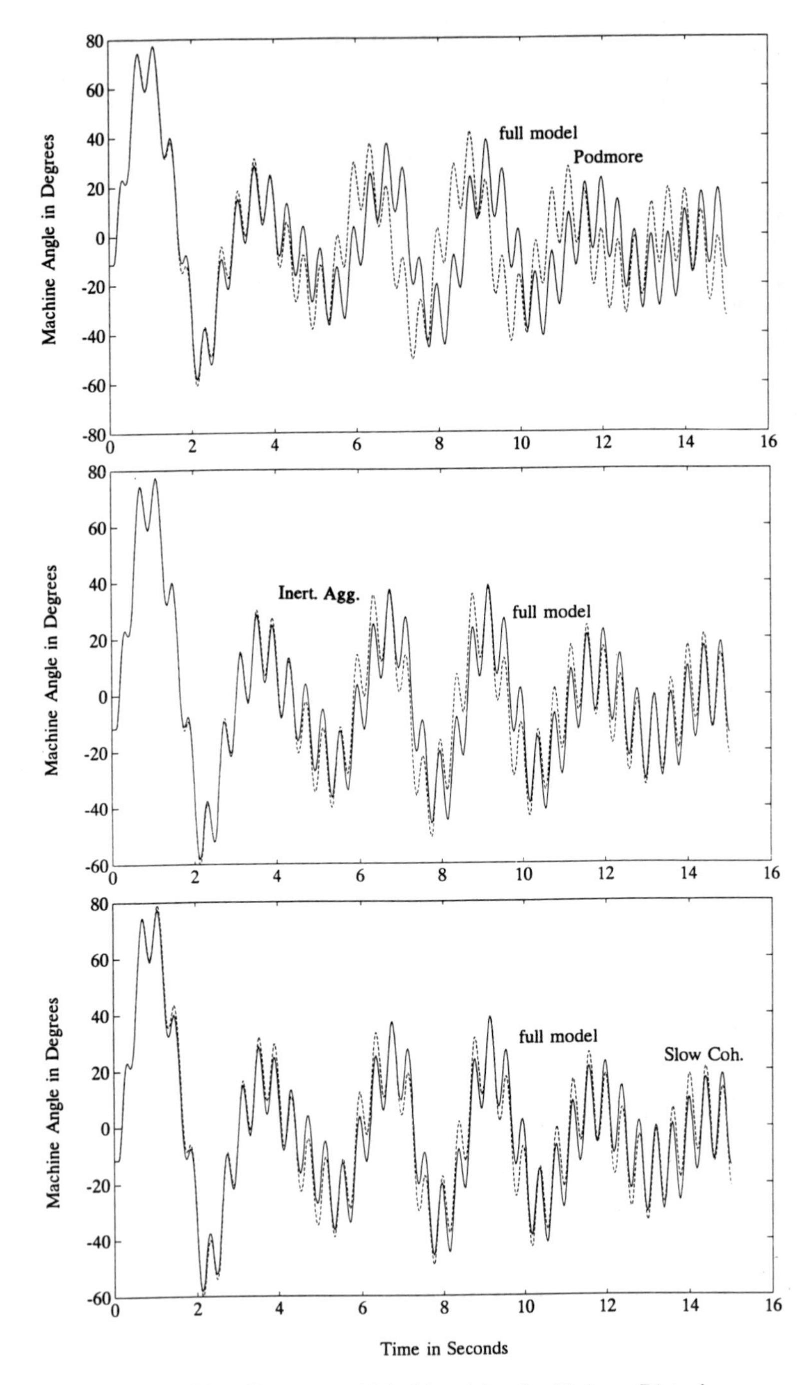

FIG. 4.5. *Time Response of Machine 4 for the Medway Disturbance*

on the parameters within an coherent area, and represents the improvement to the inertial aggregate.

In the slow coherency aggregation, the equivalent machine internal node is also linked to multiple terminal buses. Thus the approach used in the inertial aggregation to reconstruct a single generator terminal bus can also be applied here, as shown in Figure 4.3-iii.

To provide a comparison of the nonlinear aggregate models obtained from the three techniques, we apply the techniques to the 17 area tight coherency partition of the NPCC system. The disturbance considered is a 6 cycle short-circuit fault at Medway located in Area 1, which is cleared by removing the line from Medway to Sherman Road. The study region in which local machine dynamics are of interest is designated to be Areas 1 and 2. As a result, all the machines in these two areas are retained individually and no aggregation is required. The aggregation algorithms are applied to the other coherent areas to combine areas with more than one coherent machine into single equivalent machine areas. As a result, all the aggregate models contain 24 machines, of which 7 are equivalent machines.

The Medway disturbance is simulated on the full model and the three aggregate models. The responses for machines 1 and 4 are shown in Figures 4.4 and 4.5. Note that the Podmore aggregate provides good approximations up improvement would have been expected. However, the addition of loads to balance the to 4 seconds, after which the higher frequency oscillation error of the Podmore aggregate becomes evident. The inertial aggregate provides a substantial improvement over the Podmore aggregate, although it still shows a slightly higher frequency oscillation error. The slow coherency aggregate shows a small improvement over the inertial aggregate model, with a smaller higher frequency error. From the eigenvalue analysis results in Tables 3.3 and 3.4, more improvement would have been expected. However, the addition of loads to balance the power flow in the slow coherency algorithm introduces an approximation that reduces the improvement.

For a practical power system, a reduced model may consist of a mixture of individual machines, inertial aggregates and slow coherency aggregates. In general, no aggregation is performed within the study region such that the machines in the study region are kept separately. For the coherent areas external to the study region, slow coherency aggregates are constructed for all the coherent areas. Then a threshold can be established such that slow coherency aggregates resulting in unacceptable load mismatches are replaced by inertial aggregates to improve accuracy.

5. Conclusions. New algorithms for improving the slow coherency method of obtaining reduced order models of large power systems are presented. The tight coherency grouping algorithm allows the user to specify a tolerance to control the disparity between the machines in a coherent area.

The inertial and slow coherency aggregations are formulated to retain a power network structure for the reduced order models. The techniques have been applied to the eigenvalue analysis and transient stability simulation of a 48-machine system, and the results are promising.

The discussion of the slow coherency method in this paper does not include detailed machine models. Future research will focus on investigating methods such as those in [25,26], to obtain equivalent detailed generator dynamic models and control systems for the aggregate machines.

6. Acknowledgment. The contributions of Ranjit Date, Hisham Othman, Pierre Acari, and William Price to the development of the slow coherency grouping and aggregation algorithms are gratefully acknowledged. The work was supported in part by NSF under grants ECS-8714480 and ECS-9215076, and in part by General Electric Company.

Appendix

A. Derivation of the Time-Scales in (2.16). Since the network admittance matrix K_4 is nonsingular, we can solve (2.2) for

$$\Delta V = -K_4^{-1} K_3 \Delta\delta \tag{A.1}$$

and eliminate ΔV in (2.2) to obtain a linearized electromechanical model reduced to the machine internal nodes

$$M\Delta\ddot{\delta} = K\Delta\delta, \tag{A.2}$$

where

$$K = K_1 - K_2 K_4^{-1} K_3. \tag{A.3}$$

From (2.3),

$$\begin{aligned} K_4^{-1} &= (K_4^I + \epsilon K_4^E)^{-1} \\ &= (K_4^I(I + \epsilon(K_4^I)^{-1} K_4^E))^{-1}. \end{aligned} \tag{A.4}$$

For ϵ small, (A.4) can be expanded as

$$\begin{aligned} K_4^{-1} &= (I - \epsilon(K_4^I)^{-1} K_4^E + \epsilon^2((K_4^I)^{-1} K_4^E)^2 + \ldots)(K_4^I)^{-1} \\ &= (K_4^I)^{-1} + \epsilon K_{4\epsilon}^E. \end{aligned} \tag{A.5}$$

Substituting (A.5) into (A.3), the connection matrix K can also be separated into

$$K = K^I + \epsilon K^E, \tag{A.6}$$

where

$$K^I = K_1 - K_2 (K_4^I)^{-1} K_3 \tag{A.7}$$

is the block-diagonal matrix of internal connections between the machines in the same coherent areas, and

$$K^E = -K_2 K^E_{4\epsilon} K_3 \tag{A.8}$$

is the matrix of external connections between different coherent areas. Note that the each of the rows of K^I sums to zero.

Applying the transformation (2.13) to the model (A.2), (A.6), we obtain

$$\begin{aligned} M_a \ddot{y} &= \epsilon K_a y + \epsilon K_{ad} z, \\ M_d \ddot{z} &= \epsilon K_{da} y + (K_d + \epsilon K_{dd}) z, \end{aligned} \tag{A.9}$$

where

$$\begin{aligned} M_d &= (GM^{-1}G^T)^{-1}, \quad K_a = U^T K^E U, \\ K_{ad} &= U^T K^E M^{-1} G^T M_d, \quad K_{da} = M_d G M^{-1} K^E U, \\ K_{dd} &= M_d G M^{-1} K^E M^{-1} G^T M_d, \\ K_d &= M_d G M^{-1} K^I M^{-1} G^T M_d. \end{aligned} \tag{A.10}$$

Note that K_a, K_{ad}, K_{da} and K_{dd} are independent of K^I because $K^I U = 0$. System (A.9) is in the *standard singularly perturbed form* [18,1] showing that y is the slow variable and z is the fast variable.

REFERENCES

[1] J. H. Chow, editor, *Time-scale Modeling of Dynamic Networks with Applications to Power Systems,* Springer-Verlag, New York, 1982.

[2] R. J. Piwko, H. Othman, O. A. Alvarez and C. Y. Wu, "Eigenvalue and Frequency-domain Analysis of the Intermountain Power Project and WSCC Network," *IEEE Trans. Power Systems,* vol. PWRS-6, pp. 238-244, 1991.

[3] Y. Mansour, "Application of Eigenanalysis to the Western North American Power System," in *Eigenanalysis and Frequency Domain Methods for System Dynamic Performance*, IEEE Publications 90TH0292-3-PWR, pp. 97-104, 1990.

[4] L. Rouco and I. J. Pérez-Arriaga, "Multi-area Analysis of Small Signal Stability in Large Electric Power Systems by SMA", 1992 IEEE/PES Summer Power Meeting, Paper 92 SM 601-5 PWRS, Seattle, Washington.

[5] R. L. Cresap and J. F. Hauer, "Emergence of a New Swing Mode in the Western Power System," *IEEE Trans. Power Apparatus and Systems,* vol. PAS-100, pp. 2037-2045, 1981.

[6] M. Klein, G. J. Rogers and P. Kundur, "A Fundamental Study of Inter-area Oscillations in Power Systems," *IEEE Trans. Power Systems*, vol. 6, pp. 914-921, 1991.

[7] P. Kundur, G. J. Rogers, D. Y. Wong, L. Wang and M. G. Lauby, "A Comprehensive Computer Program Package for Small Signal Stability Analysis of Power Systems," *IEEE Trans. Power Systems*, vol. PWRS-5, pp. 1076-1083, 1990.

[8] N. Martins, "Efficient Eigenvalue and Frequency Response Methods Applied to Power System Small-Signal Stability Studies," *IEEE Trans. Power Systems,*, vol. 1, pp. 217-226, February 1986.

[9] J. H. Chow, R. A. Date, H. Othman and W. W. Price, "Slow Coherency Aggregation of Large Power Systems," in *Eigenanalysis and Frequency Domain Methods for System Dynamic Performance*, IEEE Publications 90TH0292-3-PWR, pp. 50-60, 1990.

[10] G. Peponides, P. V. Kokotovic and J. H. Chow, "Singular Perturbations and Time Scales in Nonlinear Models of Power Systems," *IEEE Trans. Circuits and Systems,* vol. CAS-29, pp. 758-767, 1982.

[11] K. W. Cheung and J. H. Chow, "Stability Analysis of Singularly Perturbed Systems using Slow and Fast Manifolds," *Proceedings of the 1991 American Control Conference,* pp. 1685-1690.

[12] W. W. Price, "Dynamic Equivalents in Transient Stability Studies," NPCC-10 Working Group report, 1974.

[13] R. A. Date and J. H. Chow, "Aggregation Properties of Linearized Two-time-scale Power Networks," *IEEE Trans. Circuits and Systems,* vol. 38, pp. 720-730, 1991.

[14] R. Podmore, "Development of Dynamic Equivalents for Transient Stability Studies," EPRI Report EL-456, 1977.

[15] B. R. Copeland, "Reduced Order Dynamic Equivalents for Electric Power Systems: an Analysis of the Modal-coherent and Slow-coherent Equivalency Techniques," M. S. Thesis, University of Tennessee, Knoxville, Tennessee, 1986, J. Lawler, thesis advisor.

[16] J. H. Chow and K. W. Cheung, "A Toolbox for Power System Dynamics and Control Engineering Education and Research," *IEEE Trans. Power Systems,* vol. 7, pp. 1559-1564, 1992.

[17] J. N. Little, *Matlab: User's Guide*, The Mathworks, Inc.

[18] P. V. Kokotovic, H. K. Khalil and J. O'Reilly, *Singular Perturbation Methods in Control: Analysis and Design,* Academic Press, London, 1986.

[19] R. A. Date, "Dynamic Model Reduction of Large Scale Systems: Applications to Power Systems," M.S. Thesis, Rensselaer Polytechnic Institute, Troy, New York, May, 1989.

[20] J. H. Chow and P. V. Kokotovic, "Time-scale Modeling of Sparse Dynamic Net-

works," *IEEE Trans. Automatic Control*, vol. AC-30, pp. 714-722, 1985.

[21] B. Eliasson, *Damping of Power Oscillations in Large Power Systems,* Ph.D. Thesis, Lund Institute, 1990.

[22] J. Zaborszky, K.-W. Whang, G. M. Huang, L.-J. Chiang and S.-Y. Lin, "A Clustered Dynamic Model for a Class of Linear Autonomous Systems using Simple Enumerative Sorting," *IEEE Trans. Circuits and Systems,* vol. CAS-29, pp. 747-758, 1982.

[23] J. Lawler, R. A. Schlueter, P. Rusche and D. L. Hackett, "Modal-coherent Equivalents Derived from an RMS Coherency Measure," *IEEE Trans. Power Apparatus and Systems,* vol. PAS-99, pp. 1415-1425, 1980.

[24] R. Nath, S. S. Lamba and K. S. P. Rao, "Coherency Based System Decomposition into Study and External Areas using Weak Coupling," *IEEE Trans. Power Apparatus and Systems,* vol. PAS-104, pp. 1443-1449, 1985.

[25] A. J. Germond and R. Podmore, "Dynamic Aggregation of Generating Unit Models," *IEEE Trans. Power Apparatus and Systems,* vol. PAS-97, pp. 1060-1069, 1978.

[26] S. Ahmed-Zaid, P. W. Sauer and J. R. Winkelman, "Higher Order Dynamic Equivalents for Power Systems," *Automatica,* vol. 22, pp. 489-494, 1986.

COMPUTATIONAL COMPLEXITY RESULTS IN PARAMETRIC ROBUST STABILITY ANALYSIS WITH POWER SYSTEMS APPLICATIONS

CHRISTOPHER L. DEMARCO* AND GREGORY E. COXSON*

Abstract. Reliable power system operation requires that the dynamics of a network remain well-behaved over a wide range of system parameter values and operating points. As network and load parameters vary, a minimal criterion for acceptable operation is that eigenvalues for the family of linearizations around resulting operating points maintain strictly negative real parts. This paper confirms that this type of robust stability determination is an NP-Hard decision problem when the family of system matrices for the linearized model is a matrix polytope. More specifically, it is shown that the reduction from a known NP-Complete problem to this robust stability determination can be realized by a power system model. This suggests that the worst-case characteristic of NP-Hard problems—the expectation that their cost of computation may grow exponentially with problem dimension—is likely to be observed in robust stability tests for linearized power system models.

1. Introduction. The analysis and design of engineering systems is often carried out in a context in which certain parameters affecting performance are imprecisely known. Such studies are particularly relevant to power system applications, where the large scale and geographically distributed nature of the network make precise knowledge of all relevant quantities impossible. If parameters have unknown but constant values, or if their variation is much slower than that of the underlying dynamics of the system, one may wish to study stability properties of the family of time-invariant dynamical systems that results as these parameters take values from the specified set. Assuring desirable dynamic behavior invariant with respect to such parameter changes has long been a topic of study in power systems and other applications. Probabilistic methods represent one of the most established means of examining such issues, but a huge range of possible techniques exist, depending on the description of uncertainty in the system, the type of model, and performance criteria of interest. For a perspective on uncertain dynamic systems motivated in part by large-scale problems such as power systems, the reader is referred to the pioneering text by Schweppe [1].

To predict behavior in families of linear time-invariant dynamic models, it is useful to examine the possible variation in either the system matrix for a state-space representation, or the variation of coefficients in a corresponding characteristic polynomial. This class of problems has been the focus of a large body of research over the last decade, with much of the work appearing in the literature of automatic control systems under the general heading of "robust stability" analysis and "robust control" design. An overview and

* Department of Electrical and Computer Engineering, University of Wisconsin-Madison, Madison, WI 53706-1691. email: demarco@engr.wisc.edu.

unification of recent advances in parametric robust stability for linear systems may be found in [2]. One motivation for extensive research devoted to this class of problems was the publication by Kharitonov [3] of an efficient robust stability test for an interval polynomial. An interval polynomial is a family defined by specifying a fixed degree for the polynomials of interest, along with upper and lower bounds on the coefficients of each. Kharitonov proved that all members of an interval polynomial have roots with strictly negative real part if and only if four "extreme point" polynomials in the set have this property. This result has a clear implication for computational complexity of the interval polynomial robust stability problem. Given that a necessary and sufficient test for stability of a single polynomial of degree n can be decided with computational cost that is $O(n^2)$ (consider application of a Routh array test), the cost of establishing stability of the interval polynomial family is then $O(n^2)$ as well. Control systems applications of this result were highlighted in [4] and [5], and much of the subsequent literature has built upon these works (note, however, that the results of [5] contain errors pointed out in [6] and [7]).

To extend the engineering application of these methods, many researchers have sought efficient robust stability tests for classes of linear systems with structure more general than that of an interval polynomial; see, for example, [8] and [9]. A useful perspective on computational complexity of these tests is obtained if one considers an ordering of problems. At one end of the ordering are those problems having very restrictive assumptions on how parameter variations enter the system; at the other end are families of linear systems in which the form of parameter dependence is allowed to be very general. For a suitably constructed ordering, one might expect that the worst-case cost of computation for associated robust stability problems would be nondecreasing as one progresses "up" through this ordering. Clearly, such an ordering is not uniquely defined, and we will not pursue its construction in detail here. Interested readers are referred to [10]. However, suppose such an ordering *was* constructed, and consider the following question: at what point in the ordering does the robust stability determination cease to have a polynomial-time solution? The Kharitonov result established a polynomial-time solution for the "bottom" of this ordering. The work of [9] proved that polynomial-time solutions persist while "moving up" the ordering to families of linear systems in which the coefficients of the characteristic polynomial are constrained to lie in a polytope (note that this structure may also be obtained by letting the coefficients be affine functions of independent parameters that are each constrained to lie in an interval). In contrast, the approach to be presented here will work "down" from the other direction, demonstrating a more general structure of parameter dependence for which robust stability determination is NP-Hard, and therefore cannot be expected to have a guaranteed polynomial-time solution. To do so, this work will specify families of linear systems directly through parameter dependence in a state-space representation, rather than considering the structure of the characteristic polynomial. This is arguably a more natural approach in the power systems application, where elements of the system matrix typically depend on network, generator, and control parameters in a fairly simple fashion. While the state-space representation will be the focus in the sections to follow, the main result on worst-case computational cost for robust stability determination has implications for characteristic polynomial characterizations as well. In particular, the result to be presented will imply that robust stability determination is NP-Hard when the coefficients of the characteristic polynomial are allowed to be *bilinear* functions of independent parameters, each constrained to an interval. Note that this places the line separating guaranteed polynomial-time solutions from potentially exponential-time solutions between the case of affine dependence ([9]) and the case of bilinear dependence. This division is discussed in [11].

In order to characterize worst-case computational cost of robust stability calculations in a manner that is independent of the particular algorithm employed, this work will make use of the classes of NP-Complete and NP-Hard problems. Section 2 will provide a brief overview of the required background, reviewing certain well-known NP-Complete problems, as well as the concept of a reduction that is used to establish relations between various decision problems. Section 3 will then establish a reduction from a known NP-Complete problem relating to satisfiability of certain types of logical clauses, known as "MAX 2-SAT," to the class of problems of interest, matrix polytope robust stability tests. While the conclusion that the later problem is NP-Hard is not new (see, for example, [12], and more recent extensions in [13]), the specific structure of matrix polytope used in the reduction will be important to the power systems application of this result. In particular, the main contribution of this work appears in Section 4, where it is shown that the structure of matrix polytope used in the reduction of Section 3 may be realized by a linearized power system model in which line susceptances and conductances are uncertain parameters. This will show that for any instance of the known NP-Complete problem MAX 2-SAT, one may construct an associated power system model with the following property. Testing for existence of an eigenvalue of the linearized power system model in the right half plane will be equivalent to testing for satisfiability of the original MAX 2-SAT problem. Hence, if one were to propose any general algorithm to provide a necessary and sufficient test for robust stability in such a linearized power system model, this algorithm would solve a known NP-Complete problem as a special case. As it is strongly believed that NP-Complete problems do not admit guaranteed polynomial-time solutions, this implies that the power system robust stability test to be formulated here is unlikely to have a guaranteed polynomial-time solution.

2. Basic computational complexity classes. This section will review the basic concepts of (time) complexity classes for decision problems that are necessary background for the results to follow in Section 3. These are very standard results, available in such texts as [14] and [15]. In particular, readers with an orientation towards linear systems theory and applications of linear algebra should find Chapter 5 of [15] a very accessible introduction to the theory of NP-Completeness to supplement the summary here.

For motivation, we will consider computations common to applications of linear algebra, though questions of computational complexity can be raised for algorithms of all types. In the setting of matrix/vector operations in $\mathbb{R}^n$, the cost of computation relative to problem size can be displayed in an obvious manner. Consider, for example, the problem of multiplying two arbitrary $n \times n$ real matrices. In a computer implementation, a precise characterization of the size of the data necessary to specify an "instance" of the matrix multiplication problem would depend on the finite precision representation of real numbers. However, once this is determined (say, for example, by the IEEE floating point standard), the number of bits necessary to describe the problem data is a constant times n^2, and therefore of order $O(n^2)$. The standard "textbook definition" algorithm to compute such a product would involve n^3 multiplications and $n^3 - n^2$ additions/subtractions, and therefore has computational cost that is $O(n^3)$. This provides a clear relation between the size of the data necessary to specify the problem, and the number of basic operations necessary to compute its solution with this specific algorithm. To emphasize the algorithm dependence of such a relationship, it is important to note that there exist alternate algorithms for matrix multiplication that have different order of computational cost. Strassen's algorithm [15] computes the product of two $n \times n$ real matrices in $O(n^{\log_2(7)})$ basic operations.

The simple example of matrix multiplication above motivates an obvious question. Given a well-defined, "computable" problem, what is the least upper bound (as a function of the size of data to specify an instance of such a problem) on the number of basic operations necessary to solve the problem, taken over all possible solution algorithms? Even more fundamentally, one might wish to know whether or not this upper bound on the number of operations required by the "best possible" algorithm is polynomial in problem size. The Linear Programming (LP) Problem provides an interesting example of the history of such questions. For many years, and indeed, in many applications still, the Simplex algorithm was the preferred method of solution implemented in most LP computer codes. While Simplex works extremely well in most instances, [16] demonstrated that it is possible to construct instances that force Simplex to visit every vertex of the constraint set associated with the LP Problem. Since this number of vertices can grow exponentially in relation to the size of the problem, the Simplex algorithm does not possess a polynomial upper bound on its worst-case computational cost. However, in the late 1970's, [17] provided an algorithm that could be proven to solve any instance of the LP Problem in a number of operations bounded by a polynomial in the problem size. A more practical polynomial-time algorithm has since generated considerable interest [18]. So, despite the fact that a widely used algorithm has exponential time worst-case performance, alternative algorithms show that the LP Problem can be characterized as belonging to the class of problems solvable in polynomial-time. However, this favorable conclusion for Linear Programming is clearly not shared by all computational problems of interest. There are a large number of computational problems that appear to have the property that worst-case instances can force an exhaustive search, exponential (or worse) in the size of the problem instance, even for the best known algorithms. Many of these can be grouped into an equivalence class sharing the properties that characterize them as "NP-Complete."

Many computationally challenging problems have their most natural formulation as optimization calculations. However, addressing the issue of computational complexity in this setting is challenging, with progress in optimization formulations having been achieved relatively recently (see, for example, [19] and [20]). Well-established results in computational complexity, such as the definition of the characteristics of NP-Complete problems, are formulated for computations posed as logical decision problems. However, an optimization problem seeking the minimum (maximum) value of a real cost function over some constraint set possesses an obvious associated decision problem. The problem instance is simply supplemented with the specification of a real cost threshold, K; an associated decision problem may be constructed as that of deciding whether or not there exists a feasible point having cost less than (greater than) K. It should be clear that successful computation of the global optimum would solve the decision problem as a by-product. Hence, the best possible algorithm to solve the optimization problem must involve at least as many operations as the best possible algorithm to solve the associated decision problem. Finally, note that here the original optimization problem is formulated so that the solution required is only the value of the cost function. In many applications, one may also desire the minimizer itself. With the reasonable assumption that the cost function and any constraint conditions may be evaluated in polynomial-time, the minimizer will have an interpretation in the definition of the problem class NP below.

With this background in mind, an intuitive definition of the problem class P is straightforward (for more rigorous definition based upon the Turing Machine model of computation, see [14]). The class P consists of all those logical decision problems for which there exists an algorithm guaranteed to solve any instance in a number of operations bounded by a polynomial in the size of the data required to describe the instance. In short, the class P consists of polynomial-time solvable problems. Next, one wishes to define a class that will capture those problems for which it is suspected that polynomial-time solution algorithms do not exist. However, since it is as yet beyond our power to prove non-existence of a polynomial-time solution, this class must be characterized in another way. Motivated by decision problems arising from an optimization formulation, one might consider characterizing the cost of verifying a solution, and in particular, a "yes" solution. A problem is said to belong to the class NP if a candidate "yes" solution may be verified in polynomial-time with an additional piece of information, known as a "certificate." The role of the certificate is clear when specialized to the optimization case: the certificate is simply the candidate minimizer. Suppose one asserts that the solution to the minimization-based decision problem is "yes," i.e., there does exist a feasible solution with cost less than the specified threshold K. By providing a certificate, in the form of a candidate minimizer, the proposed "yes" solution can be verified in polynomial-time, assuming, of course, that the cost function and any constraint conditions can be evaluated in polynomial-time . Note, however, that the ability to efficiently check a "yes" solution does not guarantee the ability to efficiently check a "no" solution. Those decision problems formulated in such a way that a "no" solution may be checked in polynomial-time are termed "Co-NP."

The class NP contains many challenging problems for which there are no known polynomial-time algorithms, and there is strong belief that many of these problems fundamentally do not admit guaranteed polynomial-time algorithms. Briefly stated, the prevailing belief is that the class P is not equal to the class NP. Interestingly, there exists a sub-class of problems in NP that form a type of equivalence class for the "hardest" problems in NP, problems with the property that a guaranteed polynomial-time solution for any one of these would imply P=NP. This sub-class of problems is said to have the property of being NP-Complete. To define those problems that are NP-Complete, the concept of a polynomial-time reduction is necessary.

Decision problem F is said to admit a polynomial-time reduction to decision problem H under the following conditions: there exists a polynomial-time transformation of any instance of F into an instance of H, and the solutions of the two are identical for every pair of instances. A slight modification of an example from [15] illustrates this concept. Suppose problem F was defined as follows:

Decision Problem: F
Instance: Matrix $\mathbf{A} \in \mathbb{R}^{n\times n}$, $\det(\mathbf{A}) \neq 0$; vector $\mathbf{b} \in \mathbb{R}^n$; and real scalar $K > 0$.
Question: Is $||\mathbf{A}^{-1}\mathbf{b}|| < K$?

Suppose a second type of problem, denoted H, is defined by further restrictions on the data which make up an allowable instance.
Decision Problem: H
Instance: Matrix $\mathbf{C} \in \mathbb{R}^{n\times n}$, $\det(\mathbf{C}) \neq 0$, $\mathbf{C} = \mathbf{C}^T$; vector $\mathbf{x} \in \mathbb{R}^n$; and real scalar $K > 0$.

Question: Is $||\mathbf{C}^{-1}\mathbf{x}|| < K$?

Problem F reduces to problem H by the following simple transformation of the data specifying an instance of F into that for H: let $\mathbf{C} = \mathbf{A}^T\mathbf{A}$; $\mathbf{x} = \mathbf{A}^T\mathbf{b}$. This is clearly a polynomial-time transformation, and the yes/no solution to the two decision problems will be identical in every instance. Note also that problem H also trivially reduces to problem F; no transformation of the instance is necessary. Hence, one would say that there is both a polynomial-time reduction of F to H, and a polynomial-time reduction of H to F. Note that in general, however, a reduction of one problem to another need not imply existence of a reduction in the other "direction."

A decision problem A is said to be NP-Complete if it is a member of the class NP, and any other problem B $\in$ NP can be reduced to A in polynomial-time. Hence, as suggested above, any other problem in NP can be solved indirectly by solution of an NP-Complete problem, indicating that the NP-Complete problems are "as hard" as any problem in NP. To allow for cases in which a "yes" solution may not be verifiable in polynomial-time, a problem C is said to be NP-Hard if an NP-Complete problem reduces to it in polynomial-time, independent of whether or not C $\in$ NP. The first problem demonstrated to be NP-Complete was that of satisfiability of certain types of logical sentences (commonly abbreviated as "SAT"), as established by Cook [21]. Since then, a host of problems have been established as NP-Complete. The problem that will be used here to establish NP-Hardness of the matrix polytope robust stability problem will be a specialization of SAT, known as MAX 2-SAT. As a first step in introducing MAX 2-SAT, it is worthwhile to review some simple notation for and properties of constructs involving Boolean logical variables.

Clearly, the basic element in a logical formula is a Boolean variable, x, defined as a variable whose values can be either 0 or 1, with these values understood to represent logical FALSE and TRUE respectively. A literal in a Boolean variable x may take either of two forms, x or its negation $\overline{x}$. A clause in the set of Boolean variables $\{x_1, x_2, \ldots, x_n\}$ will be understood to be a disjunction (i.e., "OR-ing") of literals of these n variables. For a given clause in $\{x_1, x_2, \ldots, x_n\}$, an n-tuple of literals will represent the particular choice of literals in the disjunction. For example, $\{x_1, \overline{x_2}, x_3\}$ will represent the disjunction $x_1 \vee \overline{x_2} \vee x_3$. Finally, a truth assignment will refer to a particular assignment of values to the elements of a set of Boolean variables. A satisfying truth assignment for a clause is one for which the clause has value 1. For example, one truth assignment for the Boolean variables in the clause $\{x_1, \overline{x_2}, x_3\}$ is

$$x_1 := 1\ ;\ \ x_2 := 1\ ;\ \ x_3 := 0$$

The value of the clause for this truth assignment is

$$x_1 \vee \overline{x_2} \vee x_3 = 1 \vee \overline{1} \vee 0 = 1;$$

this is a satisfying truth assignment for the clause.

In [21], to which the origin of NP-Completeness theory is generally attributed, it is shown that any decision problem in NP may be reduced in polynomial-time to the following problem relating to satisfiability of logical sentences, defined as:
Decision Problem: SATISFIABILITY (SAT).
Instance: A set U of n Boolean variables and collection C of m clauses over U.
Question: Is there a truth assignment for U which satisfies every clause in C?

The collection of clauses C in the definition of SAT may be understood to represent a conjunction (i.e., "AND-ing") of disjunctive clauses. One can therefore understand SAT to be the problem of deciding whether there exists a satisfying truth assignment for a given logical sentence formed as the conjunction of disjunctive clauses.

It is possible to find classes of special cases of SAT that may be solved in polynomial-time. For instance, if each clause contains a single literal, it is trivial to decide whether there exists a satisfying assignment. For the case of two literals per clause there are several known $O(n^2)$ algorithms, where n is the number of Boolean variables [22]. The case of three literals per clause is NP-Complete, indeed, [21] shows that every instance of SAT may be reformulated as an instance of SAT with three literals per clause. However, with a modification of the problem statement, it is possible to formulate a two-literals-per-clause construction that yields an NP-Complete decision problem. In particular, [23] introduces and proves NP-Complete the problem of interest here, MAXIMUM 2-SATISFIABILITY, commonly abbreviated as "MAX 2-SAT." MAX 2-SAT may be defined as follows:
Decision Problem: MAXIMUM 2-SATISFIABILITY (MAX 2-SAT).
Instance: A set U of p Boolean variables, a collection C of clauses over U such that each clause $c_i \in C$ satisfies $|c_i| = 2$, and a positive integer $K < |C|$.
Question: Is there a truth assignment for U which satisfies at least K of the clauses in C?

To summarize, the problem MAX 2-SAT provides an example of an NP-Complete problem. It follows that any algorithm for solving MAX 2-SAT is unlikely to have the property that the number of operations it performs can be bounded by a polynomial in the problem size. Moreover, if one finds a second decision problem with the property that MAX 2-SAT reduces to it in polynomial-time, this second problem can be classified as NP-Hard. Such an NP-Hard problem inherits the characteristic of being unlikely to admit a guaranteed polynomial-time solution algorithm. The section to follow will display a reduction from MAX 2-SAT to a robust stability problem.

3. NP-hardness of the matrix polytope stability problem. To facilitate our discussion, it will prove convenient to restrict attention to a particular structure of parameter dependent matrices in $\mathbb{R}^{n\times n}$. Our motivation, of course, will be to capture a simple form of dependence that might reasonably be expected to appear in the system matrices for state variable models of physical systems. For the case of k real parameters, consider the function $\mathbf{A} : \mathbb{R}^k \to \mathbb{R}^{n\times n}$ defined as:

$$\mathbf{A}(\mathbf{p}) = \mathbf{A}_0 + \sum_{i=1}^{k} p_i \mathbf{A}_i \,. \tag{3.1}$$

Suppose further that the domain of interest for this function is the k-dimensional unit hypercube:

$$Q_k := [0,1]^k \,.$$

Then the image of Q_k under the map $\mathbf{A}$, $\mathbf{A}(Q_k)$, defines a matrix polytope. With suitable scaling and offset, this formulation can represent a wide variety of system models in which parameters enter in an affine fashion.

For a matrix polytope specified in this fashion, it is reasonable to assume that the nominal matrix $\mathbf{A}_0$ is known to have all its eigenvalues in the left half plane. With this formulation, the robust stability problem of interest may be succinctly stated as a search for any unstable matrix in the polytope. In particular, we will consider the formulation as:

Decision Problem: MATRIX POLYTOPE INSTABILITY (MP-INSTAB).

Instance: A set of matrices $\mathbf{A_0}$, $\mathbf{A}_1$, ... $\mathbf{A}_k \in \mathbb{R}^{n\times n}$, with spectrum of $\mathbf{A}_0$ contained in the open left half of the complex plane.

Question: Does there exist $\hat{\mathbf{p}} \in Q_k$ such that $\mathbf{A}(\hat{\mathbf{p}})$ has at least one eigenvalue in the closed right half plane?

As discussed above, this problem of MP-INSTAB will be demonstrated to be NP-Hard by proving that any instance of MAX 2-SAT may be transformed through a polynomial-time reduction into an instance of MP-INSTAB. As we will see, the associated matrix polytopes resulting from this transformation all have a particular structure. The goal of Section 4 will then be to show that there exists a plausible scenario of parameter variation in a power system model that produces this structure. To begin, we present the following:

PROPOSITION 3.1. *MP-INSTAB is NP-Hard.*

Proof. The proof consists of showing that the NP-Complete problem MAX 2-SAT reduces to MP-INSTAB in polynomial-time. Assume that one is given an arbitrary instance of MAX 2-SAT defined by a set U of p distinct Boolean variables x_i and a collection C of k 2-literal clauses. Suppose that the jth literal in clause i is represented by $l_{ij}(x_{m_{ij}})$, where i ranges from 1 to k, j may be either 1 or 2 and the index m_{ij} is determined by the specified ordering of clauses and a predetermined ordering of the literals within the clauses. Furthermore, assume that for each i, if $j_1 \neq j_2$ then $m_{ij_1} \neq m_{ij_2}$. In other words, the two literals in a clause cannot involve the same Boolean variable.

Consider a mapping from literals to linear factors defined by

$$h(l_{ij}(x_{m_{ij}})) := \begin{cases} 1 - x_{m_{ij}} & \text{if } l_{ij}(x_{m_{ij}}) = x_{m_{ij}} \\ x_{m_{ij}} & \text{if } l_{ij}(x_{m_{ij}}) = \overline{x_{m_{ij}}}, \end{cases} \tag{3.2}$$

in which an identification is made between logical values 0 and 1 and integer values 0 and 1. Then for each clause $c_i(x_{m_{i1}}, x_{m_{i2}}) = \{l_{i1}(x_{m_{i1}}), l_{i2}(x_{m_{i2}})\}$ the product

$$h(l_{i1}(x_{m_{i1}}))h(l_{i2}(x_{m_{i2}})) \tag{3.3}$$

constitutes the characteristic function for the set of truth assignments for which the clause is not satisfied. Recall that a characteristic function has value 1 for all members of the set and 0 otherwise.

The sum over the k clauses in the conjunction C of the products 3.3 then represents, for a given truth assignment for U, the number of clauses not satisfied. Define

$$g(x_1, \ldots, x_n) := \sum_{i=1}^{k} h(l_{i1}(x_{m_{i1}}))h(l_{i2}(x_{m_{i2}})). \tag{3.4}$$

Given an $\tilde{\mathbf{x}} \in \{0,1\}^p$ representing a truth assignment, $g(\tilde{\mathbf{x}})$ takes on a non-negative integer value that is the number of clauses not satisfied. Therefore, at least K of the clauses in the collection are satisfied for some truth assignment if and only if

$$\min_{\mathbf{x} \in \{0,1\}^p} g(\tilde{x}) \leq k - K \,.$$

Since g is a multilinear function of $\mathbf{x}$, if we extend the domain of interest to $\mathbf{x} \in Q_p$, it must be true that the extreme values of g are achieved at vertices of Q_p; i.e,

$$\min_{\mathbf{x} \in \{0,1\}^p} g(\mathbf{x}) = \min_{\mathbf{x} \in Q_p} g(\mathbf{x}) \,.$$

Moreover, note that the extreme values of $g(\mathbf{x})$ over Q_p will by construction be non-negative integers between zero and k.

A matrix-valued function $\mathbf{\Pi}(\mathbf{x})$ is constructed below that is robustly stable over Q_p if and only if $g(\mathbf{x}) > k - K$ over Q_p. Moreover, this function is affine with respect to $\mathbf{x}$, so that $\mathbf{\Pi}(Q_p)$ is a matrix polytope. Consider

$$\mathbf{\Pi}(\mathbf{x}) :=$$

$$\begin{pmatrix} -1 & \frac{1}{\sqrt{k}}h(l_{11}) & \frac{1}{\sqrt{k}}h(l_{21}) & \dots & \frac{1}{\sqrt{k}}h(l_{k1}) & \sqrt{R} \\ -\frac{1}{\sqrt{k}}h(l_{12}) & -1 & 0 & \dots & 0 & 0 \\ -\frac{1}{\sqrt{k}}h(l_{22}) & 0 & -1 & \dots & 0 & 0 \\ \vdots & \vdots & \vdots & \ddots & \vdots & \vdots \\ -\frac{1}{\sqrt{k}}h(l_{k2}) & 0 & 0 & \dots & -1 & 0 \\ \sqrt{R} & 0 & 0 & \dots & 0 & -1 \end{pmatrix}, \tag{3.5}$$

defined in terms of a constant R to be specified below. Then

$$\begin{aligned} \det(\lambda I - \mathbf{\Pi}(\mathbf{x})) &= (\lambda+1)^k \left((\lambda+1)^2 - \left(R - \frac{1}{k}\sum_{i=1}^{k} h(l_{i1})h(l_{i2}) \right) \right) \\ &= (\lambda+1)^k \left((\lambda+1)^2 - (R - g(x)/k) \right). \end{aligned}$$

There are three distinct root locations for $\det(\lambda I - \mathbf{\Pi}(\mathbf{x})) = 0$, given by -1 and the pair of locations

$$-1 \pm \sqrt{R - \frac{g(\mathbf{x})}{k}}.$$

Therefore, $\mathbf{\Pi}(\hat{\mathbf{x}})$ has an eigenvalue in the closed right half plane for some $\hat{\mathbf{x}} \in Q_p$ if and only if $(R - g(\hat{\mathbf{x}})/k) \geq 1$. Setting

$$R := 1 + \frac{1}{2k} + \frac{k-K}{k} \tag{3.6}$$

results in the desired equivalence.

Finally, note that $\mathbf{\Pi}(\mathbf{x})$ may be formed from the MAX 2-SAT instance by a straightforward transformation that involves a number of operations bounded by a polynomial in the instance size. □

The proposition above implies that if one had a general, guaranteed polynomial-time algorithm to test instability for matrix polytopes of the form 3.5, then it could be used to solve MAX 2-SAT as a special case. This indicates that the existence of such an algorithm is highly unlikely.

For the development of Section 4, it is also useful to note the following points regarding the construction used in the proof above:

(i) $\mathbf{\Pi}(\mathbf{x})$ possesses purely real eigenvalues for all $\mathbf{x} \in Q_p$;

(ii) If $\mathbf{\Pi}(\mathbf{x})$ has strictly negative eigenvalues for all $\mathbf{x} \in Q_p$, these must be less than or equal to $-1 + \sqrt{1 - 1/2k}$;

(iii) If there exists an $\hat{\mathbf{x}} \in Q_p$ such that $\mathbf{\Pi}(\hat{\mathbf{x}})$ has one or more non-negative eigenvalues, then there must exist an $\hat{\hat{\mathbf{x}}} \in Q_p$ such that $\mathbf{\Pi}(\hat{\hat{\mathbf{x}}})$ has at least one eigenvalue greater than or equal to $-1 + \sqrt{1 + 1/2k}$.

4. Power system realization of the matrix polytope instability problem. Proof that a problem has the property of being NP-Hard is, by its nature, a worst-case result. For example, the result of the preceding section, that the given matrix polytope stability problem is NP-Hard, indicates only that there are some special cases in this class of computations that can not be expected to be solved in polynomial-time. A skeptical reader might naturally ask whether or not these difficult special cases ever appear in practical problems relating to physical engineering systems. In brief, are the worst-cases "physically realizable?" It is the hypothesis of this work that this is the case, and in particular, that there exist practically interesting robust stability problems in power systems that are not solvable in polynomial-time. While the judgment of whether or not the case constructed is "practically interesting" must be left to the reader, the construction below will prove that there do exist realizations of the worst-case problems in a power system model.

Consider a network consisting of n generator buses, with the structure of interconnections between buses as shown in Figure 4.1. As illustrated, transmission links exist between bus 1 and each of the other buses, with additional links from bus n to each of the buses 2 through n-1. The representation to be used is that of electro-mechanical swing dynamics, with a classical model of constant voltage behind transient reactance for each of the generators. Derivation of this simple stability model may be found in standard texts; see, for example, [24]. With bus voltage magnitudes held fixed at 1.0 per unit, the dynamic variables of interest will be bus voltage phase angles $\theta \in \mathbb{R}^n$, and generator frequencies $\omega \in \mathbb{R}^n$. It is further assumed that for each generator, the normalized damping constants, d_i, divided by normalized rotational inertias, M_i, are equal to a constant γ. This is commonly known as the "uniform damping" assumption in the literature. In this scenario, standard results establish that a minimal state-space description can be constructed with the vector $\delta \in \mathbb{R}^{n-1}$ of n-1 phase angles relative to the reference bus n,

$$\delta_i := \theta_i - \theta_n, i = 1, 2, ..., n-1,$$

and the vector $\omega_r \in \mathbb{R}^{n-1}$ of $n-1$ frequency deviations, also relative to the reference bus n,

$$\omega_{r,i} := \omega_i - \omega_n, i = 1, 2, ..., n-1.$$

As a final simplifying assumption, we shall treat the reference bus n as an infinite bus, with (effectively) infinite inertia, $M_n \to \infty$ (note that $d_n = \gamma M_n$, so that d_n will also approach ∞ to maintain machine n's damping ratio). The electro-mechanical swing dynamics then take the form (see [24], p.79])

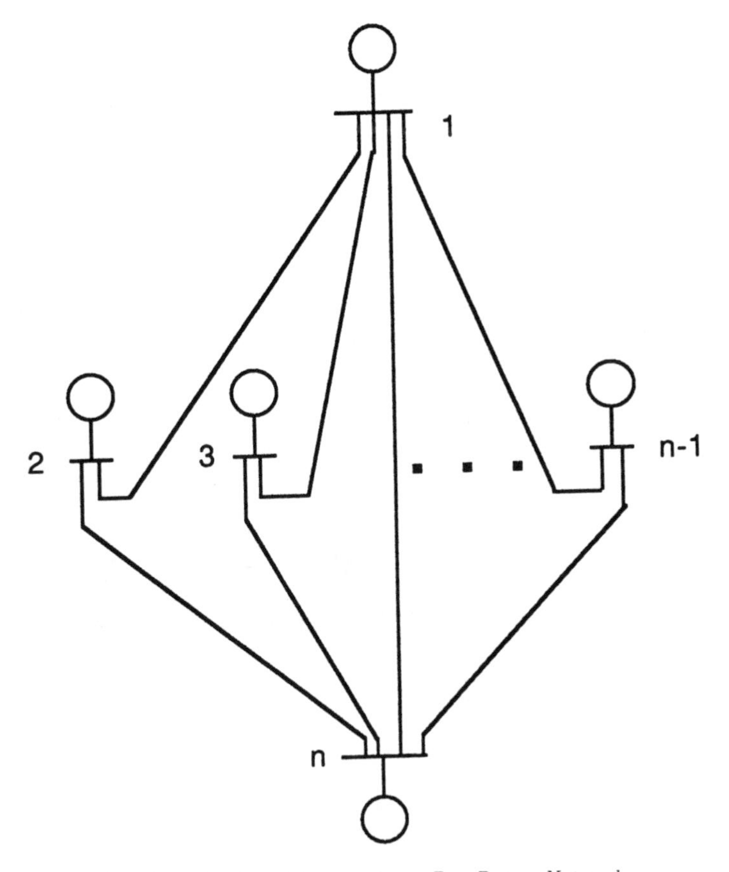

FIG. 4.1. *One-line Diagram for n-Bus Power Network*

$$\dot{\omega}_r = -\gamma\omega_r + \mathbf{M}^{-1}(\mathbf{P}^0 - \mathbf{P}_e(\delta)) \tag{4.1a}$$

$$\dot{\delta} = \omega_r \tag{4.1b}$$

where $\mathbf{P}_e : \mathbb{R}^{n-1} \to \mathbb{R}^{n-1}$, with $P_{e,i}(\delta) :=$ active electrical power absorbed by network at bus i; $\mathbf{P}^0 \in \mathbb{R}^{n-1}$, with ${P_i}^0 :=$ mechanical power delivered to generator i (constant in this model); $\mathbf{M} := diag\{M_1, M_2, ..., M_{n-1}\}$.

Linearizing about an operating point $(\delta^e, {\omega_r}^e)$, one has

$$\Delta\dot{\omega}_r = -\gamma\Delta\omega_r - \mathbf{M}^{-1}(\mathbf{J}\Delta\delta) \tag{4.2a}$$

$$\Delta\dot{\delta} = \Delta\omega_{\mathbf{r}} \tag{4.2b}$$

where

$$\mathbf{J} := \left[\frac{\mathrm{d}\mathbf{P}_e}{\mathrm{d}\delta}\right]_{\delta=\delta^e}.$$

The structure of $\mathbf{J}$ for the particular network interconnection shown in Figure 4.1 is critical to the arguments below. Its elements are formed as follows. For indices i, k running 1 to n-1, with index $k > i$, define:

$$b_{ik}^e := b_{ik} \cos(\delta_i^e - \delta_k^e), \tag{4.3a}$$

$$g_{ik}^e := g_{ik} \sin(\delta_i^e - \delta_k^e), \tag{4.3b}$$

where b_{ik} := susceptance of transmission line linking bus i to bus k, g_{ik} := conductance of transmission line linking bus i to bus k. The diagonal elements of $\mathbf{J} \in \mathbb{R}^{(n-1)x(n-1)}$ may then be written as [25]:

$$J_{11} = \left[\sum_{k=2}^{n-1} b_{1k}^e - g_{1k}^e \right] + b_{1n}^e - g_{1n}^e \tag{4.4}$$

$$J_{jj} = b_{1j}^e + g_{1j}^e + b_{jn}^e - g_{jn}^e \ ; \ \ j = 2, 3, \ldots n-1. \tag{4.5}$$

The non-zero off-diagonal terms of $\mathbf{J}$ appear only in its first row and column, and are given by:

$$J_{1k} = -b_{1k}^e + g_{1k}^e \ ; \ \ k = 2, 3, \ldots n-1, \tag{4.6}$$

$$J_{k1} = -b_{1k}^e - g_{1k}^e \ ; \ \ k = 2, 3, \ldots n-1. \tag{4.7}$$

The reader should note that the pattern of zero/non-zero entries in $\mathbf{J}$ coincides with that of the matrix $\mathbf{\Pi}(\mathbf{x})$ introduced in 3.5. To exploit this observation, our goals are two-fold. First, we will view the power flow Jacobian as a function of the linearized susceptance-related parameters, b_{ij}^e, and linearized conductance-related parameters, g_{ij}^e. We will show that given $\mathbf{\Pi}(\mathbf{x})$ of 3.5, one may construct affine functions that define b_{ij}^e and g_{ij}^e in terms of independent parameters $\mathbf{x}$ such that $\mathbf{J}(\mathbf{x}) = -\mathbf{\Pi}(\mathbf{x})$, and then relate the resulting matrix polytope to one that may be produced by variation in the underlying physical parameters of the b_{ij} and g_{ij}'s. Then we will relate the behavior of the eigenvalues of $\mathbf{J}$ to those of the overall linearized dynamics as given by 4.2.

To relate $\mathbf{J}$ to a given $\mathbf{\Pi}(\mathbf{x})$ (which in turn is constructed from a MAX 2-SAT problem having k clauses), let the number of buses $n = k + 3$. Suppose further that for all transmission lines not incident on bus n, the b_{1j}^e and g_{1j}^e parameters are defined as functions of $\mathbf{x}$, or as constants, given by:

$$\begin{bmatrix} b_{1j}^e \\ g_{1j}^e \end{bmatrix} = \begin{bmatrix} 1/2 & 1/2 \\ -1/2 & 1/2 \end{bmatrix} \begin{bmatrix} \frac{1}{\sqrt{k}} h(l_{j-1,1}(\mathbf{x})) \\ -\frac{1}{\sqrt{k}} h(l_{j-1,2}(\mathbf{x})) \end{bmatrix} \quad j = 2, 3, \ldots, n-2 \tag{4.8a}$$

and

$$\begin{bmatrix} b^e_{1,(n-1)} \\ g^e_{1,(n-1)} \end{bmatrix} = \begin{bmatrix} 1/2 & 1/2 \\ -1/2 & 1/2 \end{bmatrix} \begin{bmatrix} \sqrt{R} \\ \sqrt{R} \end{bmatrix} . \tag{4.8b}$$

The remaining parameters for transmission lines incident on bus n may then be chosen to force the diagonal elements of $\mathbf{J}$ to be unity:

$$b^e_{1n} = 1 - \left[\sum_{k=2}^{n-1} b^e_{1k} - g^e_{1k} \right] \tag{4.9a}$$

$$b^e_{jn} = 1 - b^e_{1j} - g^e_{1j} \quad , \; j = 2, 3, ..., n-1 \tag{4.9b}$$

$$g^e_{jn} = 0 \quad , \; j = 1, 2, ..., n-1 \tag{4.9c}$$

Direct calculations confirm that the above choice of b^e_{ij} and g^e_{ij} parameters yields $\mathbf{J}(\mathbf{x}) = -\mathbf{\Pi}(\mathbf{x})$. Let $\mathbf{J}(Q_k)$ denote the matrix polytope image that results as $\mathbf{x}$ ranges over the hypercube Q_k. To complete our justification of this matrix polytope structure for the power flow Jacobian, it is important to consider conditions under which variation in the underlying physical parameters of b_{ij} and g_{ij} would yield the same polytope $\mathbf{J}(Q_k)$. This reduces to the question of whether the mapping defined by (4.3), taking the b_{ij}'s and g_{ij}'s to the b^e_{ij}'s and g^e_{ij}'s, is invertible over the range of interest. If the operating point is viewed as fixed, this is simply a diagonal linear map, with full rank provided each of the relevant $\sin(\delta_i{}^e - \delta_j{}^e)$ and $\cos(\delta_i{}^e - \delta_j{}^e)$ terms is non-zero. This condition will hold generically at any operating point for which the system has non-zero power flows on each line. Finally, note that the resulting map from $\mathbf{x} \in Q_p$ to the b_{ij}'s and g_{ij}'s is a composition of a linear map and an affine map, and therefore maps polytopes into polytopes. Hence, we may make the following:

Observation 4.1: For a linearization about a fixed operating point satisfying

$$\sin(\delta_i{}^e - \delta_j{}^e) \neq 0 \;\; \forall \, i, j \text{ such that } g_{ij} \neq 0,$$

$$\cos(\delta_i{}^e - \delta_j{}^e) \neq 0 \;\; \forall \, i, j \text{ such that } b_{ij} \neq 0,$$

there exists a polytope of line parameter variations that yields a set of achievable power flow Jacobian matrices equal to the polytope $\mathbf{J}(Q_k)$.

The relationship of eigenvalues of the power flow Jacobian to those of a linearized dynamic model have been exploited in a wide variety of works on power systems. For completeness, the required relationship will be derived here, though similar results appear in [26]. For convenience,we will assume that the generator inertias for buses 1 through n-1 are all equal to 1.0 per unit, so that the matrix $\mathbf{M}$ becomes the identity. Also, it is convenient to first consider systems with zero damping. Under these assumptions, the following proposition relates the eigenvalues of $\mathbf{J}$ and those of the state matrix for the linearized dynamics of 4.2.

PROPOSITION 4.1. *Let*

$$\mathbf{A} = \begin{bmatrix} \mathbf{0}_{n-1 \times n-1} & -\mathbf{J} \\ \mathbf{I}_{n-1 \times n-1} & \mathbf{0}_{n-1 \times n-1} \end{bmatrix}$$

and assume that $\mathbf{J}$ *has n-1 distinct eigenvalues* $\lambda_i, i = 1, 2, ..., n-1$, *with associated right and left eigenvectors* $\mathbf{v}_i$ *and* $\mathbf{w}_i$ *respectively. Then for each eigenvalue of* $\mathbf{J}$, $\mathbf{A}$ *has an associated pair of eigenvalues*

$$\psi_{2i-1} = +j\sqrt{\lambda_i},$$

$$\psi_{2i} = -j\sqrt{\lambda_i},$$

with respective right eigenvectors

$$\mathbf{r}_{2i-1} = \begin{bmatrix} +j\sqrt{\lambda_i} v_i \\ v_i \end{bmatrix},$$

$$\mathbf{r}_{2i} = \begin{bmatrix} -j\sqrt{\lambda_i} v_i \\ v_i \end{bmatrix},$$

and respective left eigenvectors

$$\mathbf{s}_{2i-1} = \begin{bmatrix} w_i \\ +j\sqrt{\lambda_i} w_i \end{bmatrix},$$

$$\mathbf{s}_{2i} = \begin{bmatrix} w_i \\ -j\sqrt{\lambda_i} w_i \end{bmatrix}.$$

Proof. Direct substitution into each side of the desired equalities $\mathbf{s}_k^T \psi_k = \mathbf{s}_k^T \mathbf{A}$ and $\psi_k \mathbf{r}_k = \mathbf{A}\mathbf{r}_k$, $k = 1, 2, ...2n-2$, verifies the proposed eigenvalue/eigenvector pairs. □

To apply the results of Proposition 4.1, consider the power system model with zero damping, and a Jacobian matrix $\mathbf{J}(\mathbf{x}) = -\mathbf{\Pi}(\mathbf{x})$ using the construction of elements given in 4.4–4.9. Recall that this construction ensures that $\mathbf{J}(\mathbf{x})$ has purely real eigenvalues for all $\mathbf{x} \in Q_p$. For those values of $\mathbf{x}$ for which $\mathbf{J}(\mathbf{x})$ has all non-negative real eigenvalues, $\mathbf{A}(\mathbf{x})$ will have all its eigenvalues purely on the imaginary axis. When $\mathbf{J}(\mathbf{x})$ possesses one or more strictly negative real eigenvalues, $\mathbf{A}(\mathbf{x})$ will have an equal number of strictly positive real eigenvalues. Hence, for the construction considered thus far, robust stability of $\mathbf{\Pi}(\mathbf{x})$ over Q_p is equivalent to $\mathbf{A}(\mathbf{x})$ having its eigenvalues in the *closed* left half plane for all $\mathbf{x} \in Q_p$. To create complete equivalence between robust stability of $\mathbf{\Pi}(\mathbf{x})$ and that of $\mathbf{A}(\mathbf{x})$, damping must be introduced into the model.

In many power systems applications, the effective damping of electromechanical modes is relatively small in magnitude. With this motivation it will prove useful to consider, under the assumptions of uniform damping and $\mathbf{M} = \mathbf{I}$, the first order sensitivity in eigenvalues of $\mathbf{A}$ around $\gamma = 0$.

PROPOSITION 4.2. *Let $\mathbf{A}(\gamma)$, $\mathbf{A} : I\!R^1 \to I\!R^{2n-2 \times 2n-2}$, denote the matrix valued function mapping the uniform damping parameter γ to the state matrix of 4.2. Suppose that $\mathbf{A}(0)$ has 2n-2 distinct eigenvalues $\psi_1, \psi_2, ..., \psi_{2n-2}$. Then it follows that*

$$\left[\frac{\mathrm{d}\psi_i}{\mathrm{d}\gamma}\right]_{\gamma=0} = -0.5, \quad \forall\, i = 1, 2, ...2n-2.$$

Proof. For an isolated eigenvalue $\psi(0)$ of $A(0)$, with right and left eigenvectors $r(0)$ and $s(0)$ respectively, standard results [27] yield

$$\left[\frac{\mathrm{d}\psi_i}{\mathrm{d}\gamma}\right]_{\gamma=0} = \frac{\mathbf{s}(0)^T \left[\frac{\mathrm{d}A}{\mathrm{d}\gamma}\right]_{\gamma=0} \mathbf{r}(0)}{\mathbf{s}(0)^T \mathbf{r}(0)}$$

Applying this result to the particular structure of $\mathbf{A}(\gamma)$, $\mathbf{s}(0)$, and $\mathbf{r}(0)$ here yields:

$$\left[\frac{\mathrm{d}\psi_i}{\mathrm{d}\gamma}\right]_{\gamma=0} = \frac{[\,w_i{}^T \quad j\sqrt{\lambda_i}w_i{}^T\,] \begin{bmatrix} -\mathbf{I} & \mathbf{0} \\ \mathbf{0} & \mathbf{0} \end{bmatrix} \begin{bmatrix} j\sqrt{\lambda_i}v_i \\ v_i \end{bmatrix}}{[\,w_i{}^T \quad j\sqrt{\lambda_i}w_i{}^T\,] \begin{bmatrix} j\sqrt{\lambda_i}v_i \\ v_i \end{bmatrix}}$$

$$= \frac{-j\sqrt{\lambda_i}w_i{}^T v_i}{2j\sqrt{\lambda_i}w_i{}^T v_i} = -0.5\ .$$

□

The result of Proposition 4.2 indicates that for sufficiently small positive damping γ, eigenvalues of $\mathbf{A}$ are incrementally shifted to the left. Therefore, if $\mathbf{J}$ has all positive eigenvalues, there exists a positive γ such that $\mathbf{A}$ is guaranteed stable. Moreover, if $\mathbf{J}$ has a negative eigenvalue with magnitude greater than some known threshold, there exists a choice of γ sufficiently small to guarantee that $\mathbf{A}$ has an eigenvalue in the closed right half plane. Recalling observation (iii) following Proposition 3.1, for our application this threshold will be chosen as $(\sqrt{1 + 1/2k} - 1)$; when $\mathbf{J}(\mathbf{x}) = -\mathbf{\Pi}(\mathbf{x})$, the MAX 2-SAT problem has K or more clauses satisfied if and only if there exists some $\hat{\mathbf{x}} \in Q_p$ such that $\mathbf{J}(\hat{\mathbf{x}})$ has an eigenvalue less than or equal to $(\sqrt{1+1/2k}-1)$.

These observations may be summarized in the following proposition.

PROPOSITION 4.3. *Given an instance of MAX 2-SAT with k clauses in p Boolean variables, with the existence of a truth assignment satisfying K or more clauses to be decided, construct $\mathbf{J}(\mathbf{x}) = -\mathbf{\Pi}(\mathbf{x})$ according to 3.5, 3.6, and 4.4 through 4.9. Let*

$$\mathbf{A}(\gamma, \mathbf{x}) := \begin{bmatrix} \gamma \mathbf{I}_{n-1 \times n-1} & -\mathbf{J}(\mathbf{x}) \\ \mathbf{I}_{n-1 \times n-1} & \mathbf{0}_{n-1 \times n-1} \end{bmatrix}$$

Then there exists a $\hat{\gamma} > 0$ such that the following equivalence holds: there exists a truth assignment satisfying K or more clauses of the given MAX 2-SAT instance if and only if there exists $\hat{\mathbf{x}} \in Q_p$ such that $\mathbf{A}(\hat{\gamma}, \hat{\mathbf{x}})$ has at least one eigenvalue in the closed right half plane.

The proof follows directly from preceding discussion and Propositions 4.2 and 4.3. Coupling this result with Observation 4.1 yields the following corollary:

COROLLARY 4.4. *For a linearization about a fixed operating point, with transmission line parameters constrained to lie in a given polytope, robust stability determination for a power system model of the form 4.2 is an NP-Hard decision problem.*

5. Conclusions. The problem of maintaining acceptable dynamic performance as system parameters vary is one of considerable importance to power system operators and planners. This makes the power system a natural application for the many results on parametric robust stability analysis that have appeared over the past decade. However, power system applications also demand methods that can successfully treat models with both very large state dimension, and very large numbers of uncertain parameters. Therefore, it is particularly important that the computational cost of robust stability determination is carefully characterized in this context. As noted in the introduction, this computational cost depends on the nature of parameter dependence in the system model. For families of linear, time-invariant state-space models in which the system matrix is constrained to lie in a polytope, this paper has confirmed that robust stability determination is NP-Hard. This implies that there will exist worst-case examples in this class of problems for which the number of operations required cannot be expected to be bounded by a polynomial in problem size.

Having established that the general problem of matrix polytope robust stability determination is NP-Hard, a natural question is whether or not the worst-case examples are likely to be encountered in practical power system applications. This paper has shown that this will be the case. In particular, the results of this work demonstrate that given any instance of the known NP-Complete problem MAX 2-SAT, one can construct an associated power system model with uncertain transmission line susceptance and conductance parameters. For a linearization of this model around a fixed operating point, robust stability determination was shown to be equivalent to solving the original MAX 2-SAT problem. Hence, an algorithm that solved the robust stability problem for power system models of this form would solve a known NP-Complete problem as a special case. This offers strong evidence that the power system robust stability problem is unlikely to admit guaranteed polynomial-time solution algorithms.

REFERENCES

[1] F.C.Schweppe, Uncertain Dynamical Systems, Prentice-Hall, Englewood Cliffs, N.J. 1973.

[2] B.R.Barmish, New Tools for Robustness of Linear Systems, Macmillan Publishing Co., New York 1994.

[3] V.L.Kharitonov, *Asymptotic stability of an equilibrium position of a family of systems of linear differential equations*, Differential'nye Uraveniya **14** (1978), 2086–2088.

[4] B.R.Barmish, *Invariance of the strict Hurwitz property for polynomials with perturbed coefficients*, Proc. IEEE Conf. on Decision and Control, San Antonio, TX 1983.

[5] S.Bialas, *A necessary and sufficient condition for stability of interval matrices*, Int. Journal of Control **37** (1983), 717–722.

[6] W.C.Karl, J.P.Greschak, G.C.Verghese, *Comments on "A necessary and sufficient condition for the stability of interval matrices"*, International Journal of Control **39** (1984), 849–851.

[7] B.R.Barmish, C.V.Hollot, *Counter-example to a recent result on the stability of interval matrices by S. Bialas*, International Journal of Control **39** (1984), 1103–1104.

[8] F.J.Kraus, B.D.O.Anderson, M.Mansour, *Robust stability of polynomial with multilinear parameter dependence*, International Journal of Control **50** (1989), 1745–1762.

[9] A.Sideris, *An efficient algorithm for checking the robust stability of a polytope of polynomials*, Mathematics of Control, Signals and Systems, **4** (1991), 315–337.

[10] G.E.Coxson, *Computational complexity of robust stability and regularity in families of linear systems*, (Ph.D. dissertation) Department of Electrical and Computer Engineering, University of Wisconsin-Madison, May 1993.

[11] G.E.Coxson, C.L.DeMarco, *The computational complexity of approximation algorithms for robust stability in rank two matrix polytopes*, Proc. 1993 American Control Conference, San Francisco, CA, June 1993, 1671–1672.

[12] G.E.Coxson, C.L.DeMarco, *Testing robust stability of general matrix polytopes is an NP-hard computation*, Proc. of the 29th Allerton Conf. on Communication, Control, and Computing, Monticello, IL 1991, 105–106.

[13] A.Nemirovskii, *Several NP-hard problems arising in robust stability analysis*, (to appear in) Mathematics of Control, Signals and Systems, **6** 1993.

[14] M.Garey, Johnson, Computers and Intractability: A Guide to the Theory of NP-Completeness, W.H.Freeman and Co., NY 1979.

[15] H.S.Wilf, Algorithms and Complexity, Prentice-Hall, Inc., Englewood Cliffs, NJ 1986.

[16] V.Klee, G.J.Minty, *How good is the simplex algorithm*, Inequalities-III, O.Shisha, (ed.) Academic Press, NY 1972, 159–175.

[17] L.G.Khachian, *A polynomial algorithm in linear programming*, Soviet

Mathematics—Doklady **20** (1979), 191–194.

[18] N.Karmarkar, *A new polynomial-time algorithm for linear programming*, Combinatorica **4** (1984), 373–395.

[19] C.Papadimitriou, M.Yannakakis, *Optimization, approximation and complexity classes*, Proceedings of the 20th ACM Symposium on the Theory of Computing (1988), 229–234.

[20] S.Arora, C.Lund, R.Motwani, M.Sudan, M.Szegedy, *Proof verification and hardness of approximation problems*, Proceedings of the 33rd Annual IEEE Conference on Foundations of Computer Science, Pittsburgh, PA, October 1992, 14–23.

[21] S.A.Cook, *The complexity of theorem-proving procedures*, Proceedings of the 3rd Annual ACM Symposium on the Theory of Computing, Shaker Heights, OH, May 1971, 151–158.

[22] S.Even, A.Itai, A.Shamir, *On the complexity of timetable and multicommodity flow problems*, SIAM Journal of Computing, **5** 1976, 691–703.

[23] M.R.Garey, D.S.Johnson, L.Stockmeyer, *Some simplified NP-complete graph problems*, Theoretical Computer Science, **1** 1976, 237–267.

[24] M.A.Pai, Power System Stability, North-Holland Publishing Co., Amsterdam 1981.

[25] A.R.Bergen, Power Systems Analysis, Prentice-Hall, Englewood Cliffs, NJ 1986.

[26] Time-Scale Modeling of Dynamic Networks with Applications to Power Systems, J.H.Chow, (ed.) Springer-Verlag, Berlin 1982.

[27] G.W.Stewart, Introduction to Matrix Computations, Academic Press, NY 1973.

DAMPING AND RESONANCE IN A HIGH POWER SWITCHING CIRCUIT

IAN DOBSON*, SASAN JALALI*, AND RAJESH RAJARAMAN*

Abstract. Switching circuits with thyristor controlled reactors are used in high power systems for static VAR control and flexible AC transmission. These circuits can exhibit highly nonlinear behavior because the thyristor switch off time depends on the circuit state. This paper shows how to understand and predict damping and resonance in a basic thyristor controlled reactor circuit used for static VAR control. The circuit damping occurs even in the absence of circuit resistance. An analytic formula for the circuit Jacobian is studied to predict resonance effects and a simple test for detecting resonance involving circuit frequencies spanning an integer harmonic is proven to be correct.

1. Introduction. As static switching circuits in power systems proliferate there is an increasing need to analyze and understand these circuits. Switching circuits involving thyristor controlled reactors such as static VAR controllers and series compensators for flexible AC transmission can exhibit damping and resonances which affect the circuit design and operation. However, these effects are poorly understood. The damping effects have been observed in simulation [2] but are often neglected because the usual linear models such as average inductor models have only resistive damping. The resonance phenomena are known to occur but are not properly accounted for in the usual models. Recent work has computed the resonances using harmonic admittance matrix techniques [1,6] and has started to develop state space methods to analyze these switching circuits [10,4,13,7,8,3,12]. In particular, the stability of the circuits can be analytically computed using formulas for the Jacobian of the Poincaré map.

This paper shows how to use this Jacobian formula to compute and understand damping and resonance phenomena in a thyristor controlled reactor circuit which is used in reactive power control of power systems. We choose this circuit for our study because it is one of the simplest useful circuits in which these ideas can be introduced and developed. The prospects for extending these ideas to other high power switching circuits such as advanced static compensators for flexible AC transmission are good [9].

The thyristor controlled reactor circuit can be analyzed as a periodic succession of linear systems of varying dimension. The switching times between linear systems corresponding to thyristors switching on are fixed by the firing delay ϕ, but the switching times between the linear systems corre-

* Electrical and Computer Engineering Department, University of Wisconsin, Madison, WI 53706. email: dobson@engr.wisc.edu The authors gratefully acknowledge funding in part from EPRI under contract numbers RP 4000-29, RP 8010-30, RP 8050-03 and from NSF PYI grant ECS-9157192.

sponding to thyristors switching off depend on the state (thyristors switch off when their current decreases through zero). There are both complicated and simple aspects of these circuits: This type of circuit nonlinearity leads to novel nonlinear effects such as switching time bifurcations [7,8] and Poincaré map discontinuities [12]. On the other hand, the simple structure gives opportunities for a successful exact analysis.

Sections 2 and 3 explain the basic circuit, its modeling and its classical operation. Section 4 states the analytic formula for the Jacobian of the Poincaré map of the circuit, computes the eigenvalues as a firing delay parameter is varied and discusses the nonresistive damping inherent in the circuit. Section 5 introduces a model for the Jacobian so that the Jacobian can be better understood. The Jacobian is proved to be stable for all firing delays. (Caution: stability of the Jacobian does not preclude instability via switching time bifurcations.) Section 6 explains how to compute all the halfwave symmetric orbits of the circuit and relates the resonant blowup of these orbits to eigenvalues of the Jacobian of a half cycle map being -1. Section 7 predicts resonance when a small ambient second harmonic or third subharmonic is added to the AC voltage source and relates this to Jacobian eigenvalues on the unit circle. Section 8 gives a geometric view of the model for the Jacobian involving circular orbits on a sphere. There is a simple engineering rule of thumb involving circuit frequencies spanning an integer harmonic which is used to predict resonance in the circuit. Section 9 gives analytic conditions for circuit resonance and proves that the rule of thumb is correct.

2. System description and modeling. Figure 2.1 shows a single phase static VAR system consisting of a thyristor controlled reactor and a parallel capacitor $C = 1.5$ mF. This system is connected to an infinite bus behind a power system impedance of inductance $L_s = 0.195$ mH and resistance $R_s = 0.9$ mΩ. The controlled reactor is modeled as a series combination of an inductor $L_r = 1.66$ mH and $R_r = 31.3$ mΩ. The source voltage $u(t) = \sin \omega t$ has frequency $\omega = 2\pi 60$ rad/s and period $T = 2\pi/\omega$. The switching element of the thyristor controlled reactor consists of two oppositely poled thyristors which conduct on alternate half cycles of the

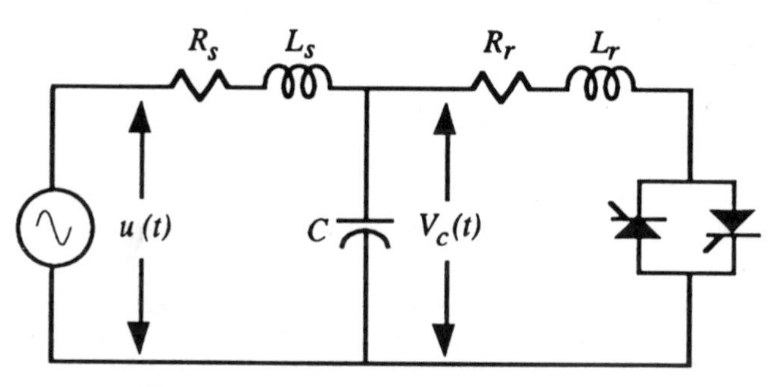

FIG. 2.1. *Single phase static VAR system.*

supply frequency.

A thyristor conducts current only in the forward direction, can block voltage in both directions, turns on when a firing signal is provided and turns off after a current zero. The thyristors are assumed ideal so that nonlinearities in the turn on/off of the thyristor are neglected. Indeed, the nonlinearity of the circuit only arises from the dependence of the switch off times of the thyristors on the system state. The thyristor firing pulses are supplied periodically and the system is controlled by varying the firing delay ϕ of the firing pulses. The system is studied with ϕ as an open loop control parameter. In practice a closed loop control would modify ϕ.

When a thyristor is on, the system state vector $x(t)$ specifies the thyristor controlled reactor current, capacitor voltage and the source current:

$$x(t) = \begin{pmatrix} I_r(t) \\ V_c(t) \\ I_s(t) \end{pmatrix}$$

During the conducting time of each of the thyristors, the system dynamics are described by the ON linear system:

$$\dot{x} = Ax + Bu \tag{2.1}$$

$$\text{where } A = \begin{pmatrix} -R_r L_r^{-1} & L_r^{-1} & 0 \\ -C^{-1} & 0 & C^{-1} \\ 0 & -L_s^{-1} & -R_s L_s^{-1} \end{pmatrix} \text{ and } B = \begin{pmatrix} 0 \\ 0 \\ L_s^{-1} \end{pmatrix}$$

During the off time of each thyristor, the circuit state is constrained to lie in the plane Z of zero thyristor current specified by $I_r = 0$. In this mode, the system state vector $y(t)$ specifies the capacitor voltage and the source current:

$$y(t) = \begin{pmatrix} V_c(t) \\ I_s(t) \end{pmatrix}$$

and the system dynamics are given by the OFF linear system

$$\dot{y} = PAQy + PBu \tag{2.2}$$

where P is the projection matrix $P = \begin{pmatrix} 0 & 1 & 0 \\ 0 & 0 & 1 \end{pmatrix}$ and $Q = P^t$ [6,8,3].

Figure 2.2 describes the system dynamics as the system state evolves over a period T. A thyristor starts conducting at time ϕ_0. This mode as described by 2.1 ends when the thyristor current goes through zero at time $\phi_0 + \sigma_1$. The non-conducting mode as described by 2.2 follows the conducting mode and continues until the next firing pulse is applied at time ϕ_1. This starts a similar on-off cycle which lasts until the next period starts at time $\phi_0 + T$.

ON (A, B)	OFF (PAQ, PB)	ON (A, B)	OFF (PAQ, PB)

ϕ_0 $\quad \phi_0 + \sigma_1$ $\quad \phi_1$ $\quad \phi_1 + \sigma_2$ $\quad \phi_0 + T$

$\longleftarrow T_1 \longrightarrow \longleftarrow T_2 \longrightarrow$

FIG. 2.2. *System dynamics from time ϕ_0 to $\phi_0 + T$.*

The state at the switch on time ϕ_0 is denoted either by the vector $y(\phi_0)$ or by the vector $x(\phi_0)$. These representations of the state at the switch on time are related by

$$x(\phi_0) = Qy(\phi_0) \tag{2.3}$$

Equation 2.3 expresses the fact that the x representation of the state at a switch on is computed from the y representation by adding a new first component which has value zero.

The state at the switch off time $\phi_0 + \sigma_1$ is similarly denoted either by $x(\phi_0 + \sigma_1)$ or $y(\phi_0 + \sigma_1)$ and these are related by

$$y(\phi_0 + \sigma_1) = Px(\phi_0 + \sigma_1) \tag{2.4}$$

The matrix P in 2.4 may be thought of as projecting the vector x onto Z, where Z denotes the plane in state space of zero thyristor current.

3. Classical analysis. The classical, idealized operation of a thyristor controlled reactor is explained in Figure 3.1. In this figure, the gray line represents the thyristor controlled reactor voltage $V_c(t)$ (c.f. Figure 2.1) and the solid line represents the thyristor current.

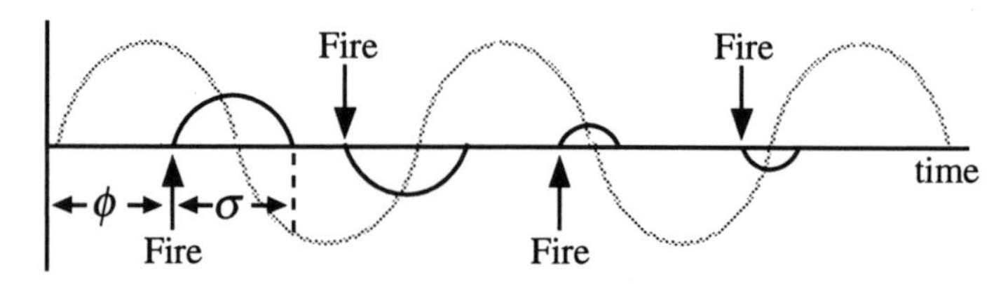

FIG. 3.1. *Classical operation of a thyristor controlled reactor.*

If the thyristors are fired at the point where the voltage $V_c(t)$ is at a peak, full conduction results. The circuit then operates as if the thyristors were shorted out, resulting in a current which lags the voltage by 90 degrees. If the firing is delayed from the peak voltage, the current becomes discontinuous with a reduced fundamental component of reactive current and a reduced thyristor conduction time σ. As the firing delay angle ϕ

ranges between 90° and 180° the thyristor conduction time σ ranges between 180° and 0°. The classical analysis assumes that the voltage $V_c(t)$ is a pure sinusoid.

The standard approach to modeling is to replace the thyristor controlled reactor by an average inductor model and then apply linear analysis techniques to the resulting circuit [5,11]. The average inductor model for a given firing delay ϕ is obtained by assuming a sinusoidal voltage across the thyristor controlled reactor and computing the fundamental component of the thyristor current. Then the value of average inductor for that firing delay ϕ is that which gives the same fundamental component of thyristor current. While this average inductor model approximation is usually effective for predicting steady state behavior, it fails to capture much of the circuit nonlinearity and it breaks down when large harmonic distortions occur. Operating conditions with large harmonic distortions are documented in [1,6,15].

4. Damping and the Poincaré map Jacobian. The Poincaré map F advances the state by one period T. In particular we choose F to advance the state $y(\phi_0)$ at turn on to $F(y) = y(\phi_0 + T)$. F can be computed by integrating the system equations 2.1 and 2.2 and taking into account the coordinate changes 2.2 and 2.4 when the switchings occur [7,3]. If the circuit has a steady state periodic orbit passing through y_0 at time ϕ_0, then y_0 is a fixed point of F:

$$F(y_0) = y_0\,.$$

The stability of the periodic orbit is the same as the stability of y_0 and is given (except in marginal cases) by the eigenvalues of DF, the Jacobian of the Poincaré map evaluated at y_0. The Poincaré map is differentiable except at switching time bifurcations where it is discontinuous [3,12]. Therefore we assume that the system is not exactly at a switching time bifurcation when computing the Jacobian.

According to [7,8,3], the Jacobian of the Poincaré map is

$$DF = e^{PAQ(T_2-\sigma_2)}Pe^{A\sigma_2}Qe^{PAQ(T_1-\sigma_1)}Pe^{A\sigma_1}Q \tag{4.1}$$

One of the interesting and useful consequences of formula 4.1 is that DF and the stability of the periodic orbit only depend on the state and the input via σ_1 and σ_2. It is remarkable that 4.1 is also the formula that would be obtained for fixed switch off times σ_1 and σ_2; the varying switch off times introduce no additional complexity in the formula, but the nonlinearity of the circuit is clear since σ_1 and σ_2 vary as a function of $y(\phi_0)$.

If the periodic orbit is assumed to be half wave symmetric, then the two conduction lengths $\sigma_1 = \sigma_2 = \sigma$ are equal and 4.1 simplifies to

$$DF = \left(e^{PAQ(T/2-\sigma)}Pe^{A\sigma}Q\right)^2 \tag{4.2}$$

It is straightforward to compute the eigenvalue locus for halfwave symmetric orbits as the conduction time σ varies over its range of 0 to 180 degrees and the results are shown in Figure 4.1.(a).

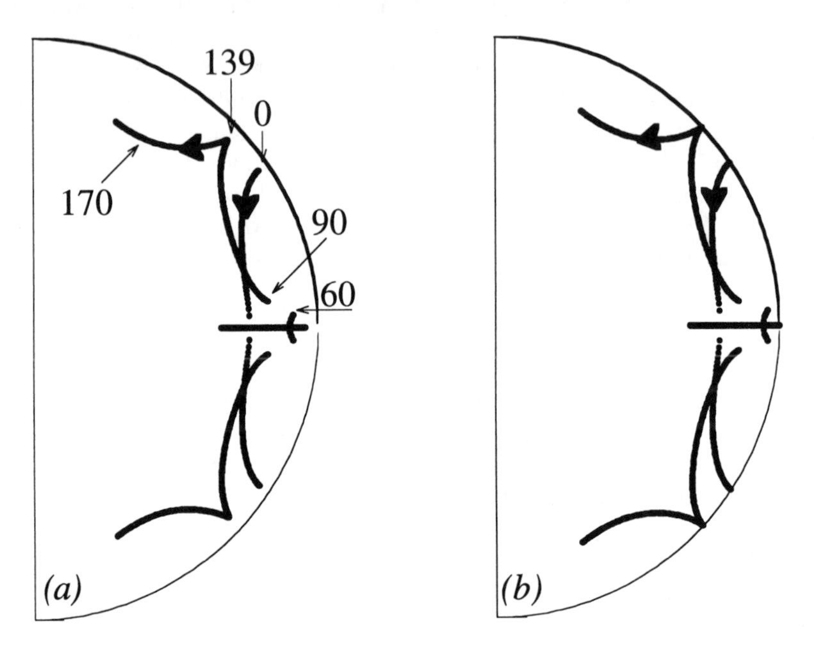

FIG. 4.1. *Eigenvalues of DH: (a) Circuit with resistance, (b) Circuit with no resistance.*

When all the eigenvalues are inside the unit circle, the circuit periodic orbit is asymptotically stable and the system damps out any small perturbations. This damping cannot be entirely attributed to the resistance in the circuit. Indeed, if the circuit resistances are set to zero then the eigenvalue locus of Figure 4.1.(b) is obtained and the damping for most values of σ is evident. This circuit damping occurs despite the circuit impedances and thyristors being lossless; the energy in a perturbation is dissipated or absorbed in the source. This nonresistive damping is not accounted for in average inductor models of the circuit.

5. Model of the Jacobian. Some insight into the properties of the Jacobian formula can be obtained by considering another dynamical system whose Poincaré map is the same Jacobian. That is, we construct a useful but fictitious system that is a model for the Jacobian. For simplicity, assume half wave symmetry and consider the map H which advances the state by half of a period. Then Section 4 gives

$$DH = e^{PAQ(T/2-\sigma)}Pe^{A\sigma}Q \tag{5.1}$$

DH is also the half period map of the dynamical system given by the circuit of Figure 5.1, where the switch is closed at time ϕ_0 and opened at time $\phi_0 + \sigma$. Note that the circuit of Figure 5.1 is the thyristor controlled reactor circuit with the source removed and the thyristor replaced by an ideal switch. The states $x_\Delta(t) = (I_{r\Delta}(t), V_{c\Delta}(t), I_{s\Delta}(t))^t$ (switch on) and $y_\Delta(t) = (V_{c\Delta}(t), I_{s\Delta}(t))^t$ (switch off) of the circuit evolve according to the following equations:

$$x_\Delta(\phi_0) = Qy_\Delta(\phi_0) \quad \text{(switch closing)} \tag{5.2.1}$$

$$\dot{x}_\Delta = Ax_\Delta \quad \text{(switch ON)} \tag{5.2.2}$$

$$y_\Delta(\phi_0 + \sigma) = Px_\Delta(\phi_0 + \sigma) \quad \text{(switch opening)} \tag{5.2.3}$$

$$\dot{y}_\Delta = PAQy_\Delta \quad \text{(switch OFF)} \tag{5.2.4}$$

and it is clear from these equations that their half cycle map is DH. In general the switch current is nonzero when the switch opens at $\phi_0 + \sigma$ and the switch opening is assumed to immediately zero the inductor current. This rather nonphysical event is described in 5.2.3 by the projection of the current $x_\Delta(\phi_0 + \sigma)$ onto the plane Z by P.

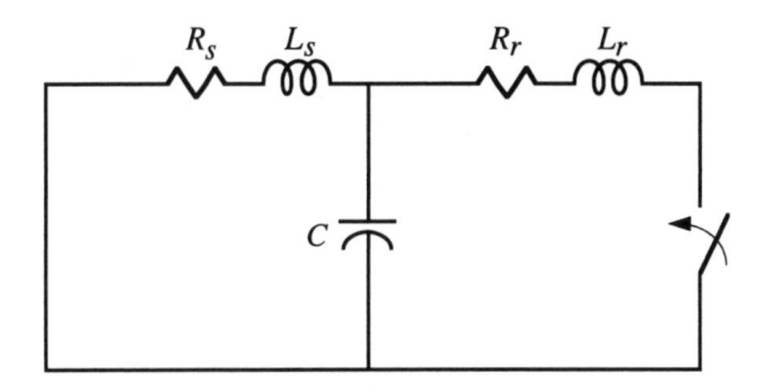

FIG. 5.1. *Circuit model for the Jacobian.*

We consider the case of zero resistances so that equations 5.2.2 and 5.2.4 are simply lossless oscillators. At the beginning of the cycle the switch turns on and the on oscillation proceeds for time σ. Then the state is projected onto Z and the off oscillation proceeds for time $T/2 - \sigma$. Since we have assumed zero resistance, the oscillators are lossless and the damping in DH is entirely accounted for by the projection onto Z.

It is straightforward to use an energy or Lyapunov method to show that DH is never unstable and that its eigenvalues always lie inside or on the unit circle. Consider the incremental energy

$$E_\Delta(t) = \frac{1}{2}L_r I_{r\Delta}^2 + \frac{1}{2}L_s I_{s\Delta}^2 + \frac{1}{2}CV_{c\Delta}^2 \tag{5.3}$$

Incremental energy E_Δ is preserved at the switch closing 5.2.1 because the reactor current $I_{r\Delta}$ is zero when the switch closes (The first component of $Qy_\Delta(\phi_0)$ is always zero). At the switch opening 5.2.4, the incremental energy E_Δ decreases by the nonnegative quantity $\frac{1}{2}L_r\,(I_{r\Delta}(\phi_0 + \sigma))^2$ because the effect of the projection P is to zero the reactor current $I_{r\Delta}$.

In the case of zero circuit resistances, equations 5.2.2 and 5.2.4 are simply lossless oscillators. Then E_Δ is preserved at switch on, is constant while the lossless oscillators act and decreases or is preserved at switch off. Since E_Δ is a Lyapunov function for the discrete time system $y_\Delta^{k+1} = DHy_\Delta^k$, $k = 0, 1, 2, 3...$, DH must be stable. If the circuit resistances are included, then E_Δ is strictly decreasing when the oscillators act and E_Δ is a strict Lyapunov function and DH is asymptotically stable.

6. Resonance. It is known that the thyristor controlled reactor circuit can exhibit a resonance for certain parameter values [1,6,7]. In particular, resonance is expected when the resonant frequency ω_{off} of the circuit when both thyristors are off and the resonant frequency ω_{on} when a thyristor is on span a harmonic of the 60 Hz fundamental frequency. Our parameter values have $\omega_{\text{off}} = 4.9 \times 60$ Hz and $\omega_{\text{on}} = 5.18 \times 60$ Hz so that the fifth harmonic is spanned. Figure 6.1 shows the amplitude of the capacitor voltage V_c of halfwave symmetric periodic orbits as σ is varied; the resonance occurs at $\sigma = 50.1^o$. This section explains and predicts the resonance by associating it with an eigenvalue of DH being −1. Figure 6.1 has a region of more complex dynamics bounded by edges at which stability of the half wave symmetric solutions is lost in switching time bifurcations. These switching time bifurcations are explained in [7,8].

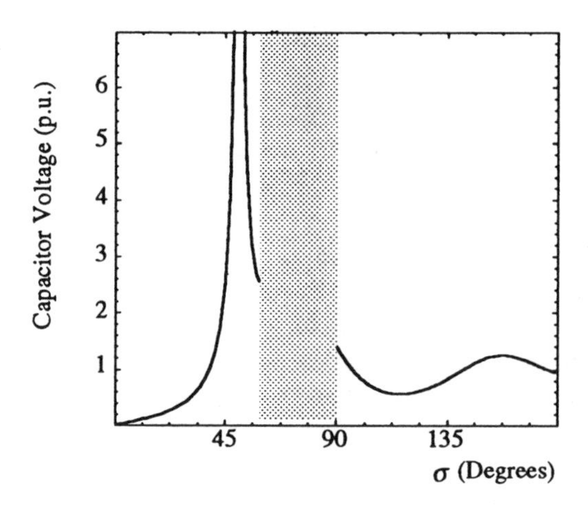

FIG. 6.1. *Capacitor voltage V_c versus firing angle σ.*

When a periodic orbit is halfwave symmetric, the system state at the half cycle is equal in magnitude and opposite in sign to the system state at the beginning of the cycle. If we write $y_0 = y(\phi_0)$ for the system state at the beginning of the cycle, then this may be written in terms of the halfwave map H as

$$H(y_0) = -y_0 \tag{6.1}$$

The conduction time σ of the conducting thyristor is given by the constraint equation:

$$0 = (I - QP)x(\phi_0 + \sigma) \tag{6.2}$$

y_0, ϕ_0 and σ corresponding to a half wave symmetric periodic orbit can be computed by solving 6.1 and 6.2 simultaneously. (Conceptually one fixes ϕ_0 and solves 6.1 and 6.2 for y_0 and σ.) Indeed, *all* halfwave symmetric solutions of the circuit can be computed by solving 6.1 and 6.2 for ϕ_0 and y_0 as σ varies from 0° to 180°. However, some of these computed solutions can be nonphysical because although the computed switch off time $\phi_0 + \sigma$ satisfies 6.2, it is not necessarily the switch off time because it need not be the *first* root of the thyristor current after ϕ_0.

Now suppose that values of ϕ_0 and σ consistent with (6.1) and (6.2) are available. We can then compute y_0 by solving (6.1) and examine the size of y_0 to try to detect when a resonance occurs. Equation (6.1) has the form

$$-y_0 = DHy_0 + g(\sigma, \phi) \tag{6.3}$$

when written in full by integrating (2.1) and (2.2) and taking the coordinate changes at the switchings into account [7,8,3]. $g(\sigma, \phi)$ is a bounded function with terms involving the integration of the input $u(t)$ over the period. Rewriting (6.3) as

$$y_0 = -(I + DH)^{-1} g(\sigma, \phi) \tag{6.4}$$

shows that y_0 becomes unbounded as an eigenvalue of DH approaches -1. (We assume that $g(\sigma, \phi)$ has a nonzero component along the right eigenvector corresponding to the eigenvalue -1.) When circuit resistance is neglected, an eigenvalue of DH can equal -1 (at $\sigma = 50.1^o$ for our parameter values) and this corresponds to an eigenvalue approaching close to -1 when the circuit resistance is included. Thus the eigenvalues of the Jacobian DH of the circuit with no resistance can be used to predict the resonance.

Note that if an eigenvalue of DH becomes -1 then an eigenvalue of the Jacobian $DF = DH^2$ becomes 1 and this may be seen on Figure 4(b). However it is not correct to argue that the eigenvalue 1 of DF implies resonance for halfwave symmetric orbits by examining the full cycle equivalent of (6.4) because the eigenvalue 1 could also arise from DH having an eigenvalue 1, in which case (6.4) is solvable for a bounded y_0. However, the next section shows that DH having eigenvalue 1 implies that the orbit corresponding to this solution is resonant if a small ambient second harmonic is added to the voltage source. The reference [14] is useful background here.

7. Resonances due to ambients. This section predicts resonance when a small ambient second harmonic or third subharmonic is added to the source voltage and eigenvalues of the halfwave Jacobian DH approach $+1$ or eigenvalues of the Jacobian DF approach certain complex roots of unity.

Suppose that the circuit has steady state periodic orbit of period T when the input is $u(t)$, the switch off times are s^1_{off} and s^2_{off} and the state at the start of the period is y_0. We assume that the two switch off times satisfy the transversality conditions $\frac{\partial f}{\partial t}(s^k_{\text{off}}-) < 0$, $k = 1, 2$ where f is the thyristor current. These transversality conditions imply that the periodic orbit is not at a switching time bifurcation and that the switching times and H and F are smooth functions of sufficiently small perturbations to the initial state or the input [3].

To analyze the resonant response to the second harmonic ambient, assume that DH does not have an eigenvalue exactly at ± 1. Let G be the map advancing the circuit state by a half period $T/2$ as a function of the initial state and the input. G has the form

$$G(y_0, u(t)) = DH y_0 + h(u(t), y_0) \tag{7.1}$$

The expression for h depends on the switch off time s_{off} which is a function of y_0 and u. However, the Jacobian simplification [7,8,3] implies that the gradient of the expression for h with respect to s_{off} vanishes. Hence if $z = O(\epsilon)$ and $v(t) = O(\epsilon)$, then

$$h(u+v, y_0+z) = \overline{h}(u+v) + O(\epsilon^2) = \overline{h}(u) + \overline{h}(v) + O(\epsilon^2)$$

where $\overline{h}$ is the expression for h evaluated with s_{off} determined by y_0 and u. Note that $\overline{h}(u)$ is linear in u and that the Jacobian simplification implies that $h_{y_0} = 0$ and $G_{y_0} = DH$.

Defining $\overline{G}(y_0, u) = DH y_0 + \overline{h}(u)$, we can deduce the following property of G which is needed in the sequel: If $z = O(\epsilon)$ and $v(t) = O(\epsilon)$, then

$$\begin{aligned} G(y_0+z, u_0+v) &= DH(y_0+z) + h(u_0+v, y_0+z) \\ &= DH y_0 + DH z + \overline{h}(u_0) + \overline{h}(v) + O(\epsilon^2) \\ &= \overline{G}(y_0, u_0) + \overline{G}(z, v) + O(\epsilon^2) \end{aligned} \tag{7.2}$$

Since y_0 and u correspond to a periodic orbit we have

$$G(y_0, u(t)) = -y_0 \tag{7.3}$$

The circuit is initially in the periodic steady state specified by y_0 and u. We suppose that a small ambient second harmonic is added to the input so that the input becomes $u(t) + \epsilon \cos 2\omega t$. If ϵ is sufficiently small, we will obtain a periodic orbit near y_0. To confirm this, define

$$K(y, \epsilon) = G(G(y, u(t) + \epsilon \cos 2\omega t), u(t+T/2) + \epsilon \cos 2\omega(t+T/2)) - y$$

Then $K(y_0, 0) = 0$ and, since DH is assumed not to have an eigenvalue exactly at ± 1, $K_y = (G_y)^2 - I = (DH)^2 - I$ is invertible. The implicit function theorem implies that for sufficiently small ϵ, there is a smooth function $\eta(\epsilon)$ such that $\eta(0) = y_0$ and $K(\eta(\epsilon), \epsilon) = 0$. We write $z = \eta(\epsilon) - y_0$ for the deviation in y caused by the ambient and compute z. That is, we want to solve $K(y_0 + z, \epsilon) = 0$ for the displacement z to determine the eigenvalue conditions under which z is large and there is a large response. We have

$$0 = K(y_0 + z, \epsilon) = \begin{array}{c} G(G(y_0 + z, u(t) + \epsilon \cos 2\omega t), \\ u(t + T/2) + \epsilon \cos 2\omega(t + T/2)) - y_0 - z \end{array}$$

Since $z = \eta(\epsilon) - y_0 = O(\epsilon)$, use property (7.2) twice to obtain

$$0 = \overline{G}(\overline{G}(z, \epsilon \cos 2\omega t), \epsilon \cos 2\omega(t + T/2)) - z + O(\epsilon^2)$$

Use the definition of $\overline{G}$ and $\cos 2\omega(t + T/2) = \cos 2\omega t$ to obtain

$$0 = (DH^2 - I)z + (I + DH)\overline{h}(\epsilon \cos 2\omega t) + O(\epsilon^2)$$

Since DH does not have eigenvalue -1, $(I + DH)$ is invertible and hence $K(y_0 + z, \epsilon) = 0$ becomes

$$z = \epsilon(I - DH)^{-1}\overline{h}(\cos 2\omega t) + (I - DH)^{-1}O(\epsilon^2) \tag{7.4}$$

Then for sufficiently small ambient (ϵ small enough), we expect a resonance when an eigenvalue of DH approaches $+1$. (We assume that $\overline{h}(\cos 2\omega t)$ has a nonzero component along the right eigenvector corresponding to the eigenvalue $+1$.)

To analyze the response to a third subharmonic ambient, assume that DF does not have an eigenvalue exactly at 1 or $e^{\pm j 2\pi/3}$. It is convenient to reuse the notation above by redefining it. Redefine G as the map advancing the circuit state by one period T as a function of the initial state and the input. G now has the form

$$G(y_0, u) = DF y_0 + h(u(t), y_0) \tag{7.5}$$

Similarly to (7.2) we can deduce that if $z = O(\epsilon)$ and $v(t) = O(\epsilon)$, then

$$G(y_0 + z, u + v) = \overline{G}(y_0, u) + \overline{G}(z, v) + O(\epsilon^2) \tag{7.6}$$

The circuit is initially in the periodic steady state specified by y_0 and u and $G(y_0, u(t)) = y_0$. A small ambient third subharmonic is added to the input so that the input becomes $u(t) + \epsilon \cos \omega_s t$ where $\omega_s = \omega/3$. If ϵ is sufficiently small, we will obtain a period 3 orbit near y_0. To confirm this, redefine

$$\begin{aligned} K(y, \epsilon) = {} & G(G(G(y, u(t) + \epsilon \cos \omega_s t), \\ & u(t) + \epsilon \cos \omega_s(t + T)), u(t) + \epsilon \cos \omega_s(t + 2T)) - y \end{aligned}$$

Then $K(y_0, 0) = 0$ and, since DF is assumed not to have an eigenvalue exactly at 1 or $e^{\pm j2\pi/3}$, $K_y = (G_y)^3 - I = DF^3 - I$ is invertible. The implicit function theorem implies that for sufficiently small ϵ, there is a smooth function $\eta(\epsilon)$ such that $\eta(0) = y_0$ and $K(\eta(\epsilon), \epsilon) = 0$. We write $z = \eta(\epsilon) - y_0$ and want to compute z from $K(y_0 + z, \epsilon) = 0$:

$$\begin{aligned}0 = K(y_0 + z, \epsilon) = G(G(G(&y_0 + z, u(t) + \epsilon \cos \omega_s t),\\ &u(t) + \epsilon \cos \omega_s (t + T)),\\ u(t) &+ \epsilon \cos \omega_s (t + 2T)) - y_0 - z\end{aligned}$$

Since $z = \eta(\epsilon) - y_0 = O(\epsilon)$, use property (7.6) three times to obtain

$$0 = \overline{G}(\overline{G}(\overline{G}(z, \epsilon \cos \omega_s t), \epsilon \cos \omega_s (t + T)), \epsilon \cos \omega_s (t + 2T)) - z + O(\epsilon^2)$$

Use the definition of $\overline{G}$ to obtain

$$\begin{aligned}0 = &(D F^3 - I)z + DF^2\overline{h}(\epsilon \cos \omega_s t)\\ &+DF\overline{h}(\epsilon \cos \omega_s (t + T)) + \overline{h}(\epsilon \cos \omega_s (t + 2T)) + O(\epsilon^2)\end{aligned}$$

Use the linearity of $\overline{h}$ to rewrite as

$$\begin{aligned}0 &= (DF^3 - I)z + \epsilon\mathbf{Re}\Big\{\big[DF^2 + DFe^{j2\pi/3} + Ie^{j4\pi/3}\big]\overline{h}(e^{j\omega_s t})\Big\}\\ &\quad +O(\epsilon^2)\\ &= (DF^3 - I)z + \epsilon(I - DF^3)\mathbf{Re}\Big\{\big[e^{j2\pi/3}I - DF\big]^{-1}\overline{h}(e^{j\omega_s t})\Big\}\\ &\quad +O(\epsilon^2)\end{aligned}$$

That is, $K(y_0 + z, \epsilon) = 0$ becomes

$$z = \epsilon\mathbf{Re}\Big\{\big[e^{j2\pi/3}I - DF\big]^{-1}\overline{h}(e^{j\omega t})\Big\} + \big[I - DF^3\big]^{-1}O(\epsilon^2) \tag{7.7}$$

Then for sufficiently small ambient (ϵ small enough), we expect a resonance when a complex pair of eigenvalues of DF approaches $e^{\pm j2\pi/3}$. (We assume that $\overline{h}(e^{j\omega t})$ has a nonzero component along the eigenspace corresponding to the eigenvalues $e^{\pm j2\pi/3}$.)

8. Geometry of the model Jacobian. This section describes the geometry of orbits of the model Jacobian to give further insight. For simplicity, we consider only the case of zero circuit resistances. In this case, section 5 described how orbits of the model Jacobian preserve incremental energy E_Δ except at switch off. While energy is preserved, the orbits remain on an ellipsoid of constant incremental energy determined by equation 5.3. It is convenient to scale coordinates so that the ellipsoid becomes a sphere and the square of the Euclidean norm becomes the incremental energy. That is, we choose coordinates

$$a = \frac{1}{\sqrt{2}}\begin{pmatrix}\sqrt{L_r}I_{r\Delta}\\ \sqrt{C}V_{c\Delta}\\ \sqrt{L_s}I_{s\Delta}\end{pmatrix} \qquad b = \frac{1}{\sqrt{2}}\begin{pmatrix}\sqrt{C}V_{c\Delta}\\ \sqrt{L_s}I_{s\Delta}\end{pmatrix}$$

so that $E_\Delta(t) = |a(t)|^2$ when the switch is on and $E_\Delta(t) = |b(t)|^2$ when the switch is off. In these coordinates, A and PAQ become the skew symmetric matrices

$$A = \begin{pmatrix} 0 & (L_rC)^{-\frac{1}{2}} & 0 \\ -(L_rC)^{-\frac{1}{2}} & 0 & (L_sC)^{-\frac{1}{2}} \\ 0 & -(L_sC)^{-\frac{1}{2}} & 0 \end{pmatrix}$$

$$PAQ = \begin{pmatrix} 0 & (L_sC)^{-\frac{1}{2}} \\ -(L_sC)^{-\frac{1}{2}} & 0 \end{pmatrix}$$

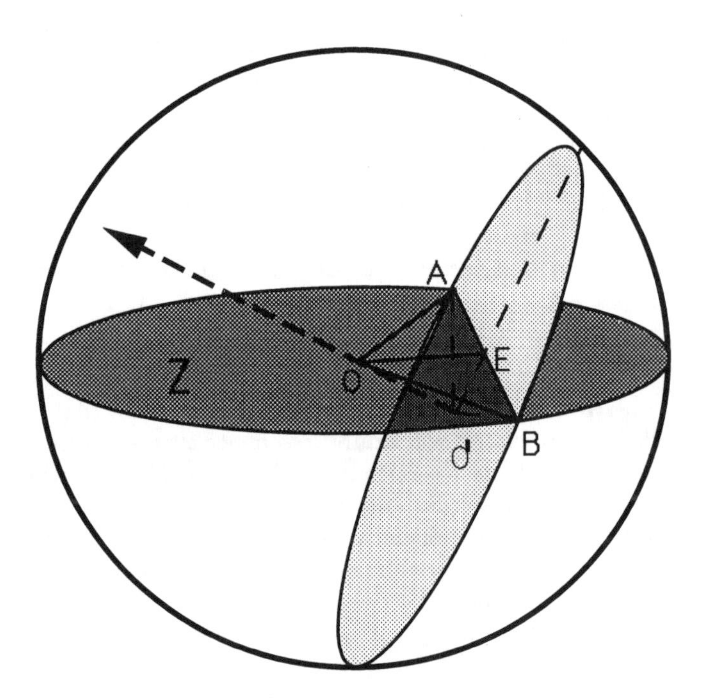

FIG. 8.1. *Dynamics of model Jacobian.*

The dynamics of the model Jacobian are shown in Figure 8.1. The off system dynamics are confined to the plane Z of zero switch current and are a circular oscillation of frequency $\omega_{\text{off}} = \sqrt{1/L_sC}$. The on system dynamics contains a circular oscillation of frequency $\omega_{\text{on}} = \sqrt{1/(L_sC) + 1/(L_rC)}$ in which the capacitor oscillates with the parallel combination of the two inductors and a constant mode with eigenvalue zero in which the loop current through the two inductors remains constant. The oscillatory mode occurs on one of an infinite number of parallel tilted planes. The vector

$$z = \begin{pmatrix} (L_s C)^{-\frac{1}{2}} \\ 0 \\ (L_r C)^{-\frac{1}{2}} \end{pmatrix} \tag{8.1}$$

is normal to all the tilted planes and is also the eigenvector of A corresponding to the zero eigenvalue (recall that eigenvectors of skew symmetric matrices corresponding to distinct eigenvalues are orthogonal). z lies along $O'O$ in Figure 8.1. Let $\theta = \angle OEO'$ in Figure 8.1 be the angle between the tilted planes and Z. Then θ is also the angle between z and Z and

$$\cos\theta = \frac{z.(1,0,0)}{|z|} = \frac{\omega_{\text{off}}}{\omega_{\text{on}}} \tag{8.2}$$

At the beginning of the cycle at ϕ_0 (end of the off switch mode), the state is on Z. The switch on frees the state to move in the vertical dimension ($I_{r\Delta}$ dimension). The on oscillation is a circular oscillation of frequency ω_{on} in one of the tilted planes which lasts for time σ. The tilted plane is selected by the position of the state when the switch is closed. The radius of the circular oscillation in the on system varies according to which tilted plane is selected. At switch off, the state is projected onto Z and the off oscillation proceeds for time $T/2 - \sigma$. The orbits remain on a sphere of constant incremental energy except that the projection at switch off to Z usually reduces the incremental energy. For our circuit parameters and $\sigma = 0^o$, an orbit makes 2.45 oscillations in the OFF system and none in the ON system whereas for $\sigma = 180^o$, an orbit makes 2.59 oscillations in the ON system and none in the OFF system.

9. Conditions for resonance. Sections 7 and 8 show that the thyristor controlled reactor circuit will exhibit resonance effects if the eigenvalues of the half map DH lie near certain roots of unity. (The condition for the resonant response to a third subharmonic ambient is given in terms of $DF = DH^2$.) When component values are specified, it is practical to check for resonance by numerically computing the eigenvalues of DH as σ varies from 0 to $T/2$ and observing whether the eigenvalues approach certain roots of unity. However, it is much better to derive the analytic results below describing whether and where DH has eigenvalues on the unit circle. Surprisingly, some of the analytic results simplify even further: Consideration of the average inductor model and case studies led in [1] to a rule of thumb for the thyristor controlled reactor circuit that if ω_{off} and ω_{on} span an odd multiple of ω, resonance will occur for some value of σ. This simple and useful rule is indeed correct and a precise version of the rule is proved from the analytic criteria. The resonance response to second harmonic ambient is predicted by a similar rule.

Theorem 9.1 describes the conditions for the eigenvalues of DH to lie on parts of the unit circle and is proved in the Appendix. The degenerate case of the ON system having eigenvectors in the plane Z is excluded.

THEOREM 9.1. *Consider the thyristor controlled reactor circuit with no resistance and assume that A has no eigenvectors in Z. Then*

(A) DH has an eigenvalue of -1 iff

(i) $\omega_{\text{on}}\sigma \neq 0 \pmod{2\pi}$

and
$$\frac{\omega_{\text{on}}}{\omega_{\text{off}}} = \tan\left(\frac{\omega_{\text{off}}(T/2-\sigma)}{2}\right)\tan\left(\frac{\omega_{\text{on}}\sigma}{2}\right)$$

or

(ii) $\omega_{\text{on}}\sigma = 0 \pmod{2\pi}$ *and* $\omega_{\text{off}}(T/2-\sigma) = \pi \pmod{2\pi}$.

(B) DH has an eigenvalue $+1$ iff

(i) $\omega_{\text{on}}\sigma \neq 0 \pmod{2\pi}$

and
$$-\frac{\omega_{\text{on}}}{\omega_{\text{off}}} = \cot\left(\frac{\omega_{\text{off}}(T/2-\sigma)}{2}\right)\tan\left(\frac{\omega_{\text{on}}\sigma}{2}\right)$$

or

(ii) $\omega_{\text{on}}\sigma = 0 \pmod{2\pi}$ *and* $\omega_{\text{off}}(T/2-\sigma) = 0 \pmod{2\pi}$.

(C) DH has eigenvalues at $e^{\pm j\pi/m}, m \geq 1$ iff

$\omega_{\text{on}}\sigma = 0 \pmod{2\pi}$

and $\omega_{\text{off}}(T/2-\sigma) = \pi/m \pmod{2\pi}$.

We examine which of these conditions are generically satisfied. For given circuit parameters, ω_{on} and ω_{off} are constants and σ is the only variable. Conditions $A(i)$ and $B(i)$ require one equation to be satisfied whereas conditions $A(ii)$, $B(ii)$ and C require two independent equations to be satisfied. Thus we expect that conditions $A(i)$ and $B(i)$ can be generically satisfied for some σ whereas conditions $A(ii)$, $B(ii)$ and C would only be satisfied for some σ for exceptional values of circuit parameters.

Theorem 9.2 further simplifies the generically occuring conditions A(i) and B(i) by reducing them to rules of thumb for resonance. A slightly more detailed result than Theorem 9.2 specifying the number of resonance conditions is stated and proved in the Appendix.

THEOREM 9.2. *Consider the thyristor controlled reactor circuit with no resistance and assume that A has no eigenvectors in Z.*
If $\omega_{\text{on}}\sigma \neq 0 \pmod{2\pi}$, then

(A) DH has an eigenvalue at -1 iff ω_{off} and ω_{on} span an odd multiple of ω.

(B) DH has an eigenvalue at $+1$ iff ω_{off} and ω_{on} span an even multiple of ω.

10. Conclusions. This paper studies damping and resonance phenomena in a basic thyristor controlled reactor circuit for static VAR control by exploiting a simple exact formula for the Jacobian of the Poincaré map. The nonresistive damping of the circuit is described and resonance effects are associated with eigenvalues of Jacobians lying near ± 1 or complex roots of unity. If circuit resistance is neglected, analytic formulas are derived which describe the conditions for eigenvalues to be at ± 1 or complex roots of unity. Hence a simple test for detecting resonance involving circuit frequencies spanning an integer harmonic is proven to be correct. This exact analysis of resonance is unusual in a nonlinear system. We expect that much of the analysis in this paper could be generalized to more complex circuits containing thyristor controlled reactors such as advanced series compensators used in flexible AC transmission. The ability to predict and understand damping and resonance in these circuits should be useful in the design of these devices.

11. Appendix: proofs of resonance conditions. Consider the thyristor controlled reactor circuit with all resistances zero. The Jacobian of the half cycle map is $DH = e^{PAQ(T/2-\sigma)}Pe^{A\sigma}Q$. Assume that A has no eigenvectors in Z and write $\beta = \omega_{\text{on}}\sigma$ for the angle traversed in the ON system. We derive conditions under which the half cycle map DH has eigenvalues on the unit circle.

Lemmas 11.1, 11.2 and 11.3 prove Theorem 9.1 by describing all the ways for DH to have eigenvalues on the unit circle.

LEMMA 11.1.

(a) *If* $\beta \neq 0 \pmod{2\pi}$, *then the ON system maps an initial vector* $a(0) = (0, r\sin(\alpha/2), r\cos(\alpha/2))$ *in* Z *to a vector* $a(\sigma)$ *in* Z *iff*

$$\tan(\beta/2)\cos\theta = \tan(\alpha/2) \tag{11.1}$$

Moreover, $a(\sigma) = (0, -r\sin(\alpha/2), r\cos(\alpha/2))$ *and* α *is the angle subtended at the origin by* $a(0)$ *and* $a(\sigma)$ *in the plane* Z.

(b) $\beta = 0 \pmod{2\pi}$ *iff the ON system maps all vectors in* Z *into* Z.

Proof. ((a)$\Rightarrow$) Suppose the ON system maps $a(0) = (0, r\sin(\alpha/2), r\cos(\alpha/2))$ to $a(\sigma) = (0, a_2, a_3)$ in Z. Recall that $z = ((L_sC)^{-\frac{1}{2}}, 0, (L_rC)^{-\frac{1}{2}})$ is the eigenvector of A with zero eigenvalue. Since the component of $a(t)$ in the direction of z is constant, $a(0).z = a(\sigma).z$ and hence $a_3 = r\cos(\alpha/2)$. The ON system preserves the Euclidean norm so that $r^2 = |a(0)|^2 = |a(\sigma)|^2 = a_2^2 + a_3^2 = a_2^2 + r^2\cos^2(\alpha/2)$ and $a_2 = \pm r\sin(\alpha/2)$. If $a_2 = r\sin(\alpha/2)$, then $a(0) = a(\sigma)$, and hence $\beta = 0 \pmod{2\pi}$ which contradicts the assumption. Therefore $a_2 = -r\sin(\alpha/2)$ and $a(\sigma) = (0, -r\sin(\alpha/2), r\cos(\alpha/2))$. In Figure 8.1, $a(0)$ is point A and $a(\sigma)$ is point B and $\alpha = \angle AOB$ is the angle subtended at the origin by $a(0)$ and $a(\sigma)$ in the plane Z. Moreover, $\theta = \angle OEO'$ and $\beta = \angle AO'B$. The angles $\angle O'EA$, $\angle EO'O$, $\angle OEA$ are right angles and $\beta/2 = \angle AO'E$ and $\alpha/2 = \angle AOE$. Now use the right angled triangles $O'EA$, $EO'O$ and OEA to obtain

$$\tan(\beta/2) = \frac{EA}{O'E} = \frac{EA}{OE\cos\theta} = \frac{\tan(\alpha/2)}{\cos\theta}$$

((a)$\Leftarrow$) It can be verified that the ON system maps $a(0) = (0,$

$r\sin(\alpha/2), r\cos(\alpha/2))$ to

$$a(\sigma) = \begin{pmatrix} r\,\mathrm{cosec}\theta\sin(\alpha/2)\sin\beta + r\cot\theta\cos(\alpha/2)(\cos\beta - 1) \\ r\sin(\alpha/2)\cos\beta - r\cos\theta\cos(\alpha/2)\sin\beta \\ r\sec\theta\sin(\alpha/2)\sin\beta + r\cos(\alpha/2)\cos\beta \end{pmatrix}$$

Using 11.1 and simplifying, $a(\sigma) = (0, -r\sin(\alpha/2), r\cos(\alpha/2))$ is in Z.

((b)$\Rightarrow$) If $\beta = 0 \pmod{2\pi}$, then $e^{A\sigma}$ is the identity.

((b)$\Leftarrow$) Suppose that the ON system maps each vector in Z into Z and $\beta \neq 0 \pmod{2\pi}$. Then by choosing any α that does not satisfy 11.1, it is clear from Lemma 11.1.(a) that $a(0) = (0, \sin(\alpha/2), \cos(\alpha/2))$ will not be mapped to Z by the ON system and a contradiction has been obtained. □

LEMMA 11.2. *Assume that $\beta \neq 0 \pmod{2\pi}$. Then*

(a) DH has an eigenvalue at -1 iff

$$\frac{\omega_{\text{on}}}{\omega_{\text{off}}} = \tan\left(\frac{\omega_{\text{off}}(T/2 - \sigma)}{2}\right)\tan\left(\frac{\omega_{\text{on}}\sigma}{2}\right), \tag{11.2}$$

(b) DH has an eigenvalue at $+1$ iff

$$-\frac{\omega_{\text{on}}}{\omega_{\text{off}}} = \cot\left(\frac{\omega_{\text{off}}(T/2 - \sigma)}{2}\right)\tan\left(\frac{\omega_{\text{on}}\sigma}{2}\right). \tag{11.3}$$

Proof. We will prove the assertion for (a); the proof for (b) is similar.

((a)$\Rightarrow$) Let $b(0) = (\sin(\alpha/2), \cos(\alpha/2))$ be the eigenvector of DH corresponding to the eigenvalue -1. $|b(T/2)| = |-b(0)| = |b(0)|$ and the preservation of the norm by the OFF system imply that $|Pa(\sigma)| = |b(\sigma)| = |b(0)|$. Also $|a(0)| = |Qb(0)| = |b(0)|$ and the preservation of the norm by the ON system imply that $|a(\sigma)| = |b(0)|$. Hence $|a(\sigma)| = |Pa(\sigma)|$ and $a(\sigma) \in Z$. $a(0) = Qb(0) = (0, \sin(\alpha/2), \cos(\alpha/2))$ and Lemma 11.1.(a) imply that $a(\sigma) = (0, -\sin(\alpha/2), \cos(\alpha/2))$. Thus $b(\sigma) = Pa(\sigma) = (-\sin(\alpha/2), \cos(\alpha/2))$. But $b(T/2) = -b(0) = (-\sin(\alpha/2), -\cos(\alpha/2))$ so that the angle swept by the OFF system is $\pi - \alpha \pmod{2\pi}$. Therefore $\omega_{\text{off}}(T/2 - \sigma) = \pi - \alpha \pmod{2\pi}$ and $\cot(\alpha/2) = \tan(\omega_{\text{off}}(T/2 - \sigma)/2)$. The result follows from 11.1 of Lemma 11.1.(a) and $\cos\theta = \omega_{\text{off}}/\omega_{\text{on}}$ from 8.2.

((a)$\Leftarrow$) Suppose 11.2 is true. Since $\beta \neq 0 \pmod{2\pi}$, we can choose $\alpha \neq 0 \pmod{2\pi}$ such that 11.1 is satisfied. Let $b(0) = (\sin(\alpha/2), \cos(\alpha/2))$. Then $a(0) = Qb(0) = (0, \sin(\alpha/2), \cos(\alpha/2))$ and Lemma 11.1.(a) implies that $a(\sigma) = (0, -\sin(\alpha/2), \cos(\alpha/2))$. Now $b(\sigma) = Pa(\sigma) = (-\sin(\alpha/2), \cos(\alpha/2))$. 11.2 and 11.1 imply that $\tan(\alpha/2)\tan(\omega_{\text{off}}(T/2 - \sigma)/2) = 1$ and hence $\omega_{\text{off}}(T/2 - \sigma) = \pi - \alpha \pmod{2\pi}$. That is, the angle swept by the OFF system is $\pi - \alpha \pmod{2\pi}$. It follows that $b(T/2) = (-\sin(\alpha/2), -\cos(\alpha/2)) = -b(0)$ so that $b(0)$ is an eigenvector of DH with eigenvalue -1. □

LEMMA 11.3. *$\beta = 0 \pmod{2\pi}$ iff both eigenvalues of the half cycle map DH are on the unit circle. In particular, the eigenvalues are located at $e^{\pm j\omega_{\text{off}}(T/2-\sigma)}$.*

Proof. ($\Rightarrow$) If $\beta = 0 \pmod{2\pi}$, then $e^{A\sigma}$ is the identity and $DH = e^{PAQ(T/2-\sigma)}$ has eigenvalues at $e^{\pm j\omega_{\text{off}}(T/2-\sigma)}$ on the unit circle.

($\Leftarrow$) If DH has both eigenvalues on the unit circle, then DH preserves the norm and $|b(0)| = |b(T/2)|$ for all $b(0)$. Suppose that the ON system maps $a(0) \in Z$ to $a(\sigma) \notin Z$. Then $|Pa(\sigma)| < |a(\sigma)|$ and $|b(T/2)| = |b(\sigma)| = |Pa(\sigma)| < |a(s)| = |a(0)| = |Qb(0)| = |b(0)|$ is a contradiction. Therefore the ON system maps all $a(0) \in Z$ into Z and the result follows from Lemma 11.1.(b). □

Theorem 9.2 part (A) is now proved by demonstrating that if ω_{off} and ω_{on} span k odd multiples of ω, then (A.2) has k roots as σ varies from 0 to $T/2$ and that if ω_{off} and ω_{on} span k even multiples of ω, then (A.3) has k roots as σ varies from 0 to $T/2$. The proof of part (B) of Theorem 9.2 demonstrates that 11.3 has as many roots as the number of even multiples of ω spanned by ω_{off} and ω_{on} and is omitted since it is very similar to the proof of part (A).

LEMMA 11.4. *Let g and h be real differentiable functions on the interval $[0,1]$. Suppose that g and h have the same sign at critical points of g and at the end points $0, 1$. Precisely, either $g(\delta)h(\delta) > 0$ or $g(\delta) = h(\delta) = 0$, when $g'(\delta) = 0$ or $\delta = 0, 1$. If all the critical points of g are isolated, the number of roots of h in $[0,1] \geq$ number of roots of g in $[0,1]$.*

Proof. Since the critical points of g are isolated, each root of g is at a critical point of g or at 0 or at 1 or occurs between successive critical points of g or between 0 and the smallest critical point or between the largest critical point and 1 or, if there are no critical points, between 0 and 1. Write δ_1 for a critical point of g (or 0) and δ_2 for the successive critical point of g (or 1).

We assert that if there is a root of g strictly between δ_1 and δ_2, then $g(\delta_1)g(\delta_2) < 0$. For suppose that $g(\delta_1)g(\delta_2) \geq 0$ and $g(\delta_r) = 0$ with $\delta_1 < \delta_r < \delta_2$. Consider the case of $g(\delta_1) \geq 0$ and $g(\delta_2) \geq 0$; the case of $g(\delta_1) \leq 0$ and $g(\delta_2) \leq 0$ is similar. Choose $\delta_m \in [\delta_1, \delta_2]$ such that $g(\delta_m) = \min\{g(\delta) \mid \delta \in [\delta_1, \delta_2]\}$. Then $g(\delta_m) \leq g(\delta_r) = 0$. Moreover, $\delta_m \neq \delta_1$ because $\delta_m = \delta_1$ implies $0 = g(\delta_r) \geq g(\delta_m) = g(\delta_1) \geq 0$ and $g(\delta_1) = g(\delta_m) = 0$. Then Rolle's theorem implies that there is a critical point of g in (δ_1, δ_m) which contradicts $\delta_2 \geq \delta_m$ being the next critical point after δ_1. Similarly, $\delta_m \neq \delta_2$. But then $\delta_m \in (\delta_1, \delta_2)$ and $g'(\delta_m) = 0$ contradicts δ_2 being the next critical point after δ_1.

Therefore if there is a root of g strictly between δ_1 and δ_2, then $g(\delta_1)g(\delta_2) < 0$. But by hypothesis, $g(\delta_1)g(\delta_2)h(\delta_1)h(\delta_2) > 0$, and hence $h(\delta_1)h(\delta_2) < 0$ and h must also have a root in (δ_1, δ_2). Further, if $g(\delta_1) = 0$, then by hypothesis, $h(\delta_1) = 0$ and if $g(\delta_2) = 0$ then $h(\delta_2) = 0$. Thus for each root of g there is a distinct root of h and the number of roots of h in $[0,1] \geq$ number of roots of g in $[0,1]$. □

LEMMA 11.5. *If ω_{off} and ω_{on} span k odd multiples of ω, then 11.2 has k roots as σ varies from 0 to $T/2$.*

Proof. First multiply both sides of 11.2 by $\pi/2$, use the following changes in notation: $\omega_{\rm off}\pi/2 = b_1\omega$, $\omega_{\rm on}\pi/2 = b_2\omega$, $\sigma = \delta T/2$, and rearrange to obtain $h(\delta) = 0$ where

$$h(\delta) = (b_1 + b_2)\cos((b_2 - b_1)\delta + b_1) + (b_2 - b_1) \cos((b_1 + b_2)\delta - b_1)\,.$$

Also define

$$g(\delta) = (b_1 + b_2)\cos((b_2 - b_1)\delta + b_1)$$

Note that $b_1 > 0$, $b_2 > 0$, $b_1 \neq b_2$ and $0 \leq \delta \leq 1$.

Let δ^* be either a critical point of h or 0 or 1. If δ^* is a critical point of h, then $h'(\delta^*) = 0$ implies that $\sin((b_1+b_2)\delta^* - b_1) = -\sin((b_2-b_1)\delta^* + b_1)$ and hence $\cos((b_1+b_2)\delta^* - b_1) = \pm\cos((b_2-b_1)\delta^* + b_1)$. Thus $h(\delta^*) = g(\delta^*)(1 \pm (b_2 - b_1)/(b_1 + b_2))$. Since $b_1 + b_2 \pm (b_2 - b_1) > 0$, either $h(\delta^*)g(\delta^*) > 0$ or $h(\delta^*) = g(\delta^*) = 0$, so that g and h have the same sign at critical points of h. If δ^* is either 0 or 1, $h(\delta^*) = g(\delta^*)(1 + (b_2 - b_1)/(b_1 + b_2))$. Thus, either $h(\delta^*)g(\delta^*) > 0$ or $h(\delta^*) = g(\delta^*) = 0$. The critical points of h are isolated and Lemma 11.4 implies that the number of roots of g in $[0,1] \geq$ number of roots of h in $[0,1]$.

Let δ_* either be a critical point of g or 0 or 1. If δ_* is a critical point of g, then $g(\delta_*) = \pm(b_1 + b_2)$ and $h(\delta_*)g(\delta_*) = (b_1 + b_2)(b_1 + b_2 \pm (b_2 - b_1)\cos((b_1 + b_2)\delta_* - b_1)) > 0$. If δ_* is either 0 or 1, then we have already shown that either $h(\delta_*)g(\delta_*) > 0$ or $h(\delta_*) = g(\delta_*) = 0$. Apply Lemma 11.4 again to conclude that the number of roots of h in $[0,1] \geq$ number of roots of g in $[0,1]$.

Thus the number of roots of h in $[0,1]$ is equal to the number of roots of g in $[0,1]$. But the argument $(b_2 - b_1)\delta + b_1$ of the cosine term in $g(\delta)$ is monotonic with respect to δ and hence the number of roots of $g(\delta) = (b_1 + b_2)\cos((b_2 - b_1)\delta + b_1)$ in $[0,1]$ is the number of odd multiples of $\pi/2$ spanned by b_1 and b_2, which is the same as the number of odd multiples of ω spanned by $\omega_{\rm off}$ and $\omega_{\rm on}$. □

REFERENCES

[1] L.J.Bohmann, R.H.Lasseter, *Harmonic interactions in thyristor controlled reactor circuits*, IEEE Trans. Power Delivery **4** (3) July (1989), 1919–1926.

[2] N.Christl, R.Hedin, P.E.Krause, S.M.McKenna, K.Sadek, P.Lützelberger, A.H.Montoya, D.Torgerson, *Advanced series compensation (ASC) with thyristor controlled impedance*, Cigré 14/37/38-05, Paris, August–Sept 1992.

[3] I.Dobson, S.G.Jalali, *Surprising simplification of the Jacobian of a diode switching circuit*, IEEE Intl. Symposium on Circuits and Systems, Chicago, IL, May 1993, 2652–2655.

[4] M.Grötzbach, R.vonLutz, *Unified modeling of rectifier-controlled DC-power supplies*, IEEE Trans. on Power Electronics, **1** (2) April (1986), 90–100.

[5] L.Gyugyi, *Power electronics in electric utilities: static var compensators*, Proceedings of the IEEE **76** (4) April (1988), 483–494.

[6] S.G.Jalali, R.H.Lasseter, *Harmonic interaction of power systems with static switching circuits*, Power Electronics Specialists Conference, MIT MA, June 1991, 330–337.

[7] S.G.Jalali, I.Dobson, R.H.Lasseter, *Instabilities due to bifurcation of switching times in a thyristor controlled reactor*, Power Electronics Specialists Conference, Toledo, Spain, July 1992, 546–552.

[8] S.G.Jalali, I.Dobson, R.H.Lasseter, G.Venkataramanan, *Switching time bifurcations in a thyristor controlled reactor*, *(research report 92-34)*, Wisconsin Electric Machines and Power Electronics Consortium, Dept. of ECE, University of Wisconsin, Madison, WI 53706.

[9] S.G.Jalali, *Harmonics and instabilities in thyristor based switching circuits*, *(Ph.D. thesis)*, University of Wisconsin at Madison 1993.

[10] J.P.Louis, *Non-linear and linearized models for control systems including static convertors*, Proceedings of the Third Intl. Federation on Automatic Control Symposium on Control in Power Electronics and Electrical Drives, Lausanne, Switzerland, Sept. 1983, 9–16.

[11] T.J.E.Miller, *(ed.) Reactive power control in electric systems*, Wiley NY, 1982.

[12] R.Rajaraman, I.Dobson, S.G.Jalali, *Nonlinear dynamics and switching time bifurcations of a thyristor controlled reactor*, IEEE Intl. Symposium on Circuits and Systems, Chicago, IL, May 1993, 2180–2183.

[13] G.C.Verghese, M.E.Elbuluk, J.G.Kassakian, *A general approach to sampled-data modeling for power electronic circuits*, IEEE Transactions on Power Electronics, **1** (2) April (1986), 76–89.

[14] J.W.Swift, K.Weisenfeld, *Suppression of period doubling in symmetric systems*, Physical Review Letters **52** (9) Feb (1984), 705–708.

[15] R.Yacamini, J.W.Resende, *Thyristor controlled reactors as harmonic sources in HVDC converter stations and AC systems*, IEE Proceedings, Part B **133** July (1986), 263–269.

DYNAMIC ANALYSIS OF VOLTAGE COLLAPSE IN POWER SYSTEMS

DAVID J. HILL* AND IAN A. HISKENS*

Abstract. An analytical framework is presented for analysis of voltage collapse as a dynamic phenomenon. The approach depends on linking static and dynamic aspects within differential-algebraic models which preserve network structure and facilitiate use of a novel approach to modelling aggregate (dynamic) load. Directions for stability/bifurcation analysis are indicated. In particular, new Lyapunov functions for large-disturbance voltage stability analysis can be derived.

1. Introduction. Development of stability theory for power systems continues to be an interesting subject. The three major problem areas are transient (angle) stability, oscillations and voltage collapse. Particularly in the nonlinear aspects of these problems, there are many questions to answer. The features which motivate this paper are:

- the interplay between static and dynamic behaviour;
- the interplay between angle and voltage behaviour;
- the network structure (which constrains power exchange) and the consequent differential-algebraic structure of models;
- the implications of using aggregate rather than device oriented nonlinear load models.

Each of these has of course received at least some attention in numerous results by researchers including the authors. No attempt will be made here to identify the detailed contributions; relevant surveys and collected works which refer to these are available [16,17,14,4]. The intention here is to present a framework for power system stability analysis with emphasis on the problem of voltage collapse analysis. The ideas follow earlier work on angle stability with related voltage behaviour [1,7,10,11,12] and voltage stability including system collapse [6,9]. The main consideration is towards integrating a nonlinear dynamic load modelling approach [6,13] into the broader stability analysis picture.

The structure of the paper is as follows. Section 2 discusses a recent approach to load modelling which captures essential dynamics. In Section 3, following a review of analytical results for angle stability to compare with, voltage stability criteria are presented. Both theories exploit system structure and load characteristics in a unified way. An earlier version of the paper was published at [8].

2. Load model structures

2.1. General form. Models for dynamical analysis of power systems typically have a consistency problem. While it is scientifically possible to

* Department of Electrical and Computer Engineering, The University of Newcastle, NSW 2308 Australia.

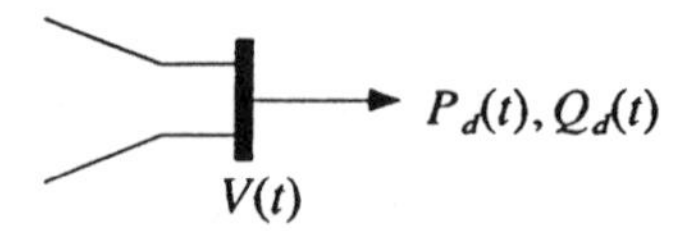

FIG. 2.1. *High Voltage Load Bus*

give quite detailed models for generators, lines, transformers and control devices, load modelling can often only be treated on an ad hoc basis. In stability analysis for instance, we need a representation of effective power demand at high voltage buses. This may include the aggregate effect of numerous load devices such as lighting, heating and motors plus some levels of transformer tap-changing and other control devices. Building up the aggregate effect by combining device characteristics may not be possible. Thus, in many cases, quite simplified aggregate load representations like impedances are used alongside detailed generator models. This section summarizes an approach for developing aggregate models of nonlinear dynamic form [6,13].

Consider a high voltage bus as in Figure 2.1. The real and reactive power demands P_d and Q_d are considered to be dynamically related to the voltage V.

Measurements in the laboratory and on power system buses [13,18] show that the load response to a step in voltage V is of the general form shown in Figure 2.2. (The responses for real and reactive power are similar qualitatively; only the real power response is shown.) The significant features of the response are as follows: 1) a step in power immediately follows a step in voltage; 2) the power recovers to a new steady-state value; 3) the recovery appears to be of exponential (sometimes underdamped) form, at least approximately; 4) the size of the step and the steady-state value are nonlinearly related to voltage. These features are easily connected to physical aspects of specific loads [6]: for recovery time constants of the order of a second, the effect of motors (transient period) is captured; on a time-scale of minutes, network and load regulating devices are modelled.

As in [6], we can propose a general load model as an implicit differential equation

$$f(P_d^{(n)}, P_d^{(n-1)}, \cdots, \dot{P}_d, V^{(m)}, V^{(m-1)}, \cdots, \dot{V}, V) = 0 \tag{2.1}$$

where $P_d^{(j)}, V^{(i)}$ denote the higher order derivatives of P_d, V respectively. (A similar equation applies for reactive power Q_d .)

An input-output version of this model is illustrated simply in Figure 2.3 where V is chosen as the input to a nonlinear dynamical system with output P_d .

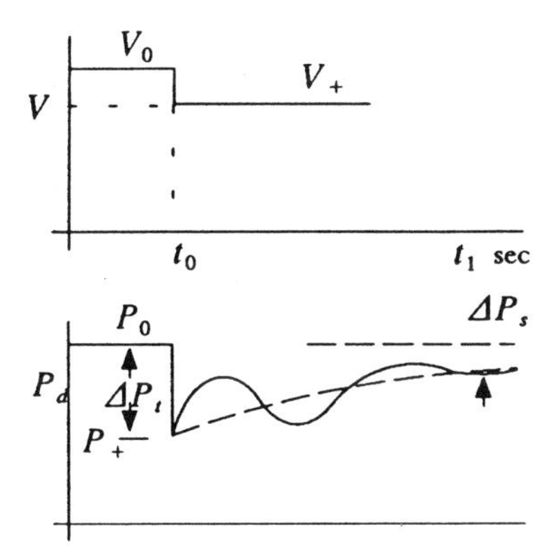

FIG. 2.2. *General Load Response*

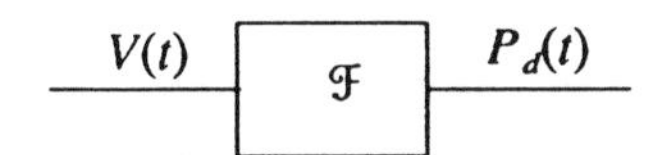

FIG. 2.3. *Input-Output Load Representation*

2.2. Linear recovery. The above discussion of Figure 2.2 suggests there are (at least) two nonlinearities in a reasonable model; one describing a steady-state relationship, i.e. the steady-state offset ΔP_s , and the other a transient one, i.e. the jump ΔP_t . Further, linear dynamics can approximately describe the transient recovery. Assuming first order dynamics, it was proposed in [6] that the load response in Figure 2.2 can be regarded as the solution of the scalar differential equation

$$T_p \dot{P}_d + P_d = P_s(V) + k_p(V)\dot{V} \tag{2.2}$$

The motivation is easy to see. Setting derivatives to zero gives the steady-state model

$$P_d = P_s(V) \tag{2.3}$$

Rewriting (2.2) as

$$T_p \frac{dP_d}{dt} + P_d = P_s(V) + T_p \frac{d}{dt}(P_t(V)) \tag{2.4}$$

where

$$P_t(V) = \frac{1}{T_p} \int_0^V k_p(\sigma)\, d\sigma + c_0 \tag{2.5}$$

c_0 a constant, clearly shows that $P_t(.)$ defines the fast changes in load according to $P_d = P_t(V)$.

For solving equation (2.2) analytically or numerically, the fact that all solutions satisfy an equivalent normal form model is used. This form is expressed as

$$\dot{x}_p = -\frac{1}{T_p} x_p + N_p(V) \tag{2.6}$$

$$P_d = \frac{1}{T_p} x_p + P_t(V) \tag{2.7}$$

where

$$N_p(V) := P_s(V) - P_t(V) \tag{2.8}$$

The solution of differential equation (2.2) to the voltage step is easily derived and

$$\Delta P_t := P_d(t_0-) - P_d(t_0+) = P_t(V_0) - P_t(V_+) \tag{2.9}$$

$$\Delta P_s := P_d(t_0-) - P_d(\infty) = P_s(V_0) - P_s(V_+) \tag{2.10}$$

So the nonlinear functions $P_s(.)$ and $k_p(.)$ (or $P_t(.)$) independently determine the steady-state and transient power increments ΔP_s and ΔP_t respectively.

Karlsson [13] has identified models of the form (2.6)-(2.8) with the special load functions:

$$P_s(V) := P_0 \left(\frac{V}{V_0}\right)^{\alpha_s} \tag{2.11}$$

$$P_t(V) := P_0 \left(\frac{V}{V_0}\right)^{a_t} \tag{2.12}$$

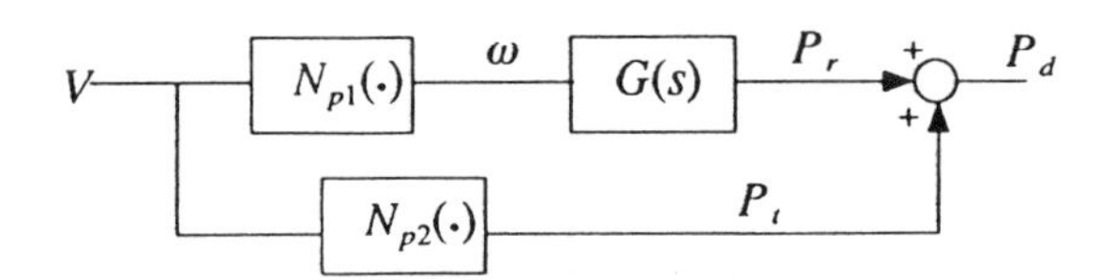

FIG. 2.4. *General Load Model with Linear Recovery Dynamics*

The steady–state model $P_d = P_0 \left(\frac{V}{V_0}\right)^{a_s}$ corresponds to the widely used static model. Now, it is convenient to note that the normal form model is easily given the block diagram representation shown in Figure 2.4 where $G(s) = \frac{1}{T_p s+1}$, $N_{p1} = N_p$ and $N_{p2} = P_t$.

In summary, the scalar nonlinear load model and its equivalent normal form (2.6)-(2.8) can be viewed as a block diagram interconnection of nonlinear functions and a linear transfer function.

For higher order dynamics

$$G(s) = \frac{b_m s^m + b_{m-1} s^m - 1 + \ldots + b_0}{s^n + a_{n-1} s^n - 1 + \ldots + a_0} \tag{2.13}$$

the response of Figure 2.2 can be obtained with more exotic recovery, i.e. multiple time-constants and/or oscillatory behavior.

Just using a second-order $G(s)$ can create the oscillatory response shown in Figure 2.2 and observed in field tests [18]. The only restriction needed in the general case is that $G(s)$ be strictly proper, i.e. $m < n$, to ensure the recovery response P_r is continuous.

Consider the second order dynamic case

$$G(s) = \frac{b_1 s + b_0}{s^2 + a_1 s + a_0} \tag{2.14}$$

In terms of the internal signal notation shown on Figure 2.4, we have

$$\ddot{P}_r + a_1 \dot{P}_r + a P_r = b_1 \dot{\omega} + b_0 \omega \tag{2.15}$$

The steady–state behavior is obtained by setting derivatives to zero; this gives

$$P_d = \frac{b_0}{a_0} N_{p1}(V) + N_{p2}(V) \tag{2.16}$$

The fast (jump) behaviour is obtained from the higher-order scalar form

$$\ddot{P}_d + a_1 \dot{P}_d + a_0 P_d = f(\ddot{V}, \dot{V}, V) \tag{2.17}$$

where

$$f(\ddot{V}, \dot{V}, V) := N'_{p2}(V).\ddot{V} + N''_{p2}(V).\dot{V}^2 + (a_1 N'_{p2}(V) + b_1 N'_{p1}(V))\dot{V} \\ + a_0 N_{p2}(V) + b_0 N_{p1}(V) \; . \tag{2.18}$$

Equality of the second–order terms gives $\frac{d}{dt}(\dot{P}_d) = \frac{d}{dt}(N'_{p2}(V).\dot{V})$ This gives the transient load as $P_d = N_{p2}(V)$.

So just as in the simple first–order case, the steady–state and fast load behaviour can be directly connected to the nonlinearities $N_{p1}(.)$ and $N_{p2}(.)$.

The first–order normal form (2.6)-(2.8) is a special case of an alternative general representation to the scalar high-order form. Translating the transfer function $G(s)$ to an equivalent state–space representation gives

$$\begin{aligned} \dot{x}_p &= F_p x_p + G_p \omega \\ P_r &= H_p^T x_p \end{aligned} \tag{2.19}$$

where x_p is an n-dimensional vector and F_p, G_p, H_p are appropriately dimensioned matrices. Combining (2.19) with the structure of Figure 2.4 gives

$$\begin{gathered} \dot{x} = Fx + GN_1(V) \\ \begin{bmatrix} P_d \\ Q_d \end{bmatrix} = H^T x + N_2(V) \end{gathered} \tag{2.20}$$

In the analysis of specific devices, it is easy to see that the P_d and Q_d models are related. To allow for this, the model (2.20) is presented in coupled form with $x = (x_p, x_q)$.

Simple MATLAB exercises show that a second order $G(s)$ gives responses close to those reported in Shackshaft et. al. [18] by appropriate choice of parameters $a_0, a_1, b_0, b_1, a_s, a_t$ (and reactive counterparts).

The model (2.20) can be easily incorporated into simulation programs for power system dynamics. The parameters which determine matrices F, G, H and nonlinear functions $N_1(.)$ and $N_2(.)$ must be obtained from measured data [13]. In [20] some comments are made about the need for second–order linear dynamics plus a nonlinearity response for load; this corresponds to $N_1 = I$ and a two pole $G(s)$ in Figure 2.4.

3. System stability analysis

3.1. System model. Consider a power system subjected to a large disturbance. We model its post–disturbance behaviour with traditional assumptions on the generator and network.

Synchronous machines are represented by a constant voltage E_i in series with transient reactance. The network is assumed to be lossless, so all lines are modeled as series reactances.

Suppose the network consists of n buses connected by transmission lines. At m of these buses there are generators. The buses which have load but no generation are labelled $i = 1, \ldots n - m$. The network is augmented with m fictitious buses representing the generator internal buses in accordance with the classical machine model. They are labelled $i + m$ where i is the bus number of the corresponding generator bus.

Let the complex voltage at the ith bus be $V_i \angle \delta_i$ where δ_i is the bus phase angle with respect to a synchronously rotating reference frame. Let $V \in \mathbb{R}^{n+m}$ denote the vector of voltage magnitudes. Using the $(n+m)$th bus as the reference, we define internodal angles $\alpha_i := \delta_i - \delta_{n+m}$. The bus frequency deviation is given by $\omega_i = \dot{\delta}_1$. Let $\alpha \in \mathbb{R}^{n+m-1}$ and $\omega_g \in \mathbb{R}^m$ denote the vectors of internodal angles and generator frequencies respectively.

Let P_{b_i} and Q_{b_i} denote the total real and reactive power leaving the ith bus via transmission lines.

Then

$$P_{bi}(\alpha, V) := \sum_{j=1}^{n+m} V_i V_j B_{ij} \sin(\alpha_i - \alpha_j) \tag{3.1}$$

$$Q_{b_i}(\alpha, V) := -\sum_{j=1}^{n+m} V_i V_j B_{ij} \cos(\alpha_i - \alpha_j) \tag{3.2}$$

where jB_{ij} are the elements of the bus admittance matrix. For convenience, we have set $V_i = E_{i-n}$ and $\alpha_{n+m} = 0$. We denote by $-P_\ell$, P_g the vectors of bus powers associated with load and generator buses respectively. Combining generator swing equations and power flow equations gives [7]

$$\dot{\gamma}_g = -S(P_g(\alpha_g, \theta, V) - P_M) \tag{3.3}$$

$$\dot{\alpha}_g = \gamma_g \tag{3.4}$$

$$0 = P_\ell(\alpha_g, \theta, V) + P_d \tag{3.5}$$

$$0 = [V]^{-1}(Q_\ell(\alpha_g, \theta, V) + Q_d) \tag{3.6}$$

where damping has been ignored, S is an inverse inertia matrix, α has been written as (α_g, θ) to identify the generator and load bus angles separately and γ is the vector of internodal velocities. $[V]$ denotes the diagonal matrix with elements of $V \in \mathbb{R}^n$ as the diagonal elements.

Combining (3.3)-(3.6) with the load model (2.20) (or special case forms in Section 2) gives a differential–algebraic model with dynamic variables (α_g, γ_g, x) and algebraic variables (θ, V).

3.2. Static loads case. Assume $P_d = P_d^o$, constant; $Q_d = Q_s(V)$.

Stability theory for this case is rather well–developed. A brief review is presented here for comparison with later results. Following an earlier result [1] showing that well-defined Lyapunov functions exist for the PV bus case, it was shown that the usual kinetic plus potential energy form can be used in this static loads case [15]. It has been pointed out to the authors that such results were already known in Russia [19] (but certainly not accessible to Western readers).

A systemmatic derivation of the Lyapunov function (and various extended versions) is provided in [7] following first integral analysis on equations (23)–(26).

Fact 1 The first integral energy function for the model (3.3)–(3.6) is given by

$$
\begin{aligned}
\mathcal{V}_g(\alpha_g, \gamma_g) &= \frac{1}{2}\gamma_g^T S^{-1}\gamma_g + \int_{\alpha_g^s}^{\alpha_g} (P_g(\rho, \theta, V) - P_M)^T d\rho \\
&= \frac{1}{2M_T}\sum_{i=n+1}^{n+m-1}\sum_{j=i+1}^{n+m} M_i M_j (\gamma_i - \gamma_j)^2 \\
&\quad -\frac{1}{2}\sum_{i=1}^{n+m}\sum_{j=1}^{n+m} B_{ij}(|V_i||V_j|\cos\alpha_{ij} - |V_i^s||V_j^s|\cos\alpha_{ij}^s) \\
&\quad +\sum_{i=1}^{n} P_{d_i}(\alpha_i - \alpha_i^s) - \sum_{i=n+1}^{n+m-1} P_{M_i}(\alpha_i - \alpha_i^s) \\
&\quad +\sum_{i=1}^{n}\int_{|V_i^s|}^{|V_i|}\frac{Q_{d_i}(\mu_i)}{\mu_i}d\mu_i
\end{aligned}
\tag{3.7}
$$

To establish this as a Lyapunov function required extending Lyapunov stability theory to differential-algebraic systems [10]. Then it is possible to show that the function V at (3.7) is a Lyapunov function showing stability of an equilibrium $(\alpha_g^o, \gamma_g^o, V^o)$. The equilibrium is given by

$$\gamma_g = 0 \tag{3.8}$$

$$P_g(\alpha_g, \theta, V) = P_M \tag{3.9}$$

and (3.5), (3.6), i.e. the usual power flow equations augmented by power balance at generator internal nodes. These can be written in the form

$$\tilde{f}_g(\alpha_g, \theta, V) = 0 \tag{3.10}$$

$$f_\ell(\alpha_g, \theta, V) = 0 \tag{3.11}$$

$$g(\alpha_g, \theta, V) = 0 \tag{3.12}$$

where $\tilde{f}_g, f_\ell, g$ are $\mathbb{R}^{m-1}, \mathbb{R}^n$ and $\mathbb{R}^n$ valued functions respectively. The power flow Jacobian has the form

$$
J = \begin{bmatrix} F_{\alpha g} & F_{\theta g} & F_{\nu g} \\ F_{\alpha \ell} & F_{\theta s} & F_{\nu s} \\ G_{\alpha \ell} & G_{\theta s} & G_{\nu s} \end{bmatrix}
$$

For later convenience, we rewrite J as

$$J = \begin{bmatrix} F_{\alpha g} & J_{\ell g} \\ J_{\alpha \ell} & J_{\ell \ell} \end{bmatrix}$$

where

$$J_{\ell \ell} = \begin{bmatrix} F_{\theta s} & F_{\nu s} \\ G_{\theta s} & G_{\nu s} \end{bmatrix}$$

The subscripts 's' refer to use of static load models.

These dynamic power flow matrices are significant in establishing two important facts. We use the notation A^* to denote the Schur complement of A within a partitioned matrix. Thus we have $F^*_{\alpha g} = F_{\alpha g} - J_{\alpha \ell} J^{-1}_{\ell \ell} J_{\ell g}$.

Fact 2 The first-integral energy function (3.7) is a locally positive definite function if $F^*_{ag}|_{(a^o_g, \theta^o, V^o)} > 0$.

If generator damping is included this property implies small disturbance (asymptotic) stability. For large-disturbance stability, we have that the small-disturbance criterion ensures a valid Lyapunov function.

Fact 3 [7,10] Suppose that $det J_{\ell\ell}|_{(\alpha^o_g, \theta^o, V^o)} \neq 0$. Then in some neighborhood of $(\alpha^o_g, \theta^o, V^o)$ the DA system is equivalent to an ordinary differential equation (ODE) model of the form

$$\dot{\gamma}_g = -S(P_g(\alpha_g, \psi(\alpha_g)) - P_M) \tag{3.13}$$

$$\dot{\alpha}_g = \gamma_g \tag{3.14}$$

where continuous function ψ relates the algebraic variables to generator angles α_g.

These observations motivate a structural picture developed in [11,12]. The impasse surfaces I on which $J_{\ell\ell}$ is singular separate open sets where the DA system model is equivalent to a local ODE model. The Lyapunov functions $\mathcal{V}$ must be seen globally as multivalued. Within a particular open set C_i , the Lyapunov stability of an equilibrium point is established using one "sheet" of this function. Discussion of regions of transient stability can proceed as for other ODE models (based on impedance loads) [5] if trajectories do not meet impasse surfaces. This depends on the load indices. In general, the critical energies must take account of the proximity of I to the stable equilibrium.

It is shown in [11] that trajectory impact on I corresponds to a short-term voltage collapse. In energy terms, this could correspond to a jump between energy "sheets"; investigations on this phenomena will require study of the perturbed DA system, i.e. with previously ignored dynamics. Related reasoning has led DeMarco and Overbye to a voltage security index [2].

In summary, we have static regularity conditions on the power flow Jacobian - see Facts 2, 3 - playing a key role in determining the large disturbance behavior. Load indices are very important and voltage dip phenomena are associated with the angle behaviour [11].

3.3. Load system stability. Assume $P_g = P_M^o$, constant.

We consider the case where the system is operating outside the transient period following a major disturbance, i.e. angle stability is not a concern. The stability analysis with first-order linear recovery load models has been considered in [9] to the extent of static regularity properties and their connections to small disturbance stability. Here we briefly review this and make some progress on large-disturbance behaviour.

System behaviour is determined by both static load and dynamic load characterisitcs. Let J_t be the Jacobian corresponding to the dynamic power flow

$$P_t(V) - P_\ell(\theta, V) = 0 \tag{3.15}$$

$$Q_t(V) - Q_\ell(\theta, V) = 0 \tag{3.16}$$

The angles α_g can be eliminated from (3.10) by the usual Implicit Function Theorem argument to give 2n power flow equations in (θ, V). The static power flow Jacobian is $J_{\ell\ell}^*$. (We drop this * for notational convenience through taking multiple Schur complements.)

For simplicity, we make a further restriction to assume slow real load recovery, i.e. $P_t \approx P_s$. For this case, set $x_p = x_p^o$, constant. The more general case is only different in detail [9].

Voltage regularity [9] requires that (for any PQ bus) $\Delta Q_i > 0$ gives all $\Delta V_j > 0$.

Fact 4 [9] Assume $detF_{\theta s} \neq 0$. The load system is at a voltage regular operating point if $G_{\nu s}^*$ is an irreducible M - matrix.

The matrix $G_{\nu s}^*$ features in much recent research on static voltage collapse proximity indicators - seen [9] for references.

Again, using the Implicit Function Theorem argument, if $F_{\theta s}$ is nonsingular, we can eliminate θ according to $\theta = \psi_p(V)$ for some differentiable function $\psi_p(.)$ in a region of the operating point. (This assumption on $F_{\theta s}$ was used in deriving the QV sensitivity.) Then the DA model becomes

$$\dot{x}_q = Q_s(V) - Q_\ell(\psi_p(V), V) \tag{3.17}$$

$$0 = -Q_\ell(\psi_p(V), V) + T_q^{-1} x_q + Q_t(V) \tag{3.18}$$

THEOREM 3.1. *[9] Assume $detF_{\theta s}(V^0) \neq 0$ and $detG_{vt}^*(V^o) \neq 0$. The load system with slow real power recovery and at operating point (x_q^0, V^0) is small disturbance stable if the matrix $G_{vs}^*(V^o)C_{qd}^{-1}(V^o)$ is stable where $C_{qd}(V^o) := -T_q(G_{vt}^*(V^o))$.*

In [6,9] various possibilities were discussed depending on relative values of G_{vs}^*, T_q and G_{vt}^* at the operating point. In particular, it is possible to operate at an irregular point with small disturbance stability.

The dynamics can be expressed in terms of the voltages. From equation (3.18), we have

$$x_q = T_q(Q_\ell(\psi_p(V), V) - Q_t(V)) \tag{3.19}$$

So $\dot{x}_q = -T_q G_{vt}^*(V)\dot{V}$ and (3.17)-(3.18) becomes

$$C_{qd}(V)\dot{V} = g_s^*(V) \tag{3.20}$$

where

$$g_s^*(V) := Q_s(V) - Q_\ell(\psi_p(V), V) \tag{3.21}$$

So the condition $detC_{qd}(V^o) \neq 0$ is seen to correspond to avoidance of an impasse surface. The relevance to voltage collapse behavior for the single load case is considered in detail in [9].

3.4. Large disturbance voltage stability. The above results in Section 3.3 enable classification of operating points according to regularity and stability. According to Lyapunov theory [21], if the operating point is hyperbolically stable, then it is asymptotically stable for the nonlinear model. We are then interested in determining the region of attraction around the operating point. Again, use is made of Lyapunov stability theory for DA systems [10].

To illustrate briefly, consider the slow real recovery model (3.20) and assume $detF_{\theta s}(V^o) \neq 0$ and $detG_{vt}^*(V^o) \neq 0$. Consider the Lyapunov function candidate

$$\mathcal{V}_\ell(V) = g_s^*(V)^T \Lambda g_s^*(V) \tag{3.22}$$

where Λ is some positive definite matrix. Differentiating $\mathcal{V}_\ell$ along the trajectories of (3.21) gives

$$\begin{aligned} \dot{\mathcal{V}}_\ell(V) &= (\dot{g}_s^*(V))^T \Lambda g_s^*(V) + (g_s^*(V))^T \Lambda \dot{g}_s^*(V) \\ &= (g_s^*(V))^T M(V;\Lambda) g_s^*(V) \end{aligned} \tag{3.23}$$

where

$$M(V;\Lambda) := (G_{vs}^* C_{qd}^{-1}(V))^T \Lambda + \Lambda (G_{vs}^*(V) C_{qd}^{-1}(V)) \ . \tag{3.24}$$

If Λ can be chosen to make $M(V;\Lambda)$ negative definite for $V \neq V^o$ for some region containing V^o , a stability result is established.

THEOREM 3.2. *Under the conditions of Theorem 3.1, the equilibrium V^o of load system (3.20) is asymptotically stable and $\mathcal{V}_\ell(.)$ is a Lyapunov function.*

Proof. Since $G_{vs}^*(V^o)C_{qd}^{-1}(V^o)$ has all eigenvalues in the open left hand plane, there exists a matrix $\Lambda_o > 0$ such that

$$(G_{vs}^*(V^o)C_{qd}^{-1}(V^o))^T \Lambda_o + \Lambda_o (G_{vs}^*(V^o)C_{qd}^{-1}(V^o)) = -R_o \tag{3.25}$$

for any matrix $R_o > 0$ [21]. Since the static and transient load flow functions are local diffeomorphisms at V^o , we have

$$M(V;\Lambda_o) < -\varepsilon I \tag{3.26}$$

for some $\varepsilon > 0$ and all V in some neighbourhood of V^o. It follows from (3.25) that using Λ_o in the Lyapunov function (3.22) gives

$$\begin{aligned} \dot{\mathcal{V}}_\ell &= (g_s^*(V))^T M(V;\Lambda_o) g_s^*(V) \\ &\leq -\varepsilon \|g_s^*(V)\|^2 \end{aligned} \tag{3.27}$$

From the LaSalle Invariance Theorem [21] and uniqueness of solution V^o within a neighbourhood, the result follows. □

In Section 3.2. Lyapunov functions were presented for voltage collapse studies in the transient period with generator dynamics; Theorem 3.2 now establishes one Lyapunov function $\mathcal{V}(.)$ for load systems.

We note that there does not appear to be any prior Lyapunov functions for nonlinear models of long-term voltage stability.

While the Lyapunov function $\mathcal{V}_g$ at (3.7) appears quite suitable for computation of stability regions, it is severely limited by relying on constant real loads for the potential energy component to be well-defined. (The usual transient energy function based on impedance loads [5] is on even shakier ground theoretically since it is only well-defined for very restricted forms of real load.) The Lyapunov function $\mathcal{V}_\ell$ at (3.22) for the load system can be easily generalised to real and reactive linear recovery loads, and does not involve any path dependence problems. The more general nth order linear recovery dynamics (2.20) can be accounted for by an additive term $x^T K x, K$ positive definite. However, the ability of these Lyapunov functions to estimate regions of stability must be tested. These issues and the question of well-defined potential energy type functions for load systems need further investigation. To illustrate for the single load case, we have the following result.

THEOREM 3.3. *For the single load version of system (3.20) under the conditions of Theorem 3.2, we have the potential energy function*

$$W(x_q, x_q^o) = -\int_{x_q^o}^{x_q} g_s^*(V) dx_q \tag{3.28}$$

is a valid Lyapunov function.

Proof. Firstly, we note that

$$\frac{\partial W}{\partial x_q} = -g_s^*(V) \quad \textit{and} \quad \frac{dx_q}{dV} = C_{qd}(V)$$

Thus

$$\frac{\partial^2 W}{\partial x_q^2} = -G_{vs}^*(V) C_{qd}^{-1}(V)$$

The conditions of Theorem 3.1 ensure that W is (locally) positive definite. Differentiating gives

$$\begin{aligned} \dot{W}(x_q, x_q^o) &= -g_s^*(V)\dot{x}_q. \\ &= -[g_s^*(V)]^2 \end{aligned} \tag{3.29}$$

The result follows from standard Lyapunov theory [21]. □

W is positive definite if the small-disturbance stability condition of Theorem 3.1 is satisfied. In the multi-load case, we again face the issue of path dependence.

The energy function (3.7) and its variants [16] can be used to provide criteria for determining critical switching times for stability enhancement actions, e.g. the equal-area criterion used for circuit breaker reclosure time. While it is tempting to think of similar equal-area type criteria for voltage stability enhancement [22], it is not clear yet how appropriate this is. Decisions on voltage defense are not clearly related to integrals such as (3.28).

4. Conclusions. This paper has presented an analytical framework for analysis of voltage collapse as a dynamic phenomenon. By giving equal emphasis to generator and load dynamics, short-term and longer-term behaviour can be studied by selecting appropriate parameter values.

Section 3 has discussed results on generator (angle) and load stability which clearly say something about voltage collapse in a static or dynamic sense. The viewpoint established is to use the power flow equations as a basis and then include generator or aggregate load (including all static and dynamic components) as the case may be. Clearly in both cases, the load characteristics and network structure represented by appropriate Jacobian matrices play a key role. For large-disturbance generator angle and voltage stability analysis, we saw many similarities:

1. Differential-algebraic models based around the power flow equations, i.e. (3.3)-(3.6) and (3.17)-(3.18) respectively;
2. Connections between static regularity, small-disturbance stability and existence of valid Lyapunov functions;
3. The possibility of multiple stable equilibria related to different 'causal regions' separated by the impasse surfaces.

However, there are many questions to consider in more detail before the theory yields a firm basis for static and dynamic voltage collapse indicators. One direction concerns geometric analysis [3,20] of the bifurcations associated with voltage collapse and the related issue of characterizing regions of stability. Lyapunov function techniques are useful in giving estimates of these regions of stability.

It has been argued [12] that in a study of system collapse we should consider both angle and voltage dynamics (especially in the initial and final stages). Thus there is motivation for further unification. One issue on which more understanding is needed is existence of well-defined potential energy functions for use in computation of stability regions with both generator and load dynamics.

REFERENCES

[1] A.R. Bergen and D.J. Hill. A structure preserving model for power system stability analysis. *IEEE Trans. Power Apparatus and Systems*, PAS-100(1):25–35, January 1981.

[2] C.L. de Marco and T.J. Overbye. An energy based security measure for assessing vulnerability to voltage collapse. *IEEE Trans. Power Systems*, Vol. 5(2):419–427, May 1990.

[3] I. Dobson and H.-D. Chiang. Towards a theory of voltage collapse in electric power systems. *Systems and Control Letters*, Vol. 13:253–262, 1989.

[4] L.H. Fink, editor. *Proceedings of Bulk Power System Voltage Phenomena II Voltage Stability and Security*, Deep Creek Lake, Maryland, August 1991.

[5] A.A. Fouad and V. Vittal. The transient energy function method. *International Journal of Electric Power and Energy Systems*, Vol. 10(4):233–246, October 1988.

[6] D.J. Hill. Nonlinear dynamic load models with recovery for voltage stability studies. *IEEE Transactions on Power Systems*, 8(1):166–176, February 1993.

[7] D.J. Hill and C.N. Chong. Energy functions for power systems based on structure preserving models. *Proc. 25th Conference on Decision and Control*, (2):1218–1223, December 1989. Athens, Greece.

[8] D.J. Hill and I.A. Hiskens. Dynamic analysis of voltage collapse in power systems. *Proceedings 31st Conference on Decision and Control*, Vol. 3:2904–2909, 1992. Tuscon, Arizona.

[9] D.J. Hill, I.A. Hiskens, and D. Popovic. Stability analysis of load systems with recovery dynamics. *International Journal Electrical Power and Energy Systems*, to appear.

[10] D.J. Hill and I.M.Y. Mareels. Stability theory for differential/algebraic systems with application to power systems. *IEEE Trans. Circuits and Systems*, CAS-37(11):1416–1423, November 1990.

[11] I.A. Hiskens and D.J. Hill. Energy functions, transient stability and voltage behaviour in power systems with nonlinear loads. *IEEE Trans. on Power Systems*, (Vol. 4 (4)):1525–1533, November 1989.

[12] I.A. Hiskens and D.J. Hill. Failure modes of a collapsing power system. *Proceedings of Bulk Power System Voltage Phenomena II Voltage Stability and Security*, August 1991. Deep Creek Lake, Maryland.

[13] D. Karlsson and D.J. Hill. Modelling and identification of nonlinear dynamic loads in power systems. *IEEE Trans on Power Systems*, to appear.

[14] Y. Mansour, editor. *Voltage Stability of Power Systems: Concepts, Analytical Tools and Industry Experience.* IEEE Task Force Report, Publication 90TH0358-2-PWR, 1991.

[15] N. Narasimhamurthi and M.T. Musavi. A generalised energy function for transient stability analysis of power systems. *IEEE Trans. Circuits and Systems*, CAS-31(7):637–645, July 1984.

[16] M.A. Pai. *Energy Function Analysis for Power System Analysis.* Kluwer Academic Publishers, Boston, 1989.

[17] M. Ribbens-Pavella and F.J. Evans. Direct methods for studying dynamics of large-scale electric power systems - a survey. *Automatica*, Vol. 21(1):1–21, January 1985.

[18] G. Shackshaft, O.C. Symons, and J.G. Hadwick. General-purpose model of power-system loads. *Proc. IEE*, Vol. 124(8), August 1977.

[19] V.P. Vasin. Energy integral for equations of electric power system transients with loads represented by steady-state characteristics. *Izvestia AN SSR Energetika i Transport*, (6):26–35, 1974. (In Russian).

[20] V. Venkatasubramanian, H. Schattler, and J. Zaborsky. A taxonomy of the dynamics of the large power system with emphasis on its voltage stability. *Proceedings of Bulk Power System Voltage Phenomena II Voltage Stability and Security*,

August 1991. Deep Creek Lake, Maryland.

[21] M. Vidyasagar. *Nonlinear Systems Analysis.* Prentice-Hall, Englewood Cliffs, New Jersey, second edition, 1993.

[22] W. Xu and Y. Mansour. Voltage stability analysis using generic dynamic load models. *IEEE Trans on Power Systems*, to appear.

EXACT CONVERGENCE OF A PARALLEL TEXTURED ALGORITHM FOR CONSTRAINED ECONOMIC DISPATCH CONTROL PROBLEMS

GARNG M. HUANG* AND SHIH-CHIEH HSIEH*

Abstract. In our earlier paper [1], a parallel textured algorithm has been developed to solve the constrained real power economic dispatch control problems and one example is used to illustrate the need of some sufficient conditions for the textured algorithm to converge to its true solution. In this paper, we provide the theoretical foundation of our claim. First, we show that for a textured decomposition, the algorithm always converges to a stationary point, which may not be a global minimum. Then, we prove that if the conditions of the exact convergence theorem are satisfied, the textured algorithm will converge to a global minimum. In addition, by using some computer runs, we demonstrate the speedup advantage of the parallel textured algorithm.

1. Introduction. Economic dispatch that minimizes the generation cost by allocating generation amounts to different generators is a ubiquitous power system operation [2]. In order to meet the increasing demand for exchanging power in a crowded network, the economic dispatch control needs to take more and more network constraints and security constraints into consideration. The corresponding problem is growing larger and complicated. On the other hand, the problem needs to be solved on-line, which demands fast convergence and reliable computation. One popular implementation of real-time (on-line) constrained economic dispatch control (CEDC) which covers line flow constraints is based on a combination of a Security Dispatch (SD) [3], which considers the line flow constraints, and a classic Economic Dispatch (ED), which does not consider the line flow constraints, since it is too time consuming to run the SD. Typically, the SD is executed every 30 minutes, and the ED is executed every 30 seconds between two consecutive SD computations. On the other hand, due to the advances of computer and communication technology, smart substations with high computing power and fast communication capacity become an inevitable trend [4]. If reliable algorithms can be developed so that the CEDC (which includes SD) can be run fast enough, we can rely on it to ensure the security and economic operations without any approximation. One way to achieve the goal is to explore the **parallelism** of the problem so that the problem can be solved parallelly or distributively at a fractional cost of using a supercomputer. In addition, the speed bottleneck caused by the physical limitation of the electronic devices can be bypassed by this approach. At the same time, the algorithm can be naturally extended to distributed control environments, which is a trend of the future. The distributed control will be more reliable since one CPU failure will not jeopardize the whole system operation.

* Dept. of Electrical Engineering, Texas A&M University, College Station, TX 77843. email: huang@ee.tamu.edu Tel: (409) 845-7476

The textured and hierarchical aggregation/disaggregation (HAD) [1], [5,8] methods are two approaches developed by the authors to explore the parallelism through network topologies. The textured method utilizes the uniformity of the network, while HAD algorithm utilizes the hierarchies of the network. These two approaches can be easily combined since at one hierarchical level, the network is relatively uniform. This paper will concentrate only on the parallel textured algorithm for constrained real power economic dispatch control problems (CEDCP).

The parallel textured decomposition method was first invented to solve the reactive power control problems [5,6], and then it has been extended to solve the systems of linear equations [9], the multicommodity flow problems [10], as well as the optimal routing problems [11,12]. The speedup advantage can be found in [5,9,12]. In [1], we extend our work [5,6] of textured decomposition method on reactive power control to the CEDCP. The extension is not trivial since the reactive power control tends to have local effect due to the PV bus configurations, but the real power tends to have global effect. In the paper, we look into the theoretical foundation of the extension in the framework of the multicommodity formulation [10,11].

The textured decomposition method first partitions a given system into a set of overlapping subsystems; then, it systematically arranges the mutually independent (nonintersected) subsystems on the same level and the subsystems overlaps from level to level to propagate the newly computed data. When the external flows in a level are fixed at the recently computed values, the independent subproblems at that level can then be solved concurrently. To fully utilize the parallelism, we require the number of the level to be minimal. Also, we require that the union of all subsystems be the whole system so that we can combine those subproblem solutions to form a global solution to the original problem.

We divide our proof into two substeps. We first show that the algorithm starting from any feasible point will converge to a solution. Then we prove that the textured algorithm satisfying the conditions of the exact convergence theorem will converge to its global minimum.

The paper is organized as follows: In section 2, the problem formulation is presented. The textured decomposition method, the decomposed subproblems, and the corresponding parallel textured algorithm are reviewed in section 3. The convergence property of the algorithm is proved in section 4. In section 5, the sufficient conditions for the algorithm to converge to its global minimum are stated in the exact convergence theorem. The exact convergence theorem is proved by mathematical induction in section 6. Section 7 provides some results of computer runs to demonstrate the speedup advantage of the parallel textured algorithm. Finally, section 8 concludes this paper.

2. Formulation of CEDCP

2.1. Conventional formulation. Following the decoupled optimal power flow approach in [13–14], we concentrate only on the real power economic dispatch control problems since the reactive part has been solved in [5–6].

Consider the system with N buses (nodes) and b transmission lines (branches). Let bus i be a generator bus for $i = 1, 2, ..., |N'|$ and bus j be a load bus for $j = |N'| + 1, |N'| + 2, ..., N$. By treating each bus as a node and each transmission line as a branch, we represent the system as a connected graph of N nodes and b branches, denoted by $G(X, U, N')$, where X is the set of bus nodes, U is the set of transmission branches, and N' is the set of generation injections. $|X| = N$ and $|U| = b$. The objective of the real power economic dispatch control problems is to find the real power generation settings which minimize the total generation cost at the same time satisfying the power flow constraints, the generation and transmission capacity constraints, as well as the security constraints. Some discussions on complete formulation can be found in [15]. To investigate the validity of the textured model to the CEDCP without messy notations, we simplify our formulation and omit the security constraints here. Including the security or other constraints will only enlarge the number of constraints and will not jeopardize the applicability of the textured model. The simplified *conventional formulation* is as follows:

Conventional Formulation

$$\text{(2.1)} \quad \text{Minimize} \; : \; \sum C_i(P_{G_i})$$

$$\text{(2.2)} \quad \text{Subject to} \; : \; P_{G_i} - P_{D_i} = \sum f_{ij}(\delta_i, \delta_j) \text{ for } i = 1, 2, ..., N,$$

$$\text{(2.3)} \quad P_{G_i,min} \leq P_{G_i} \leq P_{G_i,max} \text{ for } i = 1, 2, ..., |N'|,$$

$$\text{(2.4)} \quad f_{ij,min} \leq f_{ij} \leq f_{ij,max} \text{ for all lines } i\text{-}j,$$

where $C_i(P_{G_i})$ is the generation cost function for unit i;
P_{G_i} is the real power generation amount of unit i, and $P_{G_i} = 0$ for $i = (|N'| + 1), (|N'| + 2), ..., N$;
P_{D_i} is the load demand at bus i;
$f_{ij} = C_{ij}\left[\frac{V_i^2}{z_{ij}} \cos\xi_{ij} - \frac{V_i V_j}{z_{ij}} \cos(\delta_i - \delta_j + \xi_{ij}) + \frac{V_i^2}{z'_{ij}} \cos\xi'_{ij}\right]$ is the line flow (detected at bus i) from bus i to bus j;
$V_i = |E_i|$ is the voltage magnitude of bus i, which is treated as a constant here;
δ_i is the voltage phase angle of bus i;
$z_{ij}\angle\xi_{ij}$ is the total series impedance of line i-j;
$z'_{ij}\angle\xi'_{ij}$ is one half of shunt impedance of line i-j;

$$C_{ij} = \begin{cases} 1 & \text{if bus } i \text{ and bus } j \text{ are connected,} \\ 0 & \text{otherwise.} \end{cases}$$

In the formulation, the bus voltage magnitudes, $|E_i|$, are treated as constants and one of the buses is picked as the reference bus whose voltage phase angle, say δ_1, is fixed at zero degree. In addition, the generation cost function is assumed to be strictly convex and continuous, and it takes the form $C_i = \gamma_i + \beta_i P_{G_i} + \alpha_i P_{G_i}^2$, where α_i, β_i, and γ_i are positive real constants.

We adopt the linear programming approach [15–16] due to its well-known reliability and fast computation. In addition, many features, including piecewise linear objectives and the latest sparse methods, are included in Stott's and colleagues' programs for real power dispatch [2]. Thus, the linearization around a given operating point is used in this paper; and the linearized formulation is described as follows:

Conventional Linearized Formulation

$$\text{(2.5)} \quad \text{Minimize} \quad : \quad \sum C_i(\Delta P_{G_i})$$

$$\text{(2.6)} \quad \text{Subject to} \quad : \quad \Delta P_{G_i} = \sum \Delta f_{ij}(\Delta\delta_i, \Delta\delta_j) \text{ for } i = 1, 2, ..., N,$$

$$\text{(2.7)} \qquad \Delta P_{G_i,min} \leq \Delta P_{G_i} \leq \Delta P_{G_i,max} \text{ for } i = 1, ..., |N'|,$$

$$\text{(2.8)} \qquad \Delta f_{ij,min} \leq \Delta f_{ij} \leq \Delta f_{ij,max} \text{ for all lines } i\text{-}j,$$

where :
$$\begin{aligned}
&\Delta f_{ij} = a_{ij}(\Delta\delta_i - \Delta\delta_j),\\
&a_{ij} = C_{ij}\left[\frac{V_i V_j}{z_{ij}} \sin(\delta_i - \delta_j + \xi_{ij})\right],\\
&\Delta P_{G_i} = 0 \text{ for } i = (|N'|+1), (|N'|+2), ..., N,\\
&\Delta P_{G_i,min} = P_{G_i,min} - P_{G_i} \text{ for } i = 1, 2, ..., |N'|,\\
&\Delta P_{G_i,max} = P_{G_i,max} - P_{G_i} \text{ for } i = 1, 2, ..., |N'|,\\
&\Delta f_{ij,min} = f_{ij,min} - f_{ij}, \Delta f_{ij,max} = f_{ij,max} - f_{ij}.
\end{aligned}$$

Note that let $a_{ii} = -\sum_{j=1;\, j\neq i}^{N} a_{ij}$ for $i = 1, 2, ..., N$. Then, equation 2.6 can be rewritten as follows:

$$-\begin{bmatrix} a_{11} & a_{12} & \cdots & a_{1N} \\ a_{21} & a_{22} & \cdots & a_{2N} \\ \vdots & \vdots & \ddots & \vdots \\ a_{N1} & a_{N2} & \cdots & a_{NN} \end{bmatrix} \begin{bmatrix} \Delta\delta_1 \\ \Delta\delta_2 \\ \vdots \\ \Delta\delta_N \end{bmatrix} = \begin{bmatrix} \Delta P_{G_1} \\ \Delta P_{G_2} \\ \vdots \\ \Delta P_{G_N} \end{bmatrix}$$

Since $\Delta\delta_1 = 0$, the first column, $[a_{11}, a_{21}, ..., a_{N1}]^t$ of the above equation and the entry $\Delta\delta_1$ can be deleted. For realistic power systems, the angle ξ_{ij} is approximately equal to 90 degrees. For normal power system operations, the absolute value, $|\delta_i - \delta_j|$, of the difference between two voltage phase angles is less than 90 degrees. Thus, under normal situations, $a_{ij} \geq 0$ for all $i \neq j$ and $a_{ii} < 0$ for $i = 1, 2, ..., N$.

2.2. Reformulation of CEDCP in terms of the line flows. Motivated by the *flexible transmission system* [17] technology which can control transmission line flows directly, we would like to reformulate the problem directly in terms of the line flows. The new formulation is consistent with the multicommodity problems [10–11], in which the variables are also flows. Accordingly, the results in [11] can be adopted to this problem. Before we utilize the multicommodity formulation, we need some modification. First, we treat the real power generations as injected generation flows; secondly, we represent the incremental line flow Δf_{ji} in terms of Δf_{ij} ($\Delta f_{ji} = -\frac{a_{ji}}{a_{ij}}\Delta f_{ij}$); then, introducing the concept of **tree** in graph theory to our problem, we name $(N-1)$ incremental transmission line flows, which form a spanning tree, as **transmission tree flows** and the rest of incremental transmission line flows as **link flows**. The link flows can be expressed by these transmission tree flows without incremental generation flows since the relationship between the tree flows and the link flows are obtained from the loops, in which the incremental generation flows play no roles. Accordingly, the problem can be described by $(N-1)$ transmission tree flows and $|N'|$ incremental generation flows. Thus, the new formulation has the same number of independent equality constraints and the same number of variables as those of the conventional linearized formulation, respectively. Therefore, our new formulation is equivalent to the conventional linearized formulation.

To simplify our presentation, we keep all the tree, link, and generation variables in the following way: instead of expressing the system equations in terms of the $(N-1)$ transmission tree flows and N' incremental generation flows, we write the N equality constraints 2.6 in terms of the b incremental transmission line flow variables $\{\Delta f_{ij} | i = 1, 2, ..., (N-1); j = (i+1), (i+2), ..., N\}$ and $|N'|$ incremental generation flow variables $\{\Delta P_{G_i} | i = 1, 2, ..., |N'|\}$, and add $(b - N + 1)$ link flow equations which express the $(b - N + 1)$ link flows in terms of the $(N-1)$ transmission tree flows. Clearly, the representation is the same except that we are one step short of the substitution.

Now, for $i = 1, 2, ..., (N-1)$, let

$$\left[\Delta \underline{f}_i\right] = \begin{bmatrix} \Delta f_{i(i+1)} \\ \Delta f_{i(i+2)} \\ \vdots \\ \Delta f_{iN} \end{bmatrix} = \begin{bmatrix} a_{i(i+1)}(\Delta\delta_i - \Delta\delta_{(i+1)}) \\ a_{i(i+2)}(\Delta\delta_i - \Delta\delta_{(i+2)}) \\ \vdots \\ a_{iN}(\Delta\delta_i - \Delta\delta_N) \end{bmatrix},$$

which is an $(N-i) \times 1$ vector of incremental transmission line flows. Then, the N linearized constraints 2.6 and $(b - N + 1)$ link flow equations can be written as follows:

$$(2.9)\qquad \underbrace{\begin{bmatrix} A_1 & A_2 & \cdots & A_{N-1} & \vdots & -I_{N\times|N'|} \\ \cdots & \cdots & \cdots & \cdots & \vdots & \cdots \\ & & D & & \vdots & 0 \end{bmatrix}}_{A} \underbrace{\begin{bmatrix} \Delta \underline{f}_1 \\ \Delta \underline{f}_2 \\ \vdots \\ \Delta \underline{f}_{N-1} \\ \cdots \\ \Delta P_{G_1} \\ \Delta P_{G_2} \\ \vdots \\ \Delta P_{G_{|N'|}} \end{bmatrix}}_{\Delta f} = [0]$$

where

$$[A_i] = \begin{bmatrix} & & O_i & & \\ 1 & 1 & \cdots & \cdots & 1 \\ -\frac{a_{(i+1)i}}{a_{i(i+1)}} & 0 & 0 & \cdots & 0 \\ 0 & -\frac{a_{(i+2)i}}{a_{i(i+2)}} & 0 & \cdots & 0 \\ 0 & 0 & -\frac{a_{(i+3)i}}{a_{i(i+3)}} & \cdots & 0 \\ \vdots & \vdots & \vdots & \ddots & \vdots \\ 0 & 0 & 0 & \cdots & -\frac{a_{Ni}}{a_{iN}} \end{bmatrix} \text{ is an } N\times(N-i)$$

matrix;

$[O_i]$ is an $(i-1)\times(N-i)$ matrix with zero entries;

$[I_{N\times|N'|}] = [I_1, I_2, ..., I_{|N'|}]$, and I_i is an $N \times 1$ column vector whose i-th element is 1 and the other entries are zeros;

$[D]$ is a $(b-N+1)\times b$ matrix and can be written in the following form:

$$[D] = \begin{bmatrix} D_1 \\ D_2 \\ \vdots \\ D_{(b-N+1)} \end{bmatrix}$$

with $[D_k]$ being a row vector with b elements for $k = 1, 2, ..., (b-N+1)$, where $b = (N-1)+(N-2)+\cdots+1 = \frac{N(N-1)}{2}$. For the k-th link flow $\Delta f_{p^k q^k}(p^k < q^k)$, we can find a subset of tree flows associated with the k-th link, $\{\Delta f_{p^k q_1^k}, \Delta f_{q_1^k q_2^k}, ..., \Delta f_{q^k_{(n_k-1)} q^k_{n_k}}, \Delta f_{q^k_{n_k} q^k}\}$, such that

$$\Delta f_{p^k q^k} = a_{p^k q^k}[\frac{\Delta f_{p^k q_1^k}}{a_{p^k q_1^k}} + \frac{\Delta f_{q_1^k q_2^k}}{a_{q_1^k q_2^k}} + \cdots + \frac{\Delta f_{q^k_{(n_k-1)} q^k_{n_k}}}{a_{q^k_{(n_k-1)} q^k_{n_k}}} + \frac{\Delta f_{q^k_{n_k} q^k}}{a_{q^k_{n_k} q^k}}],$$

$(p^k < q_1^k < q_2^k < \cdots < q^k_{n_k} < q^k)$. Note that $\Delta f_{p^k q^k}$ is the l_k-th element of $[\Delta f]$, where $l_k = \frac{(p^k-1)(2N-p^k)}{2} + (q^k - p^k)$ since $l_k = (N-1)+(N-2)+$

$\cdots + (N - p^k + 1) + (q^k - p^k)$. Let $y_0 = \frac{(p^k-1)(2N-p^k)}{2} + (q_1^k - p^k)$; $y_i = \frac{(q_i^k-1)(2N-q_i^k)}{2} + (q_{i+1}^k - q_i^k)$ for $i = 1, 2, ..., (n_k - 1)$; $y_{n_k} = \frac{(q_{n_k}^k-1)(2N-q_{n_k}^k)}{2} + (q^k - q_{n_k}^k)$. And let the y_j-th element of $[D_k]$ be $d_{y_j}^k$. Then

$$d_{y_j}^k = \begin{cases} \frac{a_{p^k q^k}}{a_{p^k q_1^k}} \text{ for } j = 0, \\ \frac{a_{p^k q^k}}{a_{q_j^k q_{j+1}^k}} \text{ for } j = 1, 2, \cdots, (n_k - 1), \\ \frac{a_{p^k q^k}}{a_{q_{n_k}^k q^k}} \text{ for } j = n_k; \end{cases}$$

the l_k-th entry of $[D_k]$, $d_{l_k}^k$, is -1; and the remaining entries of $[D_k]$ are zeros.

Remark 2.1. When bus i and bus j $(i < j)$ are not connected, the coefficient C_{ij} is equal to zero, i.e., the variable Δf_{ij} and its corresponding entries of the matrix $[A]$ can all be deleted. Accordingly, the vector $[\Delta f]$ and the matrix $[A]$ can be compressed into smaller ones. Note that Δf_{ij} is the $(j - i)$-th element of $\left[\Delta \underline{f}_i\right]$ and $\left[\Delta \underline{f}_k\right]$ has $(N - k)$ variables for $k = 1, 2, ..., (i - 1)$. Thus, when $C_{ij} = 0$, the $(j - i)$-th element of $\left[\Delta \underline{f}_i\right]$, the $(j-i)$-th column of $[A_i]$ and the $[(i-1)(2N-i)/2 + (j-i)]$-th column of $[D]$ can all be eliminated. Therefore, the *System CEDCP* is considered as follows:

System CEDCP

$$\text{(2.10)} \qquad \text{Minimize} \; : \; \sum_{n \in N'} C_n(\Delta f_n)$$

$$\text{(2.11)} \qquad \text{Subject to} \; : \; [A]\,[\Delta f] = [0]$$

$$\text{(2.12)} \qquad [\Delta f_{min}] \leq [\Delta f] \leq [\Delta f_{max}]$$

where : $[A]$ is a $(b+1) \times (b + |N'|)$ matrix;
$[\Delta f] = \left[\Delta \underline{f}_1^t \; \Delta \underline{f}_2^t \; \cdots \; \Delta \underline{f}_{N-1}^t \; \Delta P_{G_1} \; \Delta P_{G_2} \; \cdots \; \Delta P_{G_{|N'|}}\right]^t$,
which is a $(b + N') \times 1$ vector;
$[0]$ is a $(b+1) \times 1$ vector with zero entries;
$[\Delta f_{min}] = [f_{min}] - [f]$; $[\Delta f_{max}] = [f_{max}] - [f]$;
$[\Delta f_{min}], [f_{min}], [\Delta f_{max}], [f_{max}]$, and
$[f]$ are all $(b + |N'|) \times 1$ vectors.

Remark 2.2. In above linearized optimization formulation (System CEDCP), the variables are the incremental flows (Δf), and the constraints 2.11–2.12 are both linear.

Here, we shall claim that the matrix $[A]$ in the above formulation is of full row rank. This will be proved in the following *Lemma 2.1.*

Lemma 2.1 The matrix $[A]$ in the equality constraints 2.11 is of full row rank.

Proof. Deleting the first row vector of the matrix $[A]$, we have a reduced $b \times (b + |N'|)$ matrix, denoted by $[A']$ as follows:

$$[A'] = \begin{bmatrix} A'' & \vdots & -I'_{(N-1)\times|N'|} \\ \cdots & \vdots & \cdots\cdots\cdots \\ D & \vdots & 0 \end{bmatrix}. \tag{2.13}$$

Here, we will show that the matrix $[A']$ is of full row rank. And then, based on this property of matrix $[A']$, *Lemma 2.1* will be proved.

Now, reorder the entries of the incremental flow variables, $[\Delta f]$, such that the first $(N-1)$ entries are the transmission tree flows; the next $(b-N+1)$ entries are the link flows, and the last $|N'|$ entries are the generation variables. Then the matrix $[A']$ becomes:

$$[A']_{new} = \begin{bmatrix} & \tilde{A} & & \vdots & -I'_{(N-1)\times|N'|} \\ \cdots & \cdots & \cdots\cdots\cdots & \vdots & \cdots\cdots \\ \hat{D} & \vdots & -I_{(b-N+1)\times(b-N+1)} & \vdots & 0 \end{bmatrix}. \tag{2.14}$$

Let $[\tilde{D}] = [\hat{D} \vdots - I_{(b-N+1)\times(b-N+1)}]$. It is sufficient to prove that $\begin{bmatrix} \tilde{A} \\ \tilde{D} \end{bmatrix}$ is of full row rank. Now use the last $(b-N+1)$ link flow equations, $[\tilde{D}\ 0][\Delta f_{new}] = [0]$, to eliminate the dependent link flows in the first $(N-1)$ equations. Then $\begin{bmatrix} \tilde{A} \\ \tilde{D} \end{bmatrix}$ becomes

$$\begin{bmatrix} \hat{A} & \vdots & 0 \\ \cdots & \vdots & \cdots\cdots\cdots \\ \hat{D} & \vdots & -I_{(b-N+1)\times(b-N+1)} \end{bmatrix},$$ where $[\hat{A}]$ is an $(N-1)\times(N-1)$ matrix.

Now claim that $[\hat{A}]$ is of full row rank. Before we show this, we observe that for any given $(N-1)$ tree flows, denoted by $[\Delta f_{tree}]$, they can be represented by the original $(N-1)$ linearly independent phase angle variables,

$\Delta\delta_2, \Delta\delta_3, \cdots, \Delta\delta_N$, as follows:

$$(2.15)\ [\Delta f_{tree}] = [T] \begin{bmatrix} \Delta\delta_2 \\ \Delta\delta_3 \\ \vdots \\ \Delta\delta_N \end{bmatrix}, \text{ where } [T] \text{ is an } (N-1)\times(N-1) \text{ matrix.}$$

Note that $[T]$ is a nonsingular matrix. This can be proved by contradiction. Suppose $[T]$ is singular. Then there exists one row which can be represented by the linear combination of some other rows. As a result, the corresponding tree flow also can be represented by the linear combination of the other corresponding tree flows. This contradicts the independent property of a tree. Hence, $[T]$ is nonsingular and of full row rank.

Also, $[\hat{A}]$ can be obtained through the conventional linearized formulation by the following way:

$$(2.16) \quad [\hat{A}] = -\begin{bmatrix} a_{22} & a_{23} & \cdots & a_{2N} \\ a_{32} & a_{33} & \cdots & a_{3N} \\ \vdots & \vdots & \ddots & \vdots \\ a_{N2} & a_{N3} & \cdots & a_{NN} \end{bmatrix} [T]^{-1} = -[A^*][T]^{-1},$$

where $[A^*]$ is a nonsingular *irreducibly diagonally dominant matrix* [18] in normal power systems since $a_{ii} = -\sum_{j=1;j\neq i}^{N} a_{ij}$ for $i = 2, 3, ..., N$, $a_{ij} \geq 0$ for all $i \neq j$, and there exists at least one diagonal term a_{ii} such that $|a_{ii}| > \sum_{j=2,j\neq i}^{N} |a_{ij}|$.

Accordingly, the rank of $[\hat{A}]$ is still (N-1) since $[T]^{-1}$ is nonsingular; that is, the rows of $[\hat{A}]$ are linearly independent. And hence $\begin{bmatrix} \hat{A} & 0 \\ \hat{D} & -I \end{bmatrix}$ is also of full row rank. Consequently, $\begin{bmatrix} \tilde{A} \\ \tilde{D} \end{bmatrix}$ is of full row rank and so is $[A']$. Therefore, the reduced matrix $[A']$ obtained by deleting one row, which contains one generation variable, from the first N rows of the matrix $[A]$ in the formulation *System OPF* is of full row rank.

Since the matrix $[A']$ is of full row rank, together with the fact that the $(b+1)$-th column of the matrix $[A']$ is a zero vector and the $(b+1)$-th entry of the first row of $[A]$ is -1, we can conclude that the matrix $[A]$ is of full row rank. □

Remark 2.3. From the proof of *Lemma 2.1*, we know that $[A^*]$ is nonsingular. Thus, when ΔP_G's are given, then $\Delta\delta$'s are uniquely determined, and so do Δf's. Conversely, when feasible Δf's are given, then $\Delta\delta$'s will be uniquely determined, and so do ΔP_G's.

3. The textured decomposition. We define the framework of the textured algorithm by introducing the following terms:

- Given a system $G = (X, U, N')$, where X, U, and N' are the sets of bus nodes, transmission lines and generations in G, respectively. Divide system G into small subsystems. A **subsystem** $G_i = (X_i, U_i, N_i')$ is a part of the system G and it satisfies:
  1. $X_i \subset X, U_i \subset U, N_i' \subset N'$.
  2. For each transmission line i-j in U_i, $i, j \in X_i$.
  3. G_i is a connected system containing at least one generator bus.
  4. For each generator bus $n \in X_i$, generation $n \in N_i'$.
- A set of subsystems $\{G_i | i = 1, 2, ..., M\}$ is called **a set of basic subsystems** if

$$\cup_{i=1}^{M} G_i = G.$$

- For the textured computation, the M basic subsystems, $\{G_i, i = 1, 2, \ldots, M\}$, are placed on L levels such that
  1. At the same level, no intersection exists between any two subsystems; hence, subsystems can be processed in parallel.
  2. At the same level, the subsystems have (approximately) equal sizes to reduce the waiting time.
  3. Each level has about the same number of subsystems which matches the number of available parallel processors to avoid idling processors.
  4. The number of levels, L, should be as small as possible to reduce sequential steps.
- An l-th level of basic subsystems, denoted by $\mathcal{G}^l = (\mathcal{X}^l, \mathcal{U}^l, \mathcal{N}^{l'})$, is a collection of basic subsystems $\{G_{l_i} | i = 1, 2, ..., n_l; n_l$ is the number of basic subsystems in the l-th level $\}$, such that $G_{l_i} \cap G_{l_j} = \emptyset$ if $i \neq j$; and $\mathcal{X}^l = \cup_{i=l_1}^{l_{n_l}} X_i, \mathcal{U}^l = \cup_{i=l_1}^{l_{n_l}} U_i, \mathcal{N}^{l'} = \cup_{i=l_1}^{l_{n_l}} N_i'$.
- $\cup_{l=1}^{L} \mathcal{G}^l = \cup_{l=1}^{L} \cup_{j=1}^{n_l} G_{l_j} = \cup_{i=1}^{M} G_i = G$.

Accordingly, two stages are used to set up a textured model.

Stage 1) **Grouping** (or **Decomposition**) : Divide a large System into M basic subsystems $\{G_i, i = 1, 2, \ldots, M\}$.

Stage 2) **Stratification** (or **Levelization**) : Arrange the basic subsystems on L levels such that L is minimum among all possible arrangements and the union of the subsystems on different levels will cover the whole large system.

Note that in the paper, we shall investigate the requirements on the subsystems which can guarantee the converged solution is optimal.

The following example illustrates the grouping and stratification concepts.

Example 3.1. Suppose we have two available processors to analyze the power flow of system X in Fig. 3.1. We can decompose the system X and

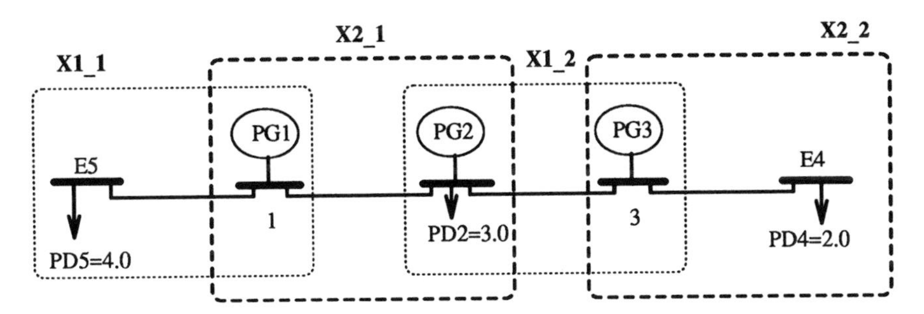

FIG. 3.1. *A 5-bus system and its textured decomposition.*

put into two levels as shown.

The subsystems $X_{1_1}, X_{1_2}, X_{2_1}$, and X_{2_2} constitute a set of basic subsystems. Note that the textured decomposition requires overlapping subsystems, through which we can patch the local solutions to form the global solutions.

3.1. Formulation of subsystem CEDCP. For a textured decomposed system, let $\{f_l\}$ be the set of line flows in the l-th level of basic subsystems and $\{e_l\}$ be the external flows in the l-th level. The external flows in a level are the union of the external flows of each subsystems in the level. An external flow of a subsystem is the flow whose branch has one terminal node in the subsystem and the other is not. When the incremental external flows are frozen, we can isolate the l-th level of basic subsystems from the rest of system. Hence, the *Level Subsystems CEDCP* is considered as follows:

Level Subsystems CEDCP

$$\text{Minimize} \quad : \quad \sum_{n \in \mathcal{N}^{l'}} C_n(\Delta f_n) \tag{3.1}$$

$$\text{Subject to} \quad : \quad [A_l]\,[\Delta f_l] = [\Delta e_l] \tag{3.2}$$

$$[\Delta f_{l,min}] \leq [\Delta f_l] \leq [\Delta f_{l,max}] \tag{3.3}$$

where all terms with subscript l are the corresponding terms in the l-th level of subsystem $\mathcal{G}^l = (\mathcal{X}^l, \mathcal{U}^l, \mathcal{N}^{l'})$.

A level of basic subsystems contains mutually nonintersected basic subsystems if the incremental external flows of each subsystems are frozen. Thus, the *Level Subsystems CEDCP* consists of n_l independent *Subsystem CEDCP* as follows:

for each $i = l_1, l_2, \ldots, l_{n_l}$, we have

Subsystem CEDCP

$$\text{(3.4)} \qquad \text{Minimize} \; : \; \sum_{n \in N_i'} C_n(\Delta f_n)$$

$$\text{(3.5)} \qquad \text{Subject to} \; : \; [A_{(i)}]\,[\Delta f_{(i)}] = [\Delta e_{(i)}]$$

$$\text{(3.6)} \qquad [\Delta f_{(i),min}] \leq [\Delta f_{(i)}] \leq [\Delta f_{(i),max}]$$

where all terms with subscript (i) are the corresponding terms in the l-th level of subsystem $G_i = (X_i, U_i, N_i')$.

3.2. The corresponding parallel textured algorithm. For a textured decomposed system, we have L levels of basic subsystems $\mathcal{G}^l, l = 1, 2, \ldots, L$, such that $G_{l_i} \cap G_{l_j} = \emptyset$ for $l_i, l_j \in \{l_1, \ldots, l_{n_l}\}$ and $l_i \neq l_j$. Let $k = max\{n_l | l = 1, 2, \ldots, L\}$. Accordingly, we describe the parallel textured algorithm as follows:

Step 0 : **for** $i = 1$ **to** k
create processor i
end for

Step 1 : Use a parallel power flow analysis to find one set of feasible flows,
$f(0)$. Initialize $m = 1$. $\{m$ denotes the m-th iteration of the outer
loop (steps 2-13), and ϵ_{outer} is the stopping criterion.$\}$

Step 2 : **while** $((m = 1)$ **or**
$(max_{u \in \{U \cup N'\}} |f_u^{new}(m-1) - f_u(m-1)| > \epsilon_{outer}))$ **do**

Step 3 : Initialize $\Delta f(0) = 0$. {It is a feasible solution for <u>*System CEDCP*</u>.} Initialize $j = 0$. $\{j$ denotes the j-th iteration of the inner loop (steps 5–7); l denotes the l-th level, and ϵ_{inner} denotes the stopping criterion for the inner loop.$\}$

Step 4 : **for** $l = 1$ **to** L **do**
for processor $i = l_1$ **to** l_{n_l} **do in parallel**
Update $[A_{(i)}]$ by using the
result of the $(m-1)$-th iteration.
end for
end for

Step 5 : **while** $((j = 0)$ **or**
$(max_{u \in \{U \cup N'\}} |\Delta f_u(j) - \Delta f_u(j-1)| > \epsilon_{inner}))$ **do**

Step 6 : $j = j + 1$.

Step 7 : **for** $l = 1$ **to** L **do**
for processor $i = l_1$ **to** l_{n_l} **do in parallel**
1. Compute the external flows $\Delta e_{(i)}$ by using the result of the $(j-1)$-th iteration.
2. Solve <u>*Subsystem CEDCP*</u> over G_i and find

the incremental line flows solution $\{\Delta \hat{f}_{(i)}\}$.
3. Update $\Delta f(j)$ by setting
$\Delta f_u = \Delta \hat{f}_u$, where $u \in \{U_i \cup N_i'\}$.

end for

end while

Step 8 : Update $f(m)$ by setting $f(m) = f(m-1) + \Delta f(j)$.

Step 9 : Compute $\{\Delta\delta_2, \Delta\delta_3, ..., \Delta\delta_N, \Delta P_{G_1}, \Delta P_{G_2}, ..., \Delta P_{G_{|N'|}}\}$ from $\{\Delta f(j)\}$.

Step 10: Update $\{\delta_p(m), P_{G_q}(m)\}$ by setting $\delta_p(m) = \delta_p(m-1) + \Delta\delta_p$ and $P_{G_q}(m) = P_{G_q}(m-1) + \Delta P_{G_q}$, where $p = 2, 3, ..., N$; $q = 1, 2, ..., |N'|$.

Step 11: Run the parallel decoupled power flow analysis by setting $P_{G_2}, P_{G_3}, ..., P_{G_{|N'|}}$ at the values obtained at step 10, and use the δ as the initial values to find the line flows, denoted by $\{f^{new}(m)\}$.

Step 12: $m = m + 1$.

Step 13: Update $f_u(m)$ by setting $f_u(m) = f_u^{new}(m-1)$, where $u \in \{U \cup N_i'\}$.

end while

Step 14: stop.

Remark 3.2. The feasible flow is a solution which satisfies the (admissible) constraints , i.e. a solution of the power flow problem. Thus, at step 1 and step 11, any nonparallel or parallel power flow program, such as Gauss-Seidel type algorithm [19, 20], can be used to find a feasible flow solution. Also, our textured model/algorithm is optimization algorithm independent because we can incorporate any optimization algorithm, with or without sparse matrix technique, to solve the *Subsystem CEDCP*. Moreover, at step 9, the phase angles can be easily computed parallelly from $\Delta f(j)$ without matrix inversion since $\Delta\delta_1 = 0$ and $\Delta\delta_i$ which connected with bus 1 can be determined concurrently by simple computations. Similarly, $\Delta\delta_j$ which connected with bus i can be obtained concurrently and so on until the problem is solved.

4. Properties of the textured algorithm. Note that the above algorithm consists of one outer loop (steps 2–13) and one inner loop (steps 5–7). The outer loop is used to enforce the nonlinear constraints, which may slightly violated due to the linearized approach. The inner loop is used to solve the linearized economic dispatch problem using the textured algorithm. The inner-loop algorithm is an *iterative descent algorithm* [21] since as each new point is generated by the algorithm, the corresponding value of the objective function decreases until the algorithm converges. In addition, each iteration j of the inner-loop algorithm can be regarded as a point-to-set mapping $M : \Omega \longrightarrow \Omega$, where Ω is defined as the feasible solution set of *System CEDCP*. (Practically, algorithms are point-to-point mappings, that is, a particular computer program for an algorithm executed twice from the same starting point will generate two copies of the same sequence. However, the more generalized concept leads to simpler convergence analysis. That is, two computer programs, designed from the same basic idea, may differ slightly in some details, and therefore may not produce identical results when given the same starting point; but both programs may be regarded as implementations of the same point-to-set mappings. See [21] for detailed examples.) In this paper, we shall concentrate on the exact convergence of the inner loop solutions. The outer loop will have minor influence on the exact convergence as long as the ΔP_G's are restarted by a reasonably small bound. Some preliminary concepts of our convergence proofs can be found in [5] and [10].

Note that the feasible solution set Ω is closed, convex, and bounded since it is defined by the equality constraints (*hyperplanes*) and the inequality constraints (*closed half spaces*). In addition, the objective function is strictly convex and continuous in ΔP_G.

To formally prove the convergence properties, we first observe that M consists of a series of *Level Subsystems CEDCP* from $l = 1$ to L, and each l-th *Level Subsystems CEDCP* is also a point-to-set mapping $M_l : \Omega \longrightarrow \Omega$. When the incremental external flows in a level of subsystems are fixed, we have

$$\Delta\hat{f} \in M_l(\Delta\tilde{f}); \quad \Delta\tilde{f}, \Delta\hat{f} \in R^{(b+|N'|)}, \tag{4.1}$$

where the value of components of $\Delta\hat{f}$ is defined as follows:

(1) $\Delta\hat{f}_u = \Delta\tilde{f}_u$ for $u \in U \setminus \mathcal{U}_l$;

(2) $\Delta\hat{f}_u$ is the solution of *Level Subsystems CEDCP* over $\mathcal{G}_l$ with $e_l = \Delta e_l(\Delta\tilde{f})$;

(3) $\Delta\hat{f}_l = \Delta\tilde{f}_l$ if $\Delta\tilde{f}_l$ also solves the above *Level Subsystems CEDCP*.

Thus, the mapping M can be represented as a composite mapping of $M_l, l = 1, 2, ..., L$, denoted by

$$M = M_L \circ M_{L-1} \circ \cdots \circ M_1 \tag{4.2}$$

By the definition of closed mapping and the definition of $M_l : \Omega \longrightarrow \Omega$, where Ω is bounded and closed, it is clear that M_l is a closed mapping for each l. The following lemma (see [21] for proof) tells us that the composite mapping of closed mappings is also closed.

LEMMA 4.1. *Let $A : X \longrightarrow Y$ and $B : Y \longrightarrow Z$ be point-to-set mappings. Suppose A is closed at x and B is closed on $A(x)$. Suppose also that if $x_k \longrightarrow x$ and $y_k \in A(x_k)$, there is a y such that, for some subsequence $\{y_{k_i}\}, y_{k_i} \longrightarrow y$. Then the composite mapping $C = BA$ is closed at x.*

By *Lemma 4.1* and the definitions of M_l and M, the following lemma and corollary can be easily verified.

LEMMA 4.2. *Let $\Delta\hat{f} \in M_l(\Delta\tilde{f})$,then $\Delta\tilde{f} \in \Omega$ implies $\Delta\hat{f} \in \Omega$ and $\sum_{i=1}^{|N'|} C_i(\Delta\hat{f}) \leq \sum_{i=1}^{|N'|} C_i(\Delta\tilde{f})$, the inequality holds if $\Delta\hat{f} \neq \Delta\tilde{f}$.*

COROLLARY 4.3. *Let $\Delta\hat{f} \in M(\Delta\tilde{f})$,then $\Delta\tilde{f} \in \Omega$ implies $\Delta\hat{f} \in \Omega$ and $\sum_{i=1}^{|N'|} C_i(\Delta\hat{f}) \leq \sum_{i=1}^{|N'|} C_i(\Delta\tilde{f})$, the inequality holds if $\Delta\hat{f} \neq \Delta\tilde{f}$.*

Thus, we have $M : \Omega \longrightarrow \Omega$ since the inner-loop algorithm is initiated by a feasible solution of *System CEDCP*.

Accordingly, the inner-loop algorithm iteratively solves the *Level Subsystems CEDCP* with the incremental external flows in a level of basic subsystems fixed at the most recently updated values. Hence, if it stops, it must stay at a stationary point for each $M_l, l = 1, 2, ..., L$, or at a stationary point of the mapping M. The stationary point is defined as a converged point of an algorithm, which is not necessary to be the global optimal solution. We define the solution set $\mathcal{F}$ of *System CEDCP* as the set of all possible converged solutions (fixed points) for our inner-loop algorithm, i.e. $\mathcal{F} = \{\Delta\hat{f} \in \Omega | \Delta\hat{f} = M(\Delta\hat{f})\}$ or $\mathcal{F} = \{\Delta\hat{f} \in \Omega | \Delta\hat{f} = M_l(\Delta\hat{f}), l = 1, 2, ..., L\}$. Before we explore the convergence characteristic of the inner-loop algorithm, we should guarantee the existence of the solution set $\mathcal{F}$ of *System CEDCP*. In particular, it is important to have the global minimum of the *System CEDCP* contained in the solution set $\mathcal{F}$; otherwise, even when the initial guess is the global minimum solution, the inner-loop algorithm will not converge to the global minimum solution. The following lemma provides the existence of the solution set $\mathcal{F}$.

LEMMA 4.4. *Let Δf^* be the global minimum of the* System CEDCP. *Then $\Delta f^* \in \mathcal{F}$ and accordingly $\mathcal{F} \neq \emptyset$.*

Proof. It is enough to show that $\Delta f^* \in \mathcal{F}$. Suppose $\Delta f^* \notin \mathcal{F}$. Since $\Delta f^* \in \Omega$, then from the definition of $\mathcal{F}$ and *Corollary 4.3*, there exists an $l \in [1, L]$, such that $\Delta\tilde{f} \in M_l(\Delta f^*)$ and $\sum_{i=1}^{|N'|} C_i(\Delta\tilde{f}) < \sum_{i=1}^{|N'|} C_i(\Delta f^*)$. This contradicts that Δf^* is the global minimum solution of *System CEDCP*. □

4.1. Convergence of the inner-loop algorithm. *Lemma 4.4* guarantees the existence of the solution set $\mathcal{F}$ of *System CEDCP*. Now, we need to show that the inner-loop algorithm will converge to the solution set $\mathcal{F}$. That is, we have to prove that as points are generated by the inner-loop algorithm, each new point will have lower cost until the solution set $\mathcal{F}$ is reached.

We apply the *Global Convergence Theorem* [21] (see appendix) to prove that the inner-loop textured algorithm will converge to the solution set $\mathcal{F}$ by checking the following three sufficient conditions.

THEOREM 4.5. *The inner-loop textured algorithm satisfies the suffi-*

cient conditions in the Global Convergence Theorem:
(1) for each j-th iteration, all points $\Delta f(j), j = 1, 2, ...$, are in a *compact set;*
(2) for cost function $\sum_{i=1}^{|N'|} C_i$ *on the feasible solution set* Ω
  (a) if $\Delta f(j) \notin \mathcal{F}$ *then* $\sum_{i=1}^{|N'|} C_i(\Delta f(j+1)) < \sum_{i=1}^{|N'|} C_i(\Delta f(j))$ *for all* $\Delta f(j+1) \in M(\Delta f(j))$,
  (b) if $\Delta f(j) \in \mathcal{F}$ *then* $\sum_{i=1}^{|N'|} C_i(\Delta f(j+1)) \leq \sum_{i=1}^{|N'|} C_i(\Delta f(j))$ *for all* $\Delta f(j+1) \in M(\Delta f(j))$;
(3) M *is closed on* $\Omega \setminus \mathcal{F}$.

Proof. Since $M : \Omega \longrightarrow \Omega$, where Ω is bounded, closed, and convex, this implies that the generated sequence is bounded in a compact set which is defined by some hyperplanes (equality constraints), capacity and flow constraints (inequality constraints). Thus, condition (1) holds.

Condition (2) is a direct consequence from *Corollary 4.3*. Now, we will prove that condition (3) is true as follows.

By *Lemma 4.1* and the definitions of $\Delta\hat{f}$ and M_l, it is enough to show that M_l, a point-to-set mapping, is closed on $\Omega \setminus \mathcal{F}$ for each $l = 1, 2, ..., L$.

Suppose $\{\Delta\tilde{f}(j)\}$ and $\{\Delta\hat{f}(j)\}$ are two sequences in $\Omega \setminus \mathcal{F}$ with $\Delta\tilde{f}(j) \longrightarrow \Delta\tilde{f}$ and $\Delta\hat{f}(j) \in M_l(\tilde{f}(j)) \longrightarrow \Delta\hat{f}$. To prove the closedness of M_l on $\Omega \setminus \mathcal{F}$, we have to show that $\Delta\hat{f} \in M_l(\Delta\tilde{f})$.

Define $Y(\Delta\tilde{f}(j))$ as **the feasible solution set** of l-th *Level Subsystems OPF* with $e_l = e_l(\Delta\ \tilde{f}(j))$ for each j-th iteration. Clearly, $M_l(\Delta\ \tilde{f}(j)) \subset Y(\Delta\tilde{f}(j)) = Y(\Delta\tilde{f})$ and $Y(\Delta\tilde{f})$ is closed, bounded, and convex. Hence $\Delta\hat{f}(j)$, $\Delta\hat{f} \in Y(\Delta\tilde{f})$, and $Y(\Delta\hat{f}(j)) = Y(\Delta\hat{f}) = Y(\Delta\tilde{f})$.

Now, we prove condition (3) by contradiction. Suppose $\Delta\hat{f} \notin M_l(\Delta\tilde{f})$, then there exists a $\Delta\bar{f} \in Y(\Delta\tilde{f})$ such that $\Delta\bar{f} \in M_l(\Delta\tilde{f})$, $\Delta\bar{f} \neq \Delta\hat{f}$. This means that $\Delta\bar{f}$ is the optimal solution when $e_l = e_l(\Delta\tilde{f})$, thus we have

$$\sum_{i=1}^{|N'|} C_i(\Delta\bar{f}) < \sum_{i=1}^{|N'|} C_i(\Delta\hat{f}).$$

Now, for every $\Delta\check{f} \in T = \{\Delta f | \Delta f = (1-\alpha)\ \Delta\hat{f} + \alpha\Delta\bar{f}, \alpha \in (0,1)\}$, it is clear that $\Delta\check{f} \in Y(\Delta\tilde{f})$ since $Y(\Delta\tilde{f})$ is convex, hence $T \subset Y(\Delta\tilde{f})$. Furthermore, since the cost function is strictly convex in ΔP_G's and from the *Remark 2.3* of *Lemma 2.1*, we have

$$\begin{aligned} \sum_{i=1}^{|N'|} C_i(\Delta\check{f}) &< (1-\alpha)\sum_{i=1}^{|N'|} C_i(\Delta\hat{f}) + \alpha\sum_{i=1}^{|N'|} C_i(\Delta\bar{f}) \\ &< (1-\alpha)\sum_{i=1}^{|N'|} C_i(\Delta\hat{f}) + \alpha\sum_{i=1}^{|N'|} C_i(\Delta\hat{f}) \\ &= \sum_{i=1}^{|N'|} C_i(\Delta\hat{f}). \end{aligned}$$

If we can prove that T does not exist, together with the knowledge of the existence of $\Delta\hat{f}$, then we can conclude that the only possibility is that

$\Delta\bar{f}$ does not exist. Thus this implies that the assumption is wrong and hence $\Delta\,\hat{f} \in M_l(\Delta\tilde{f})$.

Let S denote the set defined by the inequality constraints of System OPF, and let Int S denote the interior of S. Now, suppose $\{T\cap \text{Int}S\} \neq \emptyset$, then for every $\Delta\check{f} \in \{T\cap \text{Int } S\}$, there exists an ϵ-ball, $B(\Delta\check{f}, \epsilon)$, contained in Int S. Let $\Delta\check{f}(j) = \Delta\check{f} + (\Delta\hat{f}(j) - \Delta\hat{f})$. Note that $\Delta\check{f}, \Delta\hat{f}(j)$, and $\Delta\hat{f}$ are all in $Y(\Delta\tilde{f}(j)) \subset S$. Since $\Delta\hat{f}(j) \longrightarrow \Delta\hat{f}$, we have $\Delta\check{f}(j) \longrightarrow \Delta\check{f}$. Hence there exists a positive integer N such that $\Delta\check{f}(j) \in B(\Delta\check{f}, \epsilon)$ for all $j \geq N$. Clearly, $\Delta\check{f}(j) \in Y(\Delta\tilde{f}(j))$ for $j \geq N$. Recall that $\Delta\hat{f}(j) \in M_l(\Delta\tilde{f}(j)) \subset Y(\Delta\tilde{f}(j))$, which means that $\Delta\hat{f}(j)$ is the minimum point at j-th iteration for M_l when $\Delta\tilde{f}(j)$ is given as initial, so for any $\Delta\check{f}(j) \neq \Delta\hat{f}(j)$, $\Delta\check{f}(j) \in Y(\Delta\tilde{f}(j))$, and $j \geq N$, we have

$$\sum_{i=1}^{|N'|} C_i(\Delta\check{f}(j)) \geq \sum_{i=1}^{|N'|} C_i(\Delta\hat{f}(j)).$$

By continuity of our cost function, as j approaches infinity, the above inequality becomes

$$\sum_{i=1}^{|N'|} C_i(\Delta\check{f}) \geq \sum_{i=1}^{|N'|} C_i(\Delta\hat{f}).$$

This contradicts $\sum_{i=1}^{|N'|} C_i(\Delta\check{f}) < \sum_{i=1}^{|N'|} C_i(\Delta\hat{f})$. Hence $\{T\cap\text{Int } S\}$ should be an empty set and T should lie on the boundary of S which consists of a finite number of hyperplanes of S.

Then there exists an ϵ-ball around $\Delta\check{f} \in T$, such that $\Delta\check{f}(j) = \Delta\check{f} + (\Delta\hat{f}(j) - \Delta\hat{f}) \in B(\Delta\check{f}, \epsilon) \cap S$ for all $j \geq M$ and $|\Delta\,\hat{f}(j) - \Delta\hat{f}| < \epsilon$, where M is a positive integer. Thus all these $\Delta\check{f}(j) \in Y(\Delta\tilde{f}(j))$ for $j \geq M$ and $\Delta\check{f}(j) \longrightarrow \Delta\check{f}$. Following similar arguments as proving $\{T\cap \text{Int}S\} = \emptyset$, we can show that the intersection of T and the boundary of S is also empty.

Thus $\Delta\hat{f} \in M_l(\Delta\tilde{f})$ and M_l is closed on $\Omega \setminus \mathcal{F}$. Based on *Lemma 4.1* and the above result, therefore, we conclude that the inner-loop textured algorithm satisfies condition (3). □

The following theorem, a direct consequence of *Theorem 4.5* and the *Global Convergence Theorem*, summarizes our results.

THEOREM 4.6. *The inner-loop parallel textured algorithm starts with an initial feasible solution of the* System CEDCP *will converge to its corresponding solution set* $\mathcal{F}$.

5. The exact convergence theorem. We adopted some insights of examples in [11,22] and the Kuhn-Tucker theorem in [21] to establish the following exact convergence theorem. The theorem states that when some conditions hold on the decomposed connected basic subsystems of the system G, the textured algorithm will converge to its global minimum solution.

THEOREM 5.1. : *Let $\Delta\hat{f}$ be the global minimum of* System CEDCP. *If*

1. *The system G is connected.*
2. *$\{G_{l_i}, i = 1, 2, ..., n_l; l = 1, 2, ..., L\}$ is a set of basic subsystems and each G_{l_i} is connected.*
3. *Each G_i satisfies the following: Given the* Subsystem CEDCP *over G_i with $|N_i'|$ generator buses and $\Delta e_i = \Delta e_i(\Delta\hat{f}_i)$, where $\Delta\hat{f}_i$ is an optimal solution. Let $g(\Delta f_i) = 0$ and $q(\Delta f_i) \leq 0$ denote the equality constraints and the inequality constraints, respectively. There exists a vector $h \in R^{(|U_i|+|N_i'|)}$ such that*
   A(i) $(\nabla q_j(\Delta\hat{f}_i), h) < 0$, for all $j \in I(\Delta\hat{f}_i)$,
   A(ii) $(\nabla g_j(\Delta\hat{f}_i), h) = 0$ for $j = 1, 2, ..., (|U_i| + 1)$,
   A(iii) $\nabla g_j(\Delta\hat{f}_i), j = 1, 2, ..., (|U_i| + 1)$, are linearly independent,

   where $I(\Delta\hat{f}_i)$ is the index set of the active constraints in $\Delta\hat{f}_i$. From now on, we shall refer to this condition as the **constraint qualification assumption**.
4. *Let $\Delta\hat{f}_i, \Delta\hat{f}_j, \Delta\hat{f}_{i\wedge j}$ be the incremental flows in $\Delta\hat{f}$ corresponding to connected subsystems G_i, G_j, and $G_{i\wedge j}$ respectively. Here, $G_{i\wedge j} = G_i \cap G_j$ denotes the overlapped part of G_i and G_j, which is assumed to be connected. A* **unique K-T vector** *exists for* Subsystem CEDCP *over $G_{i\wedge j}$ with $\Delta e_{i\wedge j} = \Delta e_{i\wedge j}(\Delta\hat{f})$, where $\Delta e_{i\wedge j}$ is the vector of injected external incremental flows to $G_{i\wedge j}$. (A K-T vector is defined as the vector consists of λ's, μ's associated with the corresponding equality and inequality constraints of the problem. The vector must satisfy the K-T condition [21] of the optimization problem.)*
5. *There exists an ordering $\{k_1, k_2, ..., k_M\}$ for the indices $\{1_1, 1_2, ..., 1_{n_1}, 2_1, 2_2, ..., 2_{n_2}, ..., L_1, L_2, ..., L_{n_L}\}$, such that $\cup_{i=k_1}^{k_j} G_i$ and $(\cup_{i=k_1}^{k_j} G_i) \cap G_{k_{j+1}}$ are both connected subsystems for all $j \leq M-1$. Furthermore, $(\cup_{i=k_1}^{k_j} G_i) \cap G_{k_{j+1}} = G_{k_j} \cap G_{k_{j+1}}$.*

then the parallel textured algorithm starting from any feasible (operating) point will converge to $\Delta\hat{f}$.

Physically, the need for the first condition is straightforward. The system should be connected so that the real power can be transmitted from bus to bus. The second condition just restates the requirement of textured decomposition.

Condition 3 is usually satisfied if none or few of the capacity limits are reached. This will be discussed in *Lemma 6.1*.

Condition 4 can be replaced by the **regularity** condition at the intersection of any two overlapping subsystems since the regularity condition implies uniqueness of the K-T vector for our problem. Either condition will guarantee compatible K-T vectors at two different levels so that they can be patched as a K-T vector for the optimization problem on the union of the two subsystems.

Condition 5 is needed so that we can inductively patch the partial solution from each level to form the global overall solution.

COROLLARY 5.2. : *Let* $\Delta\hat{f}$ *be a global minimum of* System CEDCP. *If*

A). Conditions (1),(2),(3) and (5) in the above theorem are satisfied.

B). The following condition is also satisfied:

(4').Let $\Delta\hat{f}_i, \Delta\hat{f}_j, \Delta\hat{f}_{i\wedge j}$ *denote the incremental flows in* $\Delta\hat{f}$ *corresponding to connected subsystems* G_i, G_j, *and* $G_{i\wedge j}$ *respectively, where* $G_{i\wedge j}$ *is defined as that in condition (4).* $\Delta\hat{f}_{i\wedge j}$ *is a regular point of* Subsystem CEDCP *over* $G_{i\wedge j}$ *with* $\Delta e_{i\wedge j} = \Delta e_{i\wedge j}(\Delta\hat{f})$.

then the parallel textured algorithm starting from any feasible (operating) point will converge to $\Delta\hat{f}$.

6. Proof of the exact convergence theorem

6.1. Necessary and sufficient conditions for optimality of the subsystems. To show the necessary and sufficient condition for our problem, we need the *constraint qualification assumption* to apply the *generalized Kuhn-Tucker theorem* in [23] (see appendix).

For our problem, $A(iii)$ in the *constraint qualification assumption* is always true because $[A_i]$ is of full row rank. Also, $A(i)$ is always satisfied since each incremental flow can only reach either its upper bound or its lower bound and hence the gradient of the active inequality constraints are always linearly independent. Thus the constraint qualification [23] is satisfied; that is, there exists a vector $h \in R^{(|U_i|+|N_i'|)}$ such that $(\nabla q_j(\Delta\hat{f}_i), h) < 0$, for all $j \in I(\Delta\hat{f}_i)$.

LEMMA 6.1. *When the optimal solution* $\Delta\hat{f}_i$ *is regular, the* constraint qualification assumption *always holds.*

Proof. The only thing we need to prove is that there always exists a vector h with components $h_1, h_2, ..., h_{(|U_i|+|N_i'|)}$, satisfying $A(i)$ such that $A(ii)$ is also satisfied.

Let us define

$$\begin{aligned} H &= \{h|(\nabla q_j(\Delta\hat{f}_i), h) < 0, j \in I(\Delta\hat{f}_i)\} \\ &= \{h|h_j < 0, j \in I_u(\Delta\hat{f}_i); h_j > 0, j \in I(\Delta\hat{f}_i) \setminus I_u(\Delta\hat{f}_i)\}, \end{aligned}$$

where $I_u(\Delta\hat{f}_i)$ is the index set of the active constraints reaching their upper bounds in $\Delta\hat{f}_i$. Also, we define a <u>*kernel space*</u> $ker\{A_i\} = \{h|A_i h = 0\}$. Then we will prove that under the condition of regularity, there exist a vector h in H such that h is also in $ker\{A_i\}$.

Let $S = \{a^j | j \notin I(\Delta\hat{f}_i)\}$, where a^j is the j-th column vector of $[A_i]$. By the definition of regularity, the gradient vectors $\nabla g_j(\Delta\hat{f}_i), 1 \leq j \leq (|U_i| + 1)$, and $\nabla q_j(\Delta\hat{f}_i), j \in I(\Delta\hat{f}_i)$ are linearly independent. Thus, the set S must contain $(|U_i| + 1)$ linearly independent vectors. This means that the number of incremental flows in G_i which reach the limits (i.e., the number of active inequality constraints) must be less than or equal to $(|N_i'| - 1)$. Also, let $I(S)$ be an index set of $(|U_i| + 1)$ linearly independent vectors in S, and $D(S)$ be the index set of the rest of vectors in S. Consider $A_i h = 0$. Choose $h_j < 0$ for $j \in I_u(\Delta\hat{f}_i)$, $h_j > 0$ for $j \in I(\Delta\hat{f}_i) \setminus I_u(\Delta\hat{f}_i)$, and $h_j \in R$ for $j \in D(S)$. Then we can always find a solution to h_j for $j \in I(S)$ since the vectors $\{a^j | j \in I(S)\}$ are linearly independent. Clearly, such a vector h is both in H and in $ker\{A_i\}$.

Therefore, we conclude that the regularity of the optimal solution $\Delta\hat{f}_i$ implies our constraint qualification assumption. □

Remark 6.2. For normal power system operations, few line flow variables will hit their capacity limits. Thus normally the system is usually operating at a regular point and the constraint qualification assumption is usually valid. In addition, the *penalty method* [21] can be used to prevent the incremental flows from hitting their limits for stressed system conditions. The following lemma states a first-order necessary condition of optimality for the *Subsystem CEDCP*.

LEMMA 6.3. (Necessary Condition) *Let* $G_i = [X_i, U_i, N_i']$ *be a connected subsystem with* $|N_i'|$ *generator buses. Also, the* constraint qualification assumption *is satisfied. If* $\Delta\hat{f}_i$ *is the minimum solution of* Subsystem CEDCP *over* G_i *with* Δe_i, *then there exists a vector* $\lambda_i \in R^{(|U_i|+1)}$, *and a vector* $\mu_i \in R^{(|U_i|+|N_i'|)}$ *such that*

$$-\nabla_{\Delta f_i}[\sum_i C_i(\Delta\hat{f}_u)] + (A_i)^t \lambda_i + \mu_i = 0 \tag{6.1}$$

$$(\mu_i)^t \Delta\hat{f}_i = 0 \tag{6.2}$$

which are the Kuhn-Tucker optimality conditions for the Subsystem CEDCP.

Note that the component μ_{ij} in μ_i is nonpositive (nonnegative) when the corresponding flows variable $\Delta\hat{f}_{ij}$ hits its upper (lower) bound.

Proof. Directly from the *constraint qualification assumption* and the *generalized Kuhn-Tucker theorem* in [23] (see appendix), we can find a vector $\lambda_i \in R^{(|U_i|+1)}$ and a vector $\mu_i \in R^{(|U_i|+|N_i'|)}$ such that equations 6.1 and 6.2 are satisfied. □

Now, we describe the sufficient condition of optimality for the *Subsystem CEDCP* in the following lemma.

LEMMA 6.4. (Sufficient Condition) *Let $G_i = [X_i, U_i, N_i']$ be a connected subsystem with $|N_i'|$ generator buses. Also, the constraint qualification assumption is satisfied. If there exists $\lambda_i \in R^{(|U_i|+1)}$ and $\mu_i \in R^{(|U_i|+|N_i'|)}$ such that*

$$-\nabla_{\Delta f_i}[\sum_i C_i(\Delta \hat{f}_u)] + (A_i)^t \lambda_i + \mu_i = 0 \tag{6.3}$$

$$(\mu_i)^t \Delta \hat{f}_i = 0 \tag{6.4}$$

then $\Delta \hat{f}_i$ will be the minimum solution of Subsystem CEDCP *over G_i with Δe_i. Note that the component μ_{ij} in μ_i is nonpositive (nonnegative) when the corresponding flows variable $\Delta \hat{f}_{ij}$ hits its upper (lower) bound.*

Proof. The *constraint qualification assumption* and equations 6.3 and 6.4 indicate that the Kuhn-Tucker optimality conditions for the *Subsystem OPF* over G_i with $\Delta e_i = \Delta e_i(\Delta \hat{f}_i)$ is satisfied. Thus, $\Delta \hat{f}_i$ is a minimum solution for the *Subsystem OPF* over G_i with $e_i = e_i(\Delta \hat{f}_i)$. Note that the cost function is strictly convex in the subvector $\Delta \hat{P}_G$ of $\Delta \hat{f}_i$; thus, $\Delta \hat{P}_G$ is the unique solution to minimize the cost function. In addition, based on the *Remark 2.3* of *Lemma 2.1*, we can conclude that the remaining variables in the flow vector have a unique solution. Accordingly, we can conclude that $\Delta \hat{f}_i$ is the minimum solution for the *Subsystem OPF* over G_i with $e_i = e_i(\Delta \hat{f}_i)$. □

6.2. Intuition behind the exact convergence theorem. One example in [1] has demonstrated that the textured algorithm may converge to a non-optimal solution if some conditions are not satisfied. We need the exact convergence theorem to insure its convergence to a global minimum. In this subsection, we present the basic ideas of the proof of the theorem. The details will be addressed in the following subsection.

First of all, let G_i and G_j be any two connected subsystems of G such that the overlapping subsystem $G_{i\wedge j} = G_i \cap G_j$ is a connected subsystem. Let $\Delta \hat{f}_i$ and $\Delta \hat{f}_j$ be the flows in $\Delta \hat{f} \in \mathcal{F}$ corresponding to the subsystems G_i and G_j, respectively; and $\Delta \hat{f}_i$ and $\Delta \hat{f}_j$ be minimum solutions for *Subsystem CEDCP* over G_i with $\Delta e_i = \Delta e_i(\Delta \hat{f})$ and over G_j with $\Delta e_j = \Delta e_j(\Delta \hat{f})$, respectively. Also, let $\Delta \hat{f}_{i\wedge j}$ be the common flows in $\Delta \hat{f}_i$ and $\Delta \hat{f}_j$ corresponding to $G_{i\wedge j}$, and $\Delta \hat{f}_{i\vee j}$ be the flows in $\Delta \hat{f} \in \mathcal{F}$ corresponding to $G_{i\vee j} = G_i \cup G_j$, the union of subsystems G_i and G_j. We shall claim that if a unique K-T vector for *Subsystem CEDCP* over $G_{i\wedge j}$ with

$e_{i\wedge j} = e_{i\wedge j}(\Delta \hat{f})$ exists, then $\Delta \hat{f}_{i\vee j}$ is the minimum solution for *Subsystem CEDCP* over $G_{i\vee j}$ with $\Delta e_{i\vee j} = \Delta e_{i\vee j}(\Delta \hat{f})$, where $\Delta e_{i\vee j}$ is the vector of injected external flows to $G_{i\vee j}$.

Condition 5 states that for a set of M basic subsystems, an order of those subsystems exists such that for any $m \in I, m \leq M-1$, the intersection of the union of the first m subsystems and the $(m+1)$-th subsystem is connected; furthermore, this intersection is identical to the intersection of the m-th and the $(m+1)$-th subsystems. Under this condition and the first claim, we can use mathematical induction to prove that the $\Delta \hat{f} \in \mathcal{F}$ is a global minimum solution of the *Subsystem CEDCP*.

Now, we prove our first claim through the next lemma.

LEMMA 6.5. *Let $\Delta \hat{f} \in \mathcal{F}$. $\Delta \hat{f}_i$ and $\Delta \hat{f}_j$ denote the corresponding subvectors of $\Delta \hat{f}$ related to G_i and G_j. Also, the* constraint qualification assumption *is satisfied. If (i) $\Delta \hat{f}_i$ and $\Delta \hat{f}_j$ are the minimum solutions for* Subsystem CEDCP *over G_i with $\Delta e_i = \Delta e_i(\Delta \hat{f})$ and over G_j with $\Delta e_j = \Delta e_j(\Delta \hat{f})$, respectively; (ii) $G_{i\wedge j}$ is a connected subsystem; (iii) a unique K-T vector exists for* Subsystem CEDCP *over $G_{i\wedge j}$ with $\Delta e_{i\wedge j} = \Delta e_{i\wedge j}(\Delta \hat{f})$, let $\Delta \hat{f}_{i\wedge j}$ denote the common part of flows in $\Delta \hat{f}_i$ and $\Delta \hat{f}_j$ corresponding to $G_{i\wedge j}$, then $\Delta \hat{f}_{i\vee j}$, the subvector of $\Delta \hat{f}$ corresponding to the connected subsystem $G_{i\vee j} = G_i \cup G_j$, is the minimum solution for* Subsystem CEDCP *over $G_{i\vee j}$ with $\Delta e_{i\vee j} = \Delta e_{i\vee j}(\Delta \hat{f})$.*

Proof. From assumption (i) and *Lemma 6.3*, we have

$$-\nabla_{\Delta f_i}[\sum_i C_i(\Delta \hat{f}_u)] + (A_i)^t \hat{\lambda}_i + \hat{\mu}_i = 0, \tag{6.5}$$

$$(\hat{\mu}_i)^t \Delta \hat{f}_i = 0; \tag{6.6}$$

and

$$-\nabla_{\Delta f_j}[\sum_j C_j(\Delta \hat{f}_u)] + (A_j)^t \tilde{\lambda}_j + \tilde{\mu}_j = 0, \tag{6.7}$$

$$(\tilde{\mu}_j^t) \Delta \hat{f}_j = 0; \tag{6.8}$$

where $\hat{\lambda}_i \in R^{(|U_i|+1)}$, $\hat{\mu}_i \in R^{|U_i|+|N_i'|}$; $\tilde{\lambda}_j \in R^{(|U_j|+1)}$, $\tilde{\mu}_j \in R^{|U_j|+|N_j'|}$.

Furthermore, assumption (iii) implies that

$$\hat{\lambda}_{i\wedge j} = \tilde{\lambda}_{i\wedge j} \in R^{(|U_{i\wedge j}|+1)} \text{ and } \hat{\mu}_{i\wedge j} = \tilde{\mu}_{i\wedge j} \in R^{|U_{i\wedge j}|+|N_{i\wedge j}'|}, \tag{6.9}$$

where $\hat{\lambda}_{i\wedge j}, \tilde{\lambda}_{i\wedge j}, \hat{\mu}_{i\wedge j}$, and $\tilde{\mu}_{i\wedge j}$ are subvectors of $\hat{\lambda}_i, \tilde{\lambda}_j, \hat{\mu}_i$, and $\tilde{\mu}_j$, respectively, corresponding to subsystem $G_{i\wedge j}$. Equation 6.9 implies

$$-\nabla_{\Delta f_{i\wedge j}}[\sum_{i\wedge j} C_{i\wedge j}(\Delta \hat{f}_u)] + (A_{i\wedge j})^t \hat{\lambda}_{i\wedge j} + \hat{\mu}_{i\wedge j} = 0; \tag{6.10}$$

$$(\hat{\mu}^t_{i\wedge j})\Delta\hat{f}_{i\wedge j} = 0. \tag{6.11}$$

We define $\bar{\mu}_{i\vee j} \in R^{|U_i|+|U_j|-|U_{i\wedge j}|+|N'_i|+|N'_j|-|N'_{i\wedge j}|}$ and $\bar{\lambda}_{i\vee j} \in R^{|U_i|+|U_j|+1-|U_{i\wedge j}|}$ as follows:

$$\bar{\lambda}_{i\vee j}(u) = \begin{cases} \hat{\lambda}_i(u) & \text{if } u \in \{U_i\} \\ \tilde{\lambda}_j(u) & \text{if } u \in \{U_j\} \setminus \{U_i\} \end{cases} \tag{6.12}$$

and

$$\bar{\mu}_{i\vee j}(u) = \begin{cases} \hat{\mu}_i(u) & \text{if } u \in \{U_i \cup N'_i\} \\ \tilde{\mu}_j(u) & \text{if } u \in \{U_j \cup N'_j\} \setminus \{U_i \cup N'_i\} \end{cases} \tag{6.13}$$

Now, we can show that $\Delta\hat{f}_{i\vee j}$ together with $\bar{\lambda}_{i\vee j}(u)$ and $\bar{\mu}_{i\vee j}(u)$ defined in 6.12, 6.13 will satisfy the following Kuhn-Tucker condition:

$$-\nabla_{\Delta f_{i\vee j}}[\sum_{i\vee j} C_{i\vee j}(\Delta\hat{f}_u)] + (A_{i\vee j})^t\bar{\lambda}_{i\vee j} + \bar{\mu}_{i\vee j} = 0; \tag{6.14}$$

$$(\bar{\mu}^t_{i\vee j})\Delta\hat{f}_{i\vee j} = 0. \tag{6.15}$$

This is true because we can combine 6.5–6.8 and eliminate the duplicated part 6.10–6.11 to have equations 6.14 and 6.15. The proof is complete. □

Now, we will apply *Lemma 6.5* and mathematical induction method to prove the exact convergence theorem.

6.3. Proof of the exact convergence theorem. We will prove by mathematical induction that $\Delta\hat{f}_{\vee_{i=1}^m}$, a subvector of $\Delta\hat{f}$ corresponding to $\cup_{i=1}^m G_i$, is the minimum solution of *Subsystem CEDCP* over $\cup_{i=1}^m G_i$ with $\Delta e_{\vee_{i=1}^m} = \Delta e_{\vee_{i=1}^m}(\Delta\hat{f})$ for each $m = 1, 2, ..., M$, if a unique K-T vector exists at every intersection.
(i) By the definition of $\mathcal{F}$, $\Delta\hat{f}_1$ and $\Delta\hat{f}_2$ are the minimum solutions for *Subsystem CEDCP* over connected subsystems G_1 with $\Delta e_1 = \Delta e_1(\Delta\hat{f})$ and G_2 with $\Delta e_2 = \Delta e_2(\Delta\hat{f})$, respectively. By *Lemma 6.5*, if a unique K-T vector exists for *Subsystem CEDCP* over $G_{1\wedge 2}$ with $\Delta e_{1\wedge 2} = \Delta e_{1\wedge 2}(\Delta\hat{f})$ then $\Delta\hat{f}_{\vee_{i=1}^2}$ is the optimal solution for *Subsystem CEDCP* over $\cup_{i=1}^2 G_i$ with $\Delta e_{\vee_{i=1}^2} = \Delta e_{\vee_{i=1}^2}(\Delta\hat{f})$. Thus it is true for $m = 2$.
(ii) Suppose it is also true for $m = n - 1$.
(iii) Since (1) $\cup_{i=1}^{n-1} G_i$, G_n, and $(\cup_{i=1}^{n-1} G_i) \cap G_n$ are all connected subsystems; (2) $\forall m \leq M-1, (\cup_{i=j_1}^{j_m} G_i) \cap G_{j_{m+1}} = G_{j_m} \cap G_{j_{m+1}}$ (refer to the condition 5 of the exact convergence theorem); (3) $\Delta\hat{f}_{\vee_{i=1}^{n-1}}$ and $\Delta\hat{f}_n$ are the minimum solutions for *Subsystem CEDCP* over $\cup_{i=1}^{n-1} G_i$ with $\Delta e_{\vee_{i=1}^{n-1}} = \Delta e_{\vee_{i=1}^{n-1}}(\Delta\hat{f})$ and over G_n with $\Delta e_n = \Delta e_n(\Delta\hat{f})$, respectively; by *Lemma 6.5*, $\Delta\hat{f}_{\vee_{i=1}^n}$ is the minimum solution for *Subsystem CEDCP* over $\cup_{i=1}^n G_i$ with $\Delta e_{\vee_{i=1}^n} = \Delta e_{\vee_{i=1}^n}(\Delta\hat{f})$. Thus it is true for $m = n$.

In particular, when $m = M$, we have the exact convergence theorem.

6.4. Proof of the Corollary 5.2. We will prove the corollary by showing that the regularity of the optimal solution at the intersection of any two overlapping subsystems implies the uniqueness of the K-T vector at the intersection.

LEMMA 6.6. (Regularity implies uniqueness.) *Let $\Delta\hat{f} \in \mathcal{F}$. G_i, G_j are any two intersected connected subsystems under the textured decomposition model. Let $\Delta\hat{f}_i, \Delta\hat{f}_j$, and $\Delta\hat{f}_{i\wedge j}$ denote the corresponding subvector of $\Delta\hat{f}$ related to G_i, G_j, and $G_{i\wedge j} = G_i \cap G_j$. If the minimum solution $\Delta\hat{f}_{i\wedge j}$ is a regular point for* Subsystem CEDCP *over $G_{i\wedge j}$ with $\Delta e_{i\wedge j} = \Delta e_{i\wedge j}(\Delta\hat{f})$, then a unique K-T vector exists in the* Subsystem CEDCP *over $G_{i\wedge j}$ with $\Delta e_{i\wedge j} = \Delta e_{i\wedge j}(\Delta\hat{f})$.*

Proof. Let the *Subsystem OPF* over $G_{i\wedge j}$ be

$$\text{(6.16)} \qquad \text{minimize} \qquad \textstyle\sum_{u\in N'_{i\wedge j}} (\gamma_u + \beta_u \Delta f_u + \alpha_u \Delta f_u^2)$$

$$\text{(6.17)} \qquad \text{subject to} \qquad [A_{i\wedge j}][\Delta f_{i\wedge j}] = [\Delta e_{i\wedge j}(\Delta\hat{f})]$$

$$\text{(6.18)} \qquad [\Delta f_{i\wedge j,min}] \leq [\Delta f_{i\wedge j}] \leq [\Delta f_{i\wedge j,max}]$$

where α_u, β_u, and γ_u are all positive real numbers. Accordingly, we have the Lagrangian

$$\begin{aligned} \mathcal{L} = & -\sum_{u\in N'_{i\wedge j}} (\gamma_u + \beta_u \Delta f_u + \alpha_u \Delta f_u^2) + \lambda_{i\wedge j}^t [A_{i\wedge j}\Delta f_{i\wedge j} - \Delta e_{i\wedge j}(\Delta\hat{f})] \\ & + \mu_{i\wedge j}^t \Delta f_{i\wedge j}. \end{aligned}$$

By the generalized Kuhn-Tucker theorem in [23], we have

$$\text{(6.19)} \qquad \frac{\partial \mathcal{L}}{\partial \lambda_{i\wedge j}} = 0,$$

$$\text{(6.20)} \qquad \frac{\partial \mathcal{L}}{\partial \Delta f_u} = 0, u \in U_{i\wedge j} \cup N'_{i\wedge j},$$

$$\text{(6.21)} \qquad \mu_{i\wedge j}^t \Delta f_{i\wedge j} = 0.$$

Note that the k-th component of $\mu_{i\wedge j}$ is nonpositive (nonnegative) real number if the corresponding component of $\Delta\hat{f}_{i\wedge j}$ hits its upper (lower) bound.

Since the solution $\Delta\hat{f}_{i\wedge j}$ for *Subsystem OPF* over $G_{i\wedge j}$ with $\Delta e_{i\wedge j} = \Delta e_{i\wedge j}(\Delta\hat{f})$ is a regular point, the number of incremental flows in $G_{i\wedge j}$ which hit the limits (i.e. the number of active inequality constraints) must be less than or equal to $(|N'_{i\wedge j}| - 1)$. In other words, we know that there are at least $(|U_{i\wedge j}| + 1)$ incremental flows which do not reach their upper or lower bounds; this implies that the $(|U_{i\wedge j}| + |N'_{i\wedge j}|)$-dimensional vector $\mu_{i\wedge j}$ has at least $(|U_{i\wedge j}| + 1)$ zero components. Let S denote the number of zero components in $\mu_{i\wedge j}$, we have $(|U_{i\wedge j}| + 1) \leq S \leq (|U_{i\wedge j}| + |N'_{i\wedge j}|)$.

The number of variables we must solve for equations 6.19–6.21 is $(3|U_{i\wedge j}|+2|N'_{i\wedge j}|+1)$ because that both $\Delta f_{i\wedge j}$ and $\mu_{i\wedge j}$ are $(|U_{i\wedge j}|+|N'_{i\wedge j}|)$-dimensional vectors, and $\lambda_{i\wedge j}$ is an $(|U_{i\wedge j}| + 1)$-dimensional vector.

By regularity of the solution, we know that there are S zero components in $\mu_{i\wedge j}$. We reorder the index of incremental flows such that

$$\mu_{1^r}, \mu_{2^r}, \ldots, \mu_{(|U_{i\wedge j}|+|N'_{i\wedge j}|-S)^r} \neq 0,$$

$$\mu_{(|U_{i\wedge j}|+|N'_{i\wedge j}|-S+1)^r}, \mu_{(|U_{i\wedge j}|+|N'_{i\wedge j}|-S+2)^r}, \ldots, \mu_{(|U_{i\wedge j}|+|N'_{i\wedge j}|)^r} = 0.$$

The corresponding reordering is also made for $\Delta f_{i\wedge j}, A_{i\wedge j}, \lambda_{i\wedge j}$, and $\Delta e_{i\wedge j}(\Delta \hat{f})$. They are denoted by $\Delta f_{(i\wedge j)^r}, A_{(i\wedge j)^r}, \lambda_{(i\wedge j)^r}$, and $\Delta e_{(i\wedge j)^r}$, respectively. Now, we need to solve the remaining $(3|U_{i\wedge j}| + 2|N'_{i\wedge j}| + 1 - S)$ variables.

For the equations derived from 6.19 and 6.20, we put all the non-first order terms (with respect to $\Delta f_{i\wedge j}$) to the right-hand side of the equations. Thus, equations 6.19–6.21 can be rewritten as follows:

$$Cx = b, \tag{6.22}$$

where

$$\begin{aligned} x^t = {} & [\Delta f_{1^r}, \ldots, \Delta f_{(|U_{i\wedge j}|+|N'_{i\wedge j}|)^r}, \lambda_{1^r}, \ldots, \lambda_{(|U_{i\wedge j}|+1)^r}, \\ & \mu_{1^r}, \ldots, \mu_{(|U_{i\wedge j}|+|N'_{i\wedge j}|-S)^r}] \end{aligned}$$

$$b^t = [e_{1^r}(\Delta \hat{f}), \ldots, e_{(|U_{i\wedge j}|+1)^r}(\Delta \hat{f}), \beta_{1^r}, \ldots \beta_{(|U_{i\wedge j}|+|N'_{i\wedge j}|)^r}, \underbrace{0, \ldots, 0}_{|U_{i\wedge j}|+|N'_{i\wedge j}|-S}],$$

and C is a $(3|U_{i\wedge j}| + 2|N'_{i\wedge j}| + 1 - S) \times (3|U_{i\wedge j}| + 2|N'_{i\wedge j}| + 1 - S)$ matrix which can be partitioned into three parts based on 6.19–6.21, respectively, as follows:

$$C = \begin{bmatrix} & A_{(i\wedge j)^r} & \vdots & & O_1 & & \vdots & (i) \\ \cdots & \cdots & \cdots & \cdots & \cdots & \cdots & \cdots & \cdots \\ & D & \vdots & A^t_{(i\wedge j)^r} & \vdots & \begin{matrix} I_1 \\ O_2 \end{matrix} & \vdots & (ii) \\ \cdots & \cdots & \cdots & \cdots & \cdots & \cdots & \cdots & \cdots \\ I_2 & \vdots & & O_3 & & & \vdots & (iii) \end{bmatrix} \tag{6.23}$$

where

$A_{(i\wedge j)^r}$: a $(|U_{i\wedge j}|+1)\times(|U_{i\wedge j}|+|N'_{i\wedge j}|)$ matrix with rank $(|U_{i\wedge j}|+1)$;
$A^t_{(i\wedge j)^r}$: the transpose matrix of $A_{(i\wedge j)^r}$ with rank $(|U_{i\wedge j}|+1)$;
O_1 : a $(|U_{i\wedge j}|+1)\times(2|U_{i\wedge j}|+|N'_{i\wedge j}|+1-S)$ zero matrix;
O_2 : an $S\times(|U_{i\wedge j}|+|N'_{i\wedge j}|-S)$ zero matrix;
O_3 : a $(|U_{i\wedge j}|+|N'_{i\wedge j}|-S)\times(2|U_{i\wedge j}|+|N'_{i\wedge j}|+1)$ zero matrix;
I_1, I_2 : $(|U_{i\wedge j}|+|N'_{i\wedge j}|-S)\times(|U_{i\wedge j}|+|N'_{i\wedge j}|-S)$ nonsingular diagonal matrix with diagonal entries $i_{jj}=1$ or -1 for $j=1,2,...,(|U_{i\wedge j}|+|N'_{i\wedge j}|-S)$;
D : a $(|U_{i\wedge j}|+|N'_{i\wedge j}|)\times(|U_{i\wedge j}|+|N'_{i\wedge j}|)$ diagonal matrix with diagonal entries $d_{ii}=-2\alpha_{i^r}$ for $i=1,2,...,(|U_{i\wedge j}|+|N'_{i\wedge j}|)$.

Suppose $\Delta\hat{f}_{i\wedge j}$ is a regular point for *Subsystem OPF* over $G_{i\wedge j}$ with $\Delta e_{i\wedge j}=\Delta e_{i\wedge j}(\Delta\hat{f})$. By the definition of regularity, a point is a regular point if the gradients of the active constraints are linearly independent. Part (i) represents the gradients of the equality constraints while part (ii) represents those of active inequality constraints. Thus the matrix $\begin{bmatrix} & A_{(i\wedge j)^r} & \\ I_2 & \vdots & O \end{bmatrix}$ is of full row rank, and its transpose matrix is of full column rank. To determine the K-T vector, we substitute $\Delta\hat{f}_{i\wedge j}$ into the part (ii) of equation 6.22 or 6.23. Since the matrix $\begin{bmatrix} A^t_{(i\wedge j)^r} \vdots & I_1 \\ & O_2 \end{bmatrix}$ is of full column rank, there exists at most one K-T vector solution; i.e. there exists a unique K-T vector solution or no solution exists. On the other hand, by the *Kuhn-Tucker Theorem* in [21], there exists a K-T vector which satisfies equation 6.20. Therefore, there exists a unique K-T vector for the *Subsystem OPF* over $G_{i\wedge j}$ with $\Delta e_{i\wedge j}=\Delta e_{i\wedge j}(\Delta\hat{f})$. This completes the proof of *Lemma 6.6*. □

Now, by *Lemma 6.6* and the exact convergence theorem, *Corollary 5.2* follows.

7. Test results. A 228-bus system are used to demonstrate the feasibility and the speedup of the parallel textured algorithmr, which is run on a parallel nCUBE machine using 12 processors. The textured decomposition for a modified IEEE 57-bus system is shown in Fig. 7.1.

The system data are modified in [22] to include one generator for each *PV* bus. First, we generate a 114-bus system by combining two identical modified IEEE 57-bus systems with four additional transmission lines: 9-59, 29-60, 28-75, and 26-76. Then we extend to a 228-bus system which consists of two 114-bus systems by adding four additional transmission lines: 66-116, 86-117, 85-132, and 83-133.

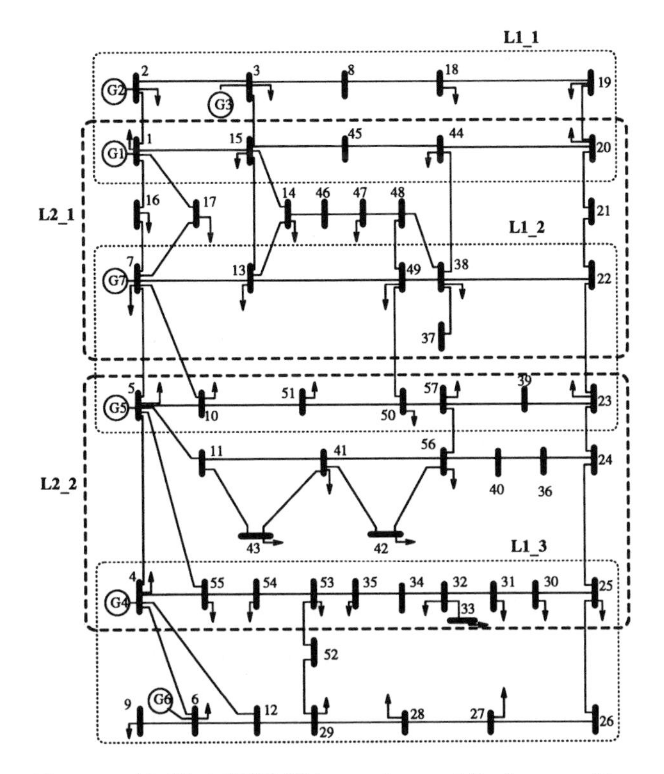

FIG. 7.1. *Modified IEEE 57-bus system and its decomposition.*

7.1. Procedure for measuring the speedup of the textured algorithm

- For each outer loop (steps 2–13), we use one processor to solve the *System CEDCP* without decomposition in the inner loop (steps 5–7), and get the CPU time for these steps. Add all these CPU times to form a total CPU time, denoted by T_1.
- Apply the textured decomposition, which satisfies the conditions of the exact convergence theorem, to the test systems. In the inner loop of the textured algorithm, apply the textured decomposed subsystems from level to level until it converges. Then repeat this process for the next outer loop until the algorithm converges. Add all the CPU times for the inner loops of all outer loops to form a total CPU time, denoted by T_2.
- Calculate the speedup ratio defined by $S = \frac{T_1}{T_2}$.

For the test systems, the **idealized speedup** between the CPU times of parallel and sequential textured algorithms is 11.5 since we use 12 processors at the first level and 11 processors at the second level. The inner iteration j is set at 2 for textured algorithm and 1 for the overall system without decomposition. The stopping criterion for solving each subsystem

TABLE 7.1

Running (including communication) time comparisons between implementations of textured and nontextured algorithms for the 228-bus system using 12 processors on nCUBE

Method used	Outer It.* (m)	Inner It.* (j)	CPU Time (sec)	Speedup	Efficiency
overall without decomposition	4	1	$T_1 =$ 16735		$E = \frac{T_2^s}{T_2^p \times 11.5}$ $\approx 64\%$
parallel textured algorithm	4	2	$T_2^p =$ 73.1	$S_p = \frac{T_1}{T_2^p}$ ≈ 229	
sequential textured algorithm	4	2	$T_2^s =$ 538.6	$S_s = \frac{T_1}{T_2^s}$ ≈ 31	

*Iterations

optimization is 10^{-8}. And the outer-loop stopping criterion is 10^{-3} for the absolute value of the difference between total generation costs of two consecutive outer iterations.

The efficiency of the parallel textured algorithm is defined by

$$E = \frac{\text{CPU time for the sequential textured algorithm}}{\text{CPU time for the parallel textured algorithm} \times \text{idealized speedup}}$$

7.2. Results and observations.

- The test results demonstrate the speedup advantage of the parallel textured algorithm. It can be also seen that even when the textured algorithm is executed sequentially, there are still some speedup. We expect that for a bigger system, we can decompose the system into many relatively smaller subsystems and can have a better speedup. As explained in [5], the inner per iteration complexity is proportional to at least the square of the problem size even when sparse techniques are used. Accordingly, for decomposed systems, the total complexity is drastically reduced since the outer loop iteration numbers are almost the same for textured systems and non-textured systems. For example, given a power system with problem size n, suppose the textured model has m ($m \geq 3$) subsystems with almost the same problem size $n_1, n_2, \ldots, n_m$, i.e. $n_i = (1+\alpha)n/m$, where α reflects the problem size of the total overlapping parts ($0 < \alpha < 1$), and to our knowledge the complexity of a available optimization algorithm is of the order n^q, where $q \geq 2$. Here, the inner loop iteration number is 2 for the textured model. Then the inner loop complexity is $n_{tex} = 2m \sum_{i=1}^{m} n_i^q = 2m[\frac{1+\alpha}{m}]^q n^q$ for the sequential textured model, and n^q for the nontextured system. Note that $n_{tex} = 2\frac{(1+\alpha)^q}{m^{q-1}}n^q \leq 2\frac{(1+\alpha)^q}{3^{q-1}}n^q = 6[\frac{1+\alpha}{3}]^q n^q$. Thus, the sequential $n_{tex} \leq 6[\frac{1+\alpha}{3}]^2 n^q < n^q$ when $\alpha \leq 20\%$. So, we can conclude that the sequential textured model can have better performance than available optimization algorithms and even better when it is parallelized. Also note that when m or q becomes larger, n_{tex} will be further reduced.

- From the exact convergence theorem, it can be seen that the generators in the overlapping parts of subsystems should have median-range fuel costs, so that these generators will have less chances to hit their generation limits.

8. Conclusion. The constrained real power economic dispatch control problem is solved in parallel by a textured decomposition based algorithm. Sufficient conditions and rigorous proofs for the proposed algorithm to converge to the global minimum solution are described in the paper. It is shown that the textured algorithm can be easily and effectively parallelized if special cares are taken for the overlapping part of the decomposed subsystems, in which it should contain enough transmission and generation capacity for allowed variations. In addition, the speedup of the textured algorithm is demonstrated by some computer runs. The result shows that even when the algorithm is executed sequentially, there still is some speedup. Moreover, we can adopt any fast available optimization algorithms into our textured model to further reduce the computational time.

Appendix

(1) To prove the convergence of our iterative descent algorithm, we need the closedness of point-to-set mappings and the Global Convergence Theorem [21], which are given in the following:

DEFINITION A.1: A point-to-set mapping A from one space X to another space Y is said to be *closed* at $x \in X$ if the assumptions

i) $x_k \to x, x_k \in X$,

ii) $y_k \to y, y_k \in A(x_k)$ imply

iii) $y \in A(x)$.

The point-to-set mapping A is said to be **closed** on X if it is closed at each point of X.

GLOBAL CONVERGENCE THEOREM A.2: *Let A be an algorithm on a set X, and suppose that, given x_0 the sequence $\{x_k\}_{k=0}^{\infty}$ is generated satisfying $x_{k+1} \in A(x_k)$.*

Let a solution set $\Gamma \subset X$ be given, and suppose

(i) all points x_k are contained in a compact set $S \subset X$,

(ii) there is a continuous function Z on X such that

(a) if $x \notin \Gamma$ then $Z(y) < Z(x)$ for all $y \in A(x)$,

(b) if $x \in \Gamma$ then $Z(y) \leq Z(x)$ for all $y \in A(x)$,

(iii) the mapping A is closed at points outside Γ.

Then the limit of any convergent subsequence of $\{x_k\}$ is a solution.

(2) A general nonlinear programming problem with equality and inequality constraints has the form as follows:

$$\text{Minimize } f(x)$$

$$\text{subject to } g(x) = 0$$

$$q(x) \leq 0$$

where $f : R^n \longrightarrow R, g : R^n \longrightarrow R^m, q : R^n \longrightarrow R^k$.

The following theorem tells us that any local minimum of a (strictly) convex objective function over a convex set is also a (the) global minimum.

THEOREM A.3: *Let Ω be a nonempty convex set in R^n, and $f : \Omega \longrightarrow R$. Consider the problem to minimize $f(x)$ subject to $x \in \Omega$, suppose that x is a local optimal solution to the problem,*

(1) If f is convex, then x is a global optimal solution.

(2) If f is strictly convex, then x is the unique global optimal solution.

DEFINITION A.4: An inequality constraint $q_i(x) \leq 0$ is said to be **active** at a feasible point x^* if $q_i(x^*) = 0$ and **inactive** at x^* if $q_i(x^*) < 0$.

DEFINITION A.5: Let x^* be a point satisfying the constraints $g(x^*) = 0$, $q(x^*) \leq 0$, and let $I(x^*)$ be the set of indices for which $q_i(x^*) = 0$. Then x^* is said to be a **regular point** of the constraints if the gradient vectors $\nabla g_i(x^*), 1 \leq i \leq m$, and $\nabla q_i(x^*), i \in I(x^*)$ are linearly independent.

ASSUMPTION A.6 [23]: Let x^* be an optimal solution of the nonlinear programming problem. Then there exists a vector $h \in R^n$ such that $(\nabla q_i(x^*), h) < 0$ for all $i \in I(x^*)$.

Note that this assumption is referred as **constraint qualification**, and a sufficient condition for this assumption to be satisfied is that the vectors $\nabla q_i(x^*), i \in I(x^*)$, are linearly independent.

THEOREM A.7 [23]: *If x^* is an optimal solution to the nonlinear programming problem, with* assumption A.6 *satisfied, then there exists a nonzero vector $\lambda \in R^{m+1}$ with $\lambda_0 \leq 0$ and a vector $\mu \in R^k$ with $\mu \leq 0$ such that*

$$\lambda_0 \nabla f(x^*) + \sum_{i=1}^{m} \lambda_i \nabla g_i(x^*) + \sum_{i=1}^{k} \mu_i \nabla q_i(x^*) = 0,$$

$$\sum_{i=1}^{k} \mu_i q_i(x^*) = 0.$$

COROLLARY A.8 [23]: *(a generalized Kuhn-Tucker theorem) If there exists a vector $h \in R^n$ such that $(\nabla q_i(x^*), h) < 0$ for all $i \in I(x^*)$, $(\nabla g_i(x^*), h) = 0$ for $i = 1, 2, ..., m$ and the vector $\nabla g_i(x^*), i = 1, 2, ..., m$ are linearly independent, then any vector $\lambda \in R^{m+1}$ satisfying the conditions of the* theorem A.7 *also satisfies $\lambda_0 < 0$.*

REFERENCES

[1] G.HUANG, S.-C.HSIEH, *A parallel textured algorithm for optimal power flow analysis*, Proceedings of IEEE ACC, Chicago, June 1992.

[2] M.HUNEAULT, F.D.GALIANA, *A survey of the optimal power flow literature*, IEEE Transactions on Power Systems **5** (2) May (1991), 762–770.

[3] R.BACKER, H.P.VAN MEETEREN, *Real-time optimal power flow in automatic generation control*, IEEE Transactions on Power Systems **3** (4) November (1988), 1518–1524.

[4] G.HUANG, A.ABUR, M.KEZUNOVIC, *The ICPS application of a new distributed approach to EMS implementation*, Proceedings of IFAC Congress, Russia 1990.

[5] J.ZABORSZKY, G.HUANG, K.W.LU, *A textured model for computationally efficient reactive power control and management*, IEEE Transactions on Power Apparatus and Systems **PAS-104** (7) July (1985), 1718–1727.

[6] G.HUANG, K.W.LU, J.ZABORSZKY, *A textured model algorithm for computationally efficient dispatch and control on the power system*, Proceedings of the 25th Conference on Decision and Control, Athens, Greece, Dec 1986, 1198–1205.

[7] W.K.TSAI, G.HUANG, J.K.ANTONIO, *Distributed iterative aggregation algorithms for box constrained minimization problems and optimal routing in data networks*, IEEE Transaction on Automatic Control **34** (1) June (1989), 34–46.

[8] J.K.ANTONIO, G.HUANG, W.K.TSAI, *A distributed shortest path algorithm for a class of hierarchically structured data networks*, IEEE Transaction on Computer, June 1992.

[9] G.HUANG, W.K.TSAI, W.LU, *Fast parallel linear equations solver based on textured decomposition*, Proceedings of the 26th Conference on Decision and Control, Los Angeles, CA 1987, 1450–1454.

[10] S.-Y.LIN, *A textured decomposition based algorithm for large scale multicommodity network problems*, Proceedings of IEEE ACC 1989.

[11] G.HUANG, W.-L.HSIEH, *A parallel textured algorithm for optimal routing in data networks*, Proceedings of IEEE Globe Com. 91, Phoenix, Arizona, Dec 1991.

[12] G.HUANG, W.L.HSIEH, S.ZHU, *Parallel implementation issues of the textured algorithm for optimal routing problem in data networks*, Proceedings of 7th International Parallel Symposium, New Port Beach, CA, April 13-16, 1993.

[13] C.H.JOLISSAINT, N.V.ARVANITIDIS, D.G.LUENBERGER, *Decomposition of real and reactive power flows: a method suitable for on-line applications*, IEEE Transaction on Power Apparatus and Systems **PAS-91** March/April (1972), 661–670.

[14] R.BILLINTON, S.S.SACHDEVA, *Real and reactive power optimization by suboptimum techniques*, IEEE Transaction on Power Apparatus and Systems **PAS-92** May/June (1973), 950–956.

[15] B.STOTT, J.L.MARINHO, *Linear programming for power system network security applications*, IEEE Transaction on Power Apparatus and Systems **PAS-98** (3) May/June (1979), 837–845.

[16] O.ALSAC, J.BRIGHT, M.PRAIS, B.STOTT, *Further developments in LP-based optimal power flow*, IEEE PES 1990 Winter Meeting, Atlanta, Georgia **90 WM 011-7 PWRS** Feb 1990.

[17] N.G.HINGORANI, *Flexible AC transmission systems (FACTS)—overview*, Panel Session on Flexible AC Transmission Systems, IEEE Winter Power Meeting 1990.

[18] E.K.Blum, Numerical Analysis and Computation Theory and Practice, Addison-Wesley, Reading, Mass. 1972.

[19] G.Huang, W.Ongsakul, *Parallel implementation of gauss-seidel type algorithm for power flow analysis on a SEQUENT parallel computer*, Proceedings of the 20th International Conference of Parallel Processing, Chicago, August 1991.

[20] G.Huang, W.Ongsakul, *Managing the bottlenecks in parallel gauss-seidel type algorithms for power flow analysis*, Proceedings of the 1993 IEEE Power Industry Computer Application Conference, Scottsdale, AZ May 4-7, 1993. (*also in*) IEEE Trans. on Power Systems, May (1994).

[21] D.G.Luenberger, Linear and Nonlinear Programming, Addison-Wesley Publishing Company 1984.

[22] S.-C.Hsieh, *A parallel textured algorithm for optimal power flow analysis*, (*thesis*) Dept. of Electrical Engineering, Texas A&M University, College Station, Texas, August 1992.

[23] M.D.Canon, C.D.Cullum,Jr., E.Polak, *Constrained minimization problems in finite-dimensional spaces*, J. SIAM Control **4** (3) (1966), 528–547.

VARIABLE STRUCTURE REGULATION OF POWER PLANT DRUM LEVEL*

HARRY G. KWATNY† AND JORDAN BERG‡

Abstract. Water level regulation in drum type power plants is a critical issue in plant operations. More effective control over all load levels would significantly improve plant operational efficiency, particularly during startup, load following, runback and sustained low load operation. Drum level control is difficult because the process dynamics are both nonlinear and nonminimum phase. Advances in multivariable nonlinear control offer new opportunities for improved control systems. In this paper, we investigate possibilities of variable structure control. Recently, Kwatny and Kim [1] gave a geometric characterization of variable structure control system design for input-output linearizable nonlinear dynamics which emphasizes the importance of 'zero dynamics' and 'relative degree,' factors relevent to the application of interest. Since variable structure controllers are switching controllers, attention is given to smoothing the motion of the mechanical control actuators.

1. Introduction. Effective water level control in drum type power plants is fundamental to safe and reliable plant operation. Recent trends in power plant automation, e.g., automated startup, rapid load following and controlled runback, highlight the difficulties associated with drum level regulation. Performance of drum level controllers is inherently limited by the nonminimum phase process characteristics–physically described as fluid shrink–swell phenomenon. Moreover, the local dynamical behavior of the circulation loop in drum type power plants varies dramatically with the plant thermal output. As a result it is almost universal practice to employ different regulators at high and low loads. Even so, drum level regulation remains typically poor at low loads and merely adequate at high loads. An analysis of the inadequacies of conventional control architectures is given by Kwatny and Berg [2].

The nonminimum phase problem must be alleviated in order to significantly improve performance. This can be done by reformulating the single–input single–output drum level control problem as a multivariable problem. Rather than simply using feedwater flow as the control input to regulate drum level, as is the prevailing custom, it is possible to use feedwater flow and throttle valve position to simultaneously regulate drum level and drum pressure. The latter two–input two–output problem is minimum phase, as we shall discuss below. There is a three–input three–output option as well.

The variation in dynamical behavior that occurs as a function of load level is due to the process nonlinearity. Rather than resort to the inte-

* This research was supported by Atlantic Electric Company.

† Raynes Professor of Mechanical Engineering, Department of Mechanical Engineering & Mechanics, Drexel University, Philadelphia, PA 19104.

‡ Research Associate, Department of Mechanical Engineering & Mechanics, Drexel University, Philadelphia, PA 19104.

gration of multiple linear controllers tuned for distinct load levels, it is preferable to seek a single nonlinear regulator that can accommodate the entire range of operating conditions. The modern geometric theory of nonlinear control provides some interesting possibilities. One alternative is the use of input-output linearization by smooth state feedback. Another is variable structure control (VSC). These methods are related in that during the sliding phase of VSC the effective control laws coincide. However, VSC has some important advantages. Linearizing control typically requires large amounts of control effort and does not directly address control constraints. VSC does so. Also, linearizing control can be quite sensitive to parametric uncertainty whereas VSC is naturally robust. On the other hand, VSC involves discontinuous control laws that can result in chattering. In the present application, which involves mechanical control actuation, chattering is objectionable. While we consider both approaches promising, at this stage we find VSC particularly attractive. In the subsequent paragraphs we describe three approaches to eliminating chattering behavior that are quite effective especially when used in combination.

In Section 2, we give a summary of the important properties and analytical methods associated with discontinuous feedback systems and in Section 3 we describe the approach used here for variable structure control system design. For additional background, the reader might consult the text [3] or the survey articles [4-6]. In Section 4 we describe the drum level control problem, present a model of the process, analyze its dynamical behavior, and describe the variable structure controller. Section 5 contains simulation results using a conventional linear control system (a detail design description is given in [2]) and the variable structure controller described herein. Concluding remarks are given in Section 6.

2. Discontinuous control systems

2.1. Variable structure control systems. Consider a nonlinear dynamical system of the form

$$\dot{x} = f(x, u) \tag{2.1}$$

where $x \in R^n$, $u \in R^m$ and f is a smooth functions of x and u. Variable structure control systems are switching controllers in which the controls u_i are discontinuous across smooth surfaces $s_i(x) = 0$, i.e.

$$u_i(x) = \begin{cases} u_i^+(x) & \text{if } s_i(x) > 0 \\ u_i^-(x) & \text{if } s_i(x) < 0 \end{cases} i = 1, .., m \tag{2.2}$$

and the control functions $u_i^+(x)$ and $u_i^-(x)$ are smooth functions of x.

The design of switching control systems of the type (2.1) often focuses on the deliberate introduction of sliding modes [3]. If there exists an open submanifold, $\mathcal{D}_s$, of any intersection of discontinuity surfaces, $s_i(x) = 0$ for $i = 1, .., p \leq m$, such that $s_i \dot{s}_i < 0$ in the neighborhood of almost every point in $\mathcal{D}_s$, then it must be true that a trajectory once entering $\mathcal{D}_s$ remains in it until a boundary is reached. $\mathcal{D}_s$ is called a *sliding manifold* and the motion in $\mathcal{D}_s$ is called a *sliding mode*. Since the control is not defined on the discontinuity surfaces, the sliding dynamics are not characterized by (2.1) & (2.2). However, sliding mode dynamics may be determined by imposing the constraint $s(x) = 0$ on the motion defined by the differential equation (2.1). Under appropriate circumstances this is sufficient to define an effective control u_{eq}, called the equivalent control, which obtains for motion constrained to lie in $\mathcal{D}_s$. If this control is smooth and unique, then the sliding behavior is well defined.

Variable structure control system design entails specification of the switching functions $s_i(x)$ and the control functions $u_i^+(x)$ and $u_i^-(x)$. The basis for design follows from the observations that the sliding mode dynamics depend on the geometry of $\mathcal{D}_s$, that is, on the switching functions $s_i(x)$, and that sliding can be induced on a desired manifold $\mathcal{D}_s$ by designing the control functions $u_i^+(x)$ and $u_i^-(x)$ to guarantee that $\mathcal{D}_s$ is attracting. Thus, control system design is a two step process: 1) design of the 'sliding mode' dynamics by the choice of switching surfaces, and 2) design of the 'reaching' dynamics by the specification of the control functions.

2.2. Basic properties of discontinuous systems. Eq. (2.1) when combined with control (2.2) is a special case of the general class of discontinuous dynamical systems

$$\dot{x} = F(x,t) \tag{2.3}$$

where for each fixed t, $F(x,t)$ is smooth ($C^k, k \geq 0$) on $\mathbb{R}^n$ except on m codimension one surfaces (codimension one regular submanifolds of $\mathbb{R}^n$) defined by $s_i(x) = 0$, $i + 1, .., m$, where $F(x,t)$ is not defined. Ordinarily, a solution to (2.3) is a curve $x(t) \subset \mathbb{R}^n$ which has the property that $dx(t)/dt = F(x(t),t)$ for each $t \in \mathbb{R}$. Such a test, however, would be impossible to apply if the prospective solution contains points on the discontinuity surfaces. Since the set of points for which $F(x,t)$ is not defined has measure zero in $\mathbb{R}^n$, one might only require that the integral curve property be satisfied only where $F(x,t)$ is defined. This is clearly inadequate because segments of trajectories which lie in a discontinuity surface would be entirely arbitrary. Filippov [7] proposed a satisfactory definition of solutions to (2.3):

DEFINITION 2.1. A curve $x(t) \subset \mathbb{R}^n$, $t \in [t_0, t_1]$, $t_1 > t_0$, is said to be a solution of (2.3) on $[t_0, t_1]$ if it is absolutely continuous on $[t_0, t_1]$ and for each $t \in [t_0, t_1]$

$$\frac{dx}{dt} \in \mathcal{F}(x(t),t) := \cap_{\delta>0} \text{ conv } F(\mathcal{G}(\delta, x(t)) - \mathcal{N}(\delta, x(t)), t)$$

where $\mathcal{G}(\delta, x)$ is the open sphere centered at x and of radius δ, $\mathcal{N}\delta, xt)$ is the subset of measure zero in $\mathcal{G}(\delta, x)$ for which F is not defined, and conv $F(\mathcal{A})$ denotes the convex closure of the set of vectors $\{F(\mathcal{A})\}$.

Remark on notation: If x is a point in $\mathbb{R}^n$ then $\mathcal{G}(\delta, x) := \{y \in \mathbb{R}^n | \, \|y - x\| < \delta\}$. If $\mathcal{A}$ is a set contained in $\mathbb{R}$ then

$$\mathcal{G}(\delta, \mathcal{A}) := \cup_{x \in \mathcal{A}} \mathcal{G}(\delta, x).$$

We call $\mathcal{G}(\delta, \mathcal{A})$ a δ-vicinity of $\mathcal{A}$.

Note that if x does not lie on a discontinuity surface, then the set $\mathcal{F}(x,t) = \{F(x,t)\}$. This definition does help characterize solutions which lie in discontinuity surfaces. Suppose a solution $x(t) \subset \mathbb{R}^n$, $t \in [t_0, t_1]$, $t_1 > t_0$, lies entirely in the intersection of some set of p discontinuity surfaces which is a regular embedded submanifold of $\mathbb{R}^n$ of dimension $n-p$, which we designate $\mathcal{M}_s$. For each $t^* \in [t_0, t_1]$, $\dot{x}(t^*)$ must belong to the set $\mathcal{F}(x(t^*), t^*)$. In addition, $\dot{x}(t^*)$ must lie in the tangent space to $\mathcal{M}_s$ at $x(t^*)$, i.e., $\dot{x}(t^*) \in T_{x(t^*)}\mathcal{M}_s$. In many important cases these two conditions uniquely define solutions which contain segments that lie in $\mathcal{M}_s$.

When we speak of solutions, or, equivalently, trajectories of discontinuous systems we shall mean solutions in the sense of Filippov. One important consequence of the definition is an extension of Lyapunov's direct stability analysis to discontinuous systems.

LEMMA 2.2. *Suppose that $\mathcal{V} : \mathbb{R}^n \to \mathbb{R}$ is a C^1 function. Then*

- *the time derivative of $\mathcal{V}(x)$ along trajectories of (2.3) satisfies the set inclusion*

$$\dot{\mathcal{V}}(x) \in \left\{ \frac{\partial \mathcal{V}}{\partial x} \xi | \xi \in \mathcal{F}(x(t), t) \right\} \tag{2.4}$$

- *if $\dot{\mathcal{V}} \leq -\rho < 0 (\geq \rho > 0$ at all points in an open set $\mathcal{P} \subset \mathbb{R}^n$ except on a set $\mathcal{N} \subset \mathcal{P}$ of measure zero where $F(x,t)$ is not defined, then $\dot{\mathcal{V}} \leq -\tilde{\rho} < 0 (\geq \tilde{\rho} > 0)$, $\tilde{\rho} < \rho$, at all points of $\mathcal{P}$.*
- *if $\dot{\mathcal{V}} \leq -\rho \|s(x)\|$, $\rho > 0$ at all points in an open set $\mathcal{P} \subset \mathbb{R}^n$ except on a set $\mathcal{N} \subset \mathcal{P}$ of measure zero where $F(x,t)$ is not defined, then $\dot{\mathcal{V}} \leq -\rho \|s(x)\|$ at all points of $\mathcal{P}$.*

Proof. The first conclusion (2.4) follows directly from the Filippov definition of a trajectory.

To prove the second, first note that at regular points the inclusion reduces to the usual $\dot{\mathcal{V}}(x) = [\partial \mathcal{V}(x)/\partial x] F(x,t)$. Consider the negative definite case. The assumption of definiteness implies that $[\partial \mathcal{V}(x)/\partial x] F(x,t) \leq -\rho$ at all regular points $x \in \mathcal{P}$. Now take any $x^* \in \mathcal{N}$. We need only show that $[\partial \mathcal{V}(x^*)/\partial x] \xi \leq -\tilde{\rho}$ for each $\xi \in \mathcal{F}(x^*, t)$. Consider the sphere $\mathcal{G}(\varepsilon, x^*)$. Since $\mathcal{V}$ is C^1, we can choose ε sufficiently small so that $[\partial \mathcal{V}(x^*)/\partial x] F(x,t) \leq -\tilde{\rho} < 0$ for any specified $\tilde{\rho} < \rho$ and all regular x in $\mathcal{G}(\varepsilon, x^*)$. By its definition $\mathcal{F}(x^*, t) \subset F(\mathcal{G}(\varepsilon, x^*) - \mathcal{N}, t)$. The conclusion follows.

To prove the third conclusion, consider a point x^* in $\mathcal{P}$. By assumption, the condition $\dot{\mathcal{V}}(x) \leq -\rho \|s(x)\|$ holds at all regular points. Suppose that x^* is not regular, then the condition holds at almost all points in a sufficiently small neighborhood $\mathcal{G}(\varepsilon, x^*)$ of x^*. Now, the smoothness of $\mathcal{V}$ and s implies the approximations $[\partial\mathcal{V}(x)/\partial x] = [\partial\mathcal{V}(x^*)/\partial x] + O(\varepsilon)$ and $\|s(x)\| = \|s(x^*)\| + O(\varepsilon)$ for all $x \in \mathcal{G}(\varepsilon, x^*)$. Thus we have at regular $x \in \mathcal{G}(\varepsilon, x^*)$

$$\frac{\partial\mathcal{V}(x^*)}{\partial x} F(x,t) \leq -\rho\|s(x^*)\| + O(\varepsilon).$$

Once again, since $\mathcal{F}(x^*, t) \subset F(\mathcal{G}(\varepsilon, x^*) - \mathcal{N}, t)$, it follows that

$$\frac{\partial\mathcal{V}(x^*)}{\partial x}\xi \leq -\rho\|s(x^*)\| + O(\varepsilon), \forall\xi \in \mathcal{F}(x^*, t)$$

The conclusion follows in the limit $\varepsilon \to 0$. □

Remark. In applications, $\dot{\mathcal{V}}$ is often relatively easy to determine, and may even be specified in design problems, at all points in a given domain other than those on the surfaces of discontinuity. The significance of the lemma is that it makes it unnecessary to actually compute $\mathcal{F}(x^*, t)$ in order to determine values of $\dot{\mathcal{V}}$ at those points.

DEFINITION 2.3. *Suppose $\mathcal{M}_s = \{x \in \mathbb{R}^n | s(x) = 0\}$ is a regular embedded manifold in $\mathbb{R}^n$ and let $\mathcal{D}_s$ be an open, connected subset of $\mathcal{M}_s$ is a sliding domain if*

(i) *for any $\varepsilon > 0$, there is a $\delta > 0$ such that trajectories of (2.3) which begin in a δ-vicinity of $\mathcal{D}_s$ remain in an ε-vicinity of $\mathcal{D}_s$ until reaching an ε-vicinity of the boundary of $\mathcal{D}_s$, $\partial\mathcal{D}_s$.*

(ii) *$\mathcal{D}_s$ must not contain any entire trajectories of the 2^m continuous systems defined in the open regions adjacent to $\mathcal{M}_s$ and partitioned by the set $\mathcal{M} := \cup_{i=1,\ldots m}\mathcal{M}$.*

Remark. This definition is due to Utkin [3]. By including (ii), it is assured that it is the switching mechanism that produces the sliding mode and the possibility of the existence of certain "pathological" sliding domains is excluded.

The definition implies that $\mathcal{D}_s$ is invariant with respect to trajectories in the sense of the following rather obvious proposition.

PROPOSITION 2.4. *If $\mathcal{D}_s$ is a sliding domain then trajectories of (2.3) which begin in $\mathcal{D}_s$ remain in $\mathcal{D}_s$ until reaching its boundary, $\partial\mathcal{D}_s$.*

Proof. Since $\mathcal{D}_s$ belongs to any δ-vicinity of itself, the definition of a sliding domain implies that trajectories which begin in $\mathcal{D}_s$ must remain in every arbitrarily small ε-vicinity of $\mathcal{D}_s$. Hence trajectories beginning in $\mathcal{D}_s$ must remain therein until reachings its boundary. □

Sufficient conditions for the existence of a sliding domain are relatively easy to formulate. One approach is as follows. Define a C^1 scalar function $\mathcal{V} : \mathcal{D} \subset \mathbb{R}^n \to \mathbb{R}$ with the following properties

$$\mathcal{V}(x) := \begin{cases} = 0 & \text{if } s(x) = 0 \\ > 0 & \text{otherwise} \end{cases} \tag{2.5}$$

Recall that $\dot{\mathcal{V}}$ is uniquely defined provided $s_i(x) \neq 0$ for $i = 1, \ldots, m$, i.e., everywhere but on $\mathcal{M} := \cup_{i=1,\ldots,m}\mathcal{M}_{s_i}$ and otherwise it is still constrained by the set inclusion of Lemma 2.2. Now the following result can be stated.

PROPOSITION 2.5. *Let $\mathcal{V}$ be given by (2.5). Suppose that*

- *$\mathcal{D}_s$ is an open, connected subset of $\mathcal{M}_s$*
- *$\mathcal{D}$ is an open connected subset of $\mathbb{R}^n$ which contains $\mathcal{D}_s$*
- *$\dot{\mathcal{V}} \leq -\rho||s(x)||$ on $\mathcal{D} - \mathcal{M}$.*

Then $\mathcal{D}_s$ is a sliding domain.

Proof. Under the stated assumptions, a trajectory cannot leave $\mathcal{D}_s$ at any point $x_0 \in \mathcal{D}_s$. This is easily proved by contradiction. Suppose a trajectory $x(t)$ does depart $\mathcal{D}_s$ from a point $x_0 \in \mathcal{D}_s$ at time t_0. Such a departure implies that there is a time $t_1 > t_0$ and sufficiently small $\varepsilon > 0$, such that the absolutely continuous trajectory segment $x(t)$, $t \in (t_0, t_1)$ is entirely contained in the set $\mathcal{G}(\varepsilon, x_o) - \mathcal{M}_s$ and along which $\dot{\mathcal{V}} > 0$. But in view of Lemma 2.2, the assumptions of the theorem imply that $\dot{\mathcal{V}} < 0$ along trajectories at all points in $\mathcal{G}(\varepsilon, x) - \mathcal{M}_s$. This is a contradiction. □

One distinguishing feature of many variable structure control systems is that trajectories beginning in a vicinity of the sliding surface each the surface in finite time. This clearly is the case if $\dot{\mathcal{V}}$ is bounded below by a negative number. However, such a bound is not necessary as the following proposition illustrates.

PROPOSITION 2.6. *Suppose that the conditions of Proposition hold and in addition $\mathcal{V}(x) = \sigma||s(x)||^2$, $\sigma > 0$ in a δ-vicinity of $\mathcal{D}_s$. Then trajectories which reach $\mathcal{D}_s$ from a δ-vicinity of $\mathcal{D}_s$ do so in finite time.*

Proof. Suppose a trajectory beginning at state x_0 in a δ-vicinity of $\mathcal{D}_s$ reaches a point $x_1 \in \mathcal{D}_s$. Then $||s(x_0)|| \leq \delta$. Since $\mathcal{V}(x) = \sigma||s(x)||^2$ we have

$$\dot{\mathcal{V}} = 2\sigma||s(x)||\frac{d||s(x)||}{dt} \leq -\rho||s(x)||$$

which in view of Lemma holds throughout the δ-vicinity of $\mathcal{D}_s$. Thus,

$$\frac{d||s(x)||}{dt} \leq -\frac{\rho}{2\sigma}$$

which implies that the trajectory reaches $\mathcal{D}_s$ in time not greater than $\delta(2\sigma/\rho)$. □

2.3. Affine systems and sliding dynamics. Consider a nonlinear dynamical system of the form

$$\dot{x} = f(x) + G(x)u \tag{2.6}$$

where $x \in \mathbb{R}^n$, $u \in \mathbb{R}^m$ and f and $G = [g_1, \ldots, g_m]$ are smooth functions of x. Systems of this type are said to be "affine" or "linear in control". Sliding mode dynamics for affine systems are best characterized by the method of equivalent control (Utkin [3]). If a trajectory lies in a sliding manifold $\mathcal{D}_s$, then $s = 0$, and all time derivatives of $s(x)$ also vanish. The control is not defined by (2.1) when $s = 0$. It must be defined by continuation. Let us denote by u_{eq} (the equivalent control) the control which must obtain while the trajectory remains in the manifold $\mathcal{D}_s$. Then, u_{eq} is defined by

$$\dot{s} = S(x)\dot{x} = S(x)\{f(x) + G(x)u_{eq}\} := 0$$

where $S(x) := \frac{\partial}{\partial x}s(x)$ and it is assumed that $\det\{S(x)G(x)\} \neq 0$, in which case we have

(2.7a) $$u_{eq} = -[S(x)G(x)]^{-1}S(x)f(x).$$

Motion in the sliding mode is then defined by

(2.7b) $$\dot{x} = [I - G(x)[S(x)G(x)]^{-1}S(x)]f(x), \qquad \text{with } s(x(0)) = 0$$

Note that the equivalent control as defined by (2.7a) depends only on the switching surfaces $s(x)$ and not on the control functions $u_i^{\pm}(x)$. If sliding occurs, it is characterized by (2.7b). Conditions for the existence of such trajectories have been given by Utkin [3].

3. VSC design for nonlinear dynamics. In the following paragraphs, we will develop a view of variable structure control system design closely associated with methods of exact linearization [8] which has evolved from work of Krener [9], Hirschorn [10] and others. The procedures outlined here are more fully developed in [1]. We consider the affine nonlinear system

(3.1a) $$\dot{x} = f(x) + G(x)u$$

(3.1b) $$y = h(x)$$

where $x \in \mathbb{R}^n$, $u \in \mathbb{R}^m$, $u \in \mathbb{R}^m$ and f, $G = [g_1, \ldots, g_m]$ and h are smooth functions of x.

3.1. Partial linearization and zero dynamics. Denote the kth Lie (directional) derivative of the scalar function $\phi(x)$ with respect to the vector field $f(x)$ by $L_f^k(\phi)$. Now, by successive differentiation of the outputs y in (3.1b) we arrive at the following definitions for the list of integers r_i, the column vector $\alpha(x)$ and the matrix $\rho(x)$:

(3.2a) $$r_i := \inf\{k | L_{g_j}(L_f^{k-1}(h_i)) \neq 0 \text{ for at least one } j\}$$

(3.2b)
$$\alpha_i(x) := L_f^{r_i}(h_i), i = 1, \ldots, m \rho_{ij}(x) := L_{g_j}(L_f^{r_i - 1}(h_i)), i, j = 1, \ldots, m.$$

Also define the vector $z \in \mathbb{R}^r, r = r_1 + \ldots + r_m$, as

(3.3a) $$z := \begin{bmatrix} z_1 \\ z_2 \\ \vdots \\ z_m \end{bmatrix}, z_i \in \mathbb{R}^{r_i}, \quad i = 1, \ldots, m$$

where

$$z_i^i(x) = L_f^{k-1}(h_i), \quad k = 1, \ldots, r_i \text{ and } i = 1, \ldots, m \tag{3.3b}$$

It is a straightforward calculation to verify that the variables z defined by (3.3) satisfy the relation

$$\dot{z} = Az + E[\alpha(x) + \rho(x)u] \tag{3.4a}$$

$$y = Cz \tag{3.4b}$$

where the only nonzero rows of E are the m rows r_1, $r_1 + r_2, \ldots, r$ and these form the identity I_m, the only nonzero columns of C are the columns $1, r_1 + 1$, $r_1 + r_2 + 1, \ldots, r - r_m + 1$ and these form the identity I_m, and

$$A = \text{ diag } (A_1, \ldots, A_m), A_i = \begin{bmatrix} 0 & I_{r_i - 1} \\ 0 & 0 \end{bmatrix} \in R^{r_i \times r_i} \tag{3.5}$$

In the following analysis, we will make use of the following elementary, but important, lemma:

LEMMA 3.1. *Suppose that $\rho(x)$ has continuous first derivatives with* $\det\{\rho(x)\} \neq 0$ *on* $\mathcal{M}_0 = \{x|z(x) = 0\}$. *Then $\partial z(x)/\partial x$ is of maximum rank on the set* $\mathcal{M}_0 = \{x|z(x) = 0\}$.

Proof. [8]. □

Lemma 2 is extremely important because it relates the invertibility of the decoupling matrix with the geometry of the set $\mathcal{M}_0$. With it, we can obtain several important results which we state here without proof.

PROPOSITION 3.2. *Suppose that $\rho(x)$ has continuous first derivatives with* $\det\{\rho(x)\} \neq 0$ *on* $\mathcal{M}_0 = \{x|z(x) = 0\}$. *Then $\mathcal{M}_0$ is a regular, $n - r$ dimensional submanifold of R^n and any trajectory segment $x(t)$, $t \in T$, T an open interval of $\mathbb{R}^1$, which satisfies $h(x(t)) = 0$ on T lies entirely in $\mathcal{M}_0$. Moreover, the control which obtains on T is*

$$u_0(x) = \rho^{-1}(x)\alpha(x) \tag{3.6}$$

and every such trajectory segment with boundary condition $x(t_0) = x_0$, $t_0 \in T$ satisfies

$$\dot{x} = f(x) - G(x)\rho^{-1}(x)\alpha(x), \quad z(x(t_0)) = 0 \tag{3.7}$$

Proof. Kwatny and Kim [1]. □

Note that the manifold $\mathcal{M}_0$ defined by $z(x) = 0$ is invariant with respect to (3.7) so that any motion beginning in it remains therein. Indeed, (3.7) defines a flow on $\mathcal{M}_0$ with all trajectories satisfying $y(t) = h(x(t)) = 0$. This justifies reference to (3.7) as the *zero output constrained dynamics* and to $\mathcal{M}_0$ as the *zero dynamics manifold*.

It is important to develop an understanding of when the construction described above will be successful (i.e., $\rho(x)$ nonsingular) and if there exists a remedy if it is not. These issues have been considered, but the theory is not complete. Dynamic extension as described in [8] can provide a useful repair, with some side effects (good and bad) which we will not discuss here. Sufficient conditions for decoupling via nonlinear dynamic feedback (of which the above construction augmented with dynamic extension is a special case) are given by Descusse and Moog [11]. It is assumed henceforth that the matrix $\rho(x)$ is nonsingular. In this case, we can apply the feedback control law

$$u = -\rho^{-1}(x)[\alpha(x) - v] \tag{3.8}$$

where v is a new control input. Thus, we have the linearized input-output model

$$\dot{z} = Az + Ev \tag{3.9a}$$

$$y = Cz \tag{3.9b}$$

Note that the control law (3.8) simultaneously linearizes the input-output relation and decouples some of the dynamics (the internal dynamics) from the output.

It is not uncommon to refer to the variables z as the linearizable coordinates. The terminology of coordinates is justified by the maximal rank condition in the following way. Let $Z : \mathbb{R}^n \to \mathbb{R}^r$ denote the map realized as the function $z(x)$. By virtue of the maximal rank assumption and the implicit function theorem we can choose local coordinates $(y_1 \dots y_n)$ on $\mathbb{R}^n$ near any point $x_0 \in \mathcal{M}_0$ such that $Z(y) = (y_1, \dots, y_r)$. In terms of these coordinates $\mathcal{M}_0$ is defined by $y_1 = 0, \dots, y_r = 0$. As a matter of fact, the first r components correspond to the level sets $z(x) = c$ which exist for all c in some neighborhood of the origin in $\mathbb{R}^n$. The remaining components $\xi := (y_{r+1}, \dots, y_n)$ provide local coordinates on $\mathcal{M}_0$. Thus, the above formal calculations make sense because the condition $\det\{\rho(x_0)\} \neq 0$ insure the existence of a local (around x_0) change of coordinates $x \to (\xi, z)$, $\xi \in \mathbb{R}^{n-r}$, $z \in \mathbb{R}^r$ such that

$$\dot{\xi} = F(\xi, z) \tag{3.10a}$$

$$\dot{z} = Az + E[\alpha(x(\xi, z)) + \rho(x(\xi, z))u]. \tag{3.10b}$$

In view of the above discussion and (3.9), it is common to refer to (3.10a) as the *internal dynamics* and (3.10b) as the *linearizable dynamics*. If z is set to zero in (3.10a) then we have local representation of the zero dynamics.

3.2. Sliding dynamics. Let us proceed to design a variable structure controller for (3.1) by selecting a switching surface which is linear in z.

PROPOSITION 3.3. *Let $s(x) = Kz(x)$ and suppose the conditions of Proposition 3.2 hold and $\partial s(x)/\partial x$ is of maximum rank on the set $\mathcal{M}_s = \{x|s(x) = 0\}$. Then $\mathcal{M}_s$ is a regular $n - m$ dimensional submanifold of $\mathbb{R}^n$ which contains $\mathcal{M}_0$. Moreover, if K is structured so that the m columns numbered $r_1, r_1 + r_2, \ldots, r$ compose an identity I_m, then for any trajectory segment $x(t)$, $t \in T$, T an open interval of $\mathbb{R}^1$, which lies entirely in $\mathcal{M}_s$, the control which obtains on T is*

$$u_{eq} = -\rho^{-1}(x)KAz - \rho^{-1}(x)\alpha(x) \tag{3.11a}$$

and every such trajectory with boundary condition $x(t_0) = x_0 \in \mathcal{M}_s$, $t_0 \in T$ satisfies

$$\dot{x} = f(x) - G(x)\rho^{-1}(x)\{\alpha(x) + KAz(x)\}, Kz(x(t_0)) = 0 \tag{3.11b}$$

Proof. Kwatny and Kim [1]. □

In this case observe that the manifold $\mathcal{M}_s$ is invariant with respect tot he dynamics (3.11b). The flow defined by (3.11b) on $\mathcal{M}_s$ is called the *sliding dynamics* and the control defined by (3.11a) is the *equivalent control*. Note that the equivalent control behaves as a linearizing feedback control. The partial state dynamics in sliding is obtained from (3.4a) and (3.11a):

$$\dot{z} = [I - EK]Az, \quad Kz(t_0) = 0 \tag{3.12}$$

PROPOSITION 3.4. *Suppose the conditions of Propositions 3.2 and 3.3 apply. Then $\mathcal{M}_0$ is an invariant manifold of the sliding dynamics (3.11b). Moreover, if K is specified as*

$$K = \text{diag}\,(k_1, \ldots, k_m), \quad k_i = [a_{i1}, \ldots a_{ir_i - 1}, 1] \tag{3.13}$$

where the m ordered sets of coefficients $\{a_{i1}, \ldots, a_{ir_i-1}\}, i = 1, \ldots, m$ each constitute a set of coefficients of a Hurwitz polynomial. Then every trajectory of (3.11b) not beginning in $\mathcal{M}_0$ approaches $\mathcal{M}_0$ exponentially.

Proof. Kwatny and Kim [1]. □

3.3. Reaching dynamics. The remaining step in VS control system design is the specification of the control functions $u_i^{\pm}$ such that the manifold $s(x) = 0$ contains a stable submanifold which insures that sliding occurs. There are many ways of approaching the reaching design problem [3]. We consider only one. Consider the function $\mathcal{V} : \mathbb{R}^n \to \mathbb{R}$

$$\mathcal{V}(x) = s^t(x)Qs(x) \tag{3.14}$$

Suppose $\dot{\mathcal{V}} < 0$ on an open set $\mathcal{D} - \mathcal{M}$ in $\mathbb{R}^n$. A sliding mode exists on a submanifold of $s(x) = 0$ which lies in $\mathcal{D}$. Upon differentiation we obtain

$$\frac{d}{dt}\mathcal{V} = 2\dot{s}^T Qs = 2[KAz + \alpha]^T QKz + 2u^T \rho^T QKz. \tag{3.15}$$

If the controls are bounded, $|u_i| \leq \bar{U}_i$, $\bar{U}_i > 0$, then obviously, to minimize the time rate of change of $\mathcal{V}$, we should choose

$$(3.16)\ u_i(x) = -\bar{U}_i \text{sign}\ (s_i^*), \quad i = 1, \ldots, m \text{ and } s^* = \rho^T(x) QKz(x).$$

It follows that $\dot{\mathcal{V}}$ is negative provided

$$|\bar{U}^T \rho^T QKz| > |[KAz + \alpha]^T QKz] \tag{3.17}$$

A useful sufficient condition is that

$$|(\rho(x)\bar{U})_i| > |[KAz(x) + \alpha(x)]_i| \tag{3.18}$$

Conditions (3.17) or (3.18) may be used to insure that the control bounds are of sufficient magnitude to guarantee sliding and to provide adequate reaching dynamics. This rather simple approach to reaching design is satisfactory when a "bang-bang" control is acceptable.

$\mathcal{A} \subset \mathcal{M}_0$ is a *stable attractor* of the zero dynamics if it is a closed invariant set and if for every neighborhood U of $\mathcal{A}$ in $\mathcal{M}_0$ there is a neighborhood V of $\mathcal{A}$ in $\mathcal{M}_0$ such that every trajectory of (3.7) beginning in V remains in U and tends to $\mathcal{A}$ as $t \to \infty$. The following proposition establishes conditions under which the variable structure controller (3.16) applied to (3.1) stabilizes $\mathcal{A}$ in R^n.

PROPOSITION 3.5. *Suppose that the conditions of Propositions 3.2, 3.3 and 3.4 apply; $\mathcal{D}$ is an open region in R^n in which (3.17) is satisfied: $\mathcal{D}_s = \mathcal{D} \cap \mathcal{M}_s$ is nonempty; and $\mathcal{A} \subset \mathcal{M}_0$ is a bounded, stable attractor of the zero dynamics which is contained in $\mathcal{D}_s \cap \mathcal{M}_0$. Then $\mathcal{A}$ is a stable attractor of the feedback system composed of (3.1) with feedback control law (3.16).*

Proof. Kwatny and Kim [1]. □

3.4. Remarks on stability. First, let us denote $\mathcal{M}_h = \{x | h(x) = 0\}$ and we assume that $\mathcal{M}_h$ is a regular submanifold of R^n of dimension $n-m$. Note that $\mathcal{M}_0$ is a submanifold of both $\mathcal{M}_h$ and $\mathcal{M}_s$ so that $\mathcal{M}_0$ lies in the intersection of $\mathcal{M}_h$ and $\mathcal{M}_s$. The relationships between these manifolds are illustrated in Figure 3.1.

Our results imply that the closed loop system behaves as follows. If the initial state is sufficiently close to $\mathcal{D}_s$, the trajectory will eventually reach $\mathcal{D}_s$ and will thereafter approximate ideal sliding. Ideal sliding is characterized by (3.11b) and sliding trajectories which remain in $\mathcal{D}_s$ approach $\mathcal{D}_0$ and eventually $\mathcal{A}$. That $\mathcal{A}$ is a stable attractor of (3.11b) is obvious. However, this only implies that trajectories of (3.11b) beginning sufficiently close to $\mathcal{A}$ approach $\mathcal{A}$. An important open problem is that of obtaining estimates of the domain of attraction.

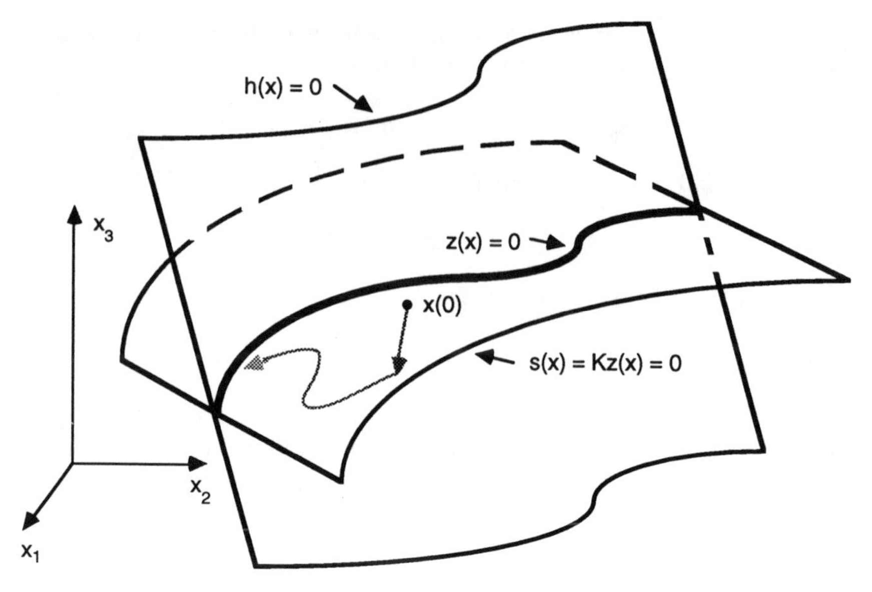

FIG. 3.1. a. *The relationship between the output constraint manifold, the sliding manifold and the zero dynamics manifold is illustrated in a three dimensional state space.*

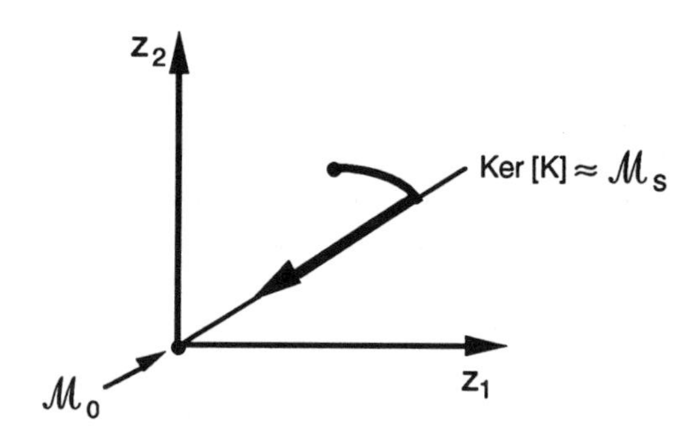

FIG. 3.1. b. *The natural projection onto the space of linearizable variables, z, shows the projections of the zero dynamics manifold, $\mathcal{M}_0$ and the sliding manifold $\mathcal{M}_s$.*

3.5. Gain scheduling via VSC. Now, let us consider a system parameterized by $\mu \in \Omega \subseteq \mathbb{R}^k$,

$$\dot{x} = f(x,\mu) + G(x,\mu)u \tag{3.19a}$$

$$y = h(x,\mu) \tag{3.19b}$$

which, using the method of the previous section can be transformed to

$$\dot{\xi} = F(\xi, z, \mu) \tag{3.20a}$$

$$\dot{z} = Az + E[\alpha(x(\xi, z), \mu) + \rho(x(\xi, z, \mu))u] \tag{3.20b}$$

$$y = Cz \tag{3.20c}$$

There are several points to be observed. The first ist hat the functions α, ρ, F are parameter dependent, however, the matrices A, C, and E are not. In the usual way, we introduce a new control variable, v, via the definition

$$u = \rho(x,\mu)^{-1}[-\alpha(x,\mu) + v] \tag{3.21}$$

so that equations (3.20) become

$$\dot{\xi} = F(\xi, z, \mu) \tag{3.22a}$$

$$\dot{z} = Az + Ev \tag{3.22b}$$

$$y = Cz \tag{3.22c}$$

Thus, we obtain a linear, parameter independent system (3.22b&c) for which a single compensator can be designed to meet the regulation performance objectives. If we proceed to design a variable structure controller in accordance with the procedures defined above, then the resultant closed loop dynamics are of the form:

$$\begin{array}{ll} \text{sliding dynamics:} & \left\{ \begin{array}{ll} \dot{\xi} & = F(\xi, (w, 0), \mu) \\ \dot{w} & = MANw \end{array} \right. \\ \\ \text{reaching dynamics:} & \dot{s} = \tilde{\rho}(\xi, w, s, \mu)(u - u_{eq}) \end{array} \tag{3.23}$$

Since we require that the equilibrium point is stable for all $\mu \in \Omega$, it follows that:

- the reaching conditions must be satisfied for all $\mu \in \Omega$,
- the zero dynamics must be stable for all $\mu \in \Omega$.

A significant observation is that:

- The projected sliding dynamics (w) are independent of μ.

3.6. Smooth versus discontinuous control. The state trajectories of ideal sliding motions are continuous functions of time contained entirely within the sliding manifold. However, the control signal, $u(t)$, is discontinuous as a consequence of the switching mechanism which generates it. In some applications this is undesirable. We discuss three remedies which have been proposed. One approach is "regularization" of the switch by replacing it with a continuous approximation. The second is "extension" of the dynamics by using additional integrators to separate an applied discontinuous pseudo-control from the actual plant inputs. Third is the "moderation" of the reaching control magnitude as errors become small.

Switch regularization entails replacing the ideal switching function, sgn $(s(x))$, with a continuous function such as

$$\text{sat}\left(\frac{1}{\varepsilon}s(x)\right) \quad \text{or} \quad \frac{s(x)}{\varepsilon+|s(x)|} \quad \text{or} \quad \tanh\left(\frac{s(x)}{\varepsilon}\right)$$

This intuitive approach is employed by Young and Kwatny [12] and Slotine and Sastry [13,14] and there are probably historical precedents. Regularization induces a boundary layer around the switching manifold whose size is $O(\varepsilon)$. Within this layer the control behaves as a high gain controller. The reaching behavior is altered significantly because the approach to the manifold is now exponential and the manifold is not reached in finite time as is the ccase with ideal switching. On the other hand within the boundary layer the trajectories are $O(\varepsilon)$ approximations to the sliding trajectories as established by Young et al [15] for linear dynamics with linear switching surfaces.

DEFINITION 3.6. *A C^k regularized switch is a function $f : \mathbb{R} \to \mathbb{R}$ which has the following properties:*

(i) *the kth derivative of f is continuous,*

(ii) *$f(x) \sim x$, as $x \to 0$,*

(iii) *$|f\left(\frac{x}{\varepsilon}\right) - \text{sgn}\ (x)| \to 0$, as $\varepsilon \to 0$ for $x \in \mathbb{R} - \{0\}$.*

The justification for this approach for linear systems is provided by the results in [15]. Some of those results have been extended to single input – single output systems has been extensively discussed by Slotine and coworkers, e.g. [13, 14]. With nonlinear systems there are subtleties and regularization can result in an unstable system.

Dynamic extension is another obvious and effective approach to control input smoothing. Our characterization of it differs very little from that of Emelyanov et al. [17].

DEFINITION 3.7. *A sliding mode is said to be of p-th order relative to an output y if the time derivatives $\dot{y}, \ddot{y}, \ldots, y^{(p-1)}$ are continuous in t but*

y^p *is not.*

The following observation is a straightforward consequence of the regular form theorem:

PROPOSITION 3.8. *Suppose (3.1) is input-output linearizable with respect to the output* $y = h(x)$ *with vector relative degree* $(r_1, \ldots, r_m)$. *Then the sliding mode corresponding to the control law (3.16) is of order* $p = \min(r_1, \ldots, r_m)$ *relative to the output* y.

We may modify the relative degree by augmenting the system with input dynamics as described. Hence, we can directly control the smoothness of the output vector y.

Control moderation involves design of the reaching control functions $u_i(x)$ such that $|u_i(x)| \rightarrow$ small as $|e(x)| \rightarrow 0$. For example,

$$u_i(x) = |e(x)| \text{ sign } (s_i(x))$$

Control moderation was used by Young and Kwatny [12] and the significance of this approach for chattering reduction in the presence of parasitic dynamics was discussed by Kwatny and Siu [18].

4. Drum level control. The issues associated with drum level control and circulation loop modeling are described in detail by Kwatny and Berg [2]. We employ the model developed therein for the present analysis. For completeness, a brief description of the model is included in the following paragraphs.

4.1. Model. The system (Fig 4.1) is comprised of a natural circulation loop whose main components are the drum which separates the steam and water, the downcomer piping which carries water from the drum to the bottom of the furnace, and the riser tubes which are exposed to the burning furnace gas and in which boiling takes place. A feedwater valve regulates the flow of water into the system and the throttle valve regulates the steam flow out of it. The heat absorbed by the fluid in the risers is a third input.

We describe the mathematical model by organizing the equations into three groups: The circulation loop thermodynamics, the circulation loop fluid mechanics, and the drum dynamics. First we establish some nomen-

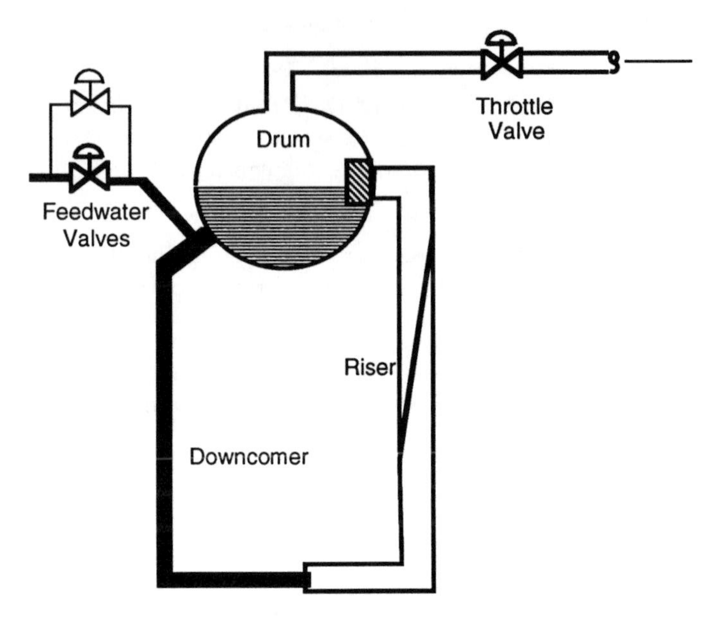

FIG. 4.1. *This illustration is a characterization of the circulation loop which identifies the major components to be modeled.*

clature:

N	number of riser sections
L_{do}, L	downcomer length and riser section length (total riser length/N)
A_{do}, A	downcomer, riser cross section areas
ω_i	mass flow rate at ith node
P_i	pressure at ith node
T_i	temperature at ith node
s_i	aggregate entropy at ith node
v_i	specific volume at ith node
$\omega_r, \omega_{dc}, \omega_s$	mass flow rates, riser, downcomer and turbine, respectively
v_{df}, v_{dg}	drum specific volume, liquid and gas, respectively
P_d	drum pressure
T_d	drum temperature
V	total drum volume
V_w	volume of water in drum
x_d	net drum quality, $x_d = V_w/V$
ω_{s0}	throttle flow at rated conditions
P_{d0}	drum pressure at rated conditions
A_t	normalized throttle valve position, at rated conditions $A_t = 1$

Circulation Loop Thermodynamics:

$$\begin{aligned}
\frac{ds_1}{dt} &= v_1\left\{\frac{1}{A_{d0}T_1}\frac{\partial q}{\partial z} - \frac{\omega_1}{A_{d0}}\left(\frac{s_1 - s_0}{L_{d0}}\right) + \frac{\omega_1 v_1}{A_{d0}T_1} f_{d0}\omega_1^2\right\} \\
\frac{ds_i}{dt} &= v_i\left\{\frac{1}{AT_i}\frac{\partial q}{\partial z} - \frac{\omega_i}{A}\left(\frac{s_i - s_{i-1}}{L}\right) + \frac{\omega_i v_i}{AT_i} f_r\omega_i^2\right\}, \quad i = 2, \ldots, N+1 \\
v &= v(s,P), T = T(s,P), \gamma_A = \left(\frac{\partial v}{\partial P}\right)_s, \gamma_B = \left(\frac{\partial v}{\partial s}\right)_P
\end{aligned}$$

Circulation Loop Fluid Dynamics:

$$\begin{aligned}
\frac{d\omega_{av}}{dt} &= \left\{-(P_d - P_0) + \sum_{i=1}^{N+1} h_i(\omega_{av}, s_i, s_{i-1}, P_{av}\right\} / \sum_{i=1}^{N+1} b_i \\
\frac{dP_{av}}{dt} &= \left\{(\omega_i - \omega_0) + \sum_{i=1}^{N+1} g_i(\omega_{av}, s_i, s_{i-1}, P_{av})\right\} / \sum_{i=1}^{N+1} b_i
\end{aligned}$$

$$a_i := L_i/A_i, \quad b_i := \gamma_A A_i L_i / v_{i-1}^2, \quad \alpha_i = \frac{a_j}{\sum_{i=1}^{N+1} a_i}, \quad \beta_i = \frac{b_i}{\sum_{i=1}^{N+1} b_i}$$

$$\begin{aligned}
h_i(\omega_{av}, s_i, s_{i-1}, P_{av}) &:= -\frac{\omega_{av}^2}{A^2}(v_i - v_{i-1}) - L\left(\frac{g}{v_i} + f_r\omega_{av}^2\right) \\
g_i(q, \omega_{av}, s_i, s_{i-1}, P_{av}) &:= -\frac{L\gamma_B}{v_{i-1}}\left(\frac{1}{AT_{i-1}}\frac{\partial q}{\partial z} - \frac{\omega_{av}}{A}\left(\frac{s_i - s_{i-1}}{L}\right)\right. \\
&\qquad \left. + \frac{\omega_{i-1}v_{i-1}}{AT_{i-1}} f_r\omega_{av}^2\right)
\end{aligned}$$

$$\begin{aligned}
P_0 &= P_{av} - \sum_{i=1}^{N}\left(\sum_{j=1}^{N+1-i}\alpha_j\right) h_i(\omega_{av}, s_i, s_{i-1}, P_{av}) \\
\omega_r &= \omega_{av} - \sum_{i=2}^{N+1}\left(\sum_{j=1}^{i-1}\beta_j\right) g_i(q, \omega_{av}, s_i, s_{i-1}, P_{av})
\end{aligned}$$

$$\omega_0 = \sqrt{|P_d - P_0|}\ \text{sign}\ (P_d - P_0)/f_{de}$$

Drum Dynamics:

$$\frac{dP_d}{dt} = -\frac{[w_e + (1 - x_r)w_r - w_{dc}]\frac{v_{df}^2}{v_{dg}} + [x_r w_r - w_s]\frac{1}{v_{df}}}{(V - V_w)\frac{v_{df}}{v_{dg}^2}\frac{\partial v_g}{\partial P_d} + V_w\frac{1}{V_{dg}}\frac{\partial v_f}{\partial P_d}}$$

$$\frac{dV_w}{dt} = \frac{[w_e + (1 - x_r)w_r - w_{dc}](V - V_w)\left(\frac{v_{df}}{v_{dg}}\right)^2\frac{\partial v_g}{\partial P_d} - V_w[x_r w_r - w_s]\frac{\partial v_f}{\partial P_d}}{(V - V_w)\frac{v_{df}}{v_{dg}^2}\frac{\partial v_g}{\partial P_d} + V_w\frac{1}{v_{dg}}\frac{\partial v_f}{\partial P_d}}$$

$$v_f = v_f(P) \text{ and } v_g = v_g(P)$$

$$\ell = f(V_w).$$

$$\omega_s = \omega_{s0}A_t\left(\frac{P_d}{P_{d0}}\right)$$

$$
\begin{aligned}
&u_1 = q,\ \ u_2 = \omega_e,\ \ u_3 = A_t \\
&\tfrac{d\omega_{av}}{dt} = f_1(\omega_{av}, s_1, s_2, s_3, P_{av}, P_d) \\
&\tfrac{ds_1}{dt} = f_2(\omega_{av}, s_1, P_{av}) + g_{21}(P_{av}, s_1)u_1 + g_{22}(\omega_{av}, P_d)u_2 \\
&\tfrac{ds_2}{dt} = f_3(\omega_{av}, s_1, s_2, P_{av}) + g_{31}(P_{av}, s_2)u_1 \\
&\tfrac{ds_3}{dt} = f_4(\omega_{av}, s_2, 8_3, P_{av}) + g_{41}(P_{av}, s_3)u_1 \\
&\tfrac{ds_4}{dt} = f_5(\omega_{av}, s_3, s_4, P_{av}) + g_{51}(P_{av}, s_4)u_1 \\
&\tfrac{dP_{av}}{dt} = f_6(\omega_{av}, s_1, s_2, s_3, s_4, P_{av}, P_d) + g_{61}(\omega_{av}, s_1, s_2, s_3, s_4, P_{av})u_1 \\
&\tfrac{dP_d}{dt} = f_7(\omega_{av}, s_1, s_2, s_3, s_4, P_{av}, P_d, V_w) + \\
&\quad g_{71}(\omega_{av}, s_1, s_2, s_3, s_4, P_{av}, P_d, V_w)u_1 + \\
&\quad g_{72}(P_d, V_w)u_2 - g_{72}(P_d, V_w)u_2 - g_{73}(P_d, V_w)u_3 \\
&\tfrac{dV_w}{dt} = f_8(\omega_{av}, s_1, s_2, s_3, s_4, P_{av}, P_d, v_w) + \\
&\quad g_{81}(\omega_{av}, s_1, s_2, s_3, s_4, P_{av}, P_d, V_w)u_1 + \\
&\quad g_{82}(P_d, V_w)u_2 - g_{83}(P_d, V_w)u_3 \\
&Y_1 = P_d, y_2 = \ell = h_2(V_2), y_3 = \omega_s = h_3(P_d) + d_3(P_d)u_3
\end{aligned}
$$

First, we describe the results of open loop studies of the circulation loop behavior. The procedure followed is:

1) trim the system at load levels ranging from near 5% to full load,
2) compute the linear perturbation equations,
3) analyze the pole-zero patterns as a function of load level.

4.2. Equilibria and perturbation dynamics. Equilibrium values are computed by specifying the desired load, drum pressure and drum level and then computing the required control inputs and the remaining state variables. A Taylor linearization at each equilibrium point yields a linear model of the perturbation dynamics. Thus, it is possible to determine the system poles and zeros and to examine how they change as a function of load. Figure 4.2 gives a sample of the results obtained from solving the equilibrium equations. Figure 4.3 is an eigenvalue plot which shows how the plant dynamics vary with load. Table 1 summarizes a complete plant modal analysis carried out at 100% load.

4.3. Zero dynamics. The linearized plant transmission zeros were also calculated as a function of load level using various combinations of inputs and outputs. These results are summarized in Figures 4.4 through 4.6. Notice that the single–input single–output results (Figure 4.4) which characterize the relation between feedwater flow and drum level show the nonminimum phase behavior typical of such systems, with a pair of zeros in the right half plane. On the other hand the multi–input multi–output results shown in Figures 4.5 and 4.6 are both minimum phase. Notice the apparent nonsmooth behavior of the zero loci. This suggests some complex behavior in the underlying nonlinear zero dynamics – perhaps bifurcations of some type – which we have not explored further.

4.4. Normal form. Figure 4.7 illustrates the circulation loop with actuators so that the command signals u_1, u_2, u_3 act through the furnace and valve drives to generate the process inputs. Because there is a direct

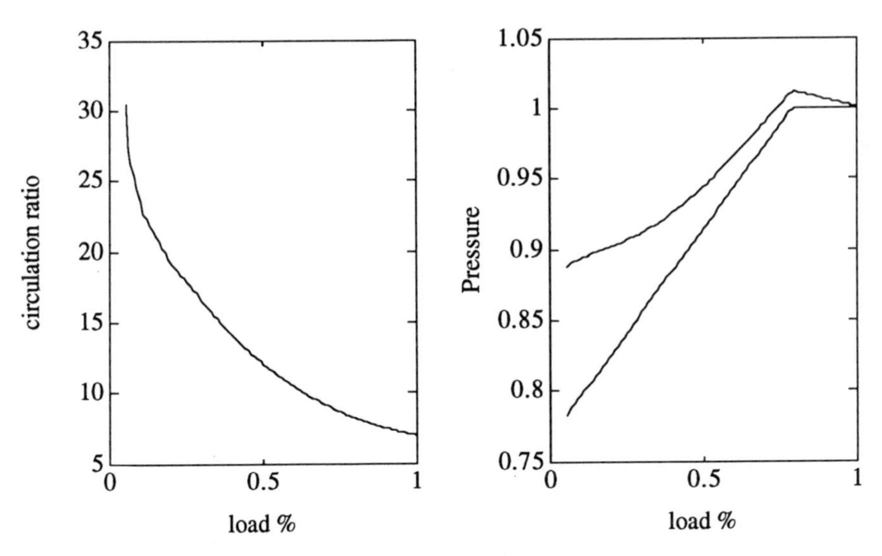

FIG. 4.2. *Typical equilibrium curves show the circulation ration, average loop pressure, and drum pressure as a function of load.*

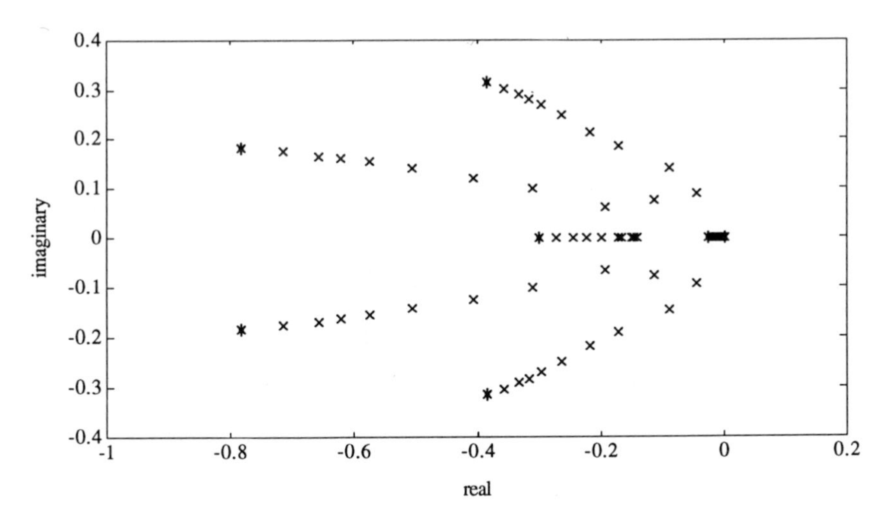

FIG. 4.3. *All but one of the eigenvalues of the circulation loop are illustrated. The missing eigenvalue is relatively far to the left at approximately* -8 *to* -20 *(depending on load level). There is an eigenvalue at the origin for all load levels as anticipated. The symbol* $(*)$ *denotes 100% load.*

TABLE 4.1
Eigenvalues & Eigenvectors at 100% Load.

	1	2&3	4&5
mode description	drum pressure-circulation flow rebalance	drum-riser mass balance oscillation	riser flow-density oscillation
eigenvalue	-7.7758	$-.7817 \pm .1835i$	$-.3854 \pm .3145i$
ω_{av}	0.5288	$.6866 \pm .5787i$	$-.9500 \pm .1684i$
s_1	-0.0066	$-.0052 \pm .0063i$	$.0034 \pm .0046i$
s_2	0.0005	$.0025 \pm .0148i$	$.0097 \pm .0058i$
s_3	0.0001	$.0186 \pm .0222i$	$.0029 \pm .0207i$
s_4	0.0001	$-.0933 \pm .0099i$	$-.0178 \pm .0280i$
P_{av}	-0.0323	$.0451 \pm .0178i$	$-.0107 \pm .0270i$
P_d	0.7644	$.0561 \pm .0283i$	$-.0321 \pm .0170i$
V_d	0.3675	$-.3925 \pm .1526i$	$.0879 \pm .2397i$

	6	7	8
mode description	circulation flow-drum level rebalance	energy-coupled drum level	drum level mass balance
eigenvalue	-0.3023	-0.0274	0.0000
ω_{av}	0.9692	0.4173	0.0000
s_1	0.0045	-0.0193	0.0000
s_2	0.0058	-0.0212	0.0000
s_3	0.0071	-0.0230	0.0000
s_4	0.0087	-0.0247	0.0000
P_{av}	-0.0267	-0.1285	0.0000
P_d	0.0187	-0.1117	0.0000
V_d	0.2436	0.8916	1.0000

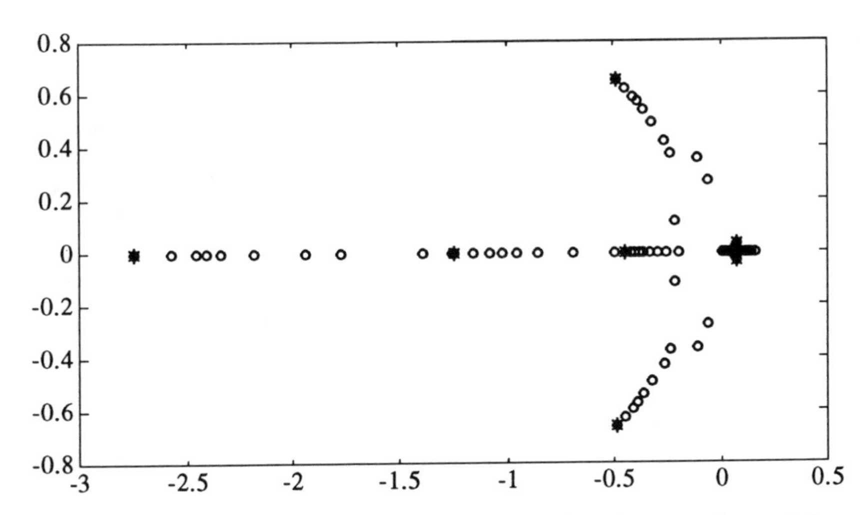

FIG. 4.4. *The zeros of the transfer function $\omega_e \rightarrow lev$ show the expected nonminimum phase characteristic of drum level dynamics.*

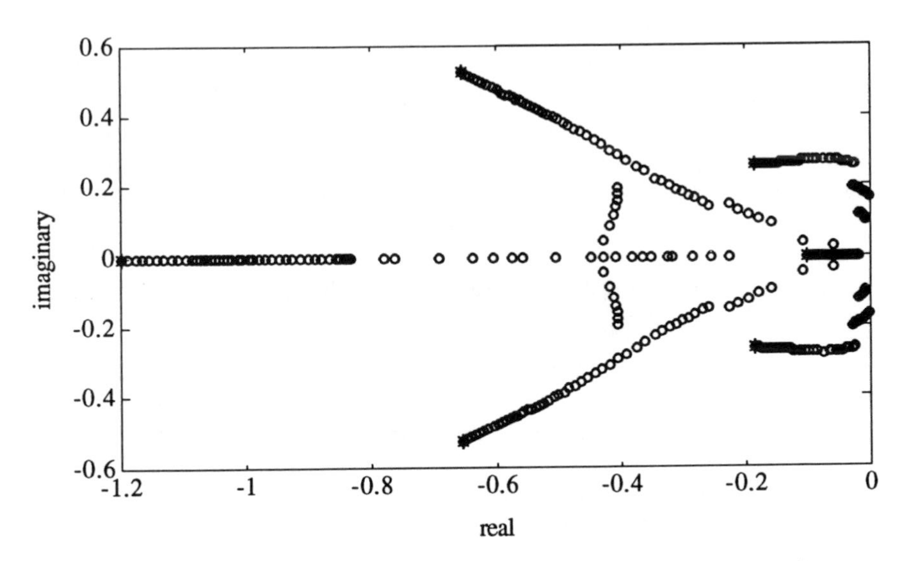

FIG. 4.5. *The transmission zeros of the 2×2 system* $([\omega_e, A_t] \rightarrow [Pd, lev])$.

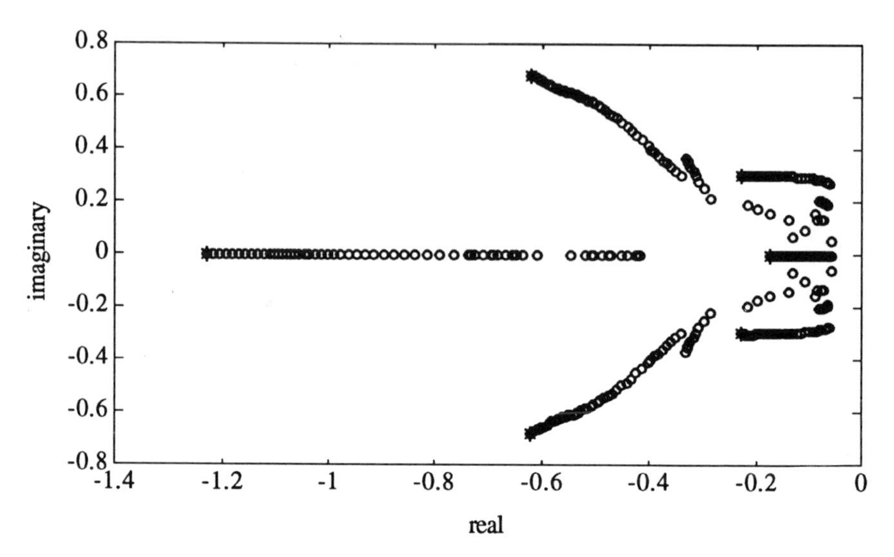

FIG. 4.6. *The transmission zeros of the* 3×3 *system* $([Q, \omega_e, A_t] \rightarrow [Pd, lev, \omega_s])$. *This system is minimum phase.*

feedthrough from throttle valve position to steam flow it is convenient (certainly not necessary) to identify the integral of stream flow as a regulated output.

It is easy to verify that if the actuators are relative degree one, then the linearized normal form, (3.9) or (3.22a&b), for the 3-input 3-output system illustrated takes the form shown in Figure 4.7. If the actuator relative degree (they can be different, of course, for each actuator) is different from one, then the number of pre-integrators would differ accordingly. The significance of the normal form is that it is the basis for designing the switching surfaces. Moreover, because of the simple structure of the normal form it is quite trivial to see how the Matrix K in Proposition 3.4 can be chosen to fix the eigenvalues associated with drum pressure P_d, drum level ℓ, and steam flow, ω_s.

4.5. The VSC regulator. The variable structure controller is designed in accordance with the discussion in Section 3. We first choose switching surfaces which produce the desired sliding dynamics:

$$\begin{aligned} s_1 &= .5z_1 + \dot{z}_1, & z_1 &= (P_d - \bar{P}_d)/P_d^* \\ s_2 &= .5z_2 + \dot{z}_2, & z_2 &= \ell \\ s_3 &= .0001z_3 + .02\dot{z}_3, & z_3 &= \int (\omega_s - \bar{\omega}_s)dt/\omega_s^* \end{aligned}$$

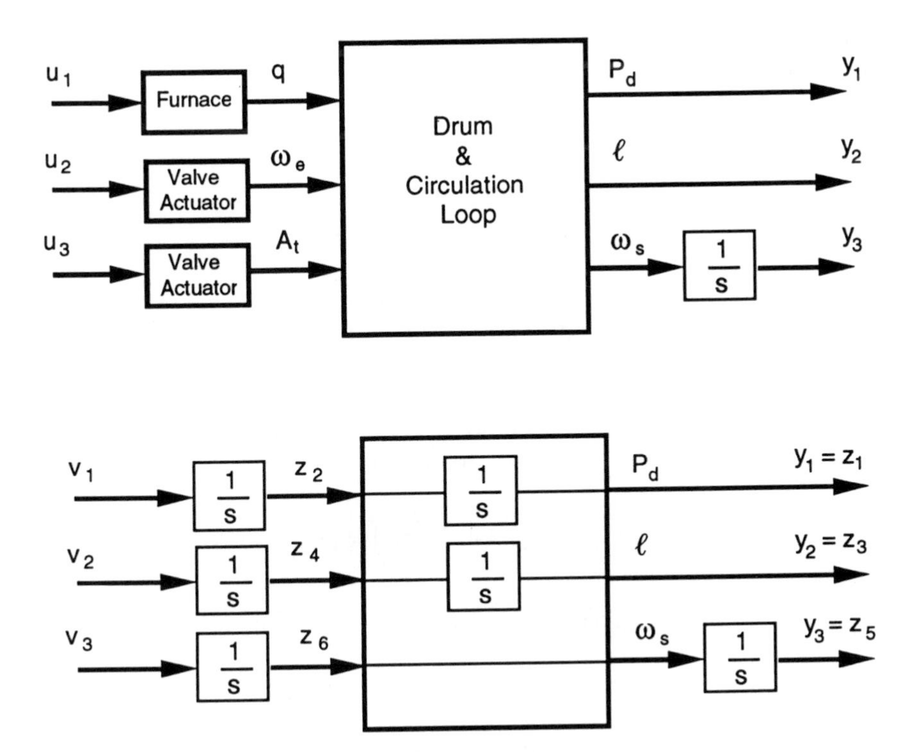

FIG. 4.7. *The drum and circulationloop is illustrated along with its linearized normal form under the assumption that the valve actuators and the furnace are of relative degree one.*

The control is chosen according to the equation (3.16):

$$u_i(x) = -\bar{U}_i \text{ sign } (s_i^*), \quad i = 1, \ldots, m \quad \text{ and } s^* = \rho^T(x) Q_s(z)$$

Calculations show that the matrix $\rho(x)$ is approximately constant over the load range, which results in an important simplification. Thus, we obtain

$$\rho \approx \begin{bmatrix} .0071 & .09436 & -.9350 \\ .0031 & .06793 & -.1461 \\ 0 & 0 & .9176 \end{bmatrix}$$

We choose

$$Q = \begin{bmatrix} 5 & 0 & 0 \\ 0 & .1 & 0 \\ 0 & 0 & .001 \end{bmatrix}$$

In addition, we take $\bar{U}_1 = .075$, $\bar{U}_2 = .15$, $\bar{U}_3 = .15$, and we employ control smoothing and control moderation to obtain the control:

$$\begin{aligned} u_1 &= -.075 \frac{|z|}{.004 + |z|} \text{ sat } \left(\frac{|z_1|}{.02} \right) \text{ sign } (s_1^*) + \bar{u}_1 \\ u_2 &= -.15 \frac{|z|}{.004 + |z|} \text{ sat } \left(\frac{|z_2|}{.2} \right) \tanh \left(\frac{s_2^*}{.0001} \right) + \bar{u}_2 \\ u_3 &= -.15 \frac{|z|}{.004 + |z|} \text{ sat } \left(\frac{|z_3|}{.02} \right) \tanh \left(\frac{s_3^*}{.005} \right) + \bar{u}_3 \end{aligned}$$

5. Simulation results. We will compare the performance of a linear controller with that of the VS controller. The design of the linear control system we use is described in detail in [2]. In structure, it is a conventional three-element feedwater controller above 30% load and a single element controller below 30% load. The linear controller can not tolerate step changes in load level of more than about 5%-10%, and somewhat smaller step changes at very low loads. In the following figures, we preset a sequence of simulation results of step load change commands spanning the range from high load to very low load. First, the linear controller performance is illustrated and then a comparable sequence of results using the variable structure controller.

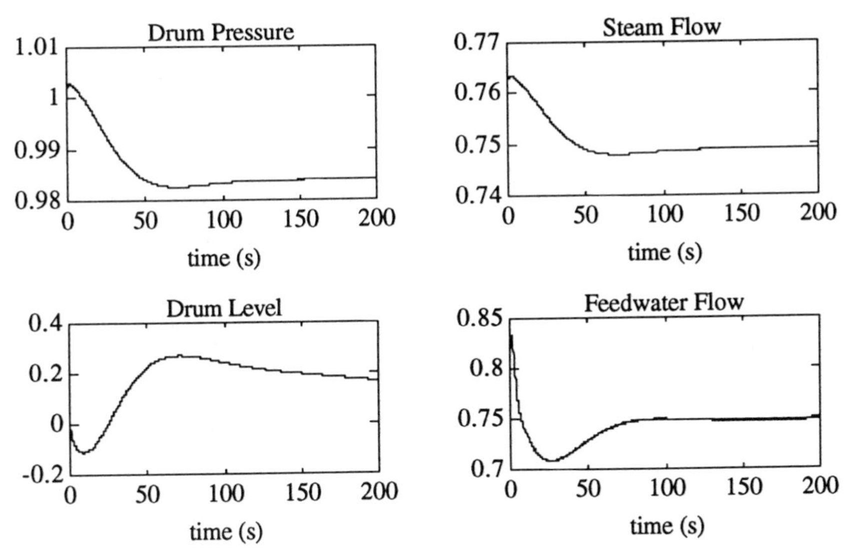

FIG. 5.1. *Linear controller response to a 5% step command changing load from 80% to 75%.*

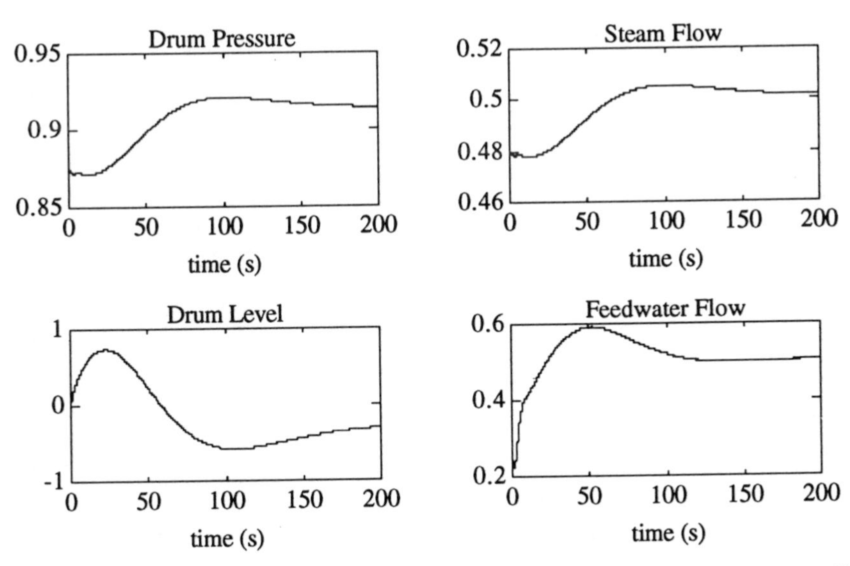

FIG. 5.2. *Linear controller response to a 10% step command changing load from 40% to 50%.*

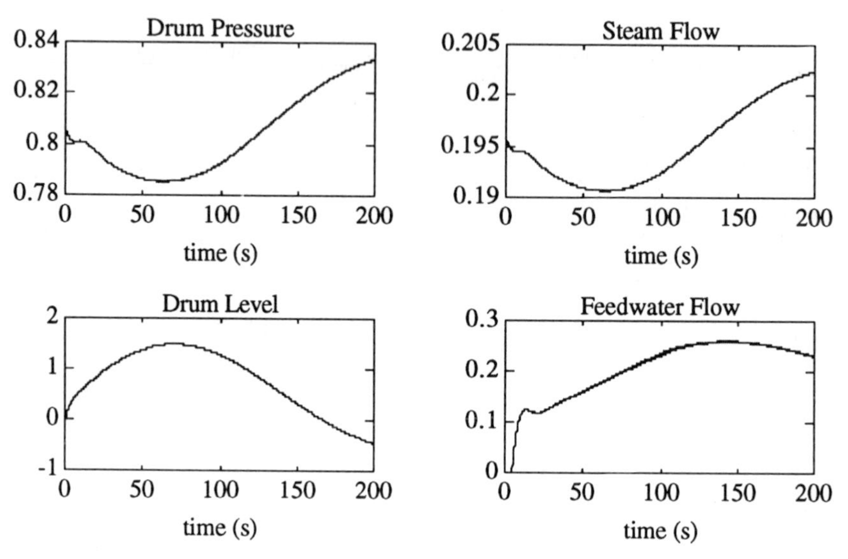

FIG. 5.3. *Linear controller response to a 5% step command changing load from 15% to 20%. Recall that below 30% load, a single element controller is used. For this particular plant, the 15%-20% range is particularly troublesome.*

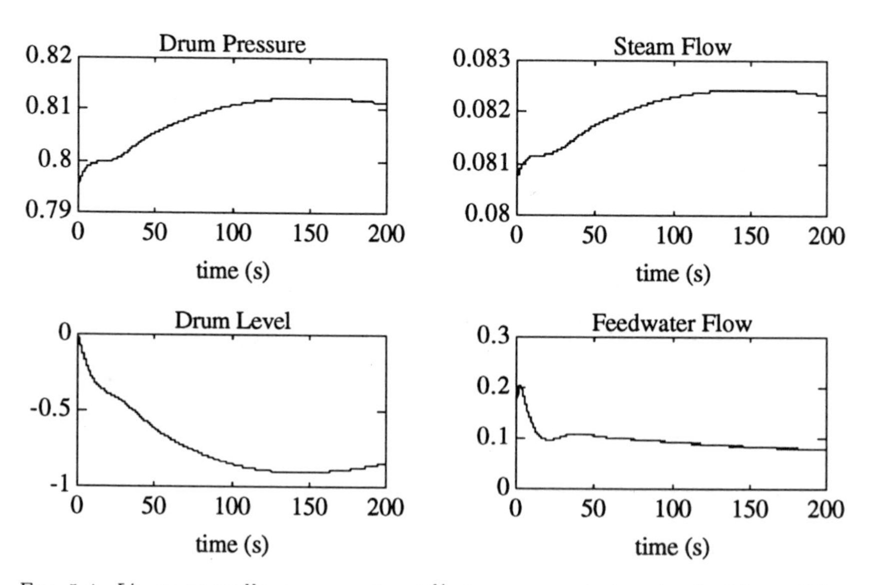

FIG. 5.4. *Linear controller response to a 2% step command changing load from 10% to 8%. At this very low load, larger step changes are not possible.*

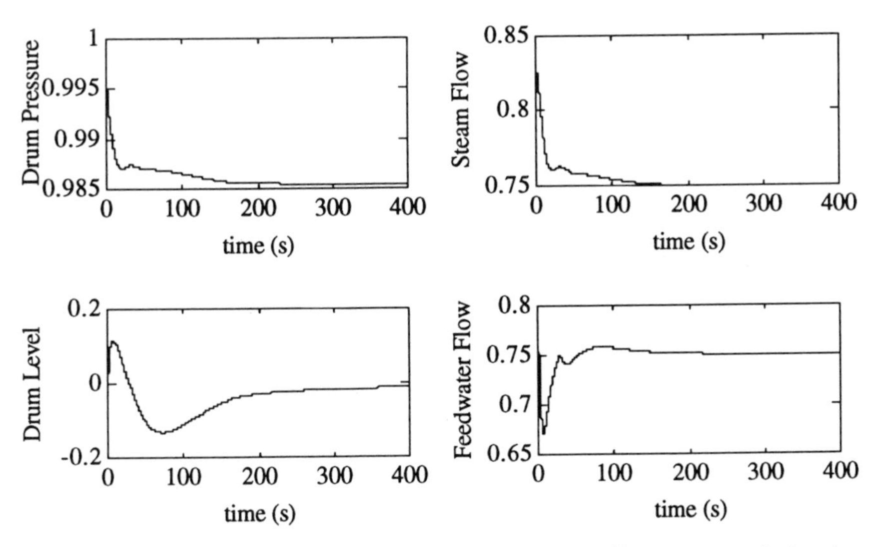

FIG. 5.5. a. *Variable structure* 3×3 *controller response to a 5% step command changing load from 80% to 75%.*

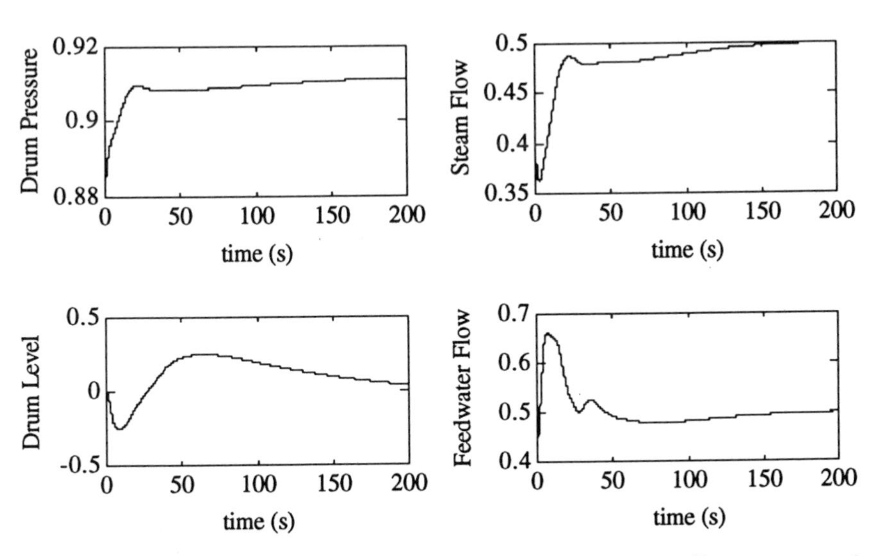

FIG. 5.5. b. *Variable structure* 3 × 3 *controller response to a 10% step command changing load from 40% to 50%.*

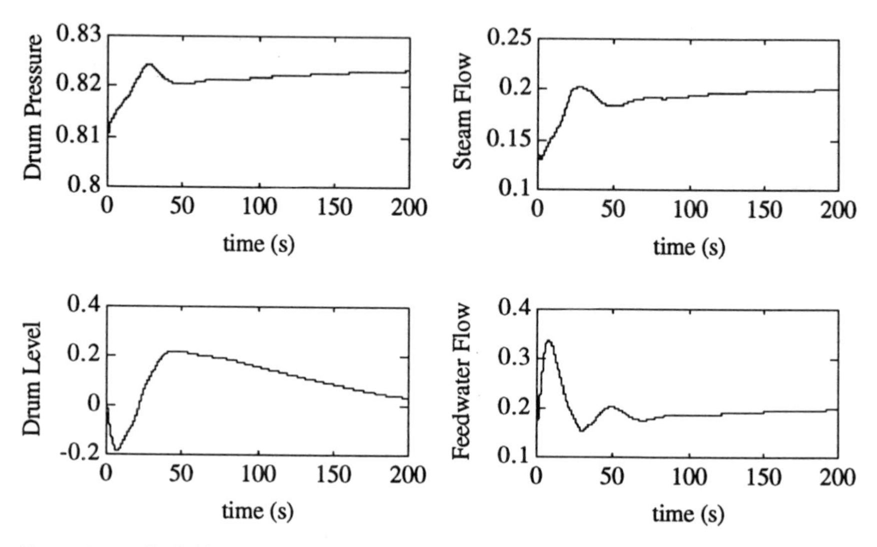

FIG. 5.6. a. *Variable structure* 3×3 *controller response to a 5% step command changing load from 15% to 20%.*

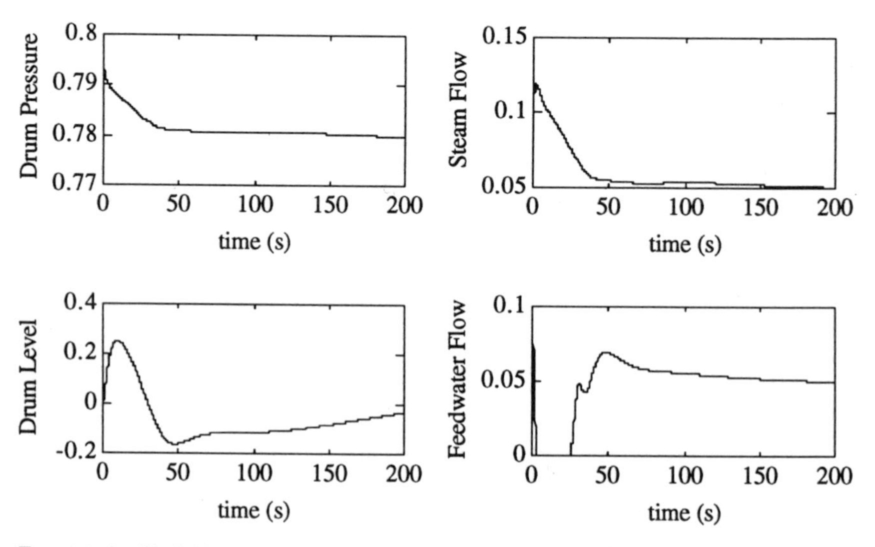

FIG. 5.6. b. *Variable structure* 3×3 *controller response to a 5% step command changing load from 10% to 5%.*

6. Conclusions. This paper has considered the application of variable structure control methods to power plant drum level control. The main issues are that the process dynamics are nonlinear and vary dramatically with load level, and the traditional single-input single-output configuration is nonminimum phase. We consider multi-input multi-output configurations in order to eliminate the inherent performance limitations of nonminimum phase systems and nonlinear control in order to achieve consistent performance over the entire load range. We choose variable structure control because of its inherent robustness, the ability to incorporate control constraints and the limited measurements needed to implement the controller. However, other nonlinear control methods, notably variants of feedback linearization, would also be appropriate.

Variable structure controllers are switching controllers and hence concern naturally arises about the possibility of control signal chattering. We employ three methods of control signal smoothing and as our simulation results show, the actuation signals are quite smooth. Overall, our results are very good and suggest that variable structure control provides a viable strategy for drum level control. Uniformly consistent performance from 5% to 100% load has been achieved. On the other hand our analysis is not comprehensive. We have set aside many practical issues which would have to be considered in a final design study.

We also note that there are important elements of variable structure theory that are not fully developed. For instance, there do not now exist systematic methods for designing the control moderation functions which we have found essential for achieving smooth, non-chattering controls. Also, the only generally applicable method for estimating the domain of attraction of a stable equilibrium point remains exhaustive computation which provides limited insight for design of the reaching controls.

REFERENCES

[1] KWATNY, H.G. AND H. KIM, *Variable Structure Regulation of Partially Linearizable Dynamics*, Systems & Control Letters, 15 1990, pp. 67–80.

[2] KWATNY, H.G. AND J. BERG, *Drum Level Regulation at All Loads: A Study of System Dynamics and Conventional Control Structures*, in Proceedings 12th IFAC World Congress, Sydney: Vol. 3 1993, pp. 405–408.

[3] UTKIN, V.I., *Sliding Modes and Their Application*, 1974 (in Russian), 1978 (in English), Moscow: MIR.

[4] UTKIN, V.I., *Variable Structure Systems with Sliding Modes: A Summary*, IEEE Transactions in Automatic Control, AC-22(April) 1977, pp. 212–222.

[5] UTKIN, V.I., *Variable Structure Systems: Present and Future*, Avtomatica i Telemekhanica, 9(September) 1983, pp. 1105–1119.

[6] DECARLO, R.A., S.H. ZAK AND G.P. MATHEWS, *Variable Structure Control of Nonlinear Multivariable Systems: A Tutorial*, Proceedings IEEE, 76(3) (1988) pp. 212–232.

[7] FILIPPOV, A.F., *A Differential Equation with Discontinuous Right Hand Side*, Matemat: Cheskii Sbornic 51(1) 1960, pp. 99–128.

[8] ISIDORI, A., *Nonlinear Control Systems*, 1989, Springer-Verlag, NY.

[9] KRENER, A.J., *On the Equivalence of Control Systems and Linearization of Nonlinear Systems*, SIAM Journal on Control and Optimization, 11 1973, p. 670.

[10] HIRSCHORN, R.M., *Invertibility of Nonlinear Control Systems*, SIAM Journal on Control and Optimization, 17(2) 1979, pp. 289–297.

[11] DESCUSSE, J. AND C.H. MOOG, *Decoupling with Dynamic Compensation for strong Invertible Affine Non-Linear Systems*, International Journal of Control, 42(6) 1985, pp. 1387–1398.

[12] YOUNG, K.D. AND H.G. KWATNY, *Variable Structure Servomechanism and its Application to Overspeed Protection Control*, Automatica, 18(4) 1982, pp. 385–400.

[13] SLOTINE, J.J.E., *Sliding Controller Design for Non-Linear Control Systems*, International Journal of Control, 40(2) 1984, pp. 421–434.

[14] SLOTINE, J.J. AND S.S. SASTRY, *Tracking Control of Non-Linear Systems Using Sliding Surfaces, with Application to Robot Manipulators*, International Journal of Control, 38(2) 1983, pp. 465–492.

[15] YOUNG, K.D., P.V. KOKOTOVIC AND V.I. UTKIN, *Singular Perturbation Analysis of High Gain Feedback Systems*, IEEE Transactions on Automatic Control, AC-22(6) 1977, pp. 931–938.

[16] MARINO, R., *High Gain Feedback Non-Linear Control Systems*, International Journal of Control, 42(6) 1985, pp. 1369–1385.

[17] EMELYANOV, S.V., S.K. KOROVIN AND L.V. LEVANTOVSKY, *A Drift Algorithm in Control of Uncertain Processes*, Problems of Control and Information Theory, 15(6) 1986, pp. 425–438.

[18] KWATNY, H.G. AND T.L. SIU, *Chattering in Variable Structure Feedback Systems*, in Proceedings 10th IFAC World Congress, Munich 1987, pp. 334–341.

ANALYSIS OF MECHANISMS OF VOLTAGE INSTABILITY IN ELECTRIC POWER SYSTEMS

CHEN-CHING LIU* AND KHÔI TIEN VU†

Abstract. This article provides a summary of a series of papers on the analysis of mechanisms of voltage instability in power systems, [1,2,3,4,5].

1. Introduction. A power system is an electrical network containing components such as generators, transmission lines, loads and voltage controllers. Power networks are usually large, containing thousands of nodes and branches. The main function of power systems is to deliver electricity to the consumers at a fixed frequency and voltage level, e.g., 60-Hz and 110-V for household consumption. Deviations from these nominal conditions may result in abnormal performance of, or even damage to, customers' equipment. However, since disturbances occur frequently in power systems, maintaining a nominal condition is a nontrivial task.

Disturbances are typically categorized as small or large, depending on their severity. An example of small disturbances is the usual load variations during the day. An example of a large disturbance is the outage of a major transmission line, or a sizable generator. Analytically, small-disturbance problems can be studied via linearization of the equations describing the dynamic behavior of the network; large-disturbance problems, however, require nonlinear techniques. The system is said to be stable if, following a disturbance, it is able to reach an acceptable steady state.

Traditionally, power system stability refers to the notion of whether synchronism of the generators can be maintained following a disturbance. Physically, this requires a balance between the mechanical power applied to each generator and its electrical power output. Any excess of power would accelerate the machine rotor, which not only affects the frequency, but might also damage the machine mechanically. The deviation of the machine's rotor from its nominal speed can be described by a second-order differential equation which resembles a pendulum equation (mass M, friction D), the driving force being the imbalance between the mechanical input and the electrical output:

$$M\ddot{\theta} + D\dot{\theta} = P_m - P_e \tag{1.1}$$

For a network of many generators, the electrical output P_e of each generator depends on the interconnection. Automatic controls added to each generator affect the terms P_e and P_m. Research effort over past decades

* Department of Electrical Engineering, FT-10, University of Washington, Seattle, WA 98195.

† Department of Electrical and Computer Engineering, Clemson University, Clemson, SC 29631.

has contributed to understanding of power system behavior governed by Eq. (1.1). Reference [6] is one of the good survey papers on this subject.

Present-day power systems are becoming more complex due to the interconnection among utilities. In addition, systems are required to meet tighter operating constraints. This is due to the fact that the demand for electricity keeps on increasing but the construction of generating plants or transmission lines is limited by many factors including environmental concerns. In such systems, long transmission lines connecting remote power sources to load centers have become common. As a result, the voltage problem is a serious concern. Several major events occurring throughout the world show that, before any significant change in frequency is observed, the voltage may decay to a very low level. A low-voltage level might trigger the protective devices, leading to a blackout or brownout. This has led to a new subject of power system stability—voltage collapse, or voltage instability.

Researchers in power system stability sometimes classify events into voltage instability and angle instability (the term *angle* is due to the fact the frequency deviation depends on the rotor position, θ in Eq. (1.1), relative to its nominal rotating frame). There are several reasons for this. One reason is that the time period of the study may be short term or long term. Based on practical experience, it is believed that frequency problems can take place over short time periods but voltage problems may require a long time to build up. The second reason involves the complexity of the problem if both angle and voltage phenomena are to be analyzed; this is due to the interaction among the many components in a power system (generator, controllers, protective devices, load, etc.). Even though different phenomena have been observed, it is clear that an unstable event may take place involving both frequency and voltage problems. To analyze such an event, one needs to have a physically sound and yet mathematically tractible model for the power system.

In some respects, the angle stability (frequency) is closely related to the rotors of the generators whereas the load side plays an important role in voltage stability problems. This guides one in selecting models for an analytical study.

This paper concentrates on the analysis of one type of stability problems, namely voltage stability. The goal is to explain how various mechanisms can interact with each other to bring about a collapse of voltage. Based on real-world phenomena, three mechanisms are deemed important: tap changing (a form of voltage controller), load behavior, and generator excitation limiting. Since the emphasis is on the load side for the voltage problem, we use a simple model for generators. That is, Eq. (1.1) is not used and the frequency is assumed to be held constant in the system model.

Our analysis begins with a single mechanism, tap changing, and is then extended to include the other two mechanisms. In each study, the analytical results rely on an investigation of the vector field representing the dynamic behavior of the nonlinear system. Invariant sets are found important in proving many of the results. Two invariant sets, namely the set of all-negative derivatives and the set of all-positive derivatives, imply monotonicity of system trajectories. These monotonic trajectories are then used in constructing (subsets of) the region of attraction.

The following standard notation in power system analysis is used. Matrices are denoted by boldface capitals, e.g., $\mathbf{B}$; vectors are indicated by an arrow, e.g., $\vec{v}$. All quantities are real numbers, except in some occasions where a phasor (complex number) is involved. An upperbar is used to distinguish a phasor from a real number; for example, $\bar{Z}$ is a phasor, but Z is a real number.

Components of a vector are denoted with a subscript index, e.g., x_i is the i-th component of vector $\vec{x}$. Inequalities are used for vectors to mean componentwise inequalities. For example, $\vec{x} \leq \vec{y}$ means $x_i \leq y_i$ for every component i of the vectors $\vec{x}$ and $\vec{y}$.

Some of the symbols appear frequently throughout the text. They are listed below together with their meaning.

V: voltage.

n: turns-ratio of a tap-changer.

E: generator voltage.

$\mathcal{P}$: the collection of all points $\vec{x}$ in the state space at which the derivative $d\vec{x}/dt$ is ≥ 0.

$\mathcal{N}$: the collection of all points $\vec{x}$ in the state space at which the derivative $d\vec{x}/dt$ is ≤ 0.

Frequently, the listed symbols are subscripted. For example, n_i is to denote the turns-ratio of the tap-changer at node i.

2. Power system model. Three mechanisms widely believed to be critical in the voltage collapse scenarios are tap changing, load dynamic and generator excitation limiting. The model for each of these three mechanisms is described in this section. For completeness, we also describe the model of transmission lines even though they are not considered a mechanism.

Tap changer: Tap changers are a special type of transformers used for voltage control (see Fig. 2.1). Typically, a transformer has two sides: the primary side which is connected to the source via transmission lines, and the secondary side serving the load. The voltages on the two sides of a transformer are proportional to the turns-ratios of their windings, i.e., $\frac{V'}{V} = \frac{1}{n}$. When the transmission-line flow is changed (e.g., due to load variations), the load voltage V may fluctuate. One way to maintain this voltage within a desired level is to make the turns-ratio $1 : n$ of the transformer adjustable. A transformer with this feature is called a tap-changer. Tap changing refers to an event in which a change in turns-ratio (or tap position) takes place.

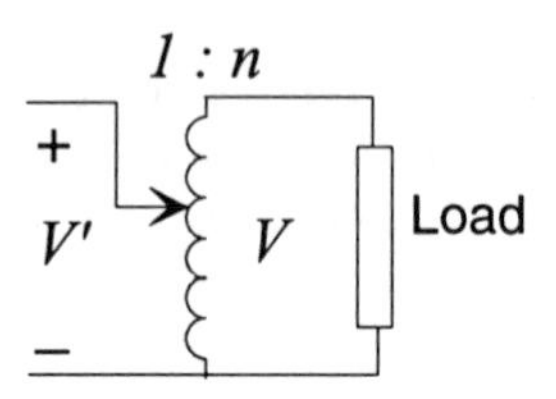

FIG. 2.1. *Schematic diagram of a tap changer.*

In practice, the turns-ratio takes on a finite number of positions, and each tap changing is a discrete event. Typically, two adjacent tap positions result in a voltage change of about 0.625%. The time required to move from one tap position to another depends largely on the design, which may vary from a few seconds to several minutes. There are two models for tap-changing dynamic.

1. Discrete model: In this model, a tap changing takes place (after some built-in time delay) if the load voltage V falls beyond a voltage range $[V_0 - \epsilon, V_0 + \epsilon]$:

$$n_{k+1} = n_k + d \cdot f(V_0 - V) \tag{2.1}$$

where n_{k+1} and n_k are the turns-ratios before and after a tap change, $d > 0$ is the size of each tap change, and

$$f(x) = \begin{cases} 1 & \text{if } x > \epsilon \\ 0 & \text{if } |x| \leq \epsilon \\ -1 & \text{if } x < -\epsilon \end{cases} \tag{2.2}$$

2. Continuous model: This model assumes a continuous range of tap positions and can be justified by the fact that the difference between two adjacent tap positions is very small. The continuous dynamic model is given by:

$$\frac{dn}{dt} = \frac{1}{T}(V_0 - V) \tag{2.3}$$

The constant $T > 0$ in the equation is to model the delay constant.

Load model: The load is the part of a power system that remains most difficult to model. This is due not only to the continuing fluctuation in demand but also to the lack of specific knowledge about how the power drawn by the load devices depends on voltage and frequency. This is particularly true when one has to deal with very low and abnormal voltages. We list two types of load models here; one is static and the other is dynamic.

1. Impedance: this is the simplest load model. For an impedance load, the power depends on voltage according to a quadratic relation:

$$\text{real power demand } P(V) = gV^2 \tag{2.4}$$

$$\text{reactive power demand } Q(V) = yV^2 \tag{2.5}$$

where g and y are positive constants.

2. Dynamic: there are a number of models for dynamic loads. In this article, we use the following model:

$$\text{real power demand } P(V, \dot{\delta}) = P_s(V) + k_P\dot{\delta} \tag{2.6}$$

$$\text{reactive power demand } Q(V, \dot{V}) = Q_s(V) + k_Q\dot{V} \tag{2.7}$$

where $\dot{\delta}$ represents the frequency deviation at the node; $\dot{V}$ is the rate of change of the voltage magnitude; P_s, Q_s, k_P, and k_Q are positive functions of V and δ. The terms P_s and Q_s are the "static" power demands since they are the power drawn by the load when a steady state is reached, i.e., when $\dot{V} = 0$ and $\dot{\delta} = 0$.

Generator excitation: In its simplest form, a generator can be pictured as a voltage source $\bar{\mathcal{E}}$ in series with an impedance jX (see Fig. 2.2). The difference between the internal source $\bar{\mathcal{E}}$ and the voltage drop across the impedance is the output voltage $\bar{E}$ of the machine. In power system operation, it is desired to keep the output voltage magnitude $E = |\bar{E}|$ within a narrow range. To achieve this objective, a feedback control loop called the excitation control system is added to the generator. This control system constantly monitors E and adjusts the internal $\mathcal{E}$ accordingly. Thus, a heavily loaded generator requires a high internal voltage. However, hardware constraints impose an upper limit on $\mathcal{E}$. When this limit is reached, the generator is said to be at its excitation limit and, as a result, the output voltage E is dictated by the power flow from the machine to the network.

The above paragraph gives a simplified and conceptual view of an excitation control system. Practically, excitation control systems have dynamics of their own, and furthermore, the event in

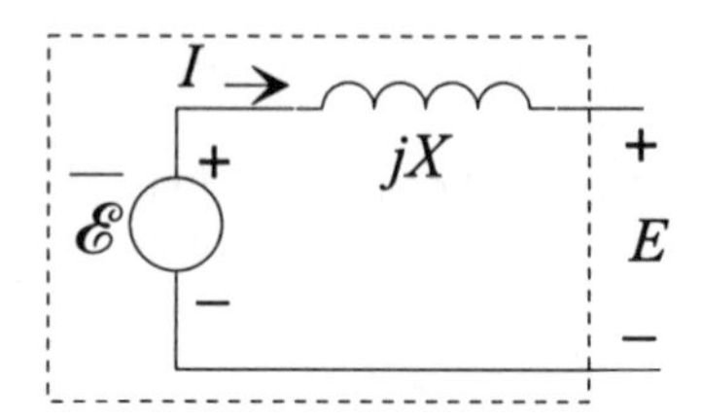

FIG. 2.2. *Simple schematic representation for generator.*

which the excitation limit is reached may undergo several stages depending on how the protection circuit is designed. Modeling an excitation system in detail therefore may not be feasible for a mathematical analysis. Fortunately, it is true that the associated dynamics are quite fast when compared with slow events such as voltage collapse. This means that it is adequate to model excitation control by an algebraic model.

The model employed in our analysis assumes that the generator has two constraints: (a) internal voltage limit $\mathcal{E}_{lim}$, and (b) generator output current limit I_{lim}. Depending on whether these limits are reached or not, we have these modes of operation:

1. Normal mode:

$$\mathcal{E} < \mathcal{E}_{lim} \text{ and } I < I_{lim} \tag{2.8}$$

In this mode, the output voltage $E = E_0$, where E_0 is a positive constant.

2. Excitation-limit mode: $\mathcal{E} = \mathcal{E}_{lim}$ or $I = I_{lim}$. (As a result, $E < E_0$.)

Transmission lines: Each line (or branch of the power network) joining two nodes i and k has an admittance $\bar{Y}_{ik} = g_{ik} - jy_{ik}$ or impedance $\bar{Z}_{ik} = r_{ik} + jx_{ik}$, where $j = \sqrt{-1}$. It is quite common in power system stability analysis to assume that lines have no resistance (the energy loss due to line resistance is small compared to the total power delivery); thus, $r_{ik} = 0$ and $g_{ik} = 0$. We use the lossless lines in our analysis, except in the simple case of Subsection 3.1 where we incorporate the line resistance.

We summarize in subsequent sections the main analytical results obtained with the above system model. As a special case, we first discuss in Section 3 only the role of tap-changer in voltage collapse. The objective is to obtain some insight into the adverse effect of the voltage controller. To do this, we use a simplified version of the model. We then consider a generalization in Section 4 where all the details of the system model are incorporated.

We emphasize that the following important assumption is made in our analysis:

ASSUMPTION 1. *For the time duration of interest, the frequency stays constant.*

3. Tap-changer analysis

3.1. Simple case. The tap-changing operation (Eqs. (2.1–2.2) or Eq. (2.3)) is designed around the normal operating voltage. Since voltage collapse typically involves a network under heavy loading, it is sensible to ask whether tap changing is still appropriate. The first detailed nonlinear analysis, given in [1], involves a simple power system (two nodes). The system diagram is shown in Fig. 3.1 whose model is a simplified version of the one presented in Section 2:

1. Tap-changer: discrete or continuous model (Eqs. (2.1–2.2) or Eq. (2.3)).
2. Load: impedance model, Eqs. (2.4–2.5), with impedance $\bar{Z} = Z\underline{/\theta}$.
3. Generator: constant voltage source, i.e., $\mathcal{E}_{lim} = \infty$ and $I_{lim} = \infty$ in Eq. (2.8). This means that the output voltage can always be maintained at the constant level of E_0.
4. Transmission line: impedance $\bar{Z}_l = Z_l\underline{/\theta_l}$.

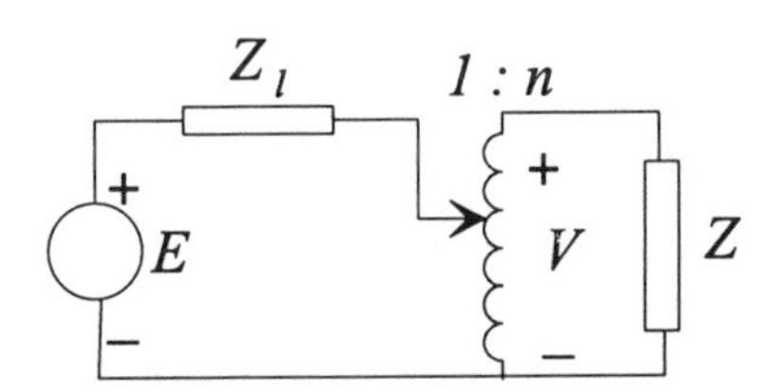

FIG. 3.1. *A two-node power system with a tap-changer.*

For a given load impedance $\bar{Z}$, the magnitude of the voltage on the load side is given by:

$$V(n) = \frac{nE_0Z}{\sqrt{n^4Z_l^2 + 2n^2ZZ_l\cos(\theta_l - \theta) + Z^2}} \tag{3.1}$$

where n represents the turns-ratio of the tap-changer.

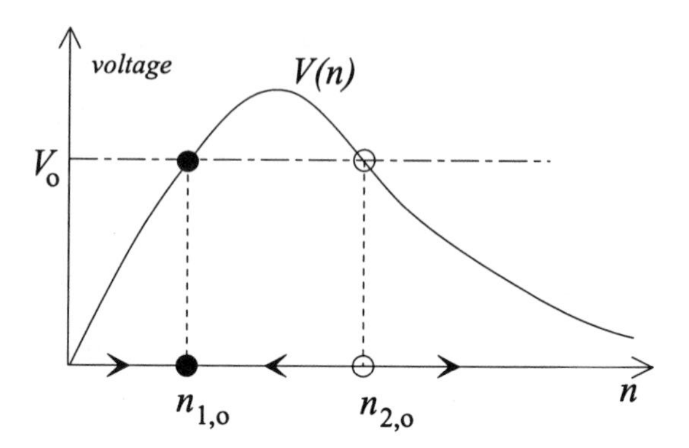

FIG. 3.2. *Magnitude of load voltage as a function of turns-ratio.*

A typical plot of V versus n is given in Fig. 3.2. The figure also shows the reference voltage level V_0. There are at most two intersection points, $n_{1,o}$ and $n_{2,o}$, between the two curves; they represent the equilibria of the system. The dynamic model of tap changing, Eqs. (2.1–2.2) or Eq. (2.3), can be used to show that $n_{1,o}$ is the only stable equilibrium and its region of attraction is $(0, n_{2,o})$. It is interesting to observe that if the initial n is outside this region ($n > n_{2,0}$) then n would keep on increasing (until a physical limit is reached), which leads to a decaying voltage.

The above analysis, though simple, provides some insights into tap changing as a mechanism of voltage instability. One, the region of attraction for tap changing is bounded. Two, outside the attraction region, tap changing is undesirable and contributes to voltage decay.

3.2. General case

3.2.1. System equations. The system, whose topology is shown in Fig. 3.3, includes the following elements:

1. Loads: there are $M + N$ load nodes; M of which have voltage regulated by tap-changing (1 through M), the remaining N nodes have no voltage regulation. Each load assumes an impedance model, Eqs. (2.4–2.5), with admittance $\bar{Y}_i = g_i + jy_i$.

2. Tap-changers: they are labeled from 1 through M and the continuous model is used, i.e.,

$$\frac{dn_i}{dt} = \frac{1}{T_i}(V_{0,i} - V_i) \qquad 1 \leq i \leq M \tag{3.2}$$

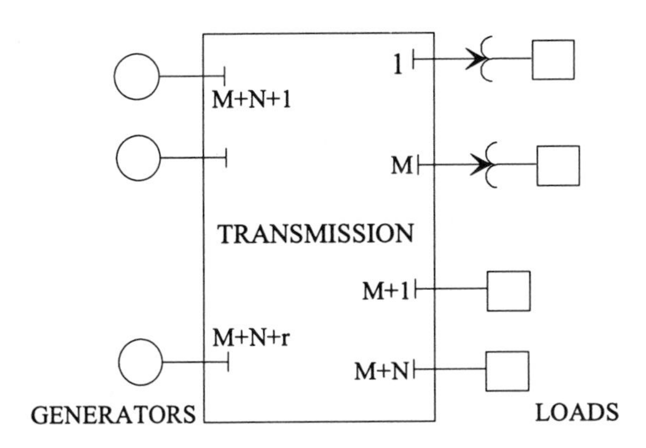

FIG. 3.3. *A multi-node power system with tap-changers.*

where V_i is the voltage at the load node i, $V_{0,i}$'s are the reference level.

3. Generators: there are r generators, labeled $M+N+1$ to $M+N+r$. Each generator k is assumed to be a perfect voltage source, i.e., $\mathcal{E}_{lim,k} = \infty$ and $I_{lim,k} = \infty$ in Eq. (2.8). This means that the output voltage can always be maintained at the constant level of E_k.

A crucial assumption in the analysis of [3] is the use of the reactive power/voltage relations in the *decoupled power flow model.* This is an implication of Assumption 1.

The system equation takes the form:

$$\frac{d\vec{x}}{dt} = \text{diag}\{\frac{1}{T_1}, \cdots, \frac{1}{T_M}\}(\vec{v}_0 - \vec{v}) \tag{3.3}$$

$$\vec{q} = \left[\mathbf{B} + \text{diag}\{x_1^2, \cdots, x_M^2\}\right] \vec{v}' \tag{3.4}$$

$$\vec{v} = \text{diag}\{n_1, \cdots, n_M\}\vec{v}' \tag{3.5}$$

where $\mathbf{B}$ is the susceptance matrix (an M-matrix [7]); $\vec{v}'$ is the nodal voltages on the primary side of tap-changers, $\vec{v}$ is on the load side, and $\vec{v}_0$ is the reference voltage levels; $\vec{q}$ consists of weighted sums of generator voltages; $T_i > 0, i = 1, \cdots, M$, are constants; $\vec{x}$ is the (scaled) tap-position, $x_i = n_i\sqrt{y_i}$.

The state space is $\mathbf{R}_+^M$, the first octant of $\mathbf{R}^M$ (turns-ratios must be positive numbers).

3.2.2. Main results. The main results of the study of tap-changer dynamics are:

1. If the system (3.3–3.5) has an equilibrium,[1] then there is a special equilibrium, denoted by $\vec{\alpha}$, which has smaller coordinates than any other equilibria.
2. The equilibrium $\vec{\alpha}$ is asymptotically stable if the linearized system does not have any 0-eigenvalue.
3. A method to approximate the region of attraction for $\vec{\alpha}$ is developed in the form of hyperboxes.

Using the technique proposed in [3], the vector field for the dynamical system (3.3–3.5) is partitioned into regions. Within each region, each derivative $\frac{dx_i}{dt}$ has a fixed sign (positive or negative). (See Fig. 3.4.) These regions are separated by hypersurfaces; each hypersurface is defined by an algebraic equation, $\frac{dx_i}{dt} = 0$ for some i, $1 \leq i \leq M$. By investigating these algebraic equations, Result 1 can be proved since their solutions are equilibrium points.

Results 2 and 3 require an investigation of the vector field. Two regions of the vector field are of particular interest: $\mathcal{P}$ (all derivatives are ≥ 0) and $\mathcal{N}$ (all derivatives are ≤ 0). It is quite routine to check that at each (nonequilibrium) point $\vec{p}$ on the boundary of $\mathcal{P}$ (resp. $\mathcal{N}$) with $\frac{dx_i}{dt}\Big|_{\vec{p}} = 0$, the second derivative $\frac{dx_i^2}{dt^2}\Big|_{\vec{p}}$ is positive (resp. negative). This implies that any trajectory starting on the boundary of either set must enter the set. In other words, $\mathcal{P}$ and $\mathcal{N}$ are invariant sets of the vector field, and system trajectories, once entering the set, must remain inside for all future time. Furthermore, since each derivative does not change sign within the set, such a trajectory if convergent will do so monotonically. The proof of Results 2 and 3 relies on such trajectories.

First, a general characterization of the system equilibria is investigated. Equilibria, according to Eqs. (3.3–3.5), are points $\vec{x}$ satisfying

$$\vec{q} = \left[\mathbf{B} + \text{diag}\{x_1^2, \cdots, x_M^2\}\right] \left[\text{diag}\{\frac{x_1}{\sqrt{y_1}}, \cdots, \frac{x_M}{\sqrt{y_M}}\}\right]^{-1} \vec{v}_0$$

where $\vec{q}$, $\mathbf{B}$, and $\vec{v}_0$ are constant vectors/matrix. Because each x_i appears only on the matrix diagonal, the above vector equation consists of M quadratic equations. To be more specific, the ith equation is a quadratic equation in x_i with coefficients in terms of x_k's, $k \neq i$. Each quadratic equation, having at most two values for x_i for a given set of coefficients, can therefore be split into two hypersurfaces $x_i = L_i(\cdot)$ and $x_i = H_i(\cdot)$. These functions L_i's and H_i's are monotonic, as illustrated in Fig. 3.4. These results are summarized in the following lemma.

LEMMA 3.1. *For each $i, i = 1, \cdots, M$, the algebraic equation $\frac{dx_i}{dt} = 0$, being quadratic in x_i with coefficients in $(x_1, \cdots, x_{i-1}, x_{i+1}, \cdots, x_M)$, can*

[1] At an equilibrium, all the load voltages are restored to their specified levels.

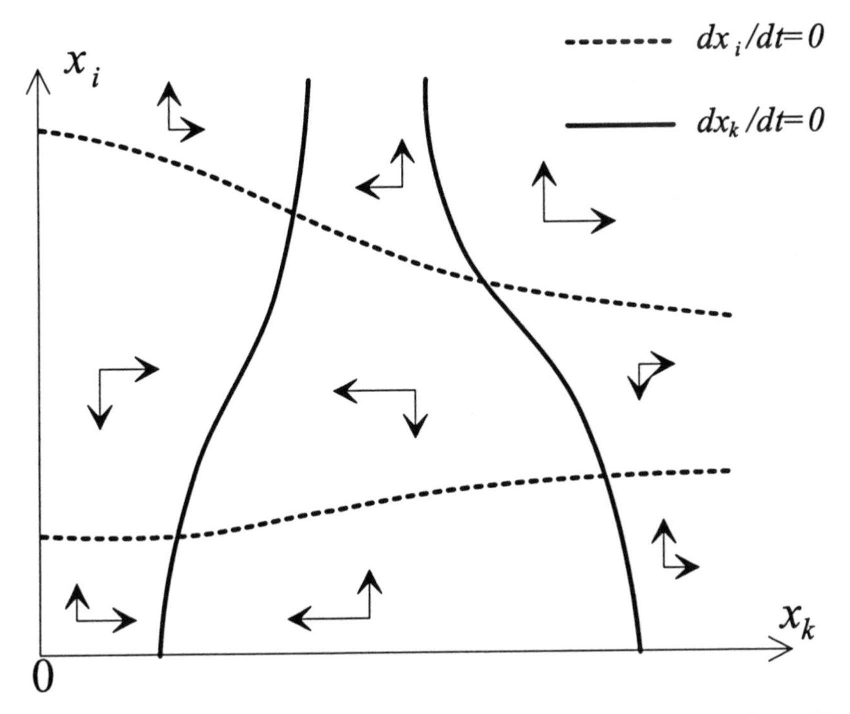

FIG. 3.4. *The quadratic nature of the system equation gives rise to a simple partition of the state space.*

be described by:

$$x_i = L_i(x_1, \cdots, x_{i-1}, x_{i+1}, \cdots, x_M) \tag{3.6}$$
$$x_i = H_i(x_1, \cdots, x_{i-1}, x_{i+1}, \cdots, x_M) \tag{3.7}$$

where $L_i(\cdot) < H_i(\cdot)$. *(L is to denote Lower and H for Higher.) Also, for each* i, $i = 1, \cdots, M$:

1. L_i *is strictly increasing in each of its argument.*
2. H_i *is strictly decreasing in each of its argument.*

Note that if $\vec{x}$ is an equilibrium then for each $i, 1 \leq i \leq M$, $\vec{x}$ must satisfy either Eq. (3.6) or Eq. (3.7). Thus, there are 2^M possible combinations. However, only one combination may yield a stable equilibrium. This result is given in Proposition 3.2. The proof can be sketched as follows. If $\vec{x}$ is an equilibrium and it satisfies $x_i = H_i(\cdot)$ for some i, then in every neighborhood of $\vec{x}$, there exists a point whose trajectory approaches a point different from $\vec{x}$. Monotonicity of H_i (Lemma 3.1) plays an important role in the selection of such a point.

PROPOSITION 3.2. *If* $\vec{x}$ *is a stable equilibrium then it must satisfy* $x_i = L_i(\cdot)$ *for each* $i = 1, \cdots, M$.

The two regions $\mathcal{N}$ and $\mathcal{P}$ of the vector field are formally defined below.

DEFINITION 3.3.

$$\mathcal{N} = \{\vec{x} \in \mathbf{R}_+^M : \frac{d\vec{x}}{dt} \leq 0\} \tag{3.8}$$

$$\mathcal{P} = \{\vec{x} \in \mathbf{R}_+^M : \frac{d\vec{x}}{dt} \geq 0\} \tag{3.9}$$

Observe that the system equilibria must be on the boundary of $\mathcal{N}$ and $\mathcal{P}$.

Each algebraic equation $\frac{dx_i}{dt} = 0$ is quadratic in x_i (Lemma 3.1), and the two roots are denoted by $L_i(\cdot)$ and $H_i(\cdot)$. Given a quadratic expression $f(x) = ax^2 + bx + c$ and an x_0, the sign of $f(x_0)$ can be determined by simply comparing x_0 with the two roots of f, i.e., whether or not x_0 is in between the roots. This observation leads to an alternative description for $\mathcal{N}$ in terms of L_i's and H_i's:

$$\mathcal{N} = \{\vec{x} : L_i(\cdot) \leq x_i \leq H_i(\cdot), \forall i\} \tag{3.10}$$

The sets $\mathcal{N}$ and $\mathcal{P}$ are shown to be invariant sets of the dynamical system (3.3–3.5). To show this, it is sufficient to pick any initial point $\vec{p}$ on the boundary of $\mathcal{P}$ or $\mathcal{N}$, and show that the corresponding trajectory does not leave the set. Let i be any index with $\frac{dx_i}{dt}\big|_{\vec{p}} = 0$. To prove that the trajectory does not leave the set, this condition can be checked: $\left.\frac{d^2x_i}{dt^2}\right|_{\vec{p}} > 0$ if $\vec{p} \in \mathcal{P}$, and $\left.\frac{d^2x_i}{dt^2}\right|_{\vec{p}} < 0$ if $\vec{p} \in \mathcal{N}$. The fact that the matrix $\mathbf{B}$ in Eq. (3.5) is an M-matrix [7] plays an important role in the derivation.

LEMMA 3.4. *$\mathcal{N}$ and $\mathcal{P}$ are invariant sets of the dynamical system (3.3–3.5).*

Due to Lemma 3.4, system trajectories upon entering either $\mathcal{N}$ or $\mathcal{P}$ will be monotonic thereafter. Convergence to an equilibrium is guaranteed if a suitable boundedness holds.

The next proposition exhibits a special equilibrium, which is also a point of the invariant set $\mathcal{N}$.

PROPOSITION 3.5. *If the set $\mathcal{N}$ is nonempty then $\mathcal{N}$ has the smallest element, denoted by $\vec{\alpha}$, i.e., $\vec{\alpha} < \vec{x}, \forall \vec{x} \in \mathcal{N}, \vec{x} \neq \vec{\alpha}$. Furthermore, $\vec{\alpha}$ is an equilibrium satisfying $\alpha_i = L_i(\cdot)$ for each i.*

The proof of Proposition 3.5 begins with a definition of $\vec{\alpha}$ by its components:

$$\alpha_i = \inf\{x_i : \vec{x} \in \mathcal{N}\} \qquad\qquad i = 1, \cdots, M$$

and then shows that $\alpha_i = L_i(\alpha_1, \cdots, \alpha_{i-1}, \alpha_{i+1}, \cdots, \alpha_M)$ for each i. (Thus, $\vec{\alpha} \in \mathcal{N}$). It can be shown that $\alpha_i \geq L_i(\alpha_1, \cdots, \alpha_{i-1}, \alpha_{i+1}, \cdots, \alpha_M)$ using Eq. (3.10) and the monotonicity of L_i's. The proof is then completed by showing that $\alpha_i > L_i(\cdot)$ for some i would contradict the definition of $\vec{\alpha}$.

Remark: Since $\vec{\alpha}$ satisfies the equations $x_i = L_i(\cdot)$, $1 \leq i \leq M$, it is a candidate for stable equilibrium according to Proposition 3.2. Furthermore, the equilibria $\vec{\alpha}$ is of importance for the following reasons:

1. $\vec{\alpha}$ corresponds to an equilibrium that yields a system voltage profile (i.e., primary side) *higher* than that due to any other equilibrium.
2. Suppose that following a disturbance, the system still has an equilibrium. Then the set $\mathcal{N}$ is nonempty (since $\mathcal{N}$ contains equilibria according to its definition). According to Prop. 3.5, the smallest-tap-setting equilibrium $\vec{\alpha}$ exists. This implies that the equilibria may drift, merge and disappear due to changes in system parameters but $\vec{\alpha}$ remains the last one to vanish.

From the preceding remarks, it is meaningful to study the stability characteristic of $\vec{\alpha}$. This stability study relies on the two subsets of $\mathcal{P}$ and $\mathcal{N}$, defined below.

DEFINITION 3.6.

$$\begin{aligned} (3.11)\quad \mathcal{P}_{\vec{\alpha}} &= \{\vec{x} : \vec{0} \leq \vec{x} \leq \vec{\alpha}\} \cap \mathcal{P} \\ (3.12)\quad \mathcal{N}_{\vec{\alpha}} &= \{\vec{x} : \vec{x} \in \mathcal{N} - \{\vec{\alpha}\} \text{ and there is no equil. } \vec{e},\ \vec{\alpha} < \vec{e} \leq \vec{x}\} \end{aligned}$$

In other words, $\mathcal{P}_{\vec{\alpha}}$ and $\mathcal{N}_{\vec{\alpha}}$ are subsets of $\mathcal{P}$ and $\mathcal{N}$, respectively, around the special equilibrium $\vec{\alpha}$. It can be checked that $\vec{0} \in \mathcal{P}$, and that $\mathcal{P}_{\vec{\alpha}}$ is an invariant set. Trajectories in $\mathcal{P}_{\vec{\alpha}}$ are monotonically increasing and converge to their upperbound $\vec{\alpha}$. Similarly, the set $\mathcal{N}_{\vec{\alpha}}$, if nonempty, is also invariant and contains trajectories converging decreasingly to $\vec{\alpha}$. In other words, when $\vec{\alpha}$ exists (this is guaranteed by an existence of any equilibrium) and $\mathcal{N}_{\vec{\alpha}} \neq \emptyset$, the vector field around $\vec{\alpha}$ is such that there are points $\vec{w} \leq \vec{\alpha}$ and $\vec{z} \geq \vec{\alpha}$ whose trajectories converge monotonically to α. Existence of such points can be used to show asymptotic stability of $\vec{\alpha}$. This result is stated in the following proposition.

PROPOSITION 3.7. *If $\mathcal{N}_{\vec{\alpha}} \neq \emptyset$ then for each $\vec{z} \in \mathcal{N}_{\vec{\alpha}}$ the hyperbox defined by*

$$B_{0\vec{z}} = \{\vec{x} : \vec{0} \leq \vec{x} \leq \vec{z}\} \tag{3.13}$$

is a region of attraction of $\vec{\alpha}$. Thus, $\vec{\alpha}$ is asymptotically stable.

The idea behind the proof of Proposition 3.7 is illustrated in Fig. 3.5. Firstly, two sequences of points $\{\vec{w}^{(k)}\}_{k=0}^{\infty}$ and $\{\vec{z}^{(k)}\}_{k=0}^{\infty}$ are constructed based on $\vec{0}$ and $\vec{z}$ in an inductive manner:

- $\vec{w}^{(0)} = \vec{0}$, $\vec{z}^{(0)} = \vec{z}$.
- $\vec{w}^{(k+1)}$ and $\vec{z}^{(k+1)}$ are obtained from $\vec{w}^{(k)}$ and $\vec{z}^{(k)}$, respectively. For each component $i, i = 1, \cdots, M$,

$$\begin{aligned} w_i^{(k+1)} &= L_i(w_1^{(k)}, \cdots, w_{i-1}^{(k)}, w_{i+1}^{(k)}, \cdots, w_M^{(k)}) \\ z_i^{(k+1)} &= L_i(z_1^{(k)}, \cdots, z_{i-1}^{(k)}, z_{i+1}^{(k)}, \cdots, z_M^{(k)}) \end{aligned}$$

Utilizing monotonicity of the functions L_i's (Lemma 3.1), it can be shown that

$$\vec{0} \leq \vec{w}^{(1)} \leq \cdots \leq \vec{w}^{(k)} \leq \cdots \leq \vec{\alpha} \leq \cdots \leq \vec{z}^{(k)} \leq \cdots \leq \vec{z}^{(1)} \leq \vec{z}$$

and that $\lim_{k\to\infty} \vec{w}^{(k)} = \lim_{k\to\infty} \vec{z}^{(k)} = \vec{\alpha}$. In other words, the hyperbox $B_k = \{\vec{x} : \vec{w}^{(k)} \leq \vec{x} \leq \vec{z}^{(k)}\}$ shrinks to the point $\vec{\alpha}$ as $k \to \infty$. The proof is completed by showing that, for any k, any trajectory starting at a point in B_k must enter B_{k+1} in finite time. □

To have a larger subset of a region of attraction, the union of hyperboxes can be used:

$$\mathcal{A} = \bigcup_{\vec{z} \in \mathcal{N}_{\vec{\alpha}}} \{\vec{x} : 0 \leq \vec{x} \leq \vec{z}\} \tag{3.14}$$

Note that in Proposition 3.7, $\mathcal{N}_{\vec{\alpha}} \neq \emptyset$ is a sufficient condition to guarantee asymptotic stability of $\vec{\alpha}$. A rather straightforward application of the Inverse Function Theorem reveals that $\mathcal{N}_{\vec{\alpha}} \neq \emptyset$ whenever the Jacobian matrix (i.e., $\partial f/\partial x$ for system $\dot{x} = f(x)$) is nonsingular at $\vec{\alpha}$. This results in the following proposition.

PROPOSITION 3.8. *If the equilibrium $\vec{\alpha}$ does not have a zero eigenvalue then it is asymptotically stable.*

If $\vec{\alpha}$ vanishes due to drifts in system parameters (a bifurcation), then the system cannot reach any steady state since there is no more equilibrium.

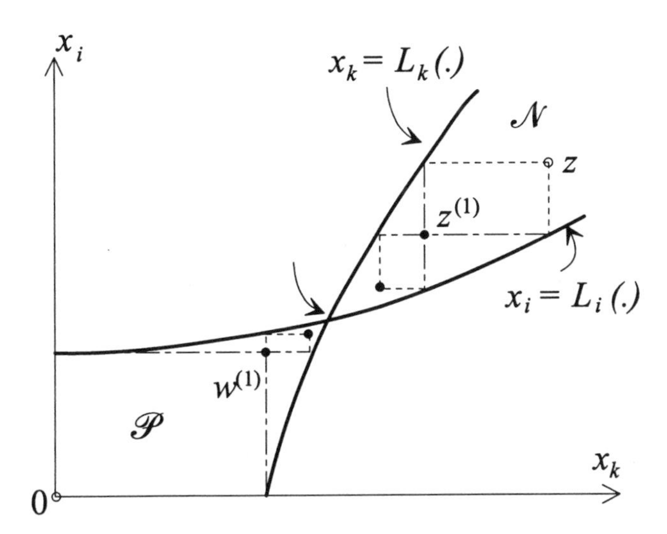

FIG. 3.5. *The points* $\{\vec{w}^{(k)}\}_{k=0}^{\infty}$ *and* $\{\vec{z}^{(k)}\}_{k=0}^{\infty}$ *define a nested sequence of boxes* $\{B_k\}_{k=0}^{\infty}$ *shrinking to* $\vec{\alpha}$. *Asymptotic stability of* $\vec{\alpha}$ *is proved by showing that if a system trajectory begins in* B_k *then it must enter* B_{k+1}.

($\vec{\alpha}$ is the last equilibrium to vanish as discussed in the preceding remark.) It follows that at least one tap-changer will keep on increasing its turns-ratio n while the corresponding load voltage decays.

4. Interaction among mechanisms. This section is an extension of Section 3. Even though the analysis presented earlier indicates that tap changing may lead to a voltage decay, it is by no means the only cause of voltage collapse scenarios that actually happened. This is because tap-changers have their physical upper-limit (i.e., n in Fig. (3.2) and $\vec{x}$ in Eqs. (3.3–3.5) are restricted), and the explanation of voltage collapse in Section 3 is not useful when the physical limits are reached. In addition, the system model is too simple (ideal generators, simplistic loads) to reflect the complex behavior of power systems. Therefore the next natural step is to improve the power system model to incorporate more dynamic mechanisms, and to study the interaction among the modeled mechanisms.

A complete analysis can be found in [4,5]. The system diagram is shown in Fig. 4.1; it is an extended version of the system in Fig. 3.3.

4.1. System model

1. Tap-changer: discrete model, Eqs. (2.1–2.2).

2. Load: each load is assumed to be dynamic, Eqs. (2.6–2.7), and is attached to a tap-changer. Due to Assumption 1, only (2.7)

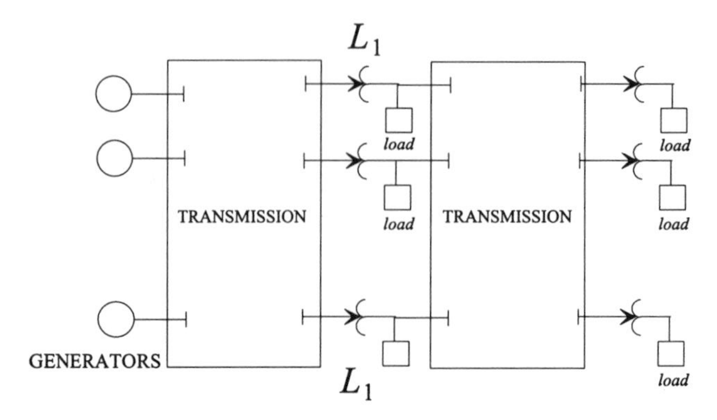

FIG. 4.1. *A power network with tap-changers and dynamic loads.*

will be used. We assume further that the "static" demand $Q_s(V)$ in (2.7) is ∞-differentiable and that $Q_s(V) \geq 0$. The total number of load nodes is denoted by M.

3. Generator: generators have finite limits $\mathcal{E}_{lim}$ and I_{lim}. Refer to Eq. (2.8).

4. Transmission lines: lossless, i.e., no resistance.

4.2. Main results. There are two types of state variables: V's (nodal voltages) due to load dynamics and n's (tap positions) due to tap changing. They are different in nature for V's vary continuously whereas n's change in discrete steps. In reality, it may take many seconds (or minutes) between two consecutive tap-changing events. Thus, we break our analysis into these steps:

- Step 1: analyze the voltage evolution between two consecutive moves of tap-changers. (Subsection 4.2.1.)
- Step 2: analyze how tap changing affects the reactive power balance and the generator excitation. (Subsection 4.2.2.)
- Step 3: analyze the effects mentioned in Step 2 on the properties established in Step 1. (Subsection 4.2.3.)

In other words, the tap-positions and generator voltages are treated as parameters. First, an analysis of system voltage evolution for a fixed set of parameters is performed (Subsec. 4.2.1); then in Subsecs. 4.2.2–4.2.3, the effect of parameters is considered.

4.2.1. Analysis of voltage stability due to load dynamics. For each load node $i, i = 1, \cdots, M$, the voltage behavior is governed by Eq. (2.7), which is shown here for ease of reference:

$$\text{reactive power demand } Q(V, \dot{V}) = Q_s(V) + k_Q \dot{V}$$

Since the demand and the incoming flow must balance each other, we have:

$$\text{reactive power flow into the load node } = Q_s(V) + k_Q\dot{V}$$

In an interconnected network, the flow into each load node i depends on the node voltage V_i and the voltages at adjacent nodes, as well as the line impedances. We can therefore write the above equation in a more explicit form for each load i:

$$\dot{V}_i = \frac{1}{k_{Q,i}}\left[-Q_{S,i}(V_i) - a_i V_i^2 + e_i(\vec{V})V_i\right] \tag{4.1}$$

where the last two terms in the square brackets represent the incoming flow. $k_{Q,i}(V_i) > 0$, $a_i > 0$ and e_i is an affine function:

$$e_i(\vec{V}) = \sum_{j\in\mathcal{A}(i)} e_{ij}V_j + e_{ig} \tag{4.2}$$

with $e_{ij} > 0, j \in \mathcal{A}(i)$, where $\mathcal{A}(i)$ denotes all load nodes adjacent to i, and $e_{ig} > 0$ if node i is adjacent to a generator (i.e., $i \in L_1$) and $e_{ig} = 0$ otherwise.

The state space is $\mathbf{R}_+^M$. Note that each point in this space represents a collection of all nodal voltages; this is different in meaning from the space in Section 3 where each point represents a collection of tap positions.

The main results of this subsection are:

1. If there is no equilibrium then the voltage must collapse.
2. Among all the equilibria, if any, there is a special equilibrium $\vec{\nu}$ that has the highest coordinate in each dimension.
3. $\vec{\nu}$ is asymptotically stable if the linearized system does not have a 0-eigenvalue.
4. $\vec{\nu}$ is the unique stable equilibrium if the reactive current of each load, $Q_s(V)/V$, is a convex function of voltage V.
5. A method to approximate the region of attraction of $\vec{\nu}$ is given based on the use of hyperboxes.
6. A subset of the voltage-collapse region is discovered.

Similar to Section 3, the analytical results here are obtained from a study of the vector field. However, the proof is more involved because the system equations are more complicated. (In Section 3, the load is assumed to be impedance, resulting in quadratic equations and simple partitions of state space (Fig. 3.4).) In this section, the static part of the load Q_s is a general function of voltage. Two most important properties of the vector field of (4.1–4.2) are given in Lemmas 4.1 and 4.3.

LEMMA 4.1. (existence of monotonic trajectories) *Let $V(\vec{x};t)$ be the system trajectory with initial state $\vec{x}$ then*

$$\dot{V}(\vec{x};0) \le 0 \implies \dot{V}(\vec{x};t) \le 0, \forall t \ge 0 \tag{4.3}$$

$$\dot{V}(\vec{x};0) \ge 0 \implies \dot{V}(\vec{x};t) \ge 0, \forall t \ge 0 \tag{4.4}$$

That is, the system trajectory is monotonic in either case.

The proof of (4.4) is sketched here; that for (4.3) is very similar. Given any $\vec{x}$ with $\dot{V}\Big|_{\vec{x}} \geq 0$ (but $\vec{x}$ is not an equilibrium), the statement (4.4) is true if it can be shown that as soon as the trajectory leaves $\vec{x}$, $\dot{V}_i$ turns strictly positive for each component i. (This behavior of the vector field can be pictured as follows. If all components of the velocity at $\vec{x}$ are non-negative, then when the trajectory drifts away from $\vec{x}$, the positive components still dominate, preventing the velocity from turning negative.) This is done by checking the higher-order derivatives of V_i. This laborious step makes use of the fact that e_i in Eqs. (4.1–4.2) has an affine form. The reader is referred to [5] for more details of the proof. □

A direct implication of Lemma 4.1 is that two invariant sets $\mathcal{N}$ and $\mathcal{P}$ exist, which is shown in the following proposition.

PROPOSITION 4.2. *The sets of all positive and all negative derivatives,*

$$\mathcal{P} = \{\vec{x} : \dot{V}_i\Big|_{\vec{x}} \geq 0 \ \forall i\} \tag{4.5}$$

$$\mathcal{N} = \{\vec{x} : \dot{V}_i\Big|_{\vec{x}} \leq 0 \ \forall i\} \tag{4.6}$$

are invariant sets of the dynamical system (4.1–4.2). Also, the set $\mathcal{P}$ is bounded from above and the set $\mathcal{N}$ is nonempty and unbounded.

Lemma 4.3 reveals that the vector field of (4.1–4.2) exhibits a partial ordering.

LEMMA 4.3. (partial ordering in the vector field) *Let $V(\vec{x};t)$ and $V(\vec{y};t)$ be the trajectories originating from different initial states $\vec{x}$ and $\vec{y}$ then*

$$\vec{x} < \vec{y} \Longrightarrow V(\vec{x};t) < V(\vec{y};t), \forall t \geq 0 \tag{4.7}$$

In some respects, the proof of this lemma is very similar to that for Lemma 4.1, except that (4.7) requires a comparison of two trajectories. Intuitively, since each component of $\vec{y}$ is no smaller (i.e., $\geq$) than the corresponding one of $\vec{x}$, those components with a $>$ force the trajectory of $\vec{y}$ to stay componentwise larger than the trajectory of $\vec{x}$. Again the special form of Eqs. (4.1–4.2) is exploited. Details can be found in [5]. □

Lemma 4.3 leads to an interesting result. For illustration, refer to Fig. 4.2. The figure shows two points $\vec{v} < \vec{w}$ and their associated trajectories. Consider the moving hyperbox with vertices $V(\vec{v};t) \leq V(\vec{w};t)$ for each $t \geq 0$, where the inequality holds at every time instant t, by Lemma 4.3. Given any point $\vec{x}$ in the original hyperbox (vertices $\vec{v}$ and $\vec{w}$) we want to know the image of $\vec{x}$ at any later time t, i.e., $V(\vec{x};t)$. Due to Lemma 4.3, the image of $\vec{x}$ will always stay within the moving hyperbox. If the hyperbox eventually contracts to a point $\vec{e}$ (this point must be an equilibrium) then $\lim_{t\to\infty} V(\vec{x};t) = \vec{e}$. In particular, if $\vec{e}$ is in the original hyperbox then $\vec{e}$ is

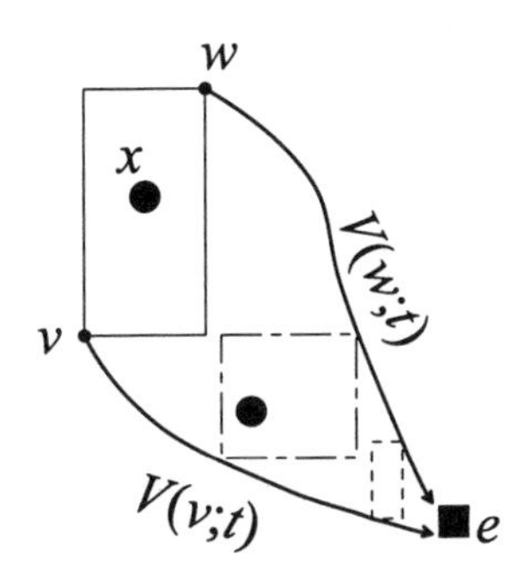

FIG. 4.2. *Illustration of Proposition 4.4.*

asymptotically stable and the hyperbox is a subset of the stability region. This result is summarized in the following proposition.

PROPOSITION 4.4. *Let $\vec{v} \leq \vec{x} \leq \vec{w}$ and $\vec{e}$ be an equilibrium. If $V(\vec{v};t) \overset{t\to\infty}{\longrightarrow} \vec{e}$ $V(\vec{w};t) \overset{t\to\infty}{\longrightarrow} \vec{e}$ then $V(\vec{x};t) \overset{t\to\infty}{\longrightarrow} \vec{e}$.*

It has widely been believed that (local) bifurcation, which results in loss of equilibria, causes a voltage collapse. This notion, though confirmed by numerical simulations, remains imprecise. When there is no steady state, how can one rule out, for example, the possibility of voltage buildup (as opposed to collapse) or strange oscillations in voltage? To resolve this issue, one needs to investigate closely the power system model. The following proposition guarantees that when the steady state is lost in the modeled system, the voltage must collapse.

PROPOSITION 4.5. (loss of equilibria implies voltage collapse) *If $\mathcal{P} = \emptyset$ then at least one nodal voltage approaches 0 (voltage collapse) regardless of the initial condition.*

In fact, Proposition 4.5 is an implication of Lemma 4.3. Given any initial point $\vec{x}$, one can always select $\vec{w} \in \mathcal{N}$ so that $\vec{x} < \vec{w}$ (existence of $\vec{w}$ is guaranteed by Prop. 4.2). Since $\mathcal{N}$ is invariant and trajectories in $\mathcal{N}$ decrease continually, the loss of equilibria implies that $V(\vec{w};t)$ must decay toward 0 for at least one component i. The same holds for $V(\vec{x};t)$ since $V(\vec{x};t) < V(\vec{w};t)$ according to Lemma 4.3. □

The following proposition is, in some sense, an extension of Prop. 3.2.

PROPOSITION 4.6. (characterization of stable equilibria) *Suppose that the equilibrium $\vec{e}$ has no 0-eigenvalue. Then, $\vec{e}$ is asymptotically stable if and only if every neighborhood of $\vec{e}$ contains points $\vec{x} \in \mathcal{P}$ and $\vec{z} \in \mathcal{N}$ so that $\vec{x} \leq \vec{e} \leq \vec{z}$.*

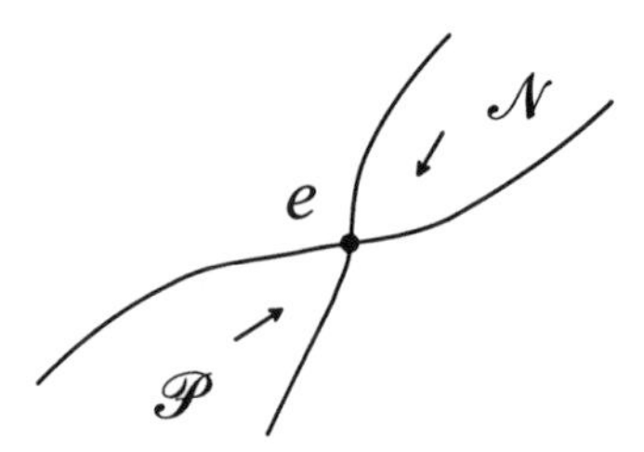

FIG. 4.3. *Regional structure of vector field around a stable equilibrium $\vec{e}$.*

In the proof of Prop. 4.6, the condition on 0-eigenvalue implies that, by using the Inverse Function Theorem, $\vec{e}$ is an isolated equilibrium and that in every neighborhood of $\vec{e}$, there are points $\vec{z} \in \mathcal{N}$ and points in $\vec{x} \in \mathcal{P}$. It can be shown that $V(\vec{z};t)$ must converge decreasingly and $V(\vec{x};t)$ increasingly to $\vec{e}$. (Monotonicity follows from Prop. 4.2.) Then use Prop. 4.4. □

Proposition 4.6 reveals that the vector field around a stable equilibrium point must have a structure illustrated in Fig. 4.3: in every neighborhood of $\vec{e}$, $\mathcal{N}$ must extend to higher and $\mathcal{P}$ to lower coordinates.

The following proposition is an extension of Proposition 3.5.

PROPOSITION 4.7. (largest-voltage equilibrium) *Suppose that an equilibrium exists for the system (4.1–4.2). Then there is an equilibrium $\vec{\nu}$ with $\vec{\nu} > \vec{a}$ for all $\vec{a} \in \mathcal{P} - \{\vec{\nu}\}$.*

The proof begins by defining each component i of $\vec{\nu}$ as

$$\nu_i = \sup_{\vec{x} \in \mathcal{P}} \{x_i\} \tag{4.8}$$

If it can be shown that $\vec{\nu} = V(\vec{\nu};t)$, then $\vec{\nu}$ is an equilibrium. An application of Lemma 4.3 and Prop. 4.2 yields $\vec{\nu} \leq V(\vec{\nu};t), \forall t \geq 0$. It can be shown that $\vec{\nu} < V(\vec{\nu};t)$ would imply the convergence of $V(\vec{\nu};t)$ to an equilibrium $\vec{e}$ ($\vec{e} \in \mathcal{P}$) with $\vec{e} > \vec{\nu}$, contradicting the definition of $\vec{\nu}$. □

Similar to Section 3, a simple criterion for stability of the special equilibrium can be obtained. This criterion is given in the following proposition.

PROPOSITION 4.8. *Assume that the system (4.1–4.2) has an equilibrium. Then $\vec{\nu}$ is asymptotically stable if and only if it does not have a 0-eigenvalue. In addition, if $Q_{s,i}(V_i)/V_i$ is a convex function for each $i, i = 1, \cdots, M$, then $\vec{\nu}$ is the unique stable equilibrium.*

Stability of $\vec{\nu}$ is proved by use of Prop. 4.6 and the fact that $\vec{\nu}$ is the largest point in $\mathcal{P}$. If, in addition, $Q_{s,i}(V_i)/V_i$ is a convex function of V_i for each i then it can be shown that $\mathcal{P}$ is a convex region. Thus, any other equilibrium $\vec{e}$, being a point of $\mathcal{P}$, can be joined to $\vec{\nu}$ by a straight line

$l \subset \mathcal{P}$. Since $\vec{v} > \vec{e}$, every point on l is componentwise greater than $\vec{e}$. This means that every neighborhood of $\vec{e}$ contains points $\vec{x}$ in $\mathcal{P}$ with $\vec{x} > \vec{e}$. Since $\mathcal{P}$ is an invariant set, the trajectory $V(\vec{x};t)$ will remain monotonically increasing. This proves that $\vec{e}$ is not a stable equilibrium point. □

The technique using hyperboxes to approximate the region of attraction of $\vec{v}$ can be applied to the system (4.1–4.2). This result, a direct implication of Prop. 4.6, is given below.

PROPOSITION 4.9. (approximating the stability region) *Suppose that the set $\mathcal{P}_{\vec{v}}$ defined by*

$$(4.9)\quad \mathcal{P}_{\vec{v}} = \{\vec{x} \in \mathcal{P} :\ \text{there is no equilibrium } \vec{e}\ , \vec{e} \neq \vec{v},\ \text{so that } \vec{x} < \vec{e}\}$$

is nonempty (equivalently, if $\vec{v}$ does not have a 0-eigenvalue) then for each $\vec{x} \in \mathcal{P}_{\vec{v}}$, the set

$$B_{\vec{x}} = \{\vec{y} : \vec{y} \geq \vec{x}\} \tag{4.10}$$

is a subset of the region of attraction to $\vec{v}$. The union

$$B_{\mathcal{P}_{\vec{v}}} = \bigcup_{\vec{x} \in \mathcal{P}_{\vec{v}}} B_{\vec{x}} \tag{4.11}$$

is a better approximation for the region of attraction.

A practical implication of Proposition 4.9 (especially Eq. (4.10)) is the existence of operating limits such as "if the voltage at each node i is greater than a corresponding threshold v_i then the system voltage stabilizes."

As pointed out in Prop. 4.5, a sufficient condition for voltage collapse is the loss of equilibria. Since this is an extreme condition, it is meaningful to investigate whether voltage collapse can occur when equilibria still exist. We first formally define the *voltage collapse region* to be the set of all $\vec{x}$ in the state space $\mathbf{R}_{+}^{M}$ so that $V_i(\vec{x};t) = 0$ for some node i and some time $t > 0$ (i.e., the voltage at some node i eventually drops to 0). The following proposition exhibits a subset of the voltage collapse region.

PROPOSITION 4.10. (a voltage collapse region) *If the set $\mathcal{C}$ defined by*

$$\mathcal{C} = \{\vec{x} : \vec{x} \in \mathcal{N} \text{ and } \vec{x} \not\geq \vec{e} \text{ for any equilibrium } \vec{e}\} \tag{4.12}$$

is nonempty then it is a subset of the voltage collapse region.

The proof of Prop. 4.10 is based on the fact that trajectories in $\mathcal{N}$ must decrease monotonically (Prop. 4.2). An equivalent description of $\mathcal{C}$ is given in the following lemma, which is a consequence of Prop. 4.10 and Lemma 4.3.

LEMMA 4.11. *The set $\mathcal{C}$ as defined by Eq. (4.12) is identical to*

$$\mathcal{C} = \{\vec{x} : \vec{x} \in \mathcal{N} \text{ and } \vec{x} \not\geq \vec{y}, \forall \vec{y} \in \mathcal{P}\} \tag{4.13}$$

The results of this subsection are valid over any time duration where there is no tap changing or change in generator voltages. In the next two subsections, the effect of tap changing and generator voltage on the obtained results is presented. Mathematically, the change in taps and/or generator voltages corresponds to a parameter change in Eqs. (4.1–4.2).

4.2.2. Effect of tap changing. DEFINITION 4.12. *The* normal excitation region *refers to the collection of all points in the state space for which no generators reach their limits I_{lim} or $\mathcal{E}_{lim}$.*

For the modeled system, the normal excitation region can be described by a set of affine inequalities:

$$\left\{(V_1, \cdots, V_M) : \sum_{i \in L_1} \frac{V_i}{n_i} b_{gi} \geq c_g, \forall g \in G\right\} \tag{4.14}$$

where G represents the collection of all generators, c_g's are constants, L_1 is the set of tap-changer nodes adjacent to the generators (see Fig. 4.1). The inequalities reflect the physical fact that the receiving-end voltages $\frac{V_i}{n_i}$ have to be kept high enough to avoid excessive transmission-line flows and burden on generators.

Due to the appearance of the turns-ratios n_i's in the set described by (4.14), the following proposition is trivial.

PROPOSITION 4.13. (effect of tap changing on normal excitation region) *Following each tap increment in group L_1, the normal excitation region shrinks.*

Physically, if the system state falls outside a shrinking normal-excitation region, at least one generator voltage is reduced since its limit has been reached. We next investigate how this reduction would affect the stability properties derived earlier.

4.2.3. Effect of generator excitation limiting. In the system equations (4.1–4.2), the term e_{ig} depends on the generator voltages E_g's. In fact, $e_{ig} = \sum_{g \in G} b_{ig} E_g$ represents the equivalent of all generator nodes that are adjacent to node i. In other words, the system (4.1–4.2) is parameterized by the generator voltages E_g's.

It is sufficient to analyze how the qualitative behavior of the dynamical system changes due to the adjustment in a single parameter. That is, we pick an arbitrary generator, and see how its voltage reduction affects the system given by Eqs. (4.1–4.2).

Given two voltage levels E and E' for the chosen generator, we use $d(E)$ and $d(E')$, respectively, to denote the corresponding dynamical systems.

Each dynamical system, $d(E)$ or $d(E')$, has associated with it the highest-voltage equilibrium $\vec{\nu}$ and the sets $\mathcal{P}_{\vec{\nu}}$, $B_{\mathcal{P}_{\vec{\nu}}}$ and $\mathcal{C}$ as described in Subsection 4.2.1. To emphasize the dependence of $\vec{\nu}$ and these sets on the generator voltage, we will use $\vec{\nu}|_{d(E)}$ versus $\vec{\nu}|_{d(E')}$, $\mathcal{P}_{\vec{\nu}}|_{d(E)}$ versus $\mathcal{P}_{\vec{\nu}}|_{d(E')}$ and so on.

The following proposition states how the highest-voltage equilibrium $\vec{\nu}$ and the sets $\mathcal{P}_{\vec{\nu}}$, etc., are affected by a reduction in the generator voltage.

PROPOSITION 4.14. *Whenever a generator voltage falls from E to $E' < E$,*

1. $\vec{\nu}|_{d(E')} < \vec{\nu}|_{d(E)}$.

2. $\mathcal{P}_{\vec{v}}|_{d(E')} \subset \mathcal{P}_{\vec{v}}|_{d(E)}$.
3. $B_{\mathcal{P}_{\vec{v}}}|_{d(E')} \subset B_{\mathcal{P}_{\vec{v}}}|_{d(E)}$.
4. $\mathcal{C}|_{d(E')} \supset \mathcal{C}|_{d(E)}$.

Remark: Combining the results of Subsections 4.2.1 through 4.2.3, we arrive at a possible scenario of voltage collapse: a disturbance, such as a line outage, takes place → tap-changers move up due to low voltages → normal-excitation region shrinks → one or more generators reach limits → stability $B_{\mathcal{P}_{\vec{v}}}$ shrinks and collapse region $\mathcal{C}$ expands → trajectory enters $\mathcal{C}$ → voltage collapse.

4.2.4. Control action. Based on the preceding analysis, a way to prevent the voltage collapse is to ensure that the collapse region does not expand (as a result of system state leaving the normal-excitation region). A method to achieve this is by controlling the tap-changers.

PROPOSITION 4.15. *Suppose that the system trajectory remains inside $\mathcal{P}_{\vec{v}}$ prior to the next tap-changing event. Then the system voltage will approach a steady state if the tap-changers are locked (deactivated).*

The important step in the proof of this result is to show that if the system state stays in $\mathcal{P}_{\vec{v}}$ then the generator voltages will not decay. Thus, the set $\mathcal{P}_{\vec{v}}$ is not shrinking, and convergence to an equilibrium is guaranteed. □

5. Conclusion. It is generally recognized that voltage collapse is a dynamic phenomenon. Computational tools for voltage collapse simulation are now available; traditional tools have been extended to include "slow" voltage control devices such as tap-changers. However, most existing analytical methods are still based on the static power flow. The purpose of our research is to investigate the dynamic mechanisms of voltage collapse and their interaction. This paper summarizes our results on the analysis of dynamic mechanisms, voltage recovery regions, and construction of voltage stability regions.

In our voltage collapse analysis, however, the generator frequency and rotor dynamics have been ignored. The ability of rotors to maintain synchronism is a primary concern in traditional stability studies. Hence, for future research, there is a need to integrate the models and methods developed for the analytical studies of rotor dynamics and voltage dynamics. It is desirable to identify a sufficient list of dynamic mechanisms and develop a method which allows the active mechanisms to be analyzed in detail and inactive mechanisms to be simplified. Analytical methods are needed for determination of control actions so that instability caused by various dynamic mechanisms can be removed.

Acknowledgement

This research is supported by the National Science Foundation through grant ECS-8657671.

REFERENCES

[1] C.C. Liu. Analysis of a voltage collapse mechanism due to effects of on-load tap-changers. *IEEE ISCAS*, 1028–1030, 1986.

[2] K. T. Vu and C.C. Liu. Analysis of dynamic voltage collapse mechanism for a three-bus power system. *Systems and Control Letters*, 11:399–407, Nov. 1988.

[3] C.C. Liu and K. Vu. Analysis of tap-changer dynamics and construction of voltage stability regions. *IEEE Trans. on Circuits and Systems*, 575–590, Apr. 1989.

[4] K.T. Vu and C.C. Liu. Dynamic mechanisms of voltage collapse. *Systems & Control Letters*, 15:329–338, Nov. 1990.

[5] K. T. Vu and C. C. Liu. Shrinking stability regions and voltage collapse in power systems. *IEEE Trans. on Circuits and Systems—I: Fund. Theory and Appl.*, 271–289, Apr. 1992.

[6] P.P. Varaiya, F.F. Wu, and R.L. Chen. Direct methods for transient stability analysis of power systems: Recent results. *Proceedings of the IEEE*, 1703–1715, Dec. 1985.

[7] J.M. Ortega and W.C. Rheinboldt. *Iterative Solution of Nonlinear Equations in Several Variables.* New York: Academic Press, 1970.

[8] I. Dobson and H. D. Chiang. Towards a theory of voltage collapse in electric power systems. *Systems and Control Letters*, 13:253–262, 1989.

[9] D.J. Hill. Nonlinear dynamic load models with recovery for voltage stability studies. *IEEE Trans. on Power Systems*, 166–176, Feb. 1993.

STRUCTURAL STABILITY IN POWER SYSTEMS

M.A. PAI*

Abstract. The concept of structural stability introduced in 1937 in the mathematical literature refers to the ability of a nonlinear dynamical system to retain its qualitative behavior under small perturbations in the vector field. Although for $n = 2$, complete characterization is possible, such is not the case for higher order dimensions. For $n \geq 3$, regions in the phase space, where such a property holds are bounded by bifurcation surfaces. Hence, bifurcation theory plays an important and complementary role. In this paper after reviewing the pertinent mathematical literature, we apply it to the dynamic and voltage stability problem in power systems.

1. Introduction. The purpose of this paper is to explain the concept of structural stability and its application to power systems. Broadly speaking, structural stability refers to the domain in the parameter space such that for small variations in the parameter vector inside the domain the phase portrait does not change qualitatively. In Lyapunov stability we analogously talk about the region of attraction of an equilibrium point such that for any initial condition in the region, the trajectory approaches the equilibrium point asymptotically. While considerable research has been done in the Lyapunov stability area, comparatively very little has been done in the structural stability area. It was partly because of the inability to formulate theorems and conditions for structural stability in the n dimensional space. However, research in bifurcation theory has thrown up an intimate connection between it and structural stability. The boundary of the structural stability region in the parameter space corresponds to the onset of bifurcations so that the phase portrait of the dynamic system on the two sides of the region are qualitatively different. Structural stability is a concept introduced in 1937 in the former USSR by Andronov and Pontryagin [1] for a two dimensional system. Its extension to higher dimensions is somewhat recent. One would like to examine stability in the sense that it deals with stability regions in the parameter space or under continuous structural variations with respect to time (topology, load demands, nature of load, etc). The parameters could be total P_{gen} or P_{load} and any other parameter such as load characteristics at a bus, P or Q at a bus, exciter gain, line switching or P_{tie}. The boundary of this region which we shall call as the structural stability limit (SSL) defines the range of parameters such that inside the region the phase portrait remains the same for minor

* Dept. of Electrical & Computer Engineering, University of Illinois, 1406 W. Green St., Urbana, IL 61801. This research was supported in part by the Institute for Mathematics and its Applications with funds provided by the National Science Foundation. The author wishes to acknowledge the support of NSF through its grant ECS 1-5-30689 and EPRI through its grant EPRI 1-5-36086 in this research . Portions of Section 3 results are from the Master's degree thesis of R. K. Ranjan at the University of Illinois, Urbana-Champaign.

variations in the parameters. More recently such a region is referred to as a 'Typal' region in Ref. [2]. In the parameter space there may be more than one such region. When the model is linear there have been applications of structural stability (known as parameter analysis) in load frequency control or adjustment in power system stabilizer (PSS) parameters. This is based on the well known D-partition technique in the former USSR [3] and the theory is well documented in Ref. [4]. But for nonlinear power systems on a global scale and involving parameters such as P_{gen}, Q_{gen} or area P_{tie} there is very little literature. There have been some papers involving maximum loadability but it is studied primarily as a static concept.

In this paper we examine this question in a dynamic sense for a nonlinear model starting from a precise definition of structural stability. It has been shown that when a system loses structural stability we have bifurcation either local or global. We concentrate on the steady state stability and voltage stability as illustrations for structural stability in the parameter space.

It is expected that this attempt at quantitatively applying structural stability to power systems will complement the work of Lyapunov stability via energy functions that has successfully been applied in power systems over the last five decades [5-7].

The paper is organized as follows. In Sec. 2, we review the pertinent mathematical literature on structural stability. Next we discuss structural stability and bifurcation theory. We explain the recently introduced concept of singularity induced bifurcation [2]. In Secs. 4 and 5 we indicate applications in multimachine systems with static nonlinear loads primarily but formulate the problem to include dynamic loads. In the final section, new research issues are discussed.

2. Structural stability. The notion of structural stability was first proposed by Andronov and Pontryagin in 1937 [1]. The basic idea is that under small perturbations, dynamical systems must preserve their topological behavior. From a simulation point of view it is nice for a dynamical system to have this property. Dynamical systems can never be modeled exactly due to measurement errors, etc, and moreover the finite precision of computers and the errors of floating point arithmetic will introduce additional sources of uncertainty. Thus the system that is being simulated is a perturbed version of the exact system. If the system is structurally stable then these errors will affect the simulation only marginally.

Mathematically, structural stability has to do with examining the change in qualitative behavior of a nonlinear dynamical system

$$\dot{x} = f(x) \tag{2.1}$$

as we change the vector field f. If the qualitative behavior remains the same for all nearby vector fields then the system 2.1 is said to be *structurally stable*. References [10-14] discuss structural stability in a rigorous mathematical framework.

2.1. Result for planar systems. The work of Andronov and Pontryagin [1,8] dealt with planar systems. They studied second order systems and conditions under which structural stability is preserved. Consider

$$\frac{dx}{dt} = P(x,y) \tag{2.2}$$

$$\frac{dy}{dt} = Q(x,y) \tag{2.3}$$

Consider a region Ω in the $x-y$ plane. We assume that the boundary of this region does not contain any equilibrium point of the system 2.2–2.3 or that the velocity vector of 2.2 and 2.3 is not tangent to the boundary. If the system 2.2–2.3 gets perturbed with $p(x,y) < \epsilon$ and $q(x,y) < \epsilon$ such that within Ω, the phase portrait of the modified system

$$\frac{dx}{dt} = P(x,y) + p(x,y) \tag{2.4}$$

$$\frac{dy}{dt} = Q(x,y) + q(x,y) \tag{2.5}$$

is qualitatively the same as 2.2 and 2.3, then we say that the system 2.2–2.3 is structurally stable. This imposes certain restrictions on the nature of equilibria of 2.2 and 2.3 and these are illustrated via the phase plane plots. Two complete characterizations for a planar system have been given in the literature, one by Debaggis [8] and the other by Peixoto [9].

2.1.1. Result of Debaggis [8]. We deal with a dynamical system

$$\dot{x}_i = P_i(x_1, x_2) \qquad i = 1,2 \tag{2.6}$$

Let G be the region where $P_i(x_1, x_2)$ has continuous first partial derivatives and let $M(x_1, x_2)$ denote a point in this region. Consider the perturbed system

$$\dot{x}_i = P_i(x_1, x_2) + p_i(x_1, x_2) \quad i = 1,2 \tag{2.7}$$

where $p_i(x_{1,}, x_2)$ belong to C^1 for all points $M(x_1, x_2)$ in G.

Definition

Let $\mathcal{P}[M_1, M_2]$ denote the Euclidean distance between two points in the plane. The system 2.4 is said to be structurally stable in the region G if for each $\epsilon > 0$ there exists a $\delta(\epsilon) > 0$ such that for all $p_i(x_1, x_2)$ which belong to class C^1 and satisfy the condition

$$\mid p_i \mid < \delta \ , \ \frac{\partial p_i}{\partial x_j} < \delta \ (i = 1,2, \ j = 1,2)$$

there exists a topological transformation T of G into itself with the additional properties:

1. $\mathcal{P}[M, T(M)] < \epsilon$.
2. T maps trajectories of 2.6 into trajectories of 2.7.

The following properties characterize a *structurally stable* system in two dimensions (proof is omitted).

1. The critical (equilibrium) points of a structurally stable system 2.6 can only be nodes, foci and saddle points.
2. A separatrix of a structurally stable system 2.6 cannot start from (end in) a saddle point and terminate on (start from) another saddle point in G (Heteroclinic orbit).
3. In a structurally stable system a separatrix which is starting from a saddle point cannot converge toward the same saddle point (Homoclinic orbit).
4. A structurally stable system 2.6 has only a finite number of limit cycles in G.
5. If $\overline{\gamma}$ is a limit cycle of a structurally stable system 2.6, then $h(\overline{\gamma}) = \frac{1}{\tau}\int_o^{\tau}\left(\frac{\partial P_1}{\partial x_1} + \frac{\partial P_2}{\partial x_2}\right) dt \neq 0$ where τ is the period of $\overline{\gamma}$. It may be noted that the limit cycle is stable (unstable) if $h(\overline{\gamma}) < 0$ (> 0).

A more concise characterization of structural stability in two dimension is due to Peixoto [9] whose theorem is stated below (proof is again omitted).

2.1.2. Peixoto's theorem [9]. Let f in 2.1 be a C^1 vector field on a compact, two-dimensional, differentiable manifold M. Then f is *structurally stable* on M if and only if

1. The number of critical points and periodic orbits is finite and each is hyperbolic.
2. There are no trajectories connecting saddle points.
3. The nonwandering set Ω consists of critical points and limit cycles only.

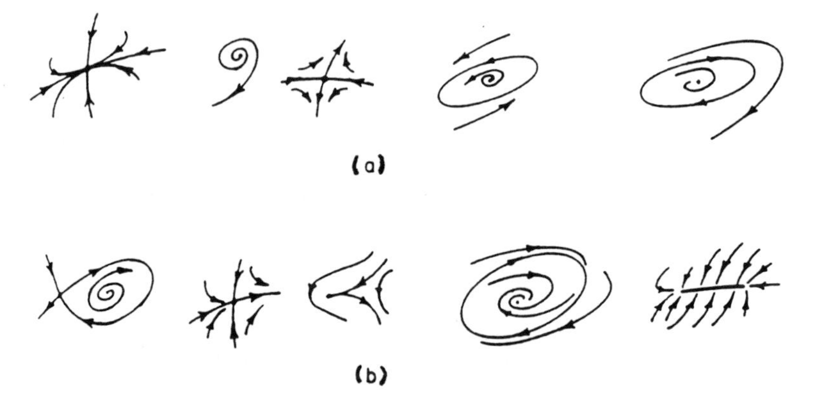

FIG. 2.1. *(a) Some structurally stable nonwandering sets on $\mathcal{R}^{\in}$, (b) some structurally unstable nonwandering sets on $\mathcal{R}^{\in}$. (Reproduced from Ref. [11]).*

Furthermore, if M is orientable, the set of structurally stable vector fields in $C^1(M)$ is an open, dense subset of $C^1(M)$. For some of the mathematical definitions in the above theorem refer to [10-14].

These two explanations of structural stability in planar systems imply that a structurally stable system will contain in its field f only sinks, saddles, sources, repelling and attracting orbits (see Fig. 2.1).

After Pexioto's theorem, it was thought that it can be extended to n dimensional systems ($n \geq 3$). Accordingly, a Morse-Smale system was defined for which

1. The number of fixed and periodic orbits is finite and each is hyperbolic. (This is the same as property (1) in Peixoto's theorem).
2. All stable and unstable manifolds intersect transversally. (This is a modification of property (2) in Peixoto's theorem).
3. The nonwandering set consists of fixed points and periodic orbits alone (again same as property (3) of Peixoto's theorem).

It was initially thought that a system of higher dimension ($n \geq 3$) is structurally stable if and only if it is Morse-Smale. But later studies indicated this to be not true. It is true that Morse-Smale systems are structurally stable but the converse is false. Smale's work on the horseshoe map showed that not all structurally stable systems are Morse-Smale [16].

With the above review of structural stability, we consider the case of a single machine infinite bus case with a classical model which is structurally unstable for a particular value of damping and which violates property (2) of Peixoto's theorem. This example is interesting in the context of transient stability. This example is due to Bansal [17] who approached the problem from a geometric point of view. Since it is structurally unstable we can find the bifurcation parameter. For the structurally unstable case, an analytical expression is derived in [28] for the critical separatrix analogous to that by Baker [18] for the phase locked loop problem.

2.2. Geometric approach to stability investigation. A single machine infinite bus system with saliency in the machine is described by

$$M\ddot{\delta} + P_d\dot{\delta} = P_1 - P_{m1}\sin\delta + P_{m2}\sin 2\delta \tag{2.8}$$

Introduce a normalized unit of time as $\tau = \frac{P_{m1}}{P_d}t$. Then eq. 2.8 becomes

$$\frac{d^2\delta}{d\tau^2} = \gamma(P_1' - \sin\delta + P_{m2}'\sin 2\delta - \frac{d\delta}{d\tau}) \tag{2.9}$$

where

$$\gamma = \frac{P_d^2}{MP_{m1}}\ ,\ P_1' = \frac{P_1}{P_{m1}}\ ,\ P_{m2}' = \frac{P_{m2}}{P_{m1}}$$

Eq. 2.9 shows that the singularities (equilibrium points) depend only on P_1' and P_{m2}' and are given by solving

$$\gamma(P_1' - \sin\delta + P_{m2}'\sin 2\delta) = 0 \tag{2.10}$$

Keeping P_1', P_{m2}' fixed and changing γ (i.e., by varying P_d or M) changes only the trajectories with the equilibrium points remaining unchanged.

Depending on the value of γ, the phase portrait will be different. Typically for a certain critical value of $\gamma = \gamma_c$, the phase portrait will be as in Fig. 2.2(a). For values of $\gamma < \gamma_c$, it changes to that in Fig. 2.2(b) and $\gamma > \gamma_c$ it looks like in Fig. 2.2(c). Thus the system is structurally unstable for $\gamma = \gamma_c$. From Peixoto's theorem at, $\gamma = \gamma_c$ it violates property (2). The trajectory joining the two critical points is termed the critical separtrix and denoted as $\omega_s(\delta)$. When the bifurcation parameter γ is changed slightly from the critical value γ_c the phase portrait changes but the number and nature of equilibria remain the same. This is not the classical definition of bifurcation which is local in nature and nature of equilibrium point changes. Hence it is called global bifurcation [19]. An analytical expression can be obtained for $\omega_s(\delta)$ through the use of a truncated Fourier analysis [17].

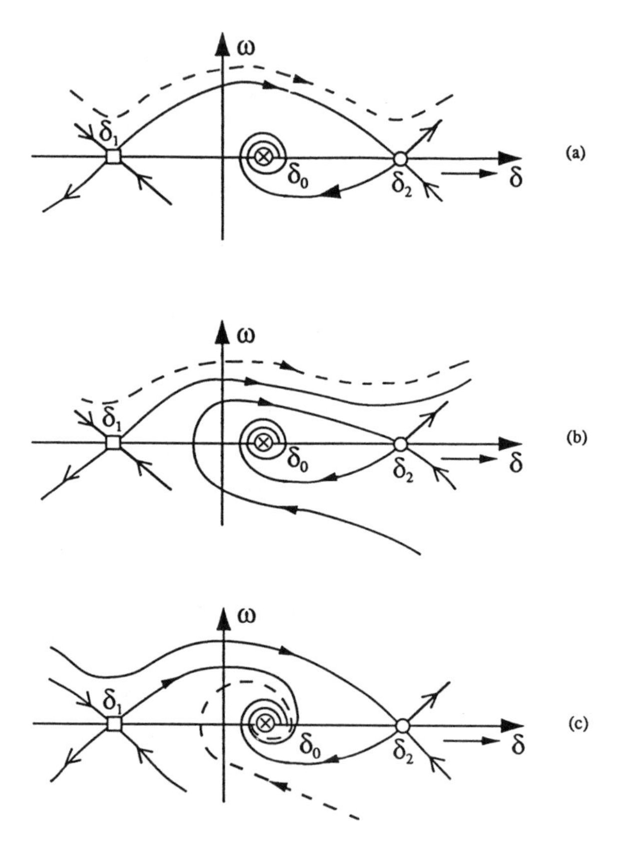

FIG. 2.2. *Typical phase portraits for Eq. (7) for different values of γ (a) $\gamma = \gamma_c$, (b) $\gamma < \gamma_c$, (c) $\gamma > \gamma_c$ (Reproduced from [17]*

2.3. Numerical example. The post-fault swing dynamics of a single machine infinite bus (SMIB) system [17] is given by

$$.0138\frac{d^2\delta}{dt^2} + D\frac{d\delta}{dt} = 0.91 - 3.02\sin\delta + .416\sin 2\delta \tag{2.11}$$

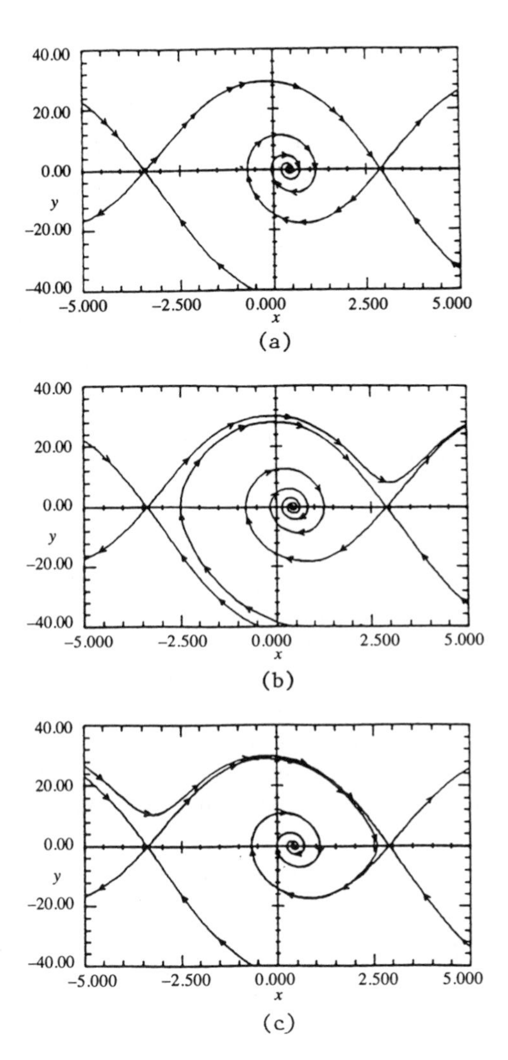

FIG. 2.3. *Phase portraits for numerical example for diff erent values of D (epsilon) (a) ϵ = .04687 (b) ϵ = .04 (c) ϵ = .05*

The phase portraits for various values of D are shown in Figs. 2.3 (a), (b), (c). The value of D for which the system is structurally unstable is obtained as 0.04687. Analytically using Fourier analysis the value agrees very well [17].

2.4. Equilibrium equivalence approach to structural stability [20]. Perhaps the only direct reference and discussion regarding structural stability in power systems appears in Ref. [20]. The definition of structural stability closely resembles our earlier definition and is given below. Consider

$$\dot{x} = f(x, \propto) \tag{2.12}$$

where $\propto$ is a parameter vector. Suppose at $\propto = \propto_o$, the system displays desirable features such as global stability or absence of chaotic motions or relatively simple structure of stability boundary. The system is said to be structurally stable if there exists a positive r such that for any parameter $\propto$ satisfying $||\propto - \propto_o|| \subset r$, the system 2.10 is *equivalent* to the system $\dot{x} = f(x, \propto_o)$. Two dynamic systems are said to be equivalent if there exists a continuous (homeomorphic) map of the phase space into itself taking trajectories of one system into another system and preserving their orientation.

The equilibrium equivalence theorem in this work refers to the retention of both the number of equilibrium points and the index of the unstable equilbrium points under change in the parameter vector $\propto$. Structural stability is discussed via this theorem. The statement of the theorems are very mathematical.

Another interesting and more intuitive interpretation of structural stability in an engineering context is contained in Ref. [21].

2.5. Structural stability in differential-algebraic dynamical systems. Bulk of the literature on structural stability concerns systems described by differential equations. However, in power systems we generally have differential equations constrained by algebraic equations (DAE systems). While Takens [22] has discussed it in a mathematical context, a recent work by V. Venkatasubramaniam, et al. [23] has discussed it in a power system context. More precisely, they consider DAE systems of the form

$$\dot{x} = f(x, y, p) \tag{2.13}$$

$$0 = g(x, y, p) \tag{2.14}$$

where p is a parameter vector, x is the state vector and y is the vector of algebraic variables. In addition to the saddle and Hopf bifurcations, they introduce the concept of singularity induced bifurcation. They have applied it to an elementary example of a single machine system. We validate this concept on the multimachine system in the next section. We discuss the effect of parameter variations on the steady state stability of a dynamic power system and observe that the system is structurally unstable at a critical value of the load. At this value of load we observe Hopf bifurcation which is a local bifurcation.

3. Application in multimachine power systems with static nonlinear loads. In normal operation a power system is structurally stable since the operator steers the system with a safe margin. Under stressed conditons this margin may get smaller. We wish to know the critical value of a parameter such as the load, which if increased incrementally will result in one or a pair of eigenvalues crossing over to the right half plane. This is the point of local bifurcation and the system is structurally unstable at this point. The analysis is done via small signal analysis of power systems, and it is the principal tool to study both low-frequency oscillations and voltage stability. In the literature these two phenomena are treated separately. In this section we take a unified approach and attempt to look at the dependence of these two phenomena on the parameters of the system, which are the loads (nonlinear dependence on voltage) and modeling of the generating unit (machine and exciter). The use of participation factors is helpful in identifying state variable participation in a critical mode.

The small-perturbation behavior of the power system in the vicinity of a steady-state operating point can be described by a set of linear, time-invariant (LTI) differential equations in the state space form as

$$\dot{x} = Ax + Bu \tag{3.1}$$

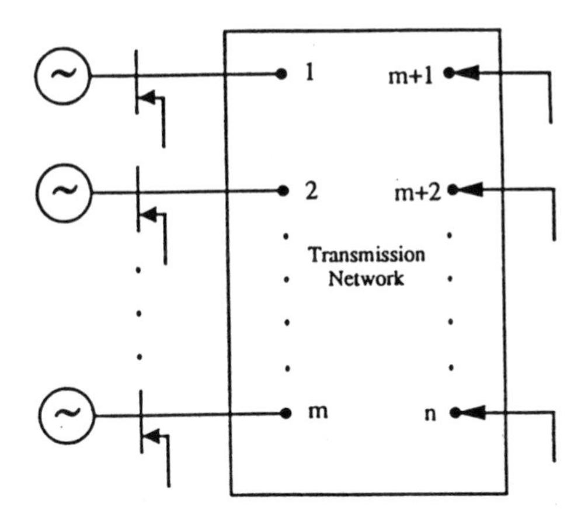

FIG. 3.1. *A general m-machine, n-bus system*

The N-dimensional state vector x represents the perturbations of the system state variables from their nominal values at the given operating condition, and the vector u represents perturbations of the system inputs such as voltage reference, desired real power or load demands. The numerical values of the matrices A and B depend on the operating condition as well as on the system parameters. The whole analysis starts with a systematic derivation of a linear model for an n-bus m-machine nonlinear differential algebraic system with nonlinear voltage-dependent loads at the network buses (Fig. 3.1). The derivation of the analytical model is not included here but is to be found in [35]. The model so obtained is flexible enough to study both low-frequency oscillations and voltage stability problems. For the former study, the system A matrix and its eigenvalues are readily obtainable for any given load model. The influence of parameters such as exciter gain and loads can be studied very easily. It is shown that the appearance of the electromechanical mode of oscillation or exciter mode oscillation depends critically on the modeling of the machine and the excitation system. For voltage stability analysis, we progressively load the system at a bus or set of buses and at each loading we monitor the eigenvalues of the linearized system. The machine is modeled by either a two-axis or a flux decay model, and the excitation system by either an IEEE Type I or a static exciter model. Thus potentially we have four types of generating unit models. Of these we limit ourselves to two models.

Model A: 2 axis machine model with an IEEE Type I exciter.

Model B: Flux decay machine model with a fast exciter.

The two exciter models are shown in Figs. 3.2 and 3.3.

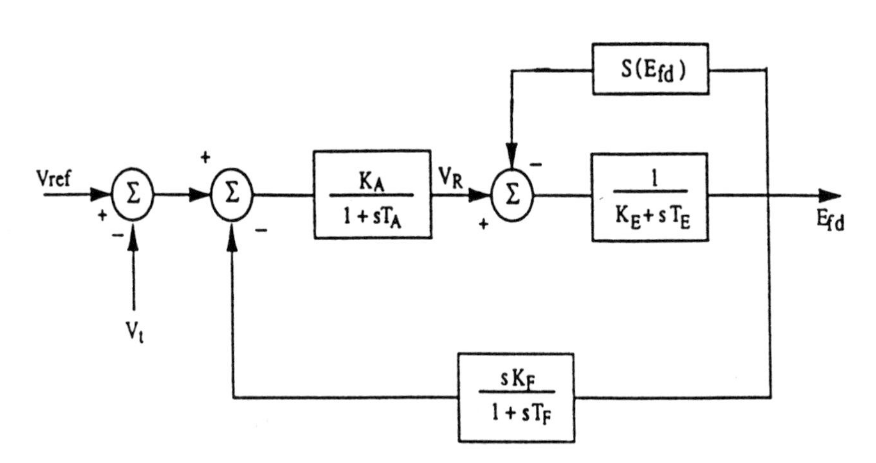

FIG. 3.2. *IEEE type-1 exciter model*

3.1. Various mathematical models. The mathematical model consists of differential equations pertaining to machine and exciter dynamics and the algebraic equations corresponding to the stator and network equations.

3.1.1. Model A (two-axis model with IEEE type I exciter) The differential equations of the machine and the exciter are given as in

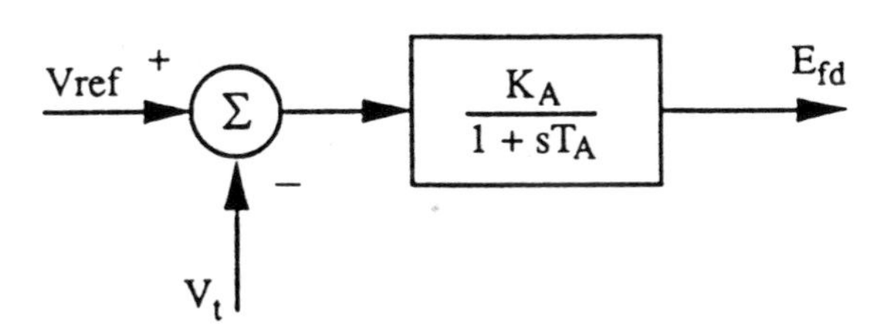

FIG. 3.3. *Static exciter model*

[28] where the various symbols are defined.

Differential equations

$$(3.2)\quad \frac{d\delta_i}{dt} = \omega_i - \omega_s$$

$$(3.3)\quad \frac{d\omega_i}{dt} = \frac{T_{Mi}}{M_i} - \frac{[E'_{qi} - X'_{di}I_{di}]I_{qi}}{M_i} - \frac{[E'_{di} + X'_{qi}I_{qi}]I_{di}}{M_i} - \frac{D_i(\omega_i - \omega_s)}{M_i}$$

$$(3.4)\quad \frac{dE'_{qi}}{dt} = -\frac{E'_{qi}}{T'_{doi}} - \frac{(X_{di} - X'_{di})I_{di}}{T'_{doi}} + \frac{E_{fdi}}{T'_{doi}}$$

$$(3.5)\quad \frac{dE'_{di}}{dt} = -\frac{E'_{di}}{T'_{qoi}} + \frac{I_{qi}}{T'_{qoi}}(X_{qi} - X'_{qi})$$

$$(3.6)\quad \frac{dE_{fdi}}{dt} = -\frac{K_{Ei} + S_E(E_{fdi})}{T_{Ei}}E_{fdi} + \frac{V_{Ri}}{T_{Ei}}$$

$$(3.7)\quad \frac{dV_{Ri}}{dt} = -\frac{V_{Ri}}{T_{Ai}} + \frac{K_{Ai}}{T_{Ai}}R_{fi} - \frac{K_{Ai}K_{Fi}}{T_{Ai}T_{Fi}}E_{fdi} + \frac{K_{Ai}}{T_{Ai}}(V_{refi} - V_i)$$

$$(3.8)\quad \frac{dR_{Fi}}{dt} = -\frac{R_{Fi}}{T_{Fi}} + \frac{K_{Fi}}{(T_{Fi})^2}E_{fdi} \qquad \text{for } i = 1, ..., m$$

The algebraic equations consist of stator algebraic equations and the network power balance equations. There are

Stator algebraic equations

$$(3.9)\quad E'_{di} - V_i \sin(\delta_i - \theta_i) - R_{si}I_{di} + X'_{qi}I_{qi} = 0$$

$$(3.10)\quad E'_{qi} - V_i \cos(\delta_i - \theta_i) - R_{si}I_{qi} - X'_{di}I_{di} = 0$$

for $i = 1, ..., m$

Network equations

$$I_{di}V_i\sin(\delta_i-\theta_i)+I_{qi}V_i\cos(\delta_i-\theta_i)+P_{Li}(V_i) \\ -\textstyle\sum_{k=1}^{n}V_iV_kY_{ik}\cos(\theta_i-\theta_k-\alpha_{ik})=0 \tag{3.11}$$

$$I_{di}V_i\cos(\delta_i-\theta_i)-I_{qi}V_i\sin(\delta_i-\theta_i)+Q_{Li}(V_i) \\ -\textstyle\sum_{k=1}^{n}V_iV_kY_{ik}\sin(\theta_i-\theta_k-\alpha_{ik})=0 \\ i=1,\ldots,m \tag{3.12}$$

$$P_{Li}(V_i)-\sum_{k=1}^{n}V_iV_kY_{ik}\cos(\theta_i-\theta_k-\alpha_{ik})=0 \tag{3.13}$$

$$\begin{aligned} Q_{Li}(V_i)-\textstyle\sum_{k=1}^{n}V_iV_kY_{ik}\sin(\theta_i-\theta_k-\alpha_{ik}) &= 0 \\ \text{for } i &= m+1,\ldots,n \end{aligned} \tag{3.14}$$

3.1.2. Model B (flux decay model and fast exciter). If the damper winding time constants T'_{qoi} are very small, then use of singular perturbations [28] makes the E'_{di} dynamics very fast so that 3.5 becomes an algebraic equation:

$$0=-E'_{di}+(X_{qi}-X'_{qi})I_{qi}$$

i.e.,

$$E'_{di}=(X_{qi}-X'_{qi})I_{qi} \tag{3.15}$$

The differential and algebraic equations of the two-axis model will be modified by substituting Eq. 3.15 in Eqs. 3.3 and 3.9. Moreover, for a simple and fast exciter, the transfer function between $(V_{ref,i}-V_i)$ and E_{fdi} is represented by a single time constant, so that the exciter model becomes

$$T_{Ai}\frac{dE_{fdi}}{dt}=-E_{fdi}+K_{Ai}(V_{refi}-V_i) \tag{3.16}$$

The overall flux decay model is represented by the following set of equations:

Differential equations

$$\frac{d\delta_i}{dt} = \omega_i - \omega_s \tag{3.17}$$

$$\frac{d\omega_i}{dt} = \frac{T_{Mi}}{M_i} - \frac{E'_{qi}I_{qi}}{M_i} - \frac{(X_{qi} - X'_{di})}{M_i} I_{di}I_{qi} - \frac{D_i(\omega_i - \omega_s)}{M_i} \tag{3.18}$$

$$\frac{dE'_{qi}}{dt} = -\frac{E'_{qi}}{T'_{doi}} - \frac{(X_{di} - X'_{di})}{T'_{doi}} I_{di} + \frac{E_{fdi}}{T'_{doi}} \tag{3.19}$$

$$\frac{dE_{fdi}}{dt} = -\frac{E_{fdi}}{T_{Ai}} + \frac{K_{Ai}}{T_{Ai}}(V_{ref,i} - V_i) \quad \text{for } i = 1, ..., m \tag{3.20}$$

The algebraic equations are

Stator algebraic equations

$$V_i \sin(\delta_i - \theta_i) + R_{si}I_{di} - X_{qi}I_{qi} = 0 \tag{3.21}$$

$$E'_{qi} - V_i \cos(\delta_i - \theta_i) - R_{si}I_{qi} - X'_{di}I_{di} = 0 \tag{3.22}$$

$$\text{for } i = 1, ..., m$$

Network equations

The network equations are the same as 3.11–3.14.

3.2. Linearization of model A. Linearizing the differential-algebraic model 3.2–3.14 we get

$$\Delta\dot{X} = A_1\Delta X + A_2\Delta I_g + A_3\Delta V_g + E\Delta U \tag{3.23}$$

$$0 = B_1\Delta X + B_2\Delta I_g + B_3\Delta V_g \tag{3.24}$$

$$0 = C_1\Delta X + C_2\Delta I_g + C_3\Delta V_g + C_4\Delta V_\ell + \Delta S_{Lg}(V) \tag{3.25}$$

$$0 = D_1\Delta V_g + D_2\Delta V_\ell + \Delta S_{L\ell}(V) \tag{3.26}$$

where

$$\begin{aligned}
\Delta x^T &= [\Delta x_1, \ldots, \Delta x_i, \ldots, \Delta x_m] \\
\Delta I_g^T &= [\Delta I_{g1}, \ldots, \Delta I_{gi}, \ldots, \Delta I_{gm}] \\
\Delta V_g^T &= [\Delta V_{g1}, \ldots, \Delta V_{gi}, \ldots, \Delta V_{gm}] \\
\Delta U^T &= [\Delta u_1, \ldots, \Delta u_i, \ldots, \Delta u_m] \\
\Delta V_\ell^T &= [\Delta V_{\ell 1}, \ldots, \Delta V_{\ell i}, \ldots, \Delta V_{\ell n}],\ \Delta S_{Lg}^T(V) = [\Delta S_{Lg1}(V_1), \ldots, \\
&\quad \Delta S_{Lgi}(V_i), \ldots, \Delta S_{Lgm}(V_m)],\ \Delta S_{L\ell}^T(V) = [\Delta S_{L\ell 1}(V_1), \ldots, \\
&\quad \Delta S_{L\ell i}(V_i), \ldots, \Delta S_{L\ell n}(V_n)] \\
\Delta x_i^T &= [\Delta\delta_i, \Delta\omega_i, \Delta E'_{qi}, \Delta E'_{di}, \Delta E_{fdi}, \Delta V_{Ri}, \Delta R_{fi}] \\
&\quad \Delta I_{gi}^T = [\Delta I_{di}, \Delta I_{qi}],\ \Delta V_{gi}^T = [\Delta\theta_{gi}, \Delta V_{gi}], \Delta u_i^T \\
&\quad = [\Delta T_{Mi}, \Delta V_{refi}] \\
\Delta V_{\ell i}^T &= [\Delta\theta_{\ell i}, \Delta V_{\ell i}],\ \Delta S_{Lgi}^T(V_i) = [\Delta P_{Lgi}(V_i), \Delta Q_{Lgi}(V_i)] \\
\Delta S_{L\ell}^T(V_i) &= [\Delta P_{L\ell i}(V_i), \Delta Q_{L\ell i}(V_i)]
\end{aligned}$$

This is the general comprehensive model of the differential-algebraic type to study both steady state and voltage stability with any type of nonlinear voltage dependent loads. The network structure is preserved and so are the stator algebraic equations for each machine. Equations 3.23–3.26 are equivalent to the model in Ref. [26] except that a machine angle is not introduced as reference. This model is quite general and can easily be expanded to include frequency or $\dot{V}$ dependence at the load buses, PSS, tap-changer dynamics, SVC's or FACTS devices. This will only augment 3.23–3.26 by either algebraic and/or differential equations. In the above model, ΔI_g is not of interest in most cases. Hence, eliminating ΔI_g from the set of Eqs. 3.23 and 3.25 using 3.24, we get following reduced set of equations,

$$\Delta\dot{X} = (A_1 - A_2B_2^{-1}B_1)\Delta X + (A_3 - A_2B_2^{-1}B_3)\Delta V_g + E\Delta U \tag{3.27}$$

$$0 = K_2\Delta X + K_1\Delta V_g + C_4\Delta V_\ell + \Delta S_{Lg}(V) \tag{3.28}$$

$$0 = D_1\Delta V_g + D_2\Delta V_\ell + \Delta S_{L\ell}(V) \tag{3.29}$$

where

$$[C_3 - C_2B_2^{-1}B_3] \triangleq K_1 \tag{3.30}$$

and

$$[C_1 - C_2B_2^{-1}B_1] \triangleq K_2 \tag{3.31}$$

More compactly Eqs. 3.27–3.29 can be put in the form

$$\begin{bmatrix} \Delta\dot{X} \\ 0 \end{bmatrix} = \begin{bmatrix} \overline{A}_1 & \overline{A}_2 \\ \overline{B}_1 & \overline{B}_2 \end{bmatrix} \begin{bmatrix} \Delta X \\ \Delta V \end{bmatrix} + \begin{bmatrix} 0 \\ \Delta S_L \end{bmatrix} + \begin{bmatrix} E \\ 0 \end{bmatrix} \Delta U \tag{3.32}$$

where $\Delta V = \begin{bmatrix} \Delta V_g \\ \Delta V_\ell \end{bmatrix}$ and $\Delta S_L = \begin{bmatrix} \Delta S_{Lg} \\ \Delta S_{L\ell} \end{bmatrix}$. Reorder the variables in the vector $\Delta V = \begin{bmatrix} \Delta V_g \\ \Delta V_\ell \end{bmatrix}$ such that the new vector is $[\Delta z^T \mid \Delta v^T] = [\Delta\theta_1, \Delta V_1, \ldots, \Delta V_m \mid \Delta\theta_2, \Delta\theta_3, \ldots, \Delta\theta_n, \Delta V_{m+1}, \ldots, \Delta V_n]$.

In this reordering of algebraic variables Δv represents those variables appearing in the standard load flow equations and Δz the remaining ones in ΔV. Also reorder the variables in $\begin{bmatrix} \Delta S_{Lg} \\ \Delta S_{L\ell} \end{bmatrix}$ to conform similarly so that $\Delta S_L^T = [\Delta P_{L1}, \Delta Q_{L1}, \cdots, \Delta Q_{Lm} \mid \Delta P_{L2}, \Delta P_{L3}, \cdots, \Delta P_{Ln}, \Delta Q_{Lm+1}, \ldots, \Delta Q_{Ln}] = [\Delta S_1^T \mid \Delta S_2^T]$. We carry out one more operation on the set of Eqs. 3.27–3.29. In any rotational system, the reference for angles is arbitrary. The order of the dynamical system in Eq. 3.27 is 7m, and can be reduced to (7m-1) by introducing relative rotor angles [26].

Selecting δ_1 as the reference, we have

$$\begin{aligned}
\delta_i' &= \delta_i - \delta_1, & i = 2,3,...,m \\
\delta_1' &= 0 \\
\dot{\delta}_i' &= \omega_i - \omega_1, & i = 2,3,...,m \\
\dot{\delta}_1' &= 0 \\
\theta_i' &= \theta_i - \delta_1, & i = 1,2,...,n
\end{aligned}$$

This implies that the differential equation corresponding to δ_1 can be deleted from Eq. 3.32 and also the column corresponding to $\Delta\delta_1$ in $\bar{A}_1$ and $\bar{B}_1$. Moreover, the entries corresponding to $\dot{\delta}_i'$, $i = 2,3,...,m$, will bring necessary changes in $\bar{A}_1$. We denote the reduced state vector as Δx. We thus have the new differential–algebraic (DAE) system as

$$(3.33)\quad \begin{bmatrix} \Delta\dot{x} \\ 0 \\ 0 \end{bmatrix} = \begin{bmatrix} A_1 & A_2 & A_3 \\ B_1 & B_2 & B_3 \\ C_1 & C_2 & C_3 \end{bmatrix} \begin{bmatrix} \Delta x \\ \Delta z \\ \Delta v \end{bmatrix} + \begin{bmatrix} 0 \\ \Delta S_1 \\ \Delta S_2 \end{bmatrix} + \begin{bmatrix} E \\ 0 \\ 0 \end{bmatrix} \Delta U$$

This model is slightly different from that in [26] in the sense that we allow for voltage dependency for the loads in the vectors ΔS_1 and ΔS_2. For the constant power case, both ΔS_1 and ΔS_2 are $\equiv 0$. Otherwise, $\Delta S_{1i} = \Delta S_{1i}(V_i)$ and $\Delta S_{2i} = \Delta S_{2i}(V_i)$. For a given voltage-dependent load, $\Delta S_{1i}(V_i)$ and $\Delta S_{2i}(V_i)$ can be computed. Only the appropriate elements of B_2, B_3, C_2 and C_3 will be modified and we obtain the system

$$(3.34)\quad \begin{bmatrix} \Delta\dot{x} \\ 0 \\ 0 \end{bmatrix} = \begin{bmatrix} \tilde{A}_1 & \tilde{A}_2 & \tilde{A}_3 \\ \tilde{B}_1 & \tilde{B}_2 & \tilde{B}_3 \\ \tilde{C}_1 & \tilde{C}_2 & \tilde{C}_3 \end{bmatrix} \begin{bmatrix} \Delta x \\ \Delta z \\ \Delta v \end{bmatrix} + \begin{bmatrix} E \\ 0 \\ 0 \end{bmatrix} \Delta U$$

Now $\tilde{C}_3$ is the load flow Jacobian J_{LF} and $\begin{bmatrix} \tilde{B}_2 & \tilde{B}_3 \\ \tilde{C}_2 & \tilde{C}_3 \end{bmatrix} = J_{AE}$ defined as the algebraic Jacobian in [26]. The system A matrix is obtained as

$$(3.35)\quad \Delta\dot{x} = A_{sys}\Delta x + E\Delta U$$

where

$$(3.36)\quad A_{sys} = \tilde{A}_1 - [\tilde{A}_2 \quad \tilde{A}_3][J_{AE}]^{-1} \begin{bmatrix} \tilde{B}_1 \\ \tilde{C}_1 \end{bmatrix}$$

This is the model used in studying low frequency oscillation, steady state stability and voltage stability. In the next section we examine the effects of increased loading on the eigenvalues of A_{sys} and the determinants of J_{AE} and J_{LF} for models A and B. Thus we do parametric study regarding load and generating unit models using the small signal stability model. In the process we illustrate both the Hopf bifurcation and the singularity induced bifurcation.

The linearized equations for model B can be derived along the same lines and hence is omitted.

3.3. Identification of critical modes: participation factor method. When a system becomes unstable due to slowly varying loads, we analyze via a linearized model and monitor the eigenvalues. Certain eigenvalues which become unstable need to be monitored more closely. The appropriate definition and determination as to which state variables significantly participate in the selected modes become very important. This requires tools for identifying state variables that are significant in producing the selected modes. Verghese et al. [29] have suggested a dimensionless measure of state variable participation (henceforth called participation factors). If v_i and w_i represent the right- and left-hand eigenvectors, respectively, for the eigenvalue λ_i of the matrix A, then the participation factor(pf) measuring the participation of the kth state variable x_k in the ith mode is defined as

$$p_{ik} = w_{ik} v_{ik}$$

This quantity is dimensionless and hence invariant under changes of scale of the variables. We can think of v_{ik} as measuring the activity of x_k in the ith mode, and w_{ik} as weighting the contribution of this activity to the mode. The step-by-step procedure for the calculation of pf is as follows:

1. For the given A_{sys}, the eigenvalues $\lambda_i's$ and the right- and the left-hand eigenvectors ($\mathcal{R}_i$ and $\mathcal{L}_i$, respectively) are calculated, for i = 1,2,...,$(7m-1)$.
2. For a given eigenvalue λ_i the corresponding pf's are calculated as follows:
   (a) The normalization constant $\mathcal{N}_i$ is calculated by finding $\sum_{k=1}^{N} abs(\mathcal{R}_{ik}) abs(\mathcal{L}_{ik})$ where N = 7m-1.
   (b) Then $p_{ik} = \frac{abs(\mathcal{R}_{ik}) abs(\mathcal{L}_{ik})}{\mathcal{N}_i}$. If $\mathcal{N}_i$ is real, then "abs" is omitted.

3.4. Bifurcation analysis in multi-machine systems. The effect of loading has been investigated on the 3-Machine, 9-Bus System (Fig. 3.4) whose data can be found in [30]. Assuming a constant power load representation, the real and/or reactive loads at a particular bus/buses was increased continuously. At each step the initial conditions of state variables were computed, after running the load flow [28], and linearization of the equations was done. Ideally the increase of load should be picked up by the generators through the economic load dispatch scheme. To simplify matters we allocate the increase in generation (real power) to the machines in proportion to the inertias. In the case of increase in reactive power, the increase is picked up by the PV buses. The A_{sys} matrix was formed and its eigenvalues were checked for stability. Also $\det J_{LF}$ and $\det J_{AE}$ were computed. The step-by-step algorithm is as follows:

(a) Increase the load at bus/buses for a particular generating unit model.

(b) If the real load is increased then distribute the load amongst various generators in proportion to their inertias.
(c) Run the load flow.
(d) Stop, if load flow fails to converge.
(e) Compute initial conditions of the state variables.
(f) Linearize the differential equations and compute the various matrices.
(g) Compute $\det J_{LF}$, $\det J_{AE}$, and the eigenvalues of A_{sys}.
(h) If A_{sys} is stable then go to step (a).
(i) Identify the states associated with the unstable eigenvalue(s) of A_{sys} using the participation factor method and go to step (a).

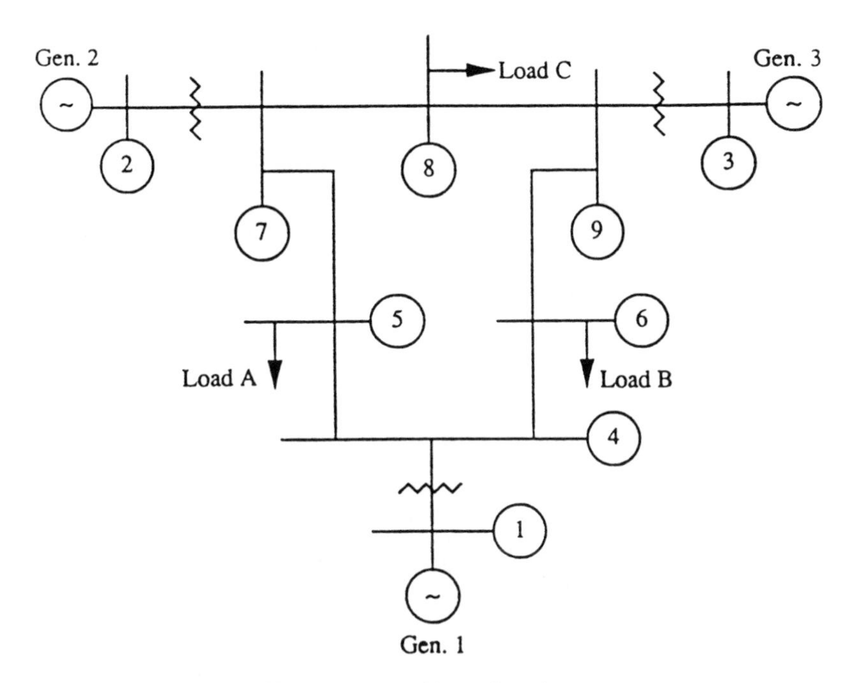

FIG. 3.4. *3 machine 9 bus system*

For the 3-machine case all the four generating units are studied. The results are summarized in Tables 3.1–3.2 for the models A and B. For the static exciter, a gain of $K_A = 50$ was assumed. In the next section we discuss the various results.

The participation factor analysis of the eigenvalues for various cases reveal the following facts:

For Model-A, when the load is increased, it is observed that the critical modes for the unstable eigenvalues are the electrical ones (E'_q and R_f).

From Table 3.1 we observe that when the load at bus 5 is increased from 4.7 pu to 4.8 pu, the complex pair of unstable eigenvalues splits into real ones which move in the opposite directions along the real axis. The one moving along the positive real axis (9.2464) is sensitive to the rotor angle mode and eventually comes back to the left-half plane via $+\infty$ when the load at bus 5 is increased from 4.8 pu to 4.9 pu. This is the point when $\det J_{AE}$ changes sign.

TABLE 3.1
Modal Behaviour of Model A for Different Loads

Load at Bus 5	sign ($\det J_{LF}$)	sign ($\det J_{AE}$)	Critical Eigenvalue(s)	Associated States
4.3	+	+	-0.1433±j2.0188	E'_{q1} & R_{f1}
4.4	+	+	0.0057±j2.2434	E'_{q1} & R_{f1}
4.5	+	+	0.3400±j2.5538	E'_{q1} & R_{f1}
4.6	+	+	1.1350±j2.8016	E'_{q1} & R_{f1}
4.7	+	+	2.5961±j2.2768	E'_{q1} & R_{f1}
4.8	+	+	9.2464, 1.8176	δ_2 & ω_2, E'_{q1} & R_{f1}
4.9	+	–	1.0542	E'_{q1} & R_{f1}
5.0	+	–	0.6298	E'_{q1} & R_{f1}
5.1	+	–	0.2463	E'_{q1} & R_{f1}
5.15	+	–	-0.6832	E'_{q1} & R_{f1}
5.2	Load flow does not converge			

TABLE 3.2
Modal Behaviour of Model B for Different Loads

Load at Bus 5	sign ($\det J_{LF}$)	sign ($\det J_{AE}$)	Critical Eigenvalue(s)	Associated States
4.4	+	+	-0.0957±j10.1407	δ_2 & ω_2
4.5	+	+	0.0308±j10.0034	δ_2 & ω_2
4.6	+	+	0.3802±j9.9008	δ_2 & ω_2
4.7	+	+	0.9344±j10.1111	δ_2 & ω_2
4.8	+	+	1.3907±j11.1963	δ_2 & ω_2
4.9	+	–	24.4174, 0.1104±j11.3605	E'_{q1} & E_{fd1}, δ_2 & ω_2
5.0	+	–	6.0978	E'_{q1} & E_{fd1}
5.1	+	–	2.5680	E'_{q1} & E_{fd1}
5.15	+	–	0.5417	E'_{q1} & E_{fd1}
5.2	Load flow does not converge			

The point when $detJ_{AE}$ changes is known in the literature as "singularity induced bifurcation" [2,23]. This is the point when the algebraic equations cannot be solved for the algebraic variables in terms of the state variables. The other unstable real eigenvalue moves to the left and is sensitive to the exciter mode. This eigenvalue returns to the left-half plane at the loading of approximately 5.15 pu at bus 5, and the system is again dynamically stable. For the load at bus 5 = 5.2 pu, the load flow does not converge. It is possible through other algebraic techniques to reach the nose of the P-V curve or the saddle node bifurcation. This phenomenon is pictorially indicated in the $P - V$ curve for model-A and also in the trajectory of critical eigenvalue(s) in s-plane (Figs. 3.5 and 3.6).

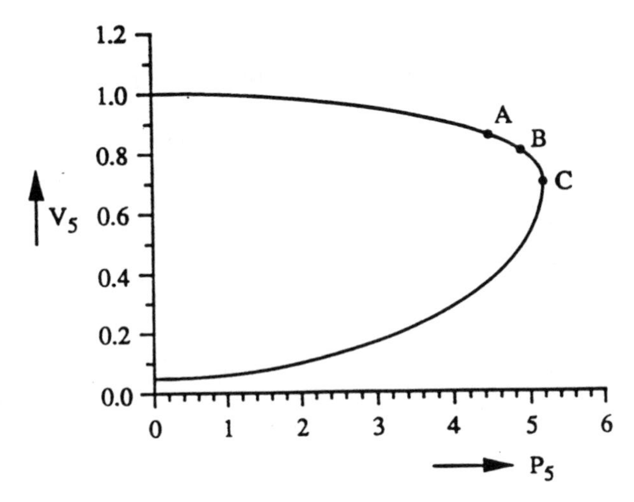

FIG. 3.5. *P-V curve for bus 5*

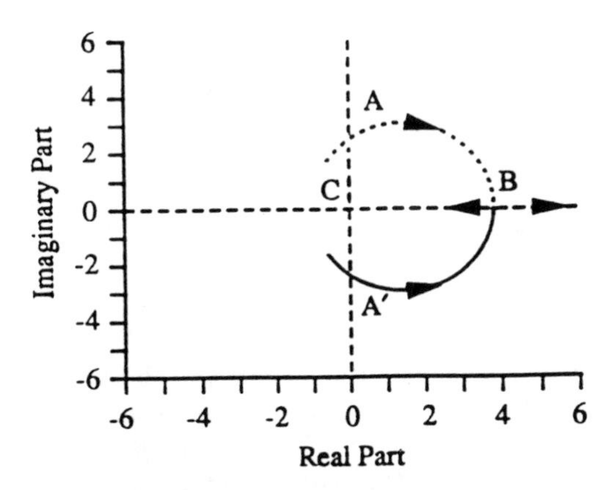

FIG. 3.6. *Critical modes of A_{sys} as a function of the load at bus-5*

Between points A and B there is Hopf-bifurcation which has been shown to be subcritical [31]. However load flow solution still exists. In this region E'_q & R_f state variables are clearly dominant initially. As we approach towards B, δ_2 & ω_2 state variables start participating substantially in the unstable eigenvalues as indicated in the Table 3.1. For Model-B which uses the fast static exciter with a single time constant, the modes which go unstable are the electromechanical ones. In Model-B, the behavior of eigenvalues between the loading of 4.8 and 5.0 is examined in detail by incrementing the load by .02. The results are shown in Table 3.3. It is observed that another pair of eigenvalues undergoes Hopf bifurcation but then returns to the left half plane while the first pair goes through a singularity induced bifurcation as in Table 3.1. The first pair of eigenvalues behaves as in Model A, namely, it undergoes singularity induced bifurcation after the Hopf bifurcation. Thus having a reduced order model and a fast exciter produces two Hopf bifurcations although not at the same time.

TABLE 3.3
Detailed modal behavior of Model-B for loading between 4.8 and 5.0 p.u.

Load at bus 5	Sign $(detJ_{LF})$	Sign $det(J_{AE})$	Critical Eigenvalue(s)	States Associated
4.8	+	+	1.3907±j11.1963	δ_2,ω_2
4.82	+	+	1.1546±j11.6623	δ_2,ω_2
			.0444±j15.503	$\delta_2,\omega_2,E'_{q2},E_{fd2}$
4.84	+	+	1.9965±j15.7308	$\delta_2\omega_2,E'_{q2},E_{fd2}$
			.5983±j11.7688	δ_2,ω_2
4.86	+	+	3.5965±j17.9188	$E'_{q1},E_{fd1},\delta_2,\omega_2$ E'_{q2},E_{fd2}
			.3174±j11.599	δ_2,ω_2
4.88	+	+	8.8996±j26.1149	$E'_{q1},E_{fd1},E'_{q2},E_{fd2}$
			.1862±j11.462	δ_2,ω_2
4.90	+	-	24.4175	E'_{q1},E_{fd1},E_{fd2}
			.1105±j11.3065	δ_2,ω_2
4.92	+	-	14.0688	E'_{q1},E_{fd1}
			.0265±j11.22	δ_2,ω_2
4.94	+	-	10.5866	E'_{q1},E_{fd1}
			.0265±j11.22	δ_2,ω_2
4.96	+	-	8.5713	E'_{q1},E_{fd1}
			.001±j11.168	δ_2,ω_2
4.98	+	-	7.1711	E_{q1},E_{fd1}
5.0	+	-	6.0978	E'_{q1},E_{fd1}

Hopf bifurcation phenomena in power systems was first discussed in Ref. [32] for a single machine case. In their studies the electro-mechanical mode was the critical one. In studies relating to voltage collapse [31] it was shown that the exciter mode may go unstable first. In this paper we have shown that both exciter modes and electromechanical modes are critical in steady state stability and voltage collapse and that they both participate in the dynamic instability. Hence decoupling the Q-V dynamics from the

$P - \delta$ dynamics as suggested in [23] may not always hold. It may be true for special system configuration/operating conditions. There is no doubt that there are some underlying dynamics such as the load dynamics as discussed in Ref. [33]. If such dynamics were represented, it will be fast dynamics and the phenomena of $detJ_{AE}$ changing sign will still exist. In conventional bifurcation theory terms one can think of solving $g(x, y) = 0$ for $y = h(x)$ and substituting this in the differential equation to get $\dot{x} = f(x, h(x))$. The change in sign of $detJ_{AE}$ is the instant when solution of y is no longer possible. However structural instability occurs much earlier. This is also tied in with the concept of implicit function theorem in singular perturbation theory. A concrete mathematical underpinning of these ideas in the context of power systems is a research issue. Recent work in "Singularity induced bifurcation," [2,23] and "Impasse surface" concepts confirms the fact that when the Jacobian of the algebraic equation becomes singular, an eigenvalue moves from $+\infty$ in the right half plane to the real axis in the left half plane [34].

4. Conclusion. In this paper we have related the structural stability phenomenon to the bifurcation analysis in power systems, particularly as it relates to voltage stability. Testing on larger systems will explain many of the observed phenomenon in power systems in a strict mathematical framework.

REFERENCES

[1] A. A. Andronov and L. Pontryagin, "Systems grossiers," Dukl. Akad. Nauk, SSSR, vol. 14, pp. 247-251, 1937.

[2] V. Venkatasubramaniam, H. Schättler and J. Zaborszky, "Voltage dynamics: study of generator with voltage control, transmission and matched MW load," *IEEE Trans. on Automatic Control*, vol. 37, no. 11, Nov. 1992, pp. 1717-1733.

[3] M. A. Aizerman and F. R. Gantmacher, *Absolute Stability of Control Systems*, A.N. SSSR, Moscow, 1963.

[4] D. D. Siljak, *Nonlinear Systems*, Wiley, New York, 1969.

[5] M. A. Pai, *Power System Stability*, N. Holland, Amsterdam, 1981.

[6] M. A. Pai, *Energy Function Analysis For Power System Stability*, Kluwer Academic Publishers, Boston, MA, 1989.

[7] A. A. Fouad and V. Vittal, *Power System Transient Stability Analysis Using The Transient Energy Function Method*, Prentice-Hall, New York, 1992.

[8] A. A. Andronov and C. E. Chaikin, *Theory of Oscillations*, Princeton University Press, Princeton, NJ, 1949.

[9] M. M. Peixoto, "Structural stability on two dimensional manifolds," *Topology*, vol. 1, pp. 101-120, 1962.

[10] L. Perko, *Differential Equations and Dynamical Systems*, Springer-Verlag, New York, 1991.

[11] J. Guckenheimer and P. Holmes, *Nonlinear Oscillations, Dynamical Systems and Bifurcation of Vector Fields*, Springer-Verlag, New York, 1983.

[12] J. Hale and H. Kocak, *Dynamics and Bifurcations*, Springer-Verlag, New York, 1991.

[13] S. N. Chow and J. Hale, *Methods of Bifurcation Theory*, Springer-Verlag, New York, 1982.

[14] V. I. Arnold, *Ordinary Differential Equations*, MIT Press, Cambridge, MA, 1973.
[15] V. I. Arnold, *Geometric Methods In The Theory Of Ordinary Differential Equations*, Springer-Verlag, New York, 1982.
[16] M. W. Hirsch and S. Smale, *Differential Equations, Dynamical Systems and Linear Algebra*, Academic Press, New York, 1974.
[17] R. K. Bansal, *Estimation of Stability Domains for the Transient Stability Investigation of Power Systems*, Ph.D. Thesis, I.I.T., Kanpur, Aug. 1975.
[18] T. S. Baker, "Analysis of the synchronization of an automatic phase control system," Tech. Report #460, Cruft Lab, Harvard University, Nov. 1964.
[19] P. Varaiya, et al., *Bifurcation and Chaos In Power system: A Survey*, EPRI Report TR-100834, Aug. 1992.
[20] L. A. Luxemburg, *Structural Stability Analysis and Its Applications To Power Systems*, Ph.D. Thesis, Texas A&M University, Dec. 1987.
[21] T. S. Parker and L. O. Chua, *Practical Numerical Algorithms For Chaotic Systems*, Springer-Verlag, New York, 1989.
[22] F. Takens, "Constrained equations: A study of implicit differential equations and their discontinuous solutions," *Structural Stability, the Theory of Catastrophes and the Application in the Sciences*, (Lecture Notes in Mathematics), vol. 525, Springer-Verlag, New York, 1976, pp. 143-234.
[23] V. Venkatasubramaniam, H. Schattler and J. Zaborszky, "A taxonomy of the dynamics of a large power system with emphasis on its voltage stability," *Proc. Bulk Voltage Stability and Security*, Deep Creek Lake, Aug. 4-7, 1991.
[24] K. R. Padiyar, M. A. Pai and C. Radhakrishna, "A versatile system model for the dynamic stability analysis of power systems including HVDC links," *IEEE Transactions on Power Apparatus and Systems*, vol. 100, April 1981, pp. 1871-1880.
[25] P. Kundur, G. J. Rogers, D. Y. Wong, L. Wang and M. G. Lauby, "A comprehensive computer program package for small signal stability analysis of power systems," *IEEE Transactions on Power Systems*, vol. 5, no. 4, Nov. 1990, pp. 1076-1083.
[26] P. W. Sauer and M. A. Pai, "Power system steady-state stability and the load flow Jacobian," *IEEE Transactions on Power Systems*, vol. 5, no. 4, Nov. 1990, pp. 1374-1383.
[27] L. Wang and A. Semlyen, "Application of sparse eigenvalue techniques to the small signal stability analysis of large power systems," *IEEE Transactions on Power Systems*, vol. 5, no. 2, May 1990, pp. 635-642.
[28] P. W. Sauer and M. A. Pai, "Modeling and simulation of multi-machine power system dynamics," *Control and Dynamic Systems: Advances in Theory and Application* (C. T. Leondes, Ed.), vol. 43, Academic Press, San Diego, CA, 1991.
[29] G. C. Verghese, I. J. Perez-Arriaga and F. C. Scheweppe, "Selective modal analysis with applications to electric power systems, Part I and II," *IEEE Transactions on Power Apparatus and Systems*, vol. PAS 101, no. 9, 1982, pp. 3117-3134.
[30] P. M. Anderson and A. A. Fouad, *Power System Control and Stability*, Iowa State University Press, Ames, IA, 1977.
[31] C. Rajagopalan, B. Lesieutre, P. W. Sauer and M. A. Pai, "Dynamic aspects of voltage/power characteristics," *IEEE Transactions on Power Systems*, vol. 7, no. 3, pp. 990-1000, 1992.
[32] E. H. Abed and P. P. Varaiya, "Nonlinear oscillations in power systems," *International Journal of Electric Power and Energy Systems*, vol. 6, no. 1, pp. 37-43, 1984.
[33] M. K. Pal, "Voltage stability conditions considering load characteristics," *IEEE Transactions on Power Systems*, vol. 7, no. 1, Feb. 1992, pp. 243-249.
[34] I. A. Hiskens and D. J. Hill, "Failure modes of a collapsing power system," *Proc. of Bulk Power System Voltage Phenomena II – Voltage Stability and Security*, Deep Creek Lake, Maryland, Aug. 1991.

[35] R. K. Ranjan, M. A. Pai and P. W. Sauer, "Analytical formulation of small signal stability analysis in power systems with nonlinear load models," Indian Academy of Sciences, Bangalore, India, *Journal of Engineering Sciences, 'Sadhana'*, Vol 18, Part 5, Sept 1993, pp. 869–889.

POWER SYSTEM LOAD MODELING*

PETER W. SAUER† AND BERNARD C. LESIEUTRE‡

Abstract. This work begins with a discussion of the overall structure of power system dynamic models and the issues of consistency in modeling the components and their interconnection. For the synchronous machine plus network and elementary load models, this is done through the use of "transient algebraic circuits". Starting with the critical assumptions associated with these interconnected circuits, the problem of load modeling is defined. These critical assumptions include the existence of an integral manifold for the electrical variables associated with the fast line frequency (i.e. 60HZ). The preservation of "frequency" dependence in loads is discussed in the context of this integral manifold. Since the existence of such an integral manifold can only be shown for elementary load models, generalization to more complex loads must include constraints to insure the solvability of the interconnected set of algebraic equations. Such constraints on traditional exponential voltage dependent load models are given. With the tremendous interest in voltage stability, there is a great need for load models which more accurately reflect the dynamic characteristics of loads. While the dynamics range from time- scales of microseconds to minutes, the phenomena of interest are usually those in the slower range. This means that it should not be necessary to reinstate the fast line frequency transients, but rather introduce some new dynamic structure associated with the slow model of these fast variables. Several models which have been proposed in the literature as well as new models based on detailed induction machine dynamic models are discussed.

1. Introduction. Load modeling has remained one of the greatest challenges of power system analysis. This is due to the large-scale nature of typical system studies and the ever increasing variety of load types. This paper begins with a discussion of the overall structure of power system dynamic models and the issues of consistency in modeling the components and their interconnection. For the synchronous machine plus network and elementary load models, this is done through the use of "transient algebraic circuits." Starting with the critical assumptions associated with these interconnected circuits, the problem of load modeling is defined. These critical assumptions include the existence of an integral manifold for the electrical variables associated with the fast line frequency (i.e. 60HZ). The preservation of "frequency" dependence in loads is discussed in the context of this integral manifold. Since the existence of such an integral manifold can only be shown for elementary load models, generalization to more complex loads must include constraints to insure the solvability of the interconnected set of algebraic equations. Such constraints on traditional exponential voltage dependent load models are given.

---

* This research was supported in part by the Institute for Mathematics and its Applications with funds provided by the National Science Foundation and in part by EPRI through contract 8010-21.

† Department of Electrical and Computer Engineering, University of Illinois at Urbana-Champaign, 1406 W. Green St., Urbana, IL 61801.

‡ Department of Electrical Engineering and Computer Science, Massachusetts Institute of Technology, 77 Massachusetts Ave., Cambridge, MA 02139.

With the tremendous interest in voltage stability, there is great need for load models which more accurately reflect the dynamic characteristics of loads. While the dynamics range from time-scales of microseconds to minutes, the phenomena of interest are usually those in the slower range. This means that it should not be necessary to reinstate the fast line frequency transients, but rather introduce some new dynamic structure associated with the slow model of these fast variables. Several models which have been proposed in the literature as well as new models based on detailed models are examined from a structural stability point of view to investigate the differences between many small machines and several large machines.

2. Modeling and simulation time-scale structure. To clearly see the many challenges of load modeling it is necessary to review the overall structure of power system dynamics from a time-scale view. Consider the following seven categories:

1. Superfast: Stator/network/load dynamics
2. Fast: Rotor electrical dynamics
3. Medium fast: Exciter/regulator dynamics
4. Medium: Shaft dynamics
5. Medium slow: Turbine/governor dynamics
6. Slow: Load dynamics
7. Super slow: System control dynamics

For a system with m balanced three-phase synchronous machines, n three-phase buses, and b-m balanced three phase resistive/inductive branches, a reasonably detailed dynamic model is [1]:

(a) 2m differential equations with 4m algebraic variables

$$\epsilon \frac{d\psi_{Di}}{dt} = R_i I_{Di} + \psi_{Qi} + V_{Di} \qquad i = 1, ..., m \tag{2.1}$$

$$\epsilon \frac{d\psi_{Qi}}{dt} = R_i I_{Qi} - \psi_{Di} + V_{Qi} \qquad i = 1, ..., m \tag{2.2}$$

(b) 2m algebraic equations with 4m algebraic variables

$$\psi_{di} = -X'_{di} I_{di} + E'_{qi} \qquad i = 1, ..., m \tag{2.3}$$

$$\psi_{qi} = -X'_{qi} I_{qi} - E'_{di} \qquad i = 1, ..., m \tag{2.4}$$

(c) 4m algebraic equations

$$(I_{Di} + jI_{Qi}) = (I_{di} + jI_{qi}) e^{j(\delta_i - \pi/2)} \qquad i = 1, ..., m \tag{2.5}$$

$$(\psi_{Di} + j\psi_{Qi}) = (\psi_{di} + j\psi_{qi}) e^{j(\delta_i - \pi/2)} \qquad i = 1, ..., m \tag{2.6}$$

Since the system is balanced three-phase, the sets of three voltages, currents and flux linkages have been reduced through the standard dqo transformation to obtain sets of two voltages, currents and flux linkages. The "o" variables are completely decoupled, and will remain zero for all transients. Thus, the "o" variable dynamics are not shown in the model. The

"DQ" subscripts are for variables in the synchronously rotating reference frame while the "dq" subscripts are for variables in the individual machine reference frames.

(d) 2(b-m) differential equations with 4(b-m) algebraic variables

$$\epsilon\frac{d\psi_{Di}}{dt} = R_i I_{Di} + \psi_{Qi} + V_{Di} \qquad i = m+1,...,b \tag{2.7}$$

$$\epsilon\frac{d\psi_{Qi}}{dt} = R_i I_{Qi} - \psi_{Di} + V_{Qi} \qquad i = m+1,...,b \tag{2.8}$$

(e) 2(b-m) algebraic equations

$$\psi_{Di} = -X_{epi} I_{Di} \qquad i = m+1,...,b \tag{2.9}$$

$$\psi_{Qi} = -X_{epi} I_{Qi} \qquad i = m+1,...,b \tag{2.10}$$

(f) b algebraic equations with b algebraic variables

$$V_i = \sqrt{V_{Di}^2 + V_{Qi}^2} \qquad i = 1,...,b \tag{2.11}$$

(g) 2b algebraic equations

$$C^t V_D = 0 \quad , \quad C^t V_Q = 0 \tag{2.12}$$

(h) 4(b-n) algebraic equations with 4(b-n) algebraic variables

$$\psi_{D\ell} = C^t \psi_D \quad , \quad \psi_{Q\ell} = C^t \psi_Q \tag{2.13}$$

$$I_D = C I_{D\ell} \quad , \quad I_Q = C I_{Q\ell} \tag{2.14}$$

(i) 5m differential equations with m control variables

$$T'_{doi}\frac{dE'_{qi}}{dt} = -E'_{qi} - (X_{di} - X'_{di})I_{di} + E_{fdi} \qquad i = 1,...,m \tag{2.15}$$

$$T'_{qoi}\frac{dE'_{di}}{dt} = -E'_{di} + (X_{qi} - X'_{qi})I_{qi} \quad i = 1,...,m \tag{2.16}$$

$$T_{Ei}\frac{dE_{fdi}}{dt} = -(K_{Ei} + S_{Ei}(E_{fdi}))E_{fdi} + V_{Ri} \quad i = 1,...,m \tag{2.17}$$

$$\begin{aligned} T_{Ai}\frac{dV_{Ri}}{dt} &= -V_{Ri} + K_{Ai}R_{fi} - \frac{K_{Ai}K_{Fi}}{T_{Fi}}E_{fdi} \\ &\quad + K_{Ai}(V_{refi} - V_i) \quad i = 1,...,m \end{aligned} \tag{2.18}$$

$$T_{Fi}\frac{dR_{fi}}{dt} = -R_{fi} + \frac{K_{Fi}}{T_{Fi}}E_{fdi} \qquad i = 1,...,m \tag{2.19}$$

(j) 5m differential equations with m control variables

$$\frac{d\delta_i}{dt} = \omega_i - \omega_s \qquad i = 1, ..., m \tag{2.20}$$

$$M_i \frac{d\omega_i}{dt} = T_{Mi} - [E'_{qi} - X'_{di} I_{di}] I_{qi} - [E'_{di} + X'_{qi} I_{qi}] I_{di} \qquad i = 1, ..., m \tag{2.21}$$

$$T_{RHi} \frac{dT_{Mi}}{dt} = -T_{Mi} + \left(1 - \frac{K_{HPi} T_{RHi}}{T_{CHi}}\right) P_{CHi} + \frac{K_{HPi} T_{RHi}}{T_{CHi}} P_{SVi} \quad i = 1, ..., m \tag{2.22}$$

$$T_{CHi} \frac{dP_{CHi}}{dt} = -P_{CHi} + P_{SVi} \qquad i = 1, ..., m \tag{2.23}$$

$$T_{SVi} \frac{dP_{SVi}}{dt} = -P_{SVi} + P_{Ci} - \frac{1}{R_{di}} \left(\frac{\omega_i}{\omega_s}\right) \quad i = 1, ..., m \tag{2.24}$$

In summary, this model contains 2b+10m differential equations in dynamic states ψ_D, ψ_Q, E'_q, E'_d, E_{fd}, V_R, R_f, δ, ω, T_M, P_{CH}, P_{SV}. Because of the loop/branch relationships, this model has only 2(b-n)+10m-1 independent differential equations. It contains 9b+4m-4n algebraic equations relating the dynamic states and the algebraic variables I_D, I_Q, V_D, V_Q, I_d, I_q, ψ_d, ψ_q, V, $I_{D\ell}$, $I_{Q\ell}$, $\psi_{D\ell}$, $\psi_{Q\ell}$. It contains 2m control variables V_{ref}, P_c.

Considering this detailed model to be exact, we now use an interesting property to introduce the concept of "transient algebraic circuits."

3. Transient algebraic circuits. The balanced three-phase nature of the synchronous machine stator plus network/load R-L branch dynamics allows the sets of two voltage, current and flux linkage equations to be written as one complex equation. For example, adding equation (2.1) plus j times (2.2) gives the following complex equation:

$$V_{Di} + jV_{Qi} = -R_i(I_{Di} + jI_{Qi}) + j(\psi_{Di} + j\psi_{Qi}) + \epsilon \frac{d}{dt}(\psi_{Di} + j\psi_{Qi}) \tag{3.1}$$

By inspection, this equation is also correct for (2.7) plus j times (2.8), so the subscript i can be shown $i = 1,...,b$. Similarily for equations (2.3), (2.4), (2.9), (2.10), using (2.5) and (2.6),

$$\psi_{Di} + j\psi_{Qi} = [-X'_{di} I_{di} + E'_{qi} + j(-X'_{qi} I_{qi} - E'_{di})] e^{j(\delta - \pi/2)} \qquad i = i, ..., m \tag{3.2}$$

$$\psi_{Di} + j\psi_{Qi} = -X_i(I_{Di} + jI_{Qi}) \qquad i = m+1, ..., b \tag{3.3}$$

By defining,

$$\overline{V}_i \triangleq V_i e^{j\theta_i} \triangleq (V_{Di} + jV_{Qi}) \qquad i = 1, ..., b \tag{3.4}$$

$$\overline{I}_i \triangleq I_i e^{j\phi_i} \triangleq (I_{Di} + jI_{Qi}) \qquad i = 1, ..., b \tag{3.5}$$

$$\overline{E}_i \triangleq E_i e^{j\gamma_i} \triangleq ((X'_{qi} - X'_{di})I_{qi} + E'_{di} + jE'_{qi})e^{j(\delta_i - \pi/2)}$$
$$(3.6) \qquad i = 1, ..., m$$

these equations can be written as,

$$\overline{V}_i = -\left(R_i + jX'_{di}\left(1 - \epsilon\frac{j}{\overline{I}_i}\frac{d}{dt}\overline{I}_i\right)\right)\overline{I}_i + \left(1 - \epsilon\frac{j}{\overline{E}_i}\frac{d}{dt}\overline{E}_i\right)\overline{E}_i$$
$$(3.7) \qquad i = 1, ..., m$$
$$(3.8) \qquad \overline{V}_i = -\left(R_i + jX_i\left(1 - \epsilon\frac{j}{\overline{I}_i}\frac{d}{dt}\overline{I}_i\right)\right)\overline{I}_i \qquad i = m+1, ..., b$$

Using

$$(3.9) \qquad \frac{-j}{\overline{I}_i}\frac{d}{dt}\overline{I}_i = \frac{d\phi_i}{dt} - j\frac{1}{I_i}\frac{dI_i}{dt} \qquad i = 1, ..., b$$
$$(3.10) \qquad \frac{-j}{\overline{E}_i}\frac{d}{dt}\overline{E}_i = \frac{d\gamma_i}{dt} - j\frac{1}{E_i}\frac{dE_i}{dt} \qquad i = 1, ..., m$$

and defining the quantities

$$(3.11) \qquad \nu_i \triangleq 1 + \epsilon\frac{d\phi_i}{dt} \qquad i = 1, ..., b$$
$$(3.12) \qquad R_{ti} \triangleq \epsilon\frac{X'_{di}}{I_i}\frac{dI_i}{dt} \qquad i = 1, ..., m$$
$$(3.13) \qquad \triangleq \epsilon\frac{X_i}{I_i}\frac{dI_i}{dt} \qquad i = m+1, ..., b$$
$$(3.14) \qquad \overline{\beta}_i = \left(1 + \epsilon\left(\frac{d\gamma_i}{dt} - j\frac{1}{E_i}\frac{dE_i}{dt}\right)\right) \qquad i = 1, ..., m$$

the machine stator plus network R-L dynamic model is,

$$(3.15) \qquad \overline{V}_i = -(R_i + R_{ti})\overline{I}_i - j\nu_i X'_{di}\overline{I}_i + \overline{\beta}_i\overline{E}_i \quad i = 1, ..., m$$
$$(3.16) \qquad \overline{V}_i = -(R_i + R_{ti})\overline{I}_i - j\nu_i X_i\overline{I}_i \quad i = m+1, ..., b$$

These complex equations motivate the concept of transient algebraic circuits [2]. The transient algebraic circuit is a circuit written using the above complex variable notation (phasors, impedance, etc.) and obeying frequency domain laws (algebraic equations ($\overline{V} = \overline{ZI}$, etc.)). The transient algebraic circuit representations of this exact model are shown in Figure 3.1.

Substitution shows that the torque of electric origin for the swing equation is,

$$(3.17) \qquad T_{ELEC_i} = Real\left(\overline{E}_i\overline{I}_i^*\right) \qquad i = 1, ..., m$$
$$(3.18) \qquad = [E'_{qi} - X'_{di}I_{di}]I_{qi} + [E'_{di} + X'_{qi}I_{qi}]I_{di} \quad i = 1, ..., m$$

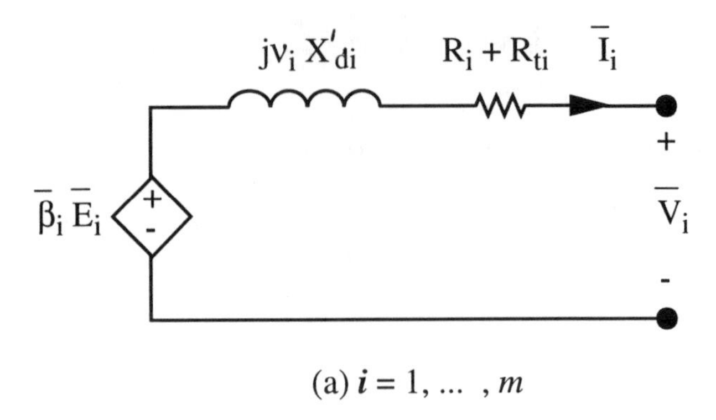

(a) $i = 1, \ldots, m$

$jv_i X_i$ $R_i + R_{ti}$ $\overline{I}_i$

- $\overline{V}_i$ +

(b) $i = m + 1, \ldots, b$

FIG. 3.1. *Exact transient algebraic circuit representations of stator/network/load dynamics*

We emphasize that these circuit representations do not imply anything about steady state or any simplifying assumptions. The dynamic model using an interconnection of these circuits replacing (1)-(10) is still an exact representation. The transient frequency ν_i, transient resistance R_{ti} and controlled source gain $\overline{\beta}_i$ are however admittedly somewhat unusual.

4. Frequency dependence. With speed governors, the steady-state speed will depend on the regulation droop R_{di}. Suppose the steady-state solution of the exact dynamic system has

$$\omega_{\underset{ss}{i}} = \omega_s + \Delta\omega \quad i = 1, \ldots, m \tag{4.1}$$

where ω_s corresponds to the nominal frequency and ω_{ss} to the steady-state frequency. All machine relative rotor angles in steady state will be

$$\delta_{\underset{ss}{i}} = \Delta\omega t + k_i \quad i = 1, \ldots, m \tag{4.2}$$

The machine variables in their respective rotor reference frames $(d_i q_i)$ will be constants whereas the system variables in the synchronously rotating reference frame $(D_i Q_i)$ will be sinusoidal with frequency $\Delta\omega$. The steady-state values of ν_i, R_{ti} and $\overline{\beta}_i$ are,

$$\nu_{\underset{ss}{i}} = 1 + \epsilon\Delta\omega \quad i = 1, \ldots, b \tag{4.3}$$

$$R_{\underset{ss}{ti}} = 0 \quad i = 1, ..., b \tag{4.4}$$

$$\overline{\beta}_{\underset{ss}{i}} = 1 + \epsilon\Delta\omega \quad i = 1, ..., m \tag{4.5}$$

When these are used with the transient algebraic circuits of Figure 3.1, they produce the correct steady-state equivalent circuit for phase a of each device.

The exact dynamic system has been written in two-time-scale form with ϵ being the small parameter which multiplies the derivatives of the fast variables. A good approximation of the slow dynamics can be obtained by simply setting ϵ to zero in all equations. When studying the system in the limit as ϵ goes to zero, the same variable symbols will be used with a subscript "o" added to signify the "zero-order" approximate system. The zero-order approximation of ν_i, R_{ti} and $\overline{\beta}_i$ are,

$$\nu_{oi} = 1 \quad i = 1, ..., b \tag{4.6}$$

$$R_{toi} = 0 \quad i = 1, ..., b \tag{4.7}$$

$$\overline{\beta}_{oi} = 1 \quad i = 1, ..., m \tag{4.8}$$

This zero-order model clearly loses all frequency dependence during transients and in steady state. An improved reduced-order slow model can be constructed to preserve the frequency dependence without resorting to integration of the fast dynamics. From equations (35)–(38), this is done by approximating the derivatives. While this can in principle be done analytically, an explicit numerical approach based on an integration time step of Δt at time t_k gives,

$$\nu_{1i}(t_k) \approx 1 + \epsilon\left(\frac{\phi_{1i}(t_k - \Delta t) - \phi_{1i}(t_k - 2\Delta t)}{\Delta t}\right) \quad i = 1, ..., b \tag{4.9}$$

$$R_{t1i}(t_k) \approx \frac{\epsilon X_i}{I_{1i}(t_k - \Delta t)}\left(\frac{I_{1i}(t_k - \Delta t) - I_{1i}(t_k - 2\Delta t)}{\Delta t}\right)$$
$$i = 1, ..., b \tag{4.10}$$

$$\overline{\beta}_{1i}(t_k) \approx 1 + \epsilon\left(\frac{\gamma_{1i}(t_k - \Delta t) - \gamma_{1i}(t_k - 2\Delta t)}{\Delta t}\right)$$
$$-j\frac{1}{E_{1i}(t_k - \Delta t)}\left(\frac{E_{1i}(t_k - \Delta t) - E_{1i}(t_k - 2\Delta t)}{\Delta t}\right)$$
$$= i = 1, ..., m \tag{4.11}$$

where the subscript "1" denotes a system where terms with ϵ to the first power are kept, but terms of order ϵ^2 are neglected.

5. Illustration. To illustrate the use of the transient algebraic circuits, it is necessary to perform a simulation with exact modeling of the R-L stator plus network and load transients. Since this is prohibitive for large systems, and since a single machine model will illustrate the result, we present the following comparison of dynamic models.

TABLE 5.1
Machine Parameters

X_d	X_q	X'_d	X'_q	T'_{do}	T'_{qo}
.8958	.8645	.1198	.1198	$6.(sec)$	$.535(sec)$
H	D	K_E	T_E	K_F	T_F
$12.(sec)$	0.	1.	$.314(sec)$	0.63	$.35(sec)$
K_A	T_A	K_{HP}	T_{RH}	T_{CH}	T_{SV}
20.	$.2(sec)$	.3	$8.(sec)$	$.05(sec)$	$.2(sec)$

The system used in this simulation used the parameters given in Table 5.1. It is a single synchronous machine having negligible stator resistance, $R_s = 0$, connected to an R-L load through a lossless transmission line with reactance equal to 0.1 pu. Initially the machine is at nominal frequency with no load and terminal voltage $V_{m1}\angle\theta_{a1} = 1\angle 0°$. At time $t = 0$ the load, $(R + jX) = (1.0 + j1.0)$, is switched in and the resulting transients are shown in Figures 5.1, 5.2, and 5.3.

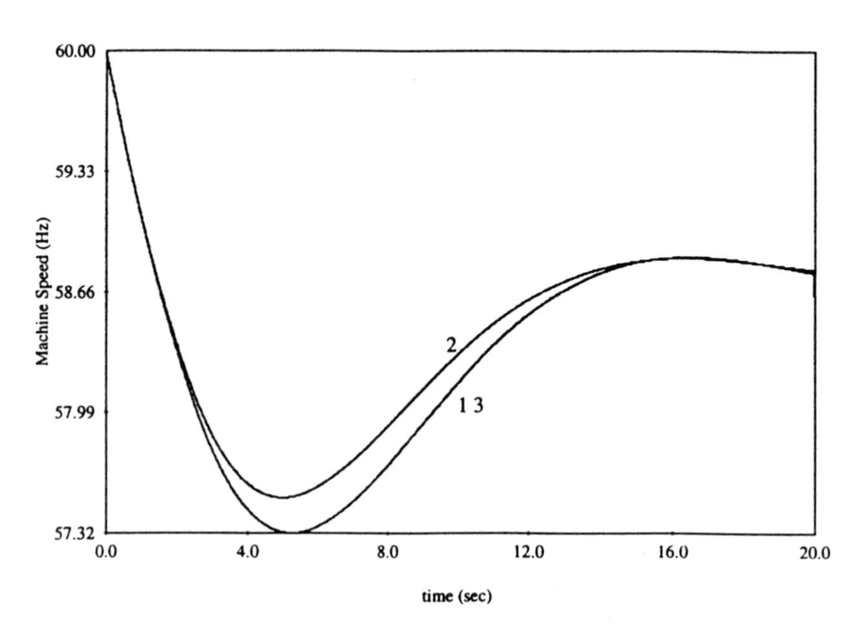

FIG. 5.1. *Machine speed (Hz) vs. time (sec)*

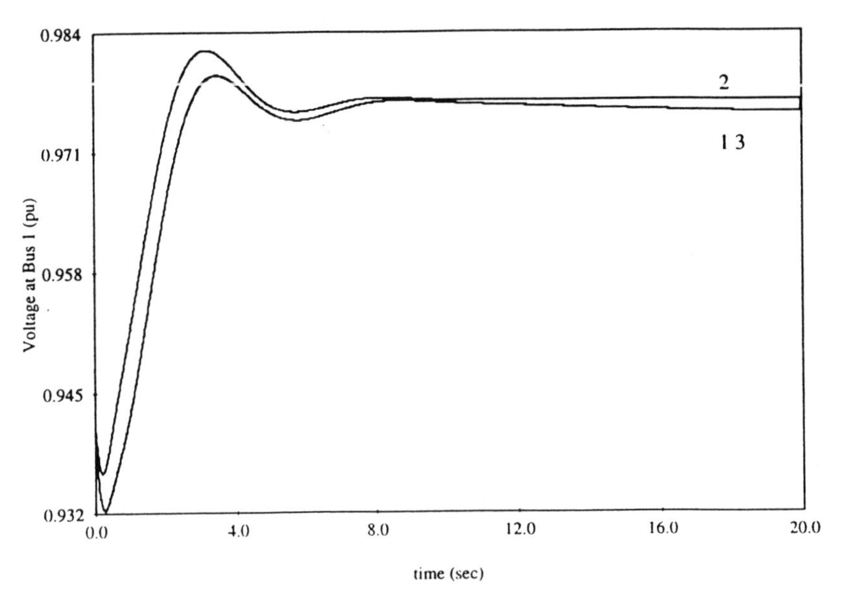

FIG. 5.2. *Voltage at bus 1 (pu) vs. time (sec)*

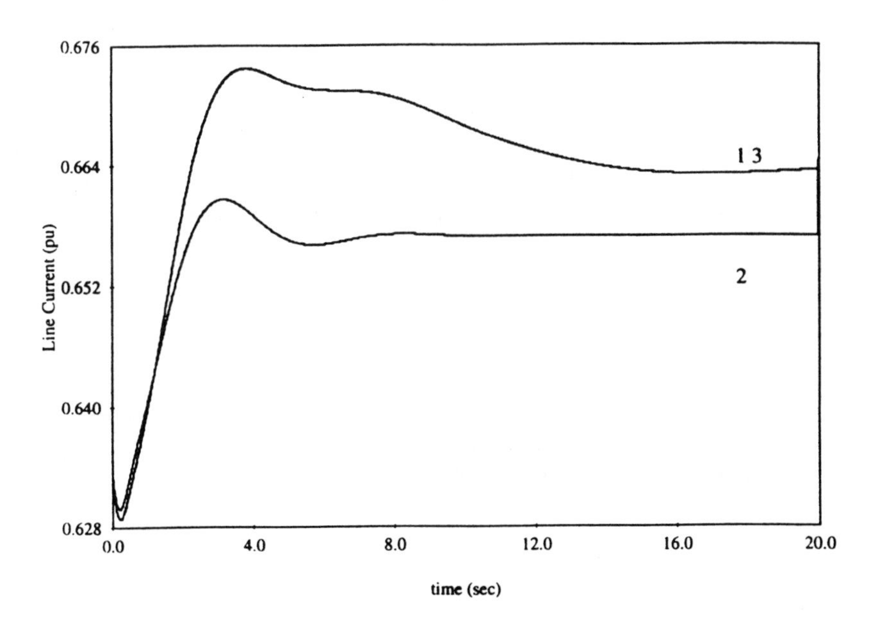

FIG. 5.3. *Line current (pu) vs. time (sec)*

This disturbance is simulated using the three different models discussed in this chapter. In Figures 5.1–5.3 the lines labeled "1," "2," and "3" correspond, respectively, to the exact model (Equations (2.1)–(2.24)), the simulation using the $\epsilon = 0$ approximation and the simulation using the $\epsilon^2 = 0, \epsilon \neq 0$ improved approximation. The machine speed in Hertz, the voltage at Bus 1, and the line current in per-unit are shown in Figures 5.1 – 5.3, respectively. All simulations use a trapezoidal rule integration scheme. The exact simulation uses a time step = 0.001 sec, and the other two simulations use a time step = 0.01 sec.

With the sudden increase in load, the machine slows down and remains below nominal frequency due to the droop in the governor regulation. This is observed in Figure 5.1. The increased load also gives rise to an increase in line current and a decrease in terminal voltage. This is shown in Figures 5.2 and 5.3. In Figures 5.2 and 5.3 the vertical scales are chosen to emphasize the long term dynamics and the steady-state error; the initial, fast transients do not appear. In the exact simulation, initially the voltage magnitude is 1.0 and the current magnitude is 0.0.

The $\epsilon^2 = 0, \epsilon \neq 0$ model does not produce any steady-state error. In this simulation it is also virtually indistinguishable from the exact model during the transient. One should note that the steady-state error is small, despite the large disturbance.

6. Static load model constraints. Power system engineers prefer to specify load constraints in terms of bus power. In terms of the synchronously rotating variables at each bus, this power is defined as,

$$(6.1) \qquad (V_{Di} + jV_{Qi})(I_{Di} - jI_{Qi}) \triangleq P_{Li} + jQ_{Li} \quad i = 1, ..., n$$

for an n bus system with injected current notation. For the case where ϵ is set to zero, this load model is typically defined as follows for the static case,

$$(6.2) \qquad P_{Li} = P_{Li}^{o} V_i^{k_{pi}} \quad i = 1, ..., n$$

$$(6.3) \qquad Q_{Li} = Q_{Li}^{o} V_i^{k_{qi}} \quad i = 1, ..., n$$

For the $\epsilon = 0$ cases, these algebraic load constraints make a set of algebraic equations which must be solved during a dynamic simulation at each time step. It has been shown [3] that for most practical power systems, these equations are always solvable provided,

$$(6.4) \qquad k_{pi} > 1 \quad i = 1, ..., n$$

$$(6.5) \qquad k_{qi} > 1 \quad i = 1, ..., n$$

6.1. Illustrations. Here we present a simple example in which the conditions above are made clear. Consider the single unity power factor

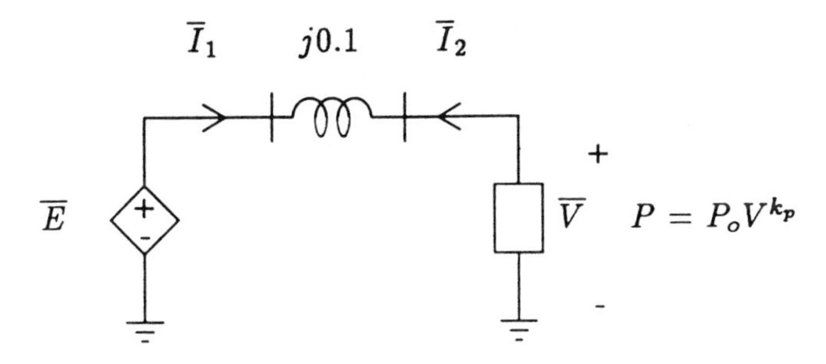

FIG. 6.1. *A single source/single load system*

load connected to a source through a lossless transmission line shown in Figure 6.1.

The network current equations describing this system are

$$\begin{bmatrix} \overline{I}_1 \\ \overline{I}_2 \end{bmatrix} = \begin{bmatrix} -j10 & j10 \\ j10 & -j10 \end{bmatrix} \begin{bmatrix} \overline{E} \\ \overline{V} \end{bmatrix}. \tag{6.6}$$

The load current is constrained by

$$-\overline{I}_2 = \frac{P_o V^{k_p}}{\overline{V}^*}. \tag{6.7}$$

Combining Equations (6.6) and (6.7) gives

$$\overline{I}_1 = -j10\overline{E} + j10\overline{V} \tag{6.8}$$

$$0 = P_o V^{k_p - 2}\overline{V} + j10\overline{E} - j10\overline{V}. \tag{6.9}$$

Using the Intermediate Value Theorem [4] we can provide bounds on the parameters to guarantee the existence of a solution to these equations. Rearranging Equation (6.9) and expressing in terms of voltage magnitudes gives

$$100E^2 = P_o^2 V^{2(k_p-1)} + 100V^2 \tag{6.10}$$

which may be written as a function in terms of the variable, V, and the known parameters and inputs, P_o, k_p, and E:

$$\begin{aligned} F(V) &\triangleq P_o^2 V^{2(k_p-1)} + 100V^2 - 100E^2 \\ &= 0. \end{aligned} \tag{6.11}$$

Consider the case when $P_o > 0$ and $k_p > 1$, then

$$F(0) = -100E^2 < 0, \tag{6.12}$$

$$F(E) = P_o^2 E^{2(k_p-1)} > 0. \tag{6.13}$$

Since $F(V)$ is continuous, by the Intermediate Value Theorem, a solution for $F(V) = 0$ must exist in the interval $0 < V < E$. Consider the case when $P > 0$ and $k_p = 1$, then

$$V^2 = E^2 - \frac{P_o^2}{100} \tag{6.14}$$

Clearly a real solution will only exist provided $P_o^2 \leq 100E^2$. We emphasize that a real solution for V always exists for any $P_o > 0$ with $k_p > 1$ but may not exist for $k_p \leq 1$. Assuming $E = 1$, the solutions for V as a function of k_p and P_o are shown in Figure 6.2. This simple example helps to illustrate the bounds on the load model such that a solution exists.

To demonstrate explicitly how these results affect dynamic analysis, we perform a nonlinear simulation on the IEEE 10-machine/39-bus system shown in [5]. Initially the system is at its base loading, and the generators are modeled using a two-axis model with IEEE type I voltage regulator and a third-order turbine/governor model.

At the base loading level, the two most heavily loaded transmission lines are those connecting bus 21 to bus 36 and bus 39 to bus 36. The simulation involves disconnecting these lines. At time $t = 0$ sec the system is in steady state. At time $t = 1$ sec the line connecting buses 21 and 36 is removed. At time $t = 2$ sec the line connecting buses 39 and 36 is removed. The bus with the lowest voltage in the system is bus 21. In Figure 6.3, the voltage at this bus is shown assuming different values for the load exponents. The trajectory labeled "$k = 0$" corresponds to the transients when all loads are modeled as constant power. The trajectory labeled "$k = 1$" corresponds to loads modeled as constant current magnitude at a constant power factor. The trajectory labeled "$k = 2$" corresponds to constant impedance loads. In the constant power case, the simulation fails to converge shortly after the second line is removed. The other two cases result in severe voltage conditions, yet the system is stable.

When the loads are modeled as constant power, the system dynamically approaches a bifurcation point of the algebraic equations. Since this corresponds to a bifurcation point of the algebraic equations at which the equations fail to exhibit a solution, this model is unacceptable. A model of any physical system must exhibit a solution. Thus, the simulations denoted by "$k = 1$" and "$k = 2$" correspond to possible power system transients. The simulation denoted by "$k = 0$" represents an impossible system transient since it fails to depict the transient after finite time.

One may argue that the constant power load model (or any model $k_p, k_q \leq 1$) is valid for some (high) voltage levels but is not valid over the complete range of possible transient conditions. These types of models are typically used to represent induction motor presence in a power system. For voltage instability studies in which low voltages transients are expected, it is necessary to include more detailed models of induction motors to capture low voltage transient induction motor phenonena.

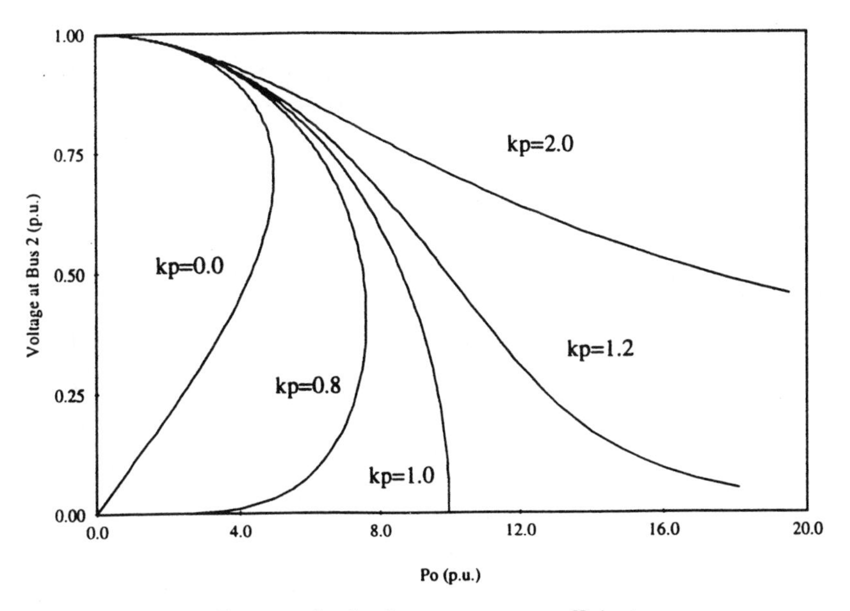

FIG. 6.2. *Load voltage vs. power coefficient*

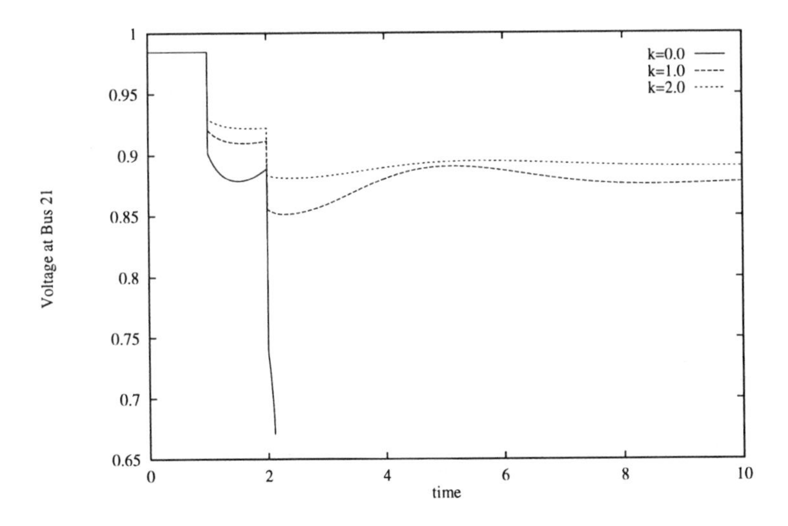

FIG. 6.3. *Low voltage after line outages*

7. Maximum loadability. With regard to voltage instabilities, maximum loadability is often viewed in terms of power/voltage curves. Consider the trivial circuit shown in Figure 7.1 with a single source, single line connected to a resistor bank. By switching in a different number of resistors one obtains data for the voltage and power consumed at the load bus. Plotting voltage vs. power gives the P/V curve shown in Figure 7.2.

There exists a theoretical, absolute maximum value of power that can be delivered to the load. Can this level of load be achieved? Are the dynamics of the system stable?

In 1975, Venikov et al. published a paper in which it was shown that an eigenvalue of the dynamic model changes sign when the load flow Jacobian becomes singular [6]. (This topic is also discussed in [7].) This corresponds to a point of maximum complex power transfer in the standard load flow. With regard to the type of P/V curve shown in Figure 7.2, this indicates that the upper half is stable, the lower half is unstable. This is intuitively appealing because an increase in P on the top half corresponds to a decrease in V, while an increase in P on the bottom half corresponds to an increase in V; however, if one views an increase in load as adding a resistor, as in Figure 7.1, an increase in load always corresponds to a decrease in voltage.

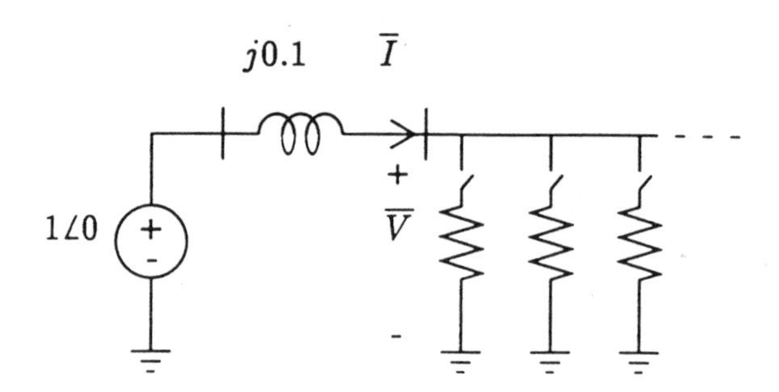

FIG. 7.1. *Single source, resistive load*

The dynamic model used by Venikov et al. is very simple. The generators are a variation of a classical model (they use a flux decay model; however, the field dynamics are eliminated by assuming a constant terminal voltage), and as in a standard load flow, the loads are considered to be constant power.

Will the same or similar result hold if more realistic generators and load models are used?

7.1. Static load models. Consider the system in which the stator/network/load model is given in Figure 7.3 and the machine is modeled by using the $\epsilon = 0$ model. Examine the steady-state stability for various

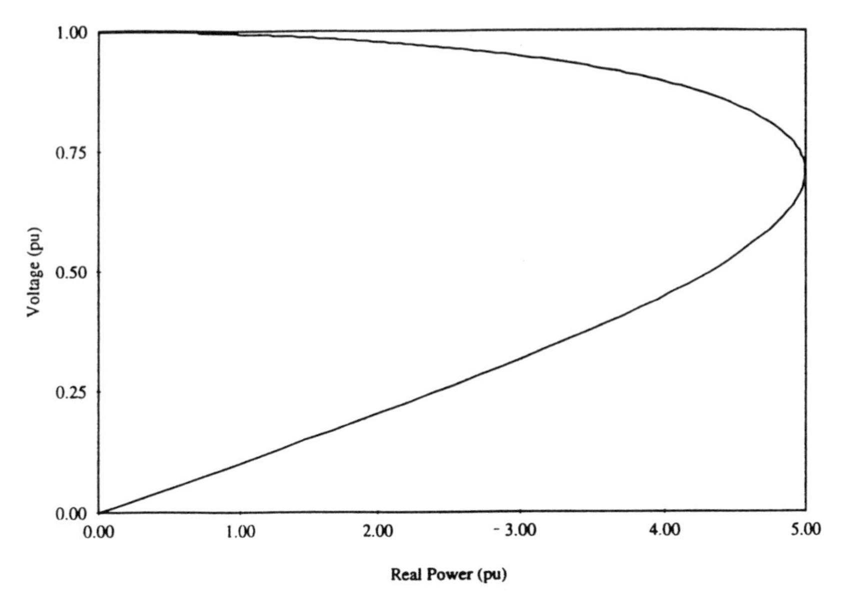

FIG. 7.2. *P/V curve*

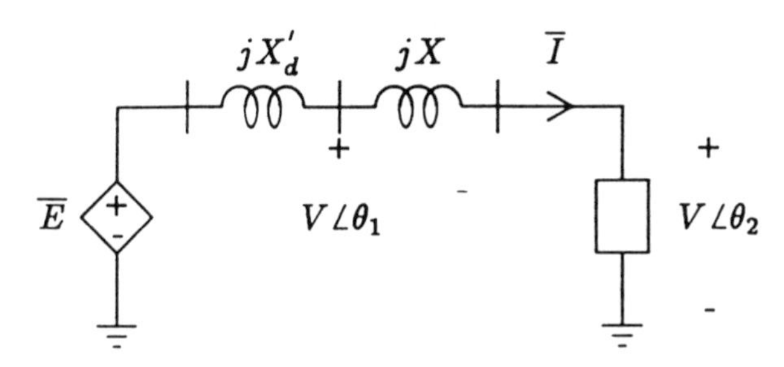

FIG. 7.3. *Stator/network/load model*

levels of load modeled by an exponential voltage relation,

$$P_L + jQ_L = P_o V^{k_p} + jQ_o V^{k_q} \tag{7.1}$$

for varying values of k_p and k_q. With each operating point computed using a terminal voltage magnitude equal to 1 pu and the reactance of the transmission line equal to 0.1 pu, the voltage vs. power curve is easily obtained for a unity power factor load ($Q_o = 0$). This is shown in Figures 7.4–7.7 for values of k_p ranging from 0 to 2; from an electrical network point of view, each point represents a valid operating point.

The regions of these curves over which the dynamic system is stable/unstable were examined by observing the eigenvalues of the linearized synchronous machine model at each point. In the cases of $0 \leq k_p < 1$ stable and unstable regions exist. Using eigenvalue sensitivity analysis for the

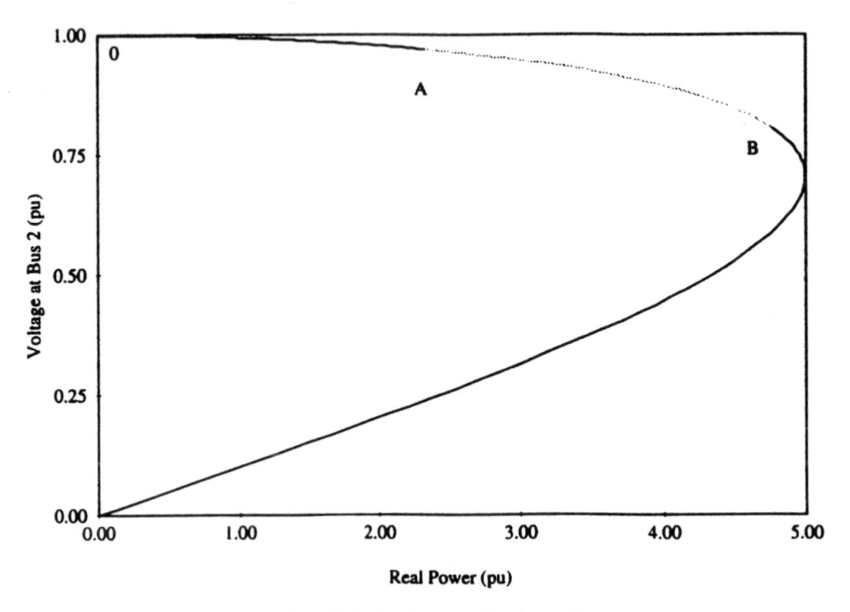

FIG. 7.4. *P-V curve for* $k_p = 0$

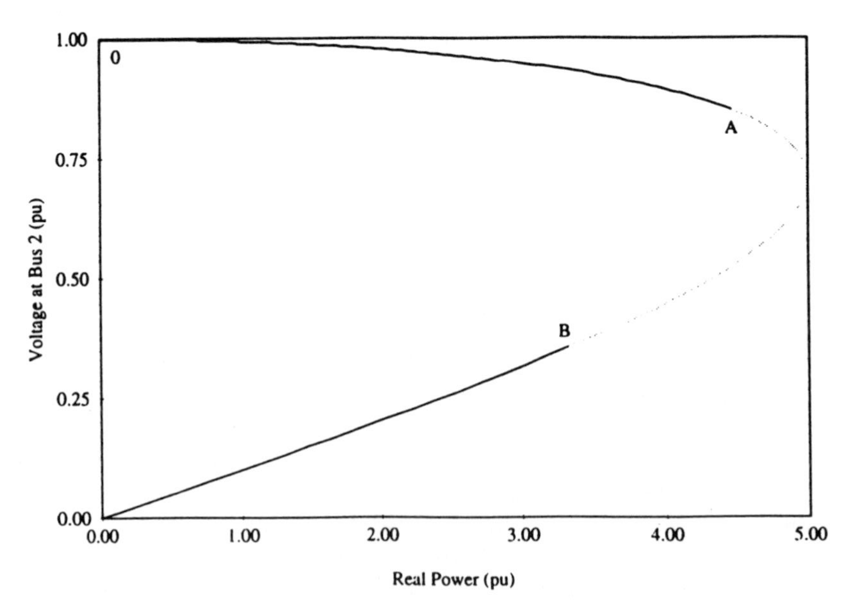

FIG. 7.5. *P-V curve for* $k_p = 0.90$

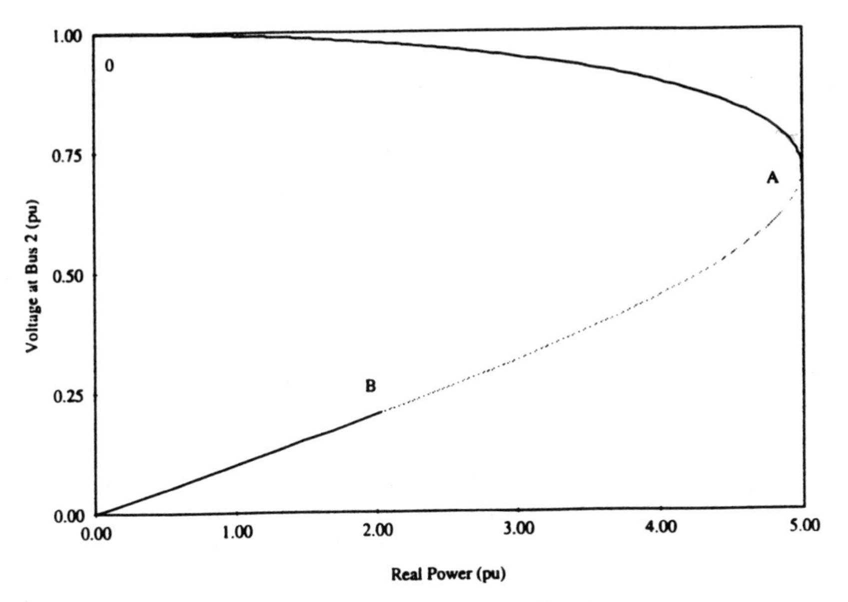

FIG. 7.6. *P-V curve for* $k_p = 0.96$

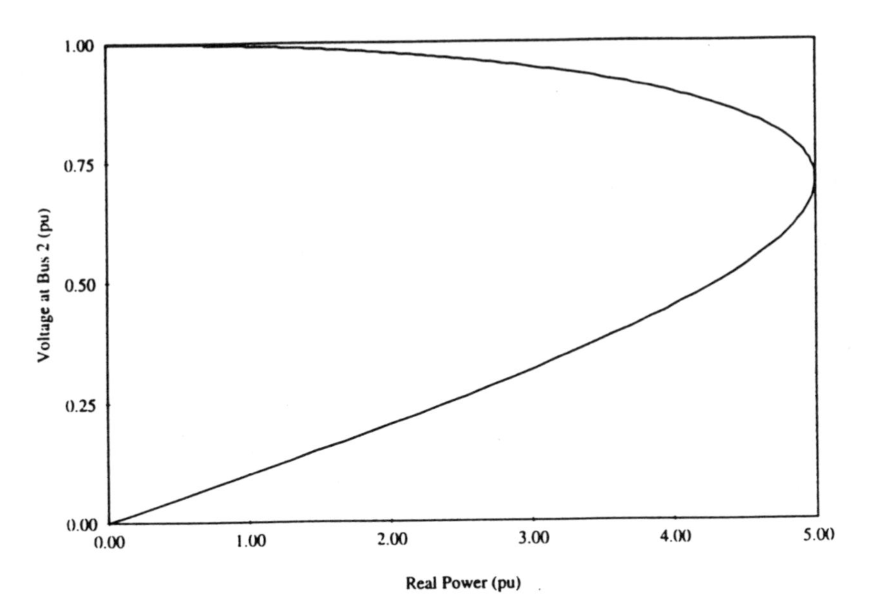

FIG. 7.7. *P-V curve for* $1 < k_p \leq 2$

constant power case ($k_p = 0$), it is determined that the variables associated with the unstable eigenvalues are E'_q, and R_f. Parametrically changing the load from 0.0 (no load) to 2.36 pu, the system is stable. In Figure 7.4 this corresponds to the portion of the curve from "O" to "A." The system loses steady-state stability at point "A." The system becomes stable again at a loading of 4.79 pu (point "B") which is before the point of maximum power transfer (power level of 5.0 pu). It is also observed that the entire lower portion of the curve in Figure 7.4 is dynamically stable. Thus, the point of maximum stable loading is $P_L = 2.36$, although if it could be reached, the system is stable for a loading at the maximum power transfer level of 5.0.

As k_p is increased, the portion of the curve that is unstable shifts from the top half to the bottom half and finally disappears completely. This is shown in Figures 7.5–7.7. In Figure 7.5, $k_p = 0.90$ and the unstable region extends from $P_L = 4.49$ on the top half (point "A") to $P_L = 3.37$ on the bottom half (point "B"). In Figure 7.6, $k_p = 0.96$ and the unstable portion is confined to the bottom half of the curve, between $P_L = 4.99$ (point "A") and $P_L = 2.03$ (point "B"). Upon further increase in k_p, the unstable portion quickly descends the curve and disappears at $k_p = 1$. In Figure 7.7 it is shown that the entire PV curve is stable for values of k_p between 1 and 2.

It is interesting to note that in each curve with $k_p < 1$ there exists a point at which the Jacobian of the algebraic equations (including the stator equations) is singular. Correspondingly, there is an eigenvalue that passes through infinity. To understand this better, consider the abstract differential/algebraic representation of the system:

$$\frac{dx}{dt} = f(x, y) \tag{7.2}$$

$$0 = g(x, y) \tag{7.3}$$

The linearized form of these equations is

$$\frac{d\Delta x}{dt} = A\Delta x + B\Delta y \tag{7.4}$$

$$0 = C\Delta x + D\Delta y \tag{7.5}$$

The reduced dynamic system is obtained by eliminating the algebraic variables,

$$\Delta y = -D^{-1}C\,\Delta x \tag{7.6}$$

Thus,

$$\frac{d\Delta x}{dt} = \tilde{A}\Delta x \tag{7.7}$$

where

$$\tilde{A} = A - BD^{-1}C \tag{7.8}$$

As the Jacobian of the algebraic equations, D, becomes singular, at least some elements in D^{-1} become unbounded. Depending upon the matrices B and C, these unbounded elements may appear in the system matrix, $\tilde{A}$, resulting in an infinite eigenvalue.

Since the presence of this eigenvalue comes from the algebraic equations, it is not affected by changes in the control parameters of the machine. For systems with load exponents $k < 1$, some points will exist at which an eigenvalue passes through infinity. Since this results in a change in the number of eigenvalues in the right-half plane, the entire P/V curve can not be made stable. It is not meaningful to view the P/V curve after this point.

7.2. Dynamic load modeling. In this section we examine bifurcations in the total dynamic system model using four different loads: a constant power model and an induction motor model with three different parameter sets.

The test systems are shown in Figures 7.8 and 7.9. Boths systems consist of a single generator connected to a single load through a lossless transmission line. In Figure 7.8 the load is constant active power.. In Figure 7.9 the load is a compensated induction motor such that at a given operating point the resistor consumes 50% of the active power, the induction motor consumes 50% of the active power, and the shunt capacitance provides 100% compensation for reactive losses in the induction motor. The generator is represented by a two-axis model with an IEEE type I voltage regulator and a third-order turbine and governor model. The induction motor is represented by the following third-order model in which the mechanical torque load is assumed to be linear to rotational speed:

$$V_D = E_D' - X' I_{Qs} \tag{7.9}$$

$$V_Q = E_Q' + X' I_{Ds} \tag{7.10}$$

$$T_o' \frac{dE_Q'}{dt} = -E_Q' + \frac{X_m^2}{X_r} I_{Ds} - s\frac{X_r}{R_r} E_D' \tag{7.11}$$

$$T_o' \frac{dE_D'}{dt} = -E_D' - \frac{X_m^2}{X_r} I_{Qs} + s\frac{X_r}{R_r} E_Q' \tag{7.12}$$

$$2H\frac{ds}{dt} = K_L(1-s) - (E_Q' I_{Qs} + E_D' I_{Ds}) \tag{7.13}$$

The parameters for the 3HP, 50HP and 500HP induction motors used in these examples are given in Table 7.1.

The dimension of even these simple power system models is high ($\geq$ 10). Detection of global bifurcations is difficult, so we begin by examining local bifurcations which can be detected from an eigenvalue analysis. In each case, the constant power load and the 3HP, 50HP, and 500HP compensated induction motor loads, the total active power of the load is increased until an eigenvalue crosses the imaginary axis of the complex plane. The

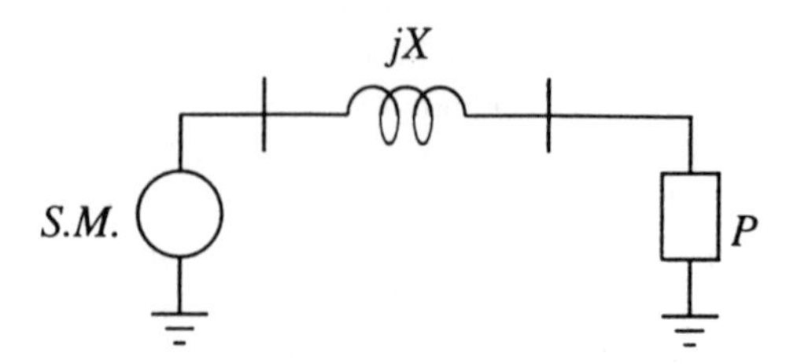

FIG. 7.8. *Single machine, constant active power load*

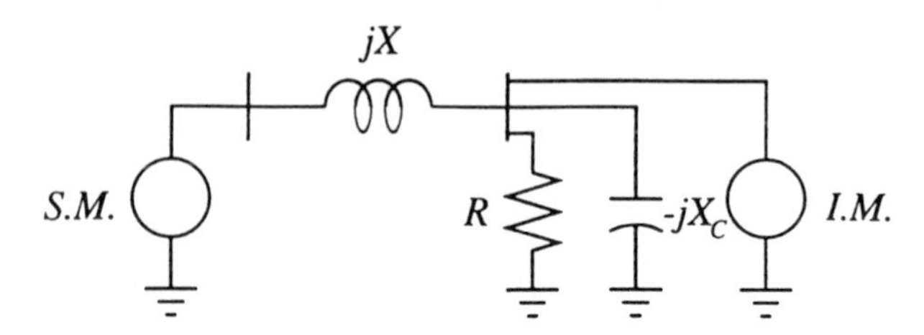

FIG. 7.9. *Compensated induction motor, unity power factor load*

eigenvalues at this point are given in Table 7.2, and the participation factors corresponding to the critical eigenvalues are shown in Table 7.3.

It is interesting to note the significance of Tables 7.2 and 7.3. Although the instability boundary occurs at different loadings for the constant power and the three different induction motor examples, qualitatively they are similar. The eigenvalues of the linearized systems have comparable values and the critical unstable eigenvalues are strongly related to the E_q' and R_f states in the generator model. The constant power model differs from the induction motor models by a large change in the dynamic form. The induction motor models differ from each other only by changes in parameter values. It is instructive to examine the qualitative behavior of the flow around the unstable points. The critical eigenvalues are a complex pair which cross the $j\omega$ axis at some point other than the origin. This is called

TABLE 7.1
Induction motor parameters

HP	R_r (pu)	X_r (pu)	X_m (pu)	X' (pu)	T_o' (sec)	H (sec)
3	0.0377	1.2429	1.2080	0.0688	0.0875	0.7065
50	0.0402	2.3590	2.3058	0.1052	0.1557	0.7915
500	0.0132	3.8943	3.8092	0.1683	0.7826	0.5269

TABLE 7.2
Eigenvalues at Critical Levels of Load

Constant P $P_L = 2.36$	3 HP $P_L = 3.16pu$	50 HP $P_L = 2.72pu$	500 HP $P_L = 2.21pu$
$0.0010 \pm j1.854$	$0.0042 \pm j1.906$	$0.0144 \pm j1.867$	$0.0275 \pm j1.770$
$-0.1852 \pm j0.2770$	$-0.2067 \pm j0.2699$	$-0.2029 \pm j0.2687$	$-0.2006 \pm j0.2698$
-2.963	-2.165	-2.069	-1.932
-4.673	-4.698	-4.701	-4.687
$-5.153 \pm j7.643$	$-5.277 \pm j7.763$	$-5.263 \pm j7.767$	$-5.244 \pm j7.775$
-20.08	-20.08	-20.08	-20.08
-0.000	0.000	0.000	0.000
N/A	-10.96	-10.44	-13.25
N/A	$-47.41 \pm j48.57$	$-44.43 \pm j40.02$	$-8.938 \pm j28.48$

TABLE 7.3
Participation Factors for the Unstable Eigenvalues

	P	3 HP	50 HP	500 HP
δ	0.000	0.036	0.032	0.013
ω	0.000	0.038	0.034	0.014
E_q'	0.439	0.287	0.293	0.227
E_d'	0.124	0.094	0.103	0.077
E_{fd}	0.101	0.061	0.062	0.046
R_f	0.267	0.159	0.162	0.122
V_r	0.070	0.043	0.043	0.032
T_m	0.000	0.005	0.004	0.002
P_{ch}	0.000	0.005	0.004	0.002
P_{sv}	0.000	0.002	0.002	0.001
E_D'		0.126	0.123	0.257
E_Q'		0.083	0.082	0.196
s		0.060	0.056	0.012

a "Hopf bifurcation." It corresponds to an intersection of an equilibrium point (a limit set) with a limit cycle (another limit set). It is expected that in the vicinity of this bifurcation either stable or unstable limit cycles should exist. It is called "subcritical" if the limit cycles are unstable and "supercritical" if the limit cycles are stable.

The Hopf bifurcation for the constant power case has been studied and is subcritical [8]. This means that, as the loading is increased towards the bifurcation point, there is an unstable limit cycle which bounds the region of attraction of the stable system. Because there are ten state variables in the model, the entire phase plane can not be visualized. Since the participation factors indicate that the critical mode is associated with the $E_q^{'}$ and R_f variables, view the $E_q^{'} - R_f$ plane, keeping in mind that this is only a cross section of the entire state space. To examine the flow in this plane, a perturbation in the value of $E_q^{'}$ is introduced and the resulting trajectories are plotted. Starting at a loading of $P_L = 2.35$ pu the flow in Figure 7.10 indicates an unstable limit cycle around the locally stable equilibrium point. Changing the loading to $P_L = 2.37$ pu, the flow in Figure 7.11 does not show any limit cycle and the equilibrium point is unstable. This is consistent with the report that this Hopf bifurcation is subcritical.

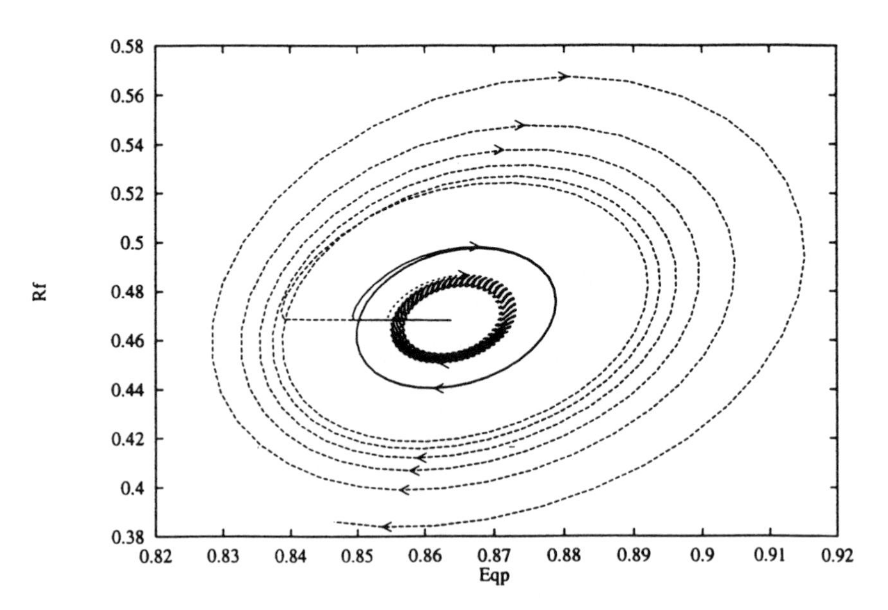

FIG. 7.10. *The constant power case,* $P_L = 2.35$

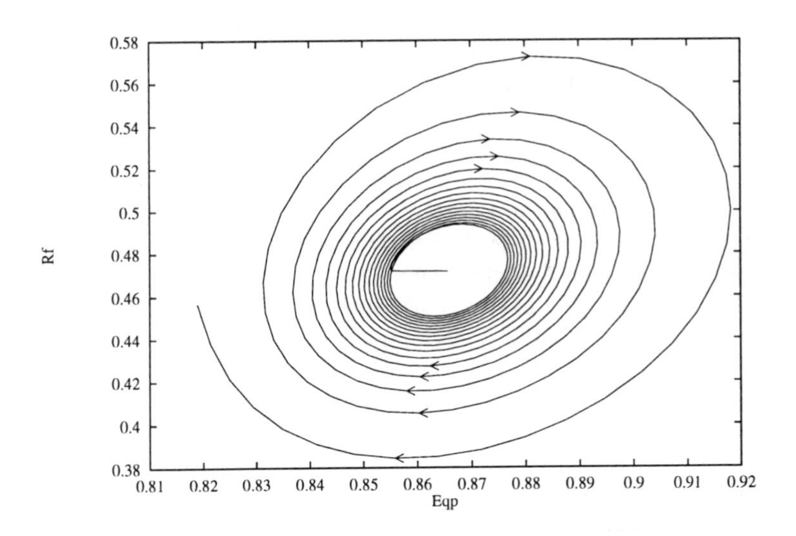

FIG. 7.11. *The constant power case,* $P_L = 2.37$

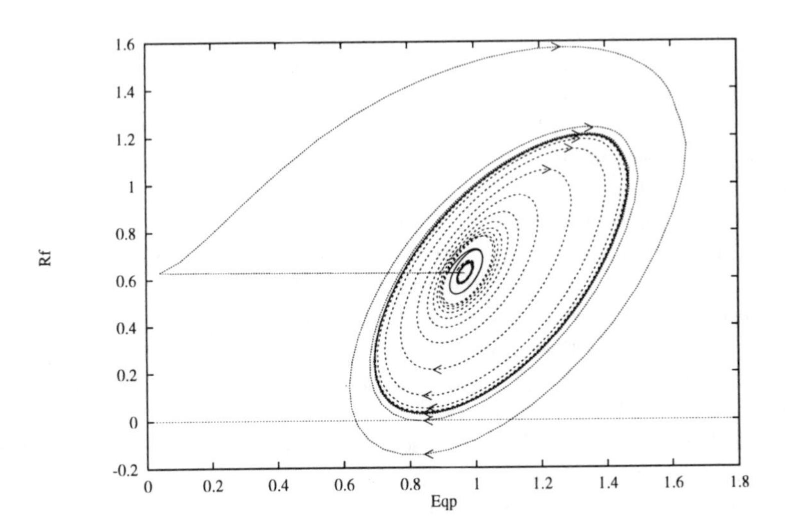

FIG. 7.12. *The 3 HP model,* $P_L = 3.15$

Now examine the 3 HP induction motor example. The participation factors indicate that again the E_q' and R_f variables greatly affect the critical mode. At a loading of $P_L = 3.15$ pu, the flow is examined by introducing a perturbation in E_q'. The trajectory in Figure 7.12 indicates that the equilibrium point is stable and is surrounded by two limit cycles, one stable and one unstable. (In Figure 7.13 the unstable limit cycle and stable trajectory in the near vicinity of the equilibrium point are shown in greater detail.) At a loading of $P_L = 3.17$ the trajectories in Figure 7.14 indicate that there is one stable limit cycle around the unstable equilibrium point. The Hopf bifurcation is still subcritical; however, the presence of the stable limit cycle makes the dynamics significantly different from those for the constant power case. Physically this means that the power system will exhibit sustained oscillations, either if the system is unstable or if a large disturbance is applied to the stable system.

Now consider an induction motor load using the 500 HP parameters. To be consistent, view the $E_q' - R_f$ plane. At loading of $P_L = 2.20$ pu, in Figure 7.15, an unstable limit cycle exists around the stable equilibrium. At a loading of $P_L = 2.22$ pu, in Figure 7.16, the equilibrium point is unstable with no limit cycles. In both Figures 7.15 and 7.16, the unstable trajectories tend toward a stable equilibrium point. This point is not properly viewed in the $E_q' - R_f$ plane because other states vary greatly. This stable equilibrium corresponds to the state of the systems after the motor stalls; the slip is high, the generator accelerates, and the system voltages are low.

The differences between the flow in the phase planes are important. One of the traditional justifications for the constant power load model is the presence of induction motors. This may even be supported, in part, by the standard stability studies in which the eigenvalues of both models indicate an instability on the top half of the P-V curve. In addition, the participation factors indicate that this instability is associated with the E_q' and R_f variables in the generator. However, the flow in state space shows that they are fundamentally different. As the constant power load is increased, the region of attraction around the equilibrium point shrinks, and after the bifurcation, the system is unstable with no limit cycles. When the eigenvalue becomes positive using the 3 HP induction motor parameters, the system exhibits stable limit cycles which appear as oscillations in the power system. The choice of load model greatly affects the dynamic behavior of the system. When the 500 HP parameters are used the system behavior more closely resembles the constant power load model than the small induction motor model. In addition to the choice of model, the choice of parameters is important.

Consider again the limit cycle that appears in the system using 3 HP induction motor parameters. The limit cycle is not present when the 500 HP parameters are used. In both cases the eigenvalues are qualitatively the same; one complex pair of eigenvalues indicates the instability. If one continuously varies the induction motor parameters from those of the 3

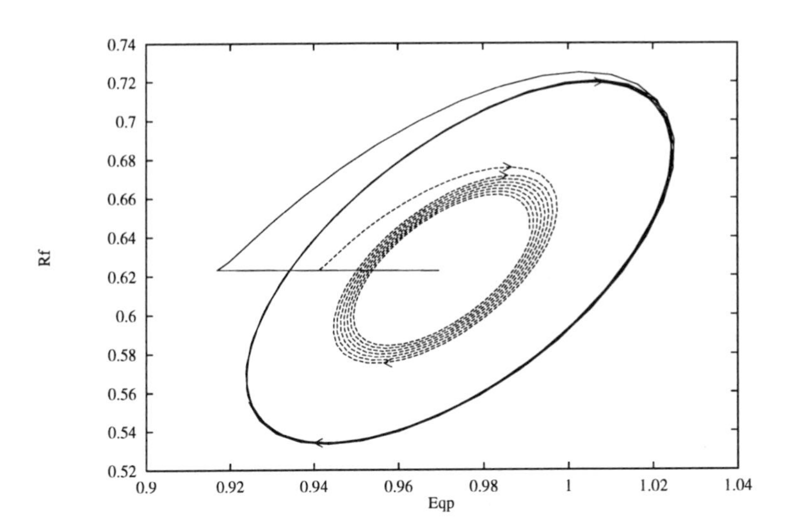

FIG. 7.13. *The 3 HP model,* $P_L = 3.15$

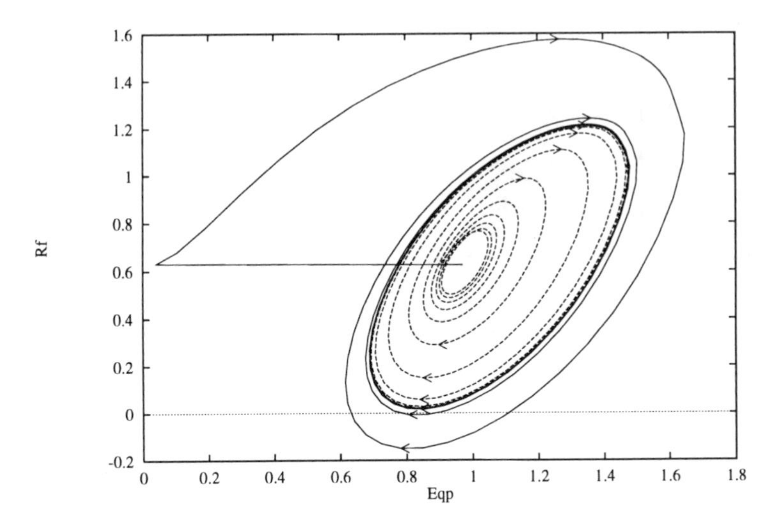

FIG. 7.14. *The 3 HP model,* $P_L = 3.17$

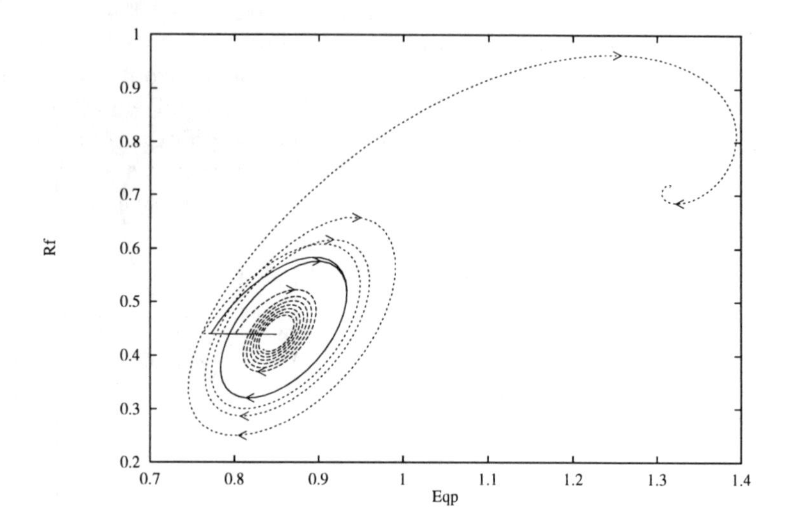

FIG. 7.15. *The 500 HP model,* $P_L = 2.20$

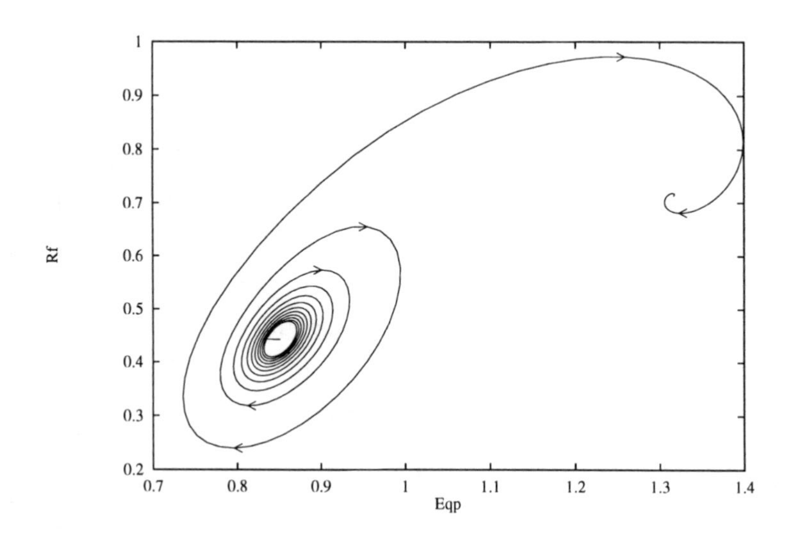

FIG. 7.16. *500 HP model,* $P_L = 2.22$

HP motor to those of the 500 HP motor, then there must be some point at which the limit cycle disappears. The importance of this is that the eigenvalues are qualitatively the same, yet the dynamic behavior in state space will fundamentally change. This may indicate the presence of a global bifurcation.

7.3. Global bifurcation with 3HP ind. motor model. To investigate a different global bifurcation, we examine the 3 HP motor load over a wide range of levels of load and observe the flow in the $E_q^{'} - R_f$ plane. At each of the different levels of load a perturbation in $E_q^{'}$ is applied and the resulting trajectories are observed. For low levels of load the operating point appears to be globally stable. At a loading of 1.0 pu shown in Figure 7.17 the trajectory approaches the equilibrium point in a damped oscillatory manner. In Figure 7.18 the trajectory is shown for a system with a level of load equal to 2.5 pu. It also exhibits oscillatory behavior although it is less damped.

The system behavior changes significantly at a loading of 2.76 pu. At a loading of 2.75 pu the flow is shown in Figure 7.19. The equilibrium point is still stable in the large; however, the flow appears compressed in an elliptical area before decaying to the equilibrium. At a level of load equal to 2.77 shown in Figure 7.20, there exist two limit cycles enclosing the equilibrium point. (In Figure 7.20, the trajectories between the limit cycles are omitted for clarity.) The outer limit cycle is stable and the inner one is unstable. As the load is increased further the unstable limit cycle decreases in size and disappears at level of load equal to 3.17 pu. This is the hopf-bifurcation noted earlier. As the loading is increased past the hopf-bifurcation the stable limit cycle persists.

The system is structurally unstable at the point at which the two limit cycles appear. It marks a qualitative change in the system behavior. This structural instability is marked by a global bifurcation. The presence of the limit cycles at this point is not reflected in any change in the local behavior around the equilibrium point. A graph summarizing this evolution is shown in Figure 7.21. In region A the system is stable in the large. At point B, two limit cycles appear as a result of a global bifurcation. At point C the hopf-bifurcation occurs where the unstable limit cycle disappears. In region D, the operating point is unstable but the behavior of the system is bounded by the stable limit cycle. The system will experience sustained oscillations in this region.

Such oscillations have been encountered in real power systems and there is concern that power systems may exhibit chaotic behavior. Research into the effects of load modeling needs to be pursued further to gain a full understanding of the critical phenomena.

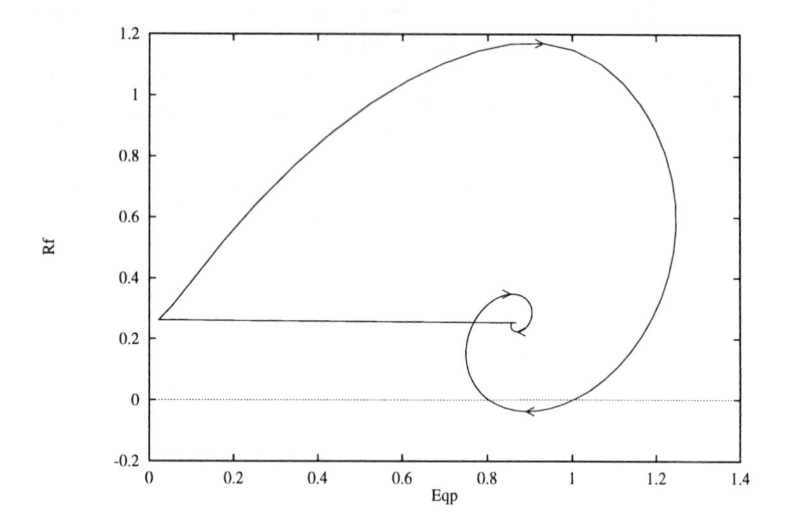

FIG. 7.17. *3 HP motor,* $P_L = 1.0$

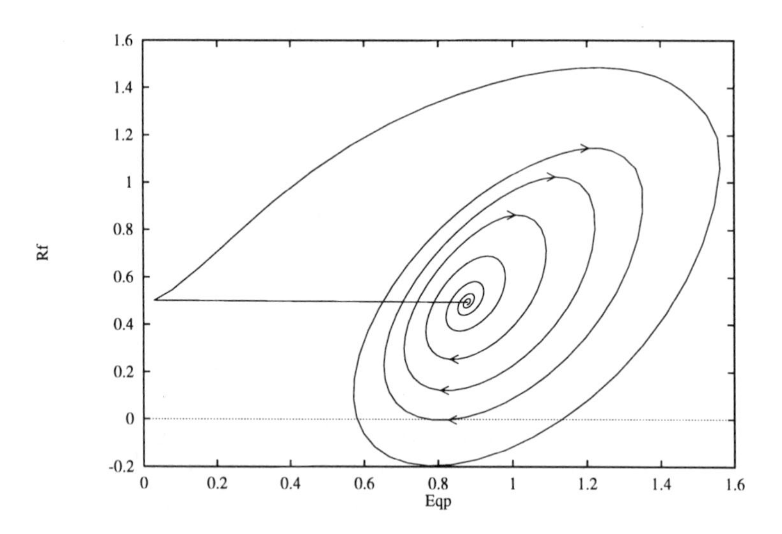

FIG. 7.18. *3 HP motor,* $P_L = 2.5$

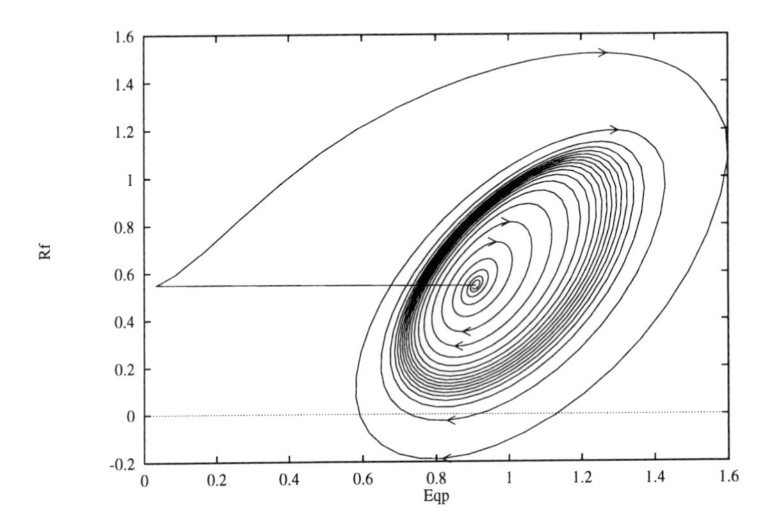

FIG. 7.19. *3 HP motor*, $P_L = 2.75$

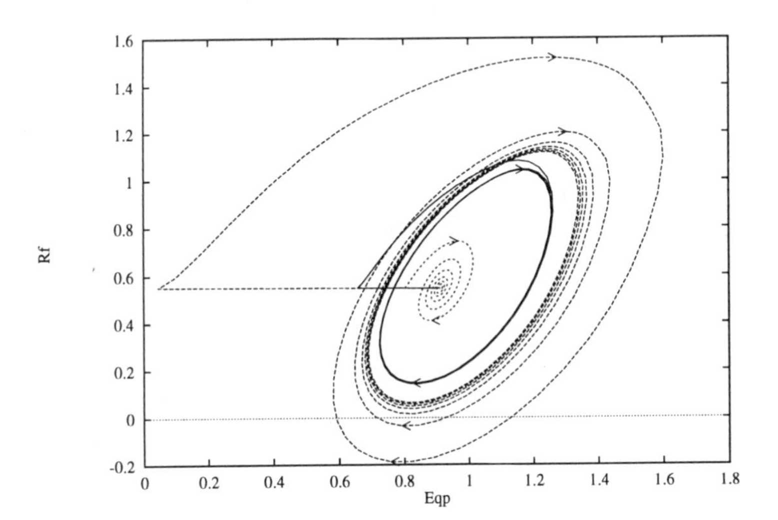

FIG. 7.20. *3 HP motor*, $P_L = 2.77$

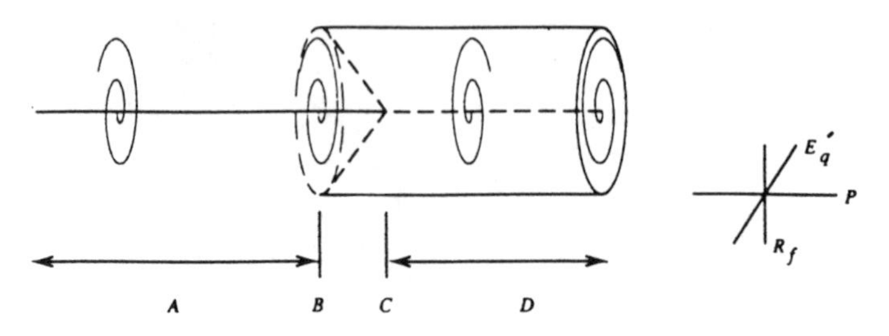

FIG. 7.21. *Evolution diagram*

8. Conclusion. The challenge of load modeling begins with the proper treatment and understanding of the time-scale nature of power system dynamics. When the fast stator/network/load dynamics are neglected in a zero-order slow model, the algebraic equations must always exhibit a solution during transients. This work has reported on the time-scale structure of power systems and the constraints which must be imposed on static load models using exponential forms involving bus voltage magnitude.

In order to extend load modeling to include dynamic properties, it is necessary to understand which dynamic property is proper for aggregate loads. This work has reported on the different dynamic properties of aggregate induction motors and the differences between small and large machine dynamics. The small machine aggregate introduces dynamics which are fundamentally different from the large machine aggregate. This difference lies in the presence of stable limit cycles surrounding unstable equilibrium points.

REFERENCES

[1] P. W. Sauer and M. A. Pai, "Modeling and simulation of multi-machine power system dynamics," *Control and Dynamic Systems: Advances in Theory and Application* (C. T. Leondes, Ed.), vol. 43, Academic Press, San Diego, CA, 1991.

[2] P. W. Sauer, B. C. Lesieutre and M. A. Pai, "Transient algebraic circuits for power system dynamic modelling," to appear, Int. Journal of Electrical Power & Energy Systems.

[3] B. C. Lesieutre, P. W. Sauer and M. A. Pai, "Sufficient conditions on static load models for network solvability," *Proceedings of the Twenty-Fourth Annual North American Power Symposium*, Reno, NV, Oct. 5-6, 1992, pp. 262-271.

[4] W. Kaplan, *Advanced Calculus*, Addison-Wesley Publishing Co., Reading, MA, 1984.

[5] B. C. Lesieutre, "Network and load modeling for power system dynamic analysis," Technical Report PAP-TR 93-2, Dept. of Electrical & Computer Engineering, University of Illinois at Urbana-Champaign, Urbana, IL, Jan. 1993.

[6] V. A. Venikov, V. A. Stroev, V. I. Idelchick and V. I. Tarasov, "Estimation of electrical power-system steady-state stability," *IEEE Trans. Power Appar. Syst.*, vol. PAS-94, pp. 1034-1040, May/June 1975.

[7] P. W. Sauer and M. A. Pai, "Power system steady-state stability and the load-flow Jacobian," *IEEE Transactions on Power Systems*, vol. 5, no. 4, November 1990, pp. 1374-1383.

[8] C. Rajagopalan, P. W. Sauer and M. A. Pai, "Analysis of voltage control systems exhibiting Hopf bifurcation," *Proceedings 28th IEEE Conference on Decision and Control*, Tampa, FL, Dec. 13-15, 1989, pp. 332-335.

PARALLEL SOLUTIONS OF LINEAR EQUATIONS BY OVERLAPPING EPSILON DECOMPOSITIONS

D.D. ŠILJAK* AND A.I. ZEČEVIĆ*

Abstract. New properties of overlapping epsilon decompositions are established, resulting in efficient algorithms for parallel solutions of large systems of linear equations. Applications to the load flow problem are presented to demonstrate the flexibility and speedups available in multiprocessor environment.

1. Introduction. The recent emergence of multiprocessor architectures has introduced new and still largely unexplored dimensions to computational problems of large systems, creating possibilities for significant speedups of the solution process. In order to adequately exploit the advantages of parallel processing, it is essential to be able to map the problem onto a set of processors in such a way that both good load balance and low inter-processor communication are achieved (Gallivan et al., 1990). With this in mind, overlapping epsilon decompositions have been proposed (Sezer and Šiljak, 1991) to identify weakly coupled (possible overlapping) blocks of a large matrix, which can then be assigned to individual processors. The coupling threshold ϵ, being directly proportional to interprocessor communication, can be selected to balance the size of the blocks (representing the processor load) with the communication overhead.

The main objective of this paper is to utilize overlapping epsilon decompositions in parallel solutions of linear equations. New properties of the decomposition algorithms are established to form a basis for efficient computations involving overlapping blocks. It is argued that the proposed decompositions provide an ideal framework for solving load flow problems on parallel machines. To this effect, calculations on the Intel iPSC/860 hypercube are presented for the IEEE 118 bus system and the 888 bus Tokyo electric power network.

In this paper we will be concerned with partitioning a matrix A as

$$A = A_0 + \epsilon A_1 \tag{1.1}$$

where A_0 is a matrix suitable for parallel inversion, A_1 has elements with absolute value less than or equal to one, and $\epsilon \geq 0$ is a sufficiently small number. Epsilon decompositions (1.1) are of particular interest in iterative methods, since the convergence of the iterates is *not* violated when the weak coupling term ϵA_1 is discarded (Zečević and Šiljak, 1994). Removal of this term is clearly desirable from a computational standpoint, observing that only the *easily invertible* matrix A_0 now remains to be factorized. An additional benefit of such a procedure is the *sparsification* of the original

* Department of Electrical Engineering, Santa Clara University, Santa Clara, CA 95053.

problem; with respect to this property, a dense (non-sparse) matrix A with decomposition (1.1) in which A_0 is *sparse* is referred to as an *epsilon sparse* matrix.

Among the possible structures of matrix A_0 a block diagonal form is considered particularly desirable, since it allows for each diagonal block to be factorized *independently* by its own designated processor. Whenever such a partitioning exists, the original matrix A is said to have an *epsilon decomposition* and the algorithm presented in (Zečević , 1993) is always capable of detecting it. This algorithm is actually designed to produce an even more general partitioning of matrix A known as an *overlapping epsilon decomposition* (Sezer and Šiljak, 1991), which allows for the use of smaller values of ϵ without violating the block diagonal structure. In the following, we will establish some properties of such decompositions and utilize them for the parallel solution of linear equations.

Let us consider an arbitrary $n \times n$ matrix A and let $X = \{x_1, x_2, ..., x_n\}$ and $Y = \{y_1, y_2, ..., y_n\}$ denote the set of its columns and rows, respectively. With this matrix we can uniquely associate a bigraph $B(X, Y; E)$ such that $|X| = |Y| = n$ and $(x_j, y_i) \in E$ if and only if $a_{ij} \neq 0$ $(i, j = 1, 2, ..., n)$.

DEFINITION 1.1. *Let $B(X, Y; E)$ have a subgraph $B^*(X, Y; E^*)$ such that all vertices in B^* have degree 1. Then the subgraph B^* uniquely defines a $1:1$ mapping $M : X \to Y$ as*

$$M(x_i) = y_j \Leftrightarrow (x_i, y_j) \in B^*. \tag{1.2}$$

Such a mapping is called a perfect matching for bigraph B (and therefore for the corresponding matrix A as well).

Finding a perfect matching is a fundamental step in initializing the algorithm for overlapping epsilon decomposition; an arbitrary choice of matching will suffice for this purpose. Our algorithm represents an improved version of the decomposition described in (Sezer and Šiljak, 1991), and consists of two main stages - the *basic decomposition* and its subsequent *modifications*. The modifications are aimed at eliminating possible redundancies and achieving a balanced block size; their execution is optional. There is also a *final stage*, in which the obtained bigraph is used to construct the expanded matrix $\tilde{A}$. Regarding the algorithm complexity, in (Zečević , 1993) it was established that the basic decomposition and modifications require $0(n^2)$ and $0(\mu^2 n^2)$ comparisons respectively, with μ denoting the maximal multiplicity of all vertices in the overlapping structure. This, however, is a worst case estimate, and in applications the dependence on the problem size was seen to be almost linear.

The following example illustrates how the algorithm is applied.

EXAMPLE. Let us consider matrix

$$A = \begin{bmatrix} * & * & 0 \\ \epsilon & * & * \\ 0 & * & \epsilon \end{bmatrix} \tag{1.3}$$

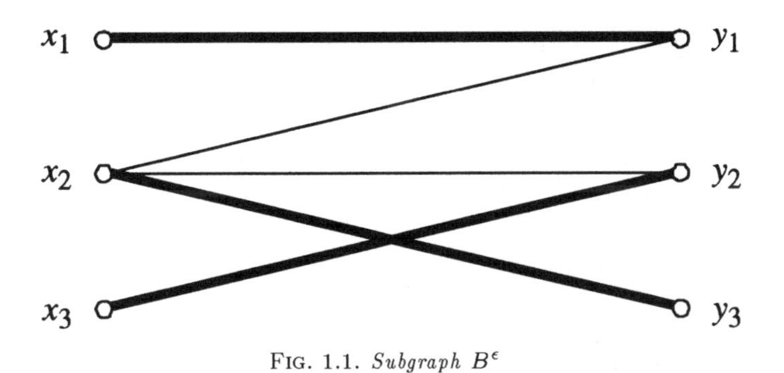

FIG. 1.1. *Subgraph* B^ϵ

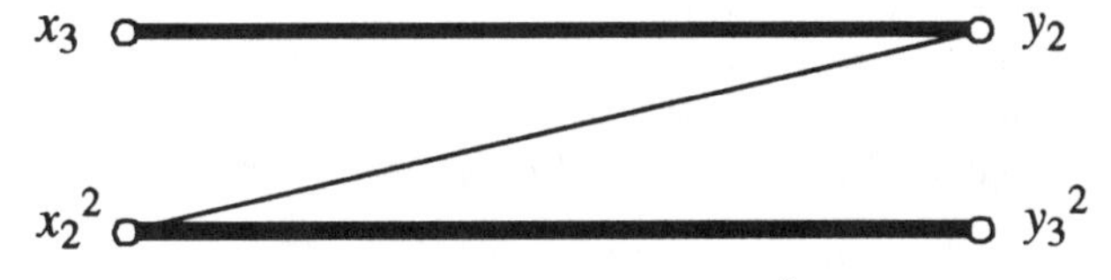

FIG. 1.2. *Expanded bigraph* $\tilde{B}^\epsilon$.

in which ϵ denotes all nonzero elements with absolute value less than or equal to some preassigned number ϵ, and $*$ denotes all other elements. Let $B(X, Y; E)$ denote the bigraph corresponding to matrix (1.3). By removing the edges of B that represent epsilon elements of the matrix, we obtain subgraph $B^\epsilon(X, Y; E^\epsilon)$ in Fig. 1.1.

Application of the overlapping epsilon decomposition algorithm to B^ϵ now results in an *expanded* bigraph $\tilde{B}^\epsilon$ shown in Fig. 1.2, in which x_i^r and y_j^r represent the $r-th$ appearance of vertices x_i and y_j, respectively. It is easily verified that bigraph $\tilde{B}^\epsilon$ indeed corresponds to the expanded matrix

$$\tilde{A} = \begin{bmatrix} * & * & | & 0 & 0 \\ 0 & * & | & \epsilon & 0 \\ - & - & - & - & - \\ \epsilon & 0 & | & * & * \\ 0 & 0 & | & \epsilon & * \end{bmatrix} = \begin{bmatrix} * & * & | & 0 & 0 \\ 0 & * & | & 0 & 0 \\ - & - & - & - & - \\ 0 & 0 & | & * & * \\ 0 & 0 & | & \epsilon & * \end{bmatrix} + \epsilon \begin{bmatrix} 0 & 0 & | & 0 & 0 \\ 0 & 0 & | & 1 & 0 \\ - & - & - & - & - \\ 1 & 0 & | & 0 & 0 \\ 0 & 0 & | & 0 & 0 \end{bmatrix} \tag{1.4}$$

and that the permutation matrices V and $\tilde{V}$

$$V = \begin{bmatrix} 1 & 0 & 0 \\ 0 & 0 & 1 \\ 0 & 1 & 0 \\ 0 & 0 & 1 \end{bmatrix}, \quad \tilde{V} = \begin{bmatrix} 1 & 0 & 0 \\ 0 & 1 & 0 \\ 0 & 0 & 1 \\ 0 & 1 & 0 \end{bmatrix} \tag{1.5}$$

are uniquely defined by the ordering of the x and y vertices in Fig. 1.2.

In this section we will consider some important properties of epsilon decompositions. They are described by several general theorems, in which it is assumed that A has *full generic rank* (extensions to the generically singular case can be found in (Zečević , 1993)); all proofs are give in the Appendix. We begin with a fundamental result relating the original $n \times n$ matrix A and the $\tilde{n} \times \tilde{n}$ ($\tilde{n} \geq n$) expanded matrix $\tilde{A}$ obtained using overlapping epsilon decomposition.

THEOREM 1.2. *The original and expanded matrix are related by*

$$VA = \tilde{A}\tilde{V}. \tag{1.6}$$

The $\tilde{n} \times n$ permutation matrices V and $\tilde{V}$ are uniquely defined by the expanded bigraph $\tilde{B}$, and $V = \tilde{V}$ whenever A has a perfect matching on the diagonal.

We next provide a theorem that is concerned with the preservation of cycles in the expanded graph; a proof is not given, since it can be found in (Sezer and Šiljak, 1991).

THEOREM 1.3. *Let the original bigraph B contain a simple alternating cycle. If a component of the expanded graph $\tilde{B}$ contains a replica of any one of the vertices appearing in the cycle, then it contains at least one replica of each of the other vertices in the cycle.*

Theorem 1.3 effectively states that cycles of B *cannot* be split in the process of overlapping epsilon decomposition. This important property now allows us to prove a theorem which justifies our strategy of initiating the basic decomposition with an *arbitrary* perfect matching.

THEOREM 1.4. *Given a generically non-singular matrix A, the block structure of the expanded matrix $\tilde{A}$ is invariant to the choice of matching.*

The final property that will be presented in this section deals with the relationship that exists between overlapping block structures and *singly bordered block diagonal* (SBBD) forms.

THEOREM 1.5. *If an $n \times n$ matrix A has an overlapping epsilon decomposition, then it can always be permuted into an $n \times n$ matrix $\tilde{A}$ that is SBBD when ϵ elements are set to zero.*

2. Parallel solution of linear equations using overlapping epsilon decompositions. We will now consider the system of linear equations

$$Ax = b \tag{2.1}$$

where A is a nonsingular matrix, and assume that there exists an overlapping epsilon decomposition

$$\tilde{A} = \tilde{A}_0 + \epsilon \tilde{A}_1 \tag{2.2}$$

related to A by

$$VA = \tilde{A}\tilde{V}. \tag{2.3}$$

In order to solve system (2.1) in parallel using p processors, we propose to apply an iterative method in the *expanded space* $\Re^{\tilde{n}}$

$$\tilde{x}_{k+1} = \tilde{x}_k - \tilde{A}_0^{-1}(\tilde{A}\tilde{x}_k - \tilde{b}). \tag{2.4}$$

In view of the fact that $\tilde{A}_0$ is a block diagonal matrix

$$\tilde{A}_0 = \text{diag}\{\tilde{A}_{11}, \tilde{A}_{22}, ..., \tilde{A}_{pp}\} \tag{2.5}$$

by (2.4) it follows that each processor can now update its *own* component x_k^i as

$$\tilde{x}_{k+1}^i = -\tilde{A}_{ii}^{-1}\left[\sum_{j \neq i} \tilde{A}_{ij}\tilde{x}_k^j - \tilde{b}^i\right], \quad i = 1, 2, ..., p \tag{2.6}$$

The diagonal blocks $\tilde{A}_{ii}$ can be factored independently by their designated processors *prior* to starting the iterations; no inter-processor communication is needed at this stage. Note, however, that computing x_{k+1}^i *does* require knowledge of $x_k^j (j \neq i)$, implying that a *multinode broadcast* is necessary after *every* iteration.

Conditions allowing for the use of the expanded block iterative method (2.6) in computing the solution of (2.1) are given by the following theorem:

THEOREM 2.1. *Let x^* be the unique solution of (2.1) and assume that A has overlapping epsilon decomposition (2.2). Then, if $\tilde{A}_0$ is invertible and satisfies*

$$\epsilon\|\tilde{A}_0^{-1}\tilde{A}_1\| < 1 \tag{2.7}$$

the block iterative method (2.6) converges to

$$\tilde{x}^* = \tilde{V}x^* \tag{2.8}$$

for any initial approximation $\tilde{x}_0 \in \Re^{\tilde{n}}$.

The iterative method described in (2.6) obviously requires the invertibility of $\tilde{A}_0$. Unfortunately, in general this cannot be guaranteed even under the assumption that A is a nonsingular matrix. If problems of this nature arise in practice, a simple way to alleviate them is to choose a different ϵ. A more rigorous approach, however, is based on the following theorem, which provides additional conditions on A that guarantee the invertibility of $\tilde{A}_0$ for *any* choice of ϵ.

THEOREM 2.2. *Let A be a positive-definite, symmetric matrix. Then, for any arbitrary overlapping epsilon decomposition (2.2) the matrix $\tilde{A}_0$ is invertible.*

Theorem 2.2 appears to restrict the applicability of our method only to positive-definite symmetric matrices. It can be argued, however, that given equation (2.1), we can solve the equivalent system (Hageman and Young, 1981)

$$A^T Ax = A^T b \tag{2.9}$$

in which $A_s \equiv A^T A$ is clearly positive-definite and symetric whenever A is nonsingular. Consequently, if an epsilon decomposition exists for A_s the block parallel iterative method (2.6) can indeed proceed as described earlier.

That (2.9) does not solve the problem entirely is shown by the following theorem, which demonstrates that although a positive-definite symmetric matrix *can* have an ordinary epsilon decomposition, the decomposition will never include overlapping.

THEOREM 2.3. *A positive-definite, symmetric matrix cannot have an overlapping epsilon decomposition.*

In view of Theorem 2.3, we now propose to scale (2.9) as

$$D^{-1}A^T Ax = D^{-1}A^T b \tag{2.10}$$

where D represents a *positive diagonal matrix* in which d_{ii} is the absolute value of the *maximal* element in row i of $A^T A$. The scaling process is justified by the following theorem.

THEOREM 2.4. *Let A be a nonsingular matrix. Then, for an arbitrary overlapping epsilon decomposition (2.2) of the scaled matrix $\overline{A} \equiv D^{-1}A^T A$ the block diagonal matrix $\tilde{A}_0$ is invertible.*

In applications to electric circuits it is of particular interest to extend Theorem 2.4 to the class of symmetric Metzler matrices (e.g., Šiljak, 1978), which are known to appear in a variety of standard computational problems.

DEFINITION 2.5. *An $n \times n$ matrix A is a Metzler matrix if*

$$a_{ij} = \begin{cases} < 0\,, & i = j \\ \geq 0\,, & i \neq j \end{cases} \tag{2.11}$$

for all $i, j \in N$.

Regarding the overlapping epsilon decomposition of scaled Metzler matrices, we can state the following simple result.

THEOREM 2.6. *Let A be a symmetric, invertible Metzler matrix satisfying*

$$|a_{ii}| \geq \sum_{j \neq i} |a_{ij}|, \quad i = 1, ..., n. \tag{2.12}$$

Then, for any scaled matrix $D^{-1}A$ with overlapping epsilon decomposition (2.2) the corresponding matrix $\tilde{A}_0$ is invertible.

The conditions of this theorem are satisfied for matrices that appear in node voltage equations, load flow analysis and other circuit related problems. It is important to note that in such applications the matrix A does not need to be multiplied by its transpose, by virtue of Theorem 2.6.

3. Application to electric power systems. Since the initial studies of parallel solutions to load-flow problems (Happ, 1969) it has been recognized that the key step in mapping a problem on a multiprocessor system is an efficient partitioning algorithm. Schemes for partitioning of power systems have been based on a wide variety of principles, such as sparsity (Duff *et al.*, 1986), coherency (Lee and Schweppe, 1973); Pai and Adgonkar, 1982), diakoptics (Happ, 1974; Kasturi and Potti, 1976), decoupling and time-scales (Medanić and Avramović, 1975; Chow, 1982; Khorasani and Pai, 1988), and overlapping subsystems (Šiljak, 1978; Ikeda and Šiljak, 1980; Zaborszky *et al.*, 1985; Brucoli *et al.*, 1987). Recent studies (Crow and Ilić, 1990; Decker *et al.*, 1991) have indicated that the most efficient parallel configurations are those that happen to group the tightly coupled variables together. These are precisely the configurations that the overlapping epsilon decomposition algorithm is designed to produce.

Although the load flow problem is actually nonlinear, the modified Newton iterative method with initial approximation x_0 and Jacobian $A(x_0)$

$$x_{k+1} = x_k - A(x_0)^{-1} f(x_k), \quad k = 0, 1, ... \tag{3.1}$$

computationally reduces to solving a system of linear equations

$$A(x_0) y = f(x_k) \tag{3.2}$$

$$
\begin{bmatrix}
-1 & .158 & .169 & .172 & 0 & 0 & 0 & 0 & 0 & 0 & 0 & 0 & 0 \\
.487 & -1 & .517 & 0 & 0 & 0 & 0 & 0 & 0 & 0 & 0 & 0 & 0 \\
.134 & .132 & -1 & .563 & 0 & .125 & 0 & .147 & 0 & 0 & 0 & 0 & 0 \\
.149 & 0 & .618 & -1 & .114 & 0 & 0 & 0 & 0 & 0 & 0 & 0 & 0 \\
0 & 0 & 0 & .229 & -1 & 0 & 0 & 0 & 0 & .236 & .183 & .352 & 0 \\
0 & 0 & .245 & 0 & 0 & -1 & .290 & .465 & 0 & 0 & 0 & 0 & 0 \\
0 & 0 & 0 & 0 & 0 & 1 & -1 & 0 & 0 & 0 & 0 & 0 & 0 \\
0 & 0 & .074 & 0 & 0 & .374 & 0 & -1 & .427 & 0 & 0 & 0 & .125 \\
0 & 0 & 0 & 0 & 0 & 0 & 0 & .702 & -1 & .298 & 0 & 0 & 0 \\
0 & 0 & 0 & 0 & .482 & 0 & 0 & 0 & .518 & -1 & 0 & 0 & 0 \\
0 & 0 & 0 & 0 & .585 & 0 & 0 & 0 & 0 & 0 & -1 & .415 & 0 \\
0 & 0 & 0 & 0 & .572 & 0 & 0 & 0 & 0 & 0 & .211 & -1 & .217 \\
0 & 0 & 0 & 0 & 0 & 0 & 0 & .567 & 0 & 0 & 0 & .433 & -1
\end{bmatrix}
$$

FIG. 3.1. *Matrix B.*

in each iteration. It has been shown (Zečević and Šiljak, 1994) that if the Jacobian has an overlapping epsilon decomposition

$$\tilde{A}(x_0) = \tilde{A}_0 + \epsilon \tilde{A}_1 \tag{3.3}$$

for a sufficiently small ϵ, it is possible to apply the simplified iterative scheme

$$\tilde{x}_{k+1} = \tilde{x}_k - \tilde{A}_0^{-1} \tilde{f}(\tilde{x}_k), \quad k = 0, 1, \ldots \tag{3.4}$$

in the expanded space $\mathfrak{R}^{\tilde{n}}$. Since $\tilde{A}_0$ is now a *block diagonal* matrix, equation (3.4) can be solved very efficiently in parallel and the solution $\tilde{x}^* \in \mathfrak{R}^{\tilde{n}}$ is seen to be related to the original solution $x^* \in \mathfrak{R}^n$ as described in (2.8) (Zečević and Šiljak, 1994).

As an illustration of the decomposition procedure, in Figs. 3.1 and 3.2 we show the susceptance matrix B and its expansion $\tilde{B}$ for $\epsilon = 0.4$; the expansion is seen to uniquely induce a decomposition of the corresponding Jacobian (Zečević and Šiljak, 1994).

More meaningful computational results have been obtained for larger systems, such as the IEEE 118 electric power transmission system (shown in Fig. 3.3) and the 888 bus Tokyo electric network. A schematic illustration of an epsilon decomposition for the 118 bus system is presented in Fig. 3.4; four processors have been considered, and the individual block sizes are indicated within the circles.

In Tables 3.1 and 3.2 we show the speedups obtained using overlapping epsilon decompositions on the Intel iPSC/860 hypercube. All speedups were calculated with respect to computations using a single processor; in these computations, the standard modified Newton method was used and

$$
\begin{bmatrix}
-1 & .517 & 0 & .487 & 0 & 0 & 0 & 0 & 0 & 0 & .0 & 0 & 0 & 0 & 0 & 0 & 0 & 0 & 0 & 0 \\
.132 & -1 & .563 & .134 & 0 & .125 & .047 & 0 & 0 & 0 & 0 & 0 & 0 & 0 & 0 & 0 & 0 & 0 & 0 & 0 \\
0 & .618 & -1 & .149 & 0 & 0 & 0 & 0 & 0 & 0 & 0 & .114 & 0 & 0 & 0 & 0 & 0 & 0 & 0 & 0 \\
.158 & .169 & .172 & -1 & 0 & 0 & 0 & 0 & 0 & 0 & 0 & 0 & 0 & 0 & 0 & 0 & 0 & 0 & 0 & 0 \\
0 & 0 & 0 & 0 & -1 & 1 & 0 & 0 & 0 & 0 & 0 & 0 & 0 & 0 & 0 & 0 & 0 & 0 & 0 & 0 \\
0 & .245 & 0 & 0 & .290 & -1 & .465 & 0 & 0 & 0 & 0 & 0 & 0 & 0 & 0 & 0 & 0 & 0 & 0 & 0 \\
0 & .074 & 0 & 0 & 0 & .374 & -1 & .427 & 0 & 0 & 0 & 0 & 0 & 0 & 0 & .125 & 0 & 0 & 0 & 0 \\
0 & 0 & 0 & 0 & 0 & 0 & .702 & -1 & .298 & 0 & 0 & 0 & 0 & 0 & 0 & 0 & 0 & 0 & 0 & 0 \\
0 & 0 & 0 & 0 & 0 & 0 & 0 & 0 & -1 & .518 & 0 & .482 & 0 & 0 & 0 & 0 & 0 & 0 & 0 & 0 \\
0 & 0 & 0 & 0 & 0 & 0 & 0 & 0 & .298 & -1 & .702 & 0 & 0 & 0 & 0 & 0 & 0 & 0 & 0 & 0 \\
0 & .074 & 0 & 0 & 0 & .374 & 0 & 0 & 0 & .427 & -1 & 0 & 0 & 0 & 0 & .125 & 0 & 0 & 0 & 0 \\
0 & 0 & .229 & 0 & 0 & 0 & 0 & 0 & .236 & 0 & 0 & -1 & .183 & .352 & 0 & 0 & 0 & 0 & 0 & 0 \\
0 & 0 & 0 & 0 & 0 & 0 & 0 & 0 & 0 & 0 & 0 & 0 & -1 & .415 & .585 & 0 & 0 & 0 & 0 & 0 \\
0 & 0 & 0 & 0 & 0 & 0 & 0 & 0 & 0 & 0 & 0 & 0 & .211 & -1 & .572 & .217 & 0 & 0 & 0 & 0 \\
0 & 0 & .229 & 0 & 0 & 0 & 0 & 0 & .236 & 0 & 0 & 0 & .183 & .352 & -1 & 0 & 0 & 0 & 0 & 0 \\
0 & 0 & 0 & 0 & 0 & 0 & 0 & 0 & 0 & 0 & 0 & 0 & 0 & 0 & 0 & -1 & .433 & 0 & .567 & 0 \\
0 & 0 & 0 & 0 & 0 & 0 & 0 & 0 & 0 & 0 & 0 & 0 & .211 & 0 & 0 & .217 & -1 & .572 & 0 & 0 \\
0 & 0 & .229 & 0 & 0 & 0 & 0 & 0 & .236 & 0 & 0 & 0 & .183 & .352 & 0 & 0 & 0 & -1 & 0 & 0 \\
0 & .074 & 0 & 0 & 0 & .374 & 0 & 0 & 0 & 0 & 0 & 0 & 0 & 0 & 0 & .125 & 0 & 0 & -1 & .427 \\
0 & 0 & 0 & 0 & 0 & 0 & 0 & 0 & .298 & 0 & .0 & 0 & 0 & 0 & 0 & 0 & 0 & 0 & .702 & -1
\end{bmatrix}
$$

FIG. 3.2. *The expanded matrix* $\tilde{B}$.

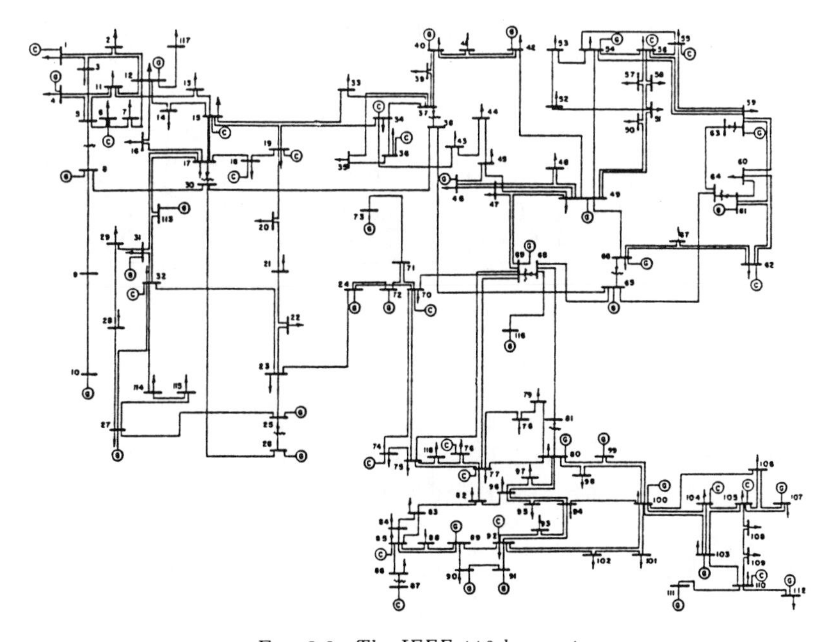

FIG. 3.3. *The IEEE 118 bus system.*

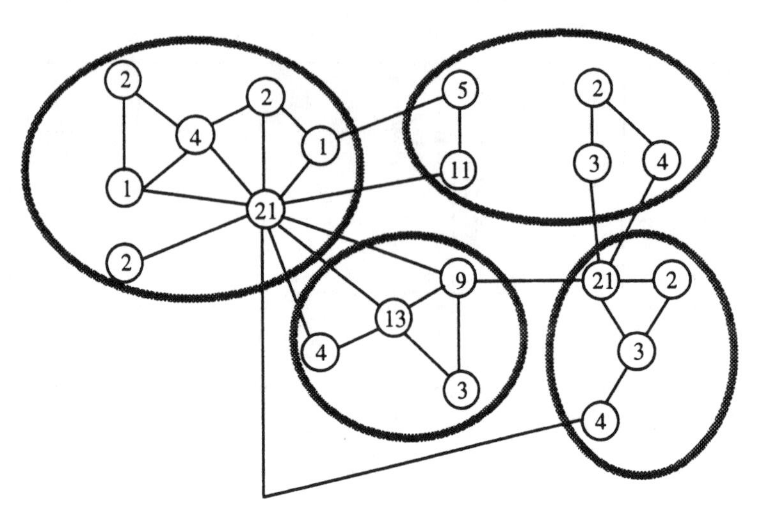

FIG. 3.4. *Decomposition of the 118 bus system for $\epsilon = 0.37$ adapted for 4 processors.*

TABLE 3.1
Computational time for the 118 bus network with optimal β.

Number of processors	Number of iterations	Computation time (seconds)	Communition time (seconds)	Total exec. time (seconds)	Speed up
single proc.	2	0.104	0	0.104	1
2	5	0.049	0.001	0.050	2.08
4	5	0.031	0.002	0.033	3.15
8	8	0.026	0.003	0.022	3.59

TABLE 3.2
Computational time for the 888 bus network with optimal β.

Number of processors	Number of iterations	Computation time (seconds)	Communition time (seconds)	Total exec. time (seconds)	Speed up
single proc.	5	0.893	0	0.893	1
2	6	0.249	0.002	0.251	3.56
4	6	0.132	0.001	0.133	6.71
8	7	0.069	0.001	0.070	12.76

the minimal degree ordering was applied to factorizing the complete (unpartitioned) Jacobian. In the parallel computations the same ordering was utilized to factorize the individual blocks, and convergence was improved with the use of modified iterates

$$x_{k+1} = x_k - A_0^{-1}[f(x_k) + \epsilon\beta A_1(x_k - x_{k-1})] \tag{3.5}$$

(β is a relaxation parameter). In all cases considered a good initial approximation was used (this is typically available in outage assessment and transient stability studies), with a mismatch of 0.001 p.u. as the convergence criterion.

Acknowledgements

The research reported herein has been supported by the National Science Foundation under Grant ECS-9114872. The authors are greatful to M. Amano, Hitachi Research Laboratory, Ibaraki-ken, Japan, for his useful comments on the paper and experimental results obtained by applying the epsilon decompositions to parallelization of load flow equations.

Appendix

Proof of Theorem 1.2. By assumption, matrix A is generically nonsingular and therefore has a perfect matching. If the matching is not on the main diagonal, it can always be placed there by an appropriate column permutation. The permuted matrix will be denoted by A_p, satisfying $A_p \equiv AP$. Application of overlapping epsilon decomposition to A_p results in a matrix $\tilde{A}$ which is *uniquely* defined by the corresponding expanded bigraph $\tilde{B}$.

Consider now an $\tilde{n} \times n$ matrix T_1 defined as

$$T_1 = \begin{array}{c} s_1 \left\{ \begin{array}{c} \\ \\ \\ \\ \end{array} \right. \\ \vdots \\ s_n \left\{ \begin{array}{c} \\ \\ \\ \\ \end{array} \right. \end{array} \begin{bmatrix} 1 & & & \\ \vdots & & 0 & \\ 1 & & & \\ & \ddots & & \\ & & & 1 \\ & 0 & & \vdots \\ & & & 1 \end{bmatrix} \tag{A.1}$$

In (A.1) $s_i (i = 1, 2, ..., n)$ is the *multiplicity* of row y_i in bigraph $\tilde{B}$ (and consequently the multiplicity of column x_i as well, since A_p has a perfect matching on the diagonal). From (A.1) it follows that the $\tilde{n} \times n$ matrix $T_1 A_p$ has rows and columns ordered as

$$Y = \{y_1^1, ..., y_1^{s_1}, y_2^1, ..., y_2^{s_2}, ..., y_n^1, ..., y_n^{s_n}\} \qquad \text{(A.2)}$$
$$X = \{x_1, x_2, ..., x_n\}$$

We can now introduce a symmetric permutation T_2 such that matrix

$$A_2 \equiv T_2^{-1} \tilde{A} T_2 \qquad \text{(A.3)}$$

has vertex ordering

$$Y = \{y_1^1, ..., y_1^{s_1}, y_2^1, ..., y_2^{s_2}, ..., y_n^1, ..., y_n^{s_n}\} \qquad \text{(A.4)}$$
$$X = \{x_1^1, ..., x_1^{s_1}, x_2^1, ..., x_2^{s_2}, ..., x_n^1, ..., x_n^{s_n}\}$$

By the overlapping epsilon decomposition algorithm, row y_p^k of matrix $\tilde{A}$ (and therefore of A_2 as well) will have element a_{pi} in some column x_i^m $(1 \leq m \leq s_i)$ and zeros in *all* other columns x_i^l $(l \neq m)$. This property allows us to write any column of matrix $T_1 A_p$ as

$$x_i = x_i^1 + x_i^2 + ... + x_i^{s_i} \quad (i = 1, 2, ..., n) \qquad \text{(A.5)}$$

where x_i^l $(l = 1, 2, ..., s_i)$ are columns of A_2. Observing (A.2), (A.3) and (A.4), it is now easily verified that

$$T_1 A_p = A_2 T_1 \qquad \text{(A.6)}$$

Invoking (A.3), and (A.6) we further obtain

$$VA = \tilde{A} \tilde{V} \qquad \text{(A.7)}$$

where $V = T_2 T_1$, and $\tilde{V} = V P^{-1}$. Clearly, when matrix A has a perfect matching on the diagonal we have $P = I$ and therefore $V = \tilde{V}$.

It should also be recalled that T_1 is uniquely defined by the vertex multiplicity and T_2 by the vertex ordering in bigraph $\tilde{B}$. Since P is determined by the perfect matching in the expanded bigraph, it follows that matrices V and $\tilde{V}$ are indeed uniquely determined by $\tilde{B}$. □

Proof of Theorem 1.4. Suppose that we have obtained a perfect matching M in which $y_i = M(x_{q(i)})$. Based on this matching we can form bigraph B corresponding to the original matrix A, and subsequently apply epsilon decomposition to obtain the expanded bigraph $\tilde{B}$ and matrix $\tilde{A}$.

Consider now a different perfect matching M_1 such that $y_i = M_1(x_{p(i)})$ $(x_{p(i)} \neq x_{q(i)})$. In the new matching the y-vertex previously matched with $x_{p(i)}$ must be *rematched*, and this reassignment proceeds successively until some y-vertex is eventually matched with $x_{q(i)}$. If all the vertices for which M and M_1 are different have *not* been visited at this point, we can continue the reassignment procedure beginning with some unvisited vertex y_m, until all such vertices have been exhausted.

Since every y-vertex *must* be rematched with some x-vertex that belongs to its incidence list, it follows that all the vertices involved in the reassignment constitute one or several cycles in the original bigraph B. As a result, it can be said that any new perfect matching corresponds to a permutation *within the cycles* of B. By Theorem 1.3 we further know that the cycles of B are not split in $\tilde{B}$, which implies that a change of perfect matching actually corresponds to a permutation *within the blocks* of the expanded matrix $\tilde{A}$ (obtained using the original matching M). It now directly follows that the initial block structure remains unchanged under a different perfect matching. □

Proof of Theorem 1.5. Let A have an overlapping epsilon decomposition $\tilde{A}$, and let $\tilde{A}$ be constructed from $\tilde{A}$ in such a way that all the *multiple* rows and columns are placed in the border, while the others remain ordered as in $\tilde{A}$. Since $\tilde{A}$ has an overlapping block structure, this procedure is guaranteed to produce a bordered block diagonal (BBD) matrix $\tilde{A}$ whenever ϵ elements are set to zero. We now show that the resulting structure will actually be SBBD. To that effect, assume conversely that some border row y_k of matrix $\tilde{A}$ has an element $a_{ki} : |a_{ki}| > \epsilon$ in a non-border column x_i, (that is, in a column x_i that was *not* multiple in $\tilde{A}$). Since y_k *was* multiple in $\tilde{A}$ (appearing in blocks $l, m, ...$) it now follows by the overlapping epsilon decomposition algorithm that x_i would have to appear in those same blocks. This, however, is in contradiction with the assumption that x_i was a *non-multiple* column in $\tilde{A}$. Consequently, all the elements in the border rows and non-border columns of $\tilde{A}$ *must* be smaller in magnitude than ϵ, which results in an SBBD structure when ϵ elements are set to zero. □

Proof of Theorem 2.1. We can write equation (2.6) as

$$\tilde{x}_{k+1} = \tilde{G}\tilde{x}_k + \tilde{h} \tag{A.8}$$

where

$$\begin{aligned} \tilde{G} &\equiv I - \tilde{A}_0^{-1}\tilde{A} = -\epsilon\tilde{A}_0^{-1}\tilde{A}_1 \\ \tilde{h} &\equiv \tilde{A}_0^{-1}\tilde{b} \end{aligned} \tag{A.9}$$

By the conditions of this theorem, mapping

$$\tilde{f}(\tilde{x}) \equiv \tilde{G}\tilde{x} + \tilde{h} \tag{A.10}$$

satisfies

$$||\tilde{f}(\tilde{x}) - \tilde{f}(\tilde{y})|| \leq \epsilon||\tilde{A}_0^{-1}\tilde{A}_1||||\tilde{x} - y|| \equiv \alpha||\tilde{x} - \tilde{y}|| \quad (\alpha < 1) \tag{A.11}$$

for any $\tilde{x}, \tilde{y} \in \mathfrak{R}^{\tilde{n}}$ and is therefore a contraction. Since $\tilde{f}$ maps $\mathfrak{R}^{\tilde{n}} \to \mathfrak{R}^{\tilde{n}}$, it follows that there is a *unique* fixed point $\tilde{x}^*$ in $\mathfrak{R}^{\tilde{n}}$ to which (2.6) converges for any initial approximation $\tilde{x}_0 \in \mathfrak{R}^{\tilde{n}}$.

On the other hand, since

$$Ax^* = b \tag{A.12}$$

we have

$$VAx^* = \tilde{A}\tilde{V}x^* = Vb = \tilde{b} \tag{A.13}$$

and therefore

$$\tilde{V}x^* = G\tilde{V}x^* + \tilde{h} \tag{A.14}$$

It now follows that $\tilde{V}x^*$ is a fixed point for $\tilde{f}$, and by uniqueness $\tilde{x}^* = \tilde{V}x^*$. □

Proof of Theorem 2.2. It should first be observed that a positive definite matrix A *always* has a perfect matching on the diagonal, since $a_{ii} > 0$ $(i = 1, 2, ..., n)$ as principal minors. This holds for epsilon decompositions as well, since our algorithm is designed to preserve *all nonzero diagonal* elements, in order to avoid a non-symmetric permutation whenever possible.

Let $\tilde{A}_{ii}$ be an arbitrary diagonal block of $\tilde{A}_0$ and let y_j^r and x_p^q be a row and column in this block. Then, by the overlapping epsilon decomposition algorithm, element a_{jp} of matrix A appears in row y_j^r and column x_p^q of $\tilde{A}_{ii}$. Since only *one* replica of any row or column can appear in $\tilde{A}_{ii}$, it follows that $\det\tilde{A}_{ii}$ will actually be a *minor* of the original matrix A. Furthermore, since A has a perfect matching on the diagonal, $\tilde{A}$ is obtained as a result of symmetric permutations and det $\tilde{A}_{ii}$ will actually be a *principal minor* of A. Recalling that all principal minors of a positive-definite matrix must be positive, we now have that $\tilde{A}_{ii}$ is nonsingular. □

Proof of Theorem 2.3. Assume conversely, that a positive-definite symmetric matrix A *has* an overlapping epsilon decomposition $\tilde{A}$. As demonstrated in Theorem 1.5, A can then be permuted into a matrix $\tilde{A}$ that is SBBD when ϵ elements are set to zero. We point out that in this case we have a symmetric permutation, in which *any* border column x_j will *always* be matched to the corresponding border row y_j (since A is positive-definite and therefore has a perfect matching on the diagonal).

It should now be observed that in $\tilde{A}$ there must exist a border column x_j with an element $a_{ij} : |a_{ij}| > \epsilon$ in some row y_i that does *not* belong to the border. If this were not the case, all border columns (and border rows as well, by symmetry of A) would have elements smaller in magnitude than ϵ everywhere except maybe in the border block. Consequently, setting ϵ elements to zero in $\tilde{A}$ would result in a purely block diagonal matrix, implying that A has an epsilon decomposition with *no* overlapping, which is contrary to our assumption.

On the other hand, this implies that row y_j in the border of $\tilde{A}$ has an element $|a_{ji}| = |a_{ij}| > \epsilon$, and it now follows that setting ϵ elements to zero in $\tilde{A}$ would *not* result in an SBBD form, which is a contradiction. □

Proof of Theorem 2.4. Suppose that the scaled matrix $\tilde{A} = D^{-1}A_s$ (where $A_s = A^T A$) has an overlapping epsilon decomposition (2.2) and consider an arbitrary block $\tilde{A}_{ii}$ of $\tilde{A}_0$. We should first note that $\tilde{A}$ has a perfect matching on the diagonal, since matrix A_s has one. As demonstrated in the proof of Theorem 2.2, det $\tilde{A}_{ii}$ then represents a *principal minor* of $\tilde{A}$. Clearly, any principal minor of $\tilde{A}$ can be written as

$$\tilde{A}\begin{pmatrix} i_1 & \dots & i_p \\ i_1 & \dots & i_p \end{pmatrix} = d_{i_1 i_1}^{-1} \dots d_{i_p i_p}^{-1} \, A_s \begin{pmatrix} i_1 & \dots & i_p \\ i_1 & \dots & i_p \end{pmatrix} \tag{A.15}$$

Furthermore, such a minor must be *nonzero* since A_s is positive-definite and D is nonsingular, so $\tilde{A}_{ii}$ is invertible. □

Proof of Theorem 2.6. Since A is a symmetric Metzler matrix satisfying (2.12), it follows by definition that $sI - A$ is an M matrix (e.g., Šiljak, 1978) for any real $s > 0$. Consequently, $sI - A$ is positive definite for any $s > 0$, and by continuity it follows that $-A$ must be a positive semi-definite matrix. However, since A is nonsingular by assumption, matrix $-A$ is actually positive definite. The rest of the proof is identical to that of Theorem 2.4. □

REFERENCES

[1] Brucoli, M., M. La Scala, F. Torelli, and M. Trovato, *Overlapping decompositions for small disturbance stability analysis of interconnected power networks*, Large Scale Systems, 13, pp. 115–129, 1987.

[2] Chow, J.H., *Time Scale Modeling of Dynamic Networks with Applications to Power Systems*, New York: Springer-Verlag, 1982.

[3] Crow, M.L., and M. Ilić, *The parallel implementation of the waveform relaxation method for transient stability simulations*, IEEE Transactions on Power Systems, 5, pp. 922–929, 1990.

[4] Decker, I. C., D.M. Falcão and E. Kaszkurewicz, *Parallel implementation of a power system dynamic simulation methodology using the conjugate gradient method*, IEEE Power Industry Computer Application Conference, Baltimore, MD, pp. 245–252, 1991.

[5] Duff, I.S., A.M. Erisman and J.K. Reid, *Direct Methods for Sparse Matrices*, Oxford: Clarendon Press, 1986.

[6] Gallivan, K.A., R.J. Plemmons, and A.H. Sameh, *Parallel algorithms for dense linear algebra computations*, SIAM Review, 32, pp. 54-135, 1990.

[7] Hageman, L.A. and D.M. Young, *Applied Iterative Methods*, New York: Academic Press, 1981.

[8] Happ, H.H., *Multicontroller configurations and diakoptics: Dispatch of real power in power pools*, IEEE Transactions on Power Apparatus and Systems, 88, pp. 764–772, 1969.

[9] Happ, H.H., *Diakoptics - the solution of system problems by tearing*, Proceedings of the IEEE, 62, pp. 930–940, 1974.

[10] Ikeda, M., and D.D. Šiljak, *Overlapping decompositions, expansions and contractions of dynamic systems*, Large Scale Systems, 1, pp. 29–38, 1980.

[11] Kasturi, R. and M.S.N. Potti, *Precise Newton-Raphson load flow - an exact method using ordered elimination*, IEEE Transactions on Power Apparatus and Systems, 95, pp. 1244–1253, 1976.

[12] Khorasani, K. and M.A. Pai, *Two time scale decomposition and stability analysis of power systems*, IEE Proceedings, 135, pp. 205, 1988.

[13] Lee, T.Y. and F.C. Schweppe, *Distance measures and coherency recognition for transient stability equivalents*, IEEE Transition on Power Apparatus and Systems, 92 1550-1557, 1973.

[14] Medanić, J. and B. Avramović, *Solution of load-flow problems in power systems by ϵ-coupling method*, IEE Proceedings, 122, pp. 801–805, 1975.

[15] Pai, M.A. and R.P. Adgonkar, *Electromechanical distance measure for decomposition of power systems*, Electrical Power and Energy Systems, 6, pp. 249–254, 1982.

[16] Sezer, M.E. and D.D. Šiljak, *Nested epsilon decomposition of linear systems: Weakly coupled and overlapping blocks*, SIAM Journal of Matrix Analysis and Applications, 12, pp. 521–533, 1991.

[17] Šiljak, D.D., *Large Scale Dynamic Systems: Stability and Structure*, New York: North-Holland, 1978.

[18] Zaborszky, J., G. Huang and K.W. Lu, *A textured model for computationally efficient reactive power control and management*, IEEE Transactions on Power Apparatus and Systems, 104, pp. 1718–1727, 1985.

[19] Zečević , A.I. and D.D. Šiljak, *A block-parallel Newton method via overlapping epsilon decompositions*, SIAM Journal of Matrix Analysis and Applications, (to appear), 1994.

[20] Zečević , A.I., *New Decomposition Methods for Parallel Computations of Large Systems*, Ph.D. Thesis, Santa Clara University, 1993.

INSECTS, FISH AND COMPUTER-BASED SUPER-AGENTS

SAROSH TALUKDAR* AND PEDRO DE SOUZA*

Abstract. This paper reports on progress towards systematic methods for designing computer-based super-agents and their organizations. (A super-agent is an aggregation of lesser agents, an organization is a set of rules or a grammar for constructing a family of super-agents.) The material is divided into two parts. The first develops some technical apparatus (a structural space and a taxonomy) for describing super-agents (old and new, natural and synthetic). The second illustrates the use of this apparatus in designing members of a new family of super-agents, called A-Teams. (An A-Team is characterized by autonomous agents that work iteratively and in parallel on populations of solutions. As such, it combines some of the best features of several natural and synthetic organizations, including cellular communities, insect societies, simulated annealing and genetic algorithms.) While this still leaves us some distance from the goal of systematic design techniques, the way is now much clearer.

1. Introduction. For social insects (ants, termites, certain bees and certain wasps), the construction of a nest is a massive engineering project requiring the cooperation of many workers [5]. Invariably, the nest is sophisticated in architecture and customized for its surroundings: features that are remarkable for three reasons.

First, there are no leaders to decide what is to be done, nor any centralized coordination of effort. (Certainly, there is a "queen", but her function is strictly reproductive: she does not lead the other colony members, nor issue orders. Although different castes exist within the colony, drones, soldiers and workers, for example, there is no hierarchical relationship among them [12]. Rather, the insects act as autonomous agents.)

Second, the workers use only local information (in space as well as time). There are no blueprints to show what the final result should be, and as Wilson notes, an individual worker "...cannot possibly perceive the actions of more than a minute fraction of its nest mates; nor can it monitor the physiological condition of the colony as a whole" [7]. Moreover, the construction process may require many worker-lifetimes. But the initial builders cannot communicate instructions to their successors: each brood of workers dies before the next hatches.

Third, the individual insect has only modest capabilities for learning, reasoning, and abstracting knowledge. It can remember the locations of important landmarks, such as its nest, and can be trained to walk through mazes [8], but seems to have little ability to generalize its knowledge or apply it to new situations. (For instance, an ant, that has learned to run a maze forwards, treats the problem of running it backwards, as being entirely new [9].) It is unlikely that a few such individuals could design an entire nest and communicate the result to their nest mates. Nor is it likely

* Engineering Design Research Center, Carnegie Mellon University, Pittsburgh, PA 15213.

that complete designs for every possible building site could be genetically stored in each of the insects: the range of site-variations would seem to be too large. Where then does the overall design come from?

There are other examples of collections with skills that seem qualitatively different from those of their individual members. The aggregates of transistors that form microprocessors and the aggregates of neurons that form neural nets, are just two of them. From where do such aggregates obtain the skills their individual members lack? It can only be (we conjecture) from the organizational schemes they use.

Organizations

An organization's reason-for-being is three-fold: first, the capabilities of any single agent are limited; second, there are tasks that require capabilities far in excess of these limits; and third, agents without a common purpose tend to work, if at all, at cross purposes. An organization provides a collection of agents with a common purpose and may, as in the cases of insects, transistors and neurons, amplify their capabilities.

There are two broad classes of organizations: hierarchies and non-hierarchies. The former have proved useful in a variety of settings including corporations, governments and some families. They have also been the subject of a good deal of research—the literature stretches past Sun Tzu's the Art of War (circa 500 BC) to at least as far as the Book of Exodus (circa 1500 BC).

A hierarchy is a multi-layered pyramid. A common purpose is dictated from above. Each layer controls the one below, providing it with tasks and watching to see that these tasks are properly performed. More specifically, the overall task(s) is selected by the agents in the top layer, decomposed into sub-tasks and passed on to the layer below. This division of labor continues, layer by layer, until the lowest layer is reached, by which time the sub-tasks have become small enough to be performed by the agents there. Often, but not always, there is a reverse flow of results, each layer collecting and integrating results from the layer below, before passing them on to the one above.

The number of layers is selected to keep the branching factor suitably small (so no agent is called on to control more underlings than it can). In principle, the addition of layers can be continued indefinitely to provide an arbitrarily large work force. In practice, corporate hierarchies have grown to include hundreds of thousands of members, and national hierarchies, such as the Peoples' Republic of China, as many as a billion members.

Of course, increases in size are useful only if they make possible the performance of bigger and better tasks.

The net capability of a work force to perform tasks depends not only on the inherent capabilities of its agents, but also on how they interact (cooperate). In other words, net capability is invariably different from the sum of the inherent capabilities, and is sensitive to the organization used, especially to the mechanisms for cooperation provided by the organization.

Two organizations with identical sets of agents can have very different net capabilities.

Over the years, organization designers have developed a great variety of mechanisms for enhancing vertical (inter-layer) and lateral (intra-layer) cooperation in hierarchies. Some examples are: matrix management, plebiscites, committees, task forces, brain-storming sessions, quality circles, and ombudsmen. However, there are two failings that are so fundamental to hierarchies that they cannot be eliminated by cooperation alone, no matter how effective it may be. These failings are:

- Supervisory overheads. An agent cannot be added to a lower level without assigning it a supervisor. If the agent is of a new type, this supervisor must be modified or retrained, often at great expense in time and effort.
- Fragility. The effects of poor decisions made by a supervisor can propagate to its subordinates and thence, to their subordinates, and so on. Also the removal of a key supervisor can be crippling. Thus, the supervisors are potential weak-points in a hierarchy.

These failings can be eliminated only by eliminating supervisors, that is, by adopting a special type of non-hierarchy called a hetrarchy: a single-layer, leaderless organization of autonomous agents, such as are used by insect societies, flocks of birds and neural nets.

Commonalty of purpose in a hetrarchy is obtained through one of two processes: embedding this purpose in the agents, or arranging for them to be drawn to it as their work progresses. Neither process is as efficient as the transmittal of purpose through the control structure of a hierarchy, nor as well understood. In other words, non-hierarchies suffer from the following relative failings:

- Inefficiency. A hetrarchy relies on an over abundance of agents that makes their efficient use unnecessary. For instance, during ant colony migrations, "many workers exit the nest carrying eggs, larvae and pupae in their mandibles, while other workers are busy carrying them back again. Still other workers run back and forth carrying nothing at all." Nevertheless, there are so many ants doing the right thing that the migration is successfully accomplished.
- Strangeness. Hetrarchies are difficult to understand and sometimes, counter-intuitive. How, for instance, can collections of autonomous insects involved in nest building, each with access only to information that is local in space and time, do anything of global worth?

The problem

To summarize the material above, the problem of designing an organization involves a choice between a hierarchy and a hetrarchy, the former having the disadvantages of supervisory overhead and fragility, the latter, of inefficiency and strangeness. For several years, we have been working on a version of this problem, specialized by task and agent-type as follows:

(P1):

given: a space of optimization tasks and a space of agents with optimization skills,

design: a good organization for the agents to use in performing the tasks;

What exactly is a space of optimization tasks? What distinguishes good organizations from bad ones? We will deal with these issues of terminology in Section 2. Meanwhile, we have some more general issues to discuss.

Goals

Our long term goal is to develop a computational framework for the systematic design of organizations that combine both human and computer-based agents. This paper reports on progress towards the intermediate goal of a framework for computer-based agents (essentially procedures or algorithms).

Our interest is in engineering tasks. We have chosen optimization as the field in which to ground our work because its terminology comes as close to providing a domain-independent language for engineering tasks as exists. (In principle, all intellectual tasks can be formulated as optimization tasks. However, computationally useful specifications of their components—objectives and constraints—are not always available.)

Why bother?

In the past, software engineers have been forced to use very simple organizations. The reason has been the difficulty in connecting large, computer-based agents (large programs such as solid modelers, optimizers, and circuit simulators), and the expense in maintaining these connections, once they have been made. The result has been software systems with very limited structures: chains with the output of one agent becoming the input of the next, and minimal networks, such as a single, shared memory, called a blackboard, from which all the agents read, and to which they all write, while under the direction of a single supervisor).

It is still difficult to connect large programs that are written in different languages and use different data structures. But it is becoming rapidly less so. Meanwhile, the computing power available in networks of distributed machines is growing even more rapidly. Together, these developments make possible more complex organizations which in turn, make possible much higher performance.

An approach

Our approach to (P1) has two concurrent and ongoing tracks: a comparative study of existing organizations and a growing series of applications from which we hope to "learn through doing."

The first track is devoted to examining a variety of organizations, ranging from cellular communities to scientific societies. Our interest is in questions like: what are the best features of this organization? how does this organization differ from that one? and can the differences be quantified? To answer such questions we need new technical apparatus: generic models and taxonomies of organizations, a space covering most if not all the organizational possibilities, and metrics for measuring the distances (differences) between members of this space. In much of Sections 2 through 7 we report on progress in developing this apparatus. The rest of these sections describe how (P1), a vague and ill posed problem, can be made less so. To do this we will use an iterative process, each iteration producing a formulation with a little more precision and detail.

The second track is devoted to building organizations for a number of important and difficult optimization tasks including large traveling salesman problems, high-rise building design, robot design, real-time control, circuit diagnosis, and train scheduling. Section 8 describes some of these efforts and summarizes the insights they have yielded. More specifically, it illustrates how a subset of organizations, called asynchronous teams, may be realized. These teams are open and effective. Their structures are characterized by autonomous agents that work iteratively and in parallel on populations of solutions. As such, they combine some of the best features of several natural and synthetic organizations, including cellular communities, insect societies, simulated annealing and genetic algorithms.

2. Open and effective super-agents

Some terminology

Some terms in (P1), such as "organization" and "good," can have very broad meanings. Here, the meanings of these and certain other terms will be restricted as follows:

- an optimization task is one specified in terms of two types of criteria: objectives to be minimized and constraints to be satisfied.
- a space is a set. Often this set is large and its elements are arranged with the aid of axes or metrics. Henceforth, we will also use the terms: family, template, region, fuzzy set and population to mean sets of one sort or another. Specifically, a family is a related set of super-agents, a template is a set of graphs; a region is a subset of a space; a fuzzy set allows for partial membership; and a population is a set of data-objects in a memory.
- an organization is a set of rules by which lesser agents aggregate to form greater or super-agents, just as birds, fish and people aggregate to form flocks, schools and corporations. As such, an organization is a super-agent grammar (it is to a super-agent what a grammar is to a sentence). Henceforth, we will refer to the set of super-agents produced by the rules of an organization as a family or language of super-agents.
- a good organization is one that yields super-agents that are both effective and open.

- an effective super-agent is one that performs its tasks well (some details will be provided shortly).
- an open super-agent is one to which new agents can be easily added. (The rationale for openness is this: besides being effective now, we would like our super-agents to be effective in the future, that is, over expanded and as yet undetermined sets of tasks that might require agents still to be developed. The more open a super-agent, the better it is able absorb such new agents and thereby, to adapt to uncertain futures.)
- cooperation is the exchange of data, whether the exchange is beneficial or not.

In dealing with instances of (P1) that have practical significance, difficulties appear in three forms: conflicts, uncertainty and complexity.

- Conflicts occur when either the constraints cannot be simultaneously satisfied or the objectives, simultaneously minimized. Conflict handling requires interaction with the customer to determine which constraints are to be relaxed and which tradeoffs among the objectives are to be preferred.
- Uncertainty is equivalent to a lack of knowledge about the criteria (constraints and objectives). Either these criteria are imprecisely known to begin with, or they change randomly during the solution process. Uncertainties can be handled in two ways: a) by inserting safety margins large enough to cover anticipated variations in the criteria or b) by using feedback to adjust the computations to the latest and best estimates of the criteria. The latter approach is usually more practical.
- Complexity refers to the rate of growth of computational effort with task size. This rate is often so large that algorithms which are both rigorous and practical (algorithms which always find optimal solutions in reasonable amounts of time) are unavailable and likely to remain so into the foreseeable future. Instead, one is forced to use heuristics which are faster but fail to find optimal solutions in many, if not most, cases. This leaves two approaches to obtaining improvements: a) invent better heuristics or b) organize existing heuristic and rigorous algorithms so thay can cooperate and together, find better solutions than any of them could separately. Since inventions are difficult to schedule, we prefer the latter approach.

Thus, to be effective a super-agent must allow for interactions with customers (to deal with conflicts), must allow for iteration and feedback (to deal with uncertainties), and must promote cooperation among optimization algorithms (to deal with complexity).

A second problem formulation

Many natural super-agents are both open and effective. A school of Silversides (a sort of fish) is one example. The school is open to all Silversides (any Silverside can enter or leave whenever it wishes). Also, the school is effective over the task of protecting its members from Barracuda (the odds against a Silverside being eaten by a Barracuda increase at least linearly, and perhaps superlinearly, with the size of its school [1]).

In contrast to schools of fish, the organizations used by computer-based super-agents tend to bring openness and effectiveness into conflict. In other words, increases in openness are obtained at the expense of effectiveness and vice-versa. Real-time control systems and program-libraries are two extreme examples. In the former, the agents (programs such as state estimators, optimizers and simulators) tend to be so tightly bound that new agents cannot be added without tearing the system apart; in the latter, agents are unconnected and consequently, unable to cooperate [2].

One would much prefer super-agents that are effective over their current task-assignments and open to a wide variety of new agents so they can handle future and more difficult task-assignments. By interpreting the term "good" in (P1) to mean "open and effective" and by shifting our attention from the "organization" in (P1) to the family of super-agents it produces, we obtain a second, more specific formulation of our design problem, namely:

(P2):

given: T, a space (set) of optimization tasks, and
A, a space (set) of agents,
design: K*, a family (set) of super-agents, such that:
the members of K* are effective over T, and
the members of K* are open to A.

3. Super-agent structure. What exactly are super-agents? Where might one find open and effective super-agents? This section answers these questions, first, by developing structural models of super-agents (a structural model emphasizes how something is built as opposed to a behavioral model which emphasizes what it does), and second, by assembling a space where a wide variety of super-agents live. The focus is on computer-based agents, though the models can be extended to insects, fish and even more intelligent agents like people. Of course, these extensions may be accompanied by some losses of fidelity, especially if the agents being modeled are much more intelligent than today's computer programs.

An assumption

We assume that all communications among a finite set of agents can be modeled by a finite set of shared memories. (In support of this assumption we note that If the agents are computer-based, they can communicate in only two ways: by passing messages among themselves or by leaving information in memories they share. Since message-passing can be simulated by shared memories and vice-versa, only one of these mechanisms need be included in a model.)

Data flows, agents and control flows

As a result of the above assumption, one can think of a super-agent as a network of memories and agents. It is convenient to represent this network by a directed graph, called a data flow, whose nodes denote memories and whose arcs denote agents, as in Fig. 3.1. Each agent reads data from the memory at its tail, processes these data, and writes the results into the memory at its head. As such, each agent can be thought of as a bundle of processes. These processes can be divided into two fuzzy subsets: production processes and control processes. The production processes actually do the reading, calculating and writing. The control processes decide what is to be read, calculated and written, and when. If an agent contains all the control processes for its production processes, that is, if all the agent's input-selection and scheduling decisions are made internally, then the agent is said to be autonomous. If however, some of the control processes for agent-a reside in agent-b, then a is said to be supervised by b. We denote supervisory relationships among agents by a directed graph, called a control flow, whose nodes represent agents and whose arcs represent the supervisory relations among them, as in Fig. 3.2. Note: if the control flow is null (a graph with no edges), then all the agents are autonomous; if the data flow is null, then the agents cannot cooperate (exchange information).

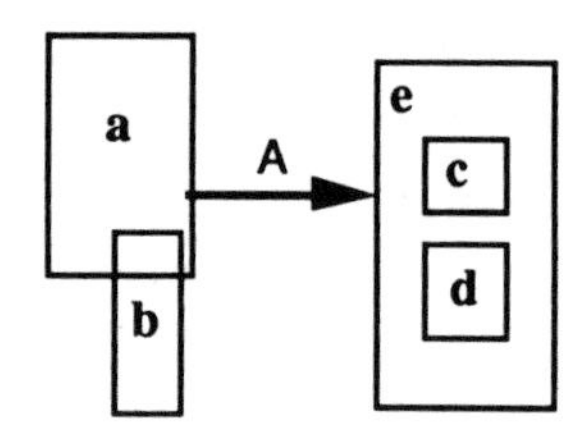

FIG. 3.1. *A data flow signifying that agent-A reads from memory-a and writes to memory-e. Also, memory-a has an intersection with memory-b, and e includes c and d.*

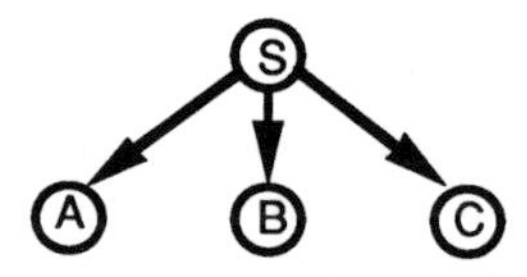

FIG. 3.2. *A control flow signifying that agent-S supervises agents A, B and C. In other words, S contains processes that select inputs for, or schedule the activities of, processes in A, B and C.*

Super-agents

Together, a data flow and a control flow describe the structure of a super-agent. In other words, a super-agent, k, can be thought of as a double:

$$k = (Df, Cf)$$

where Df is a data flow and Cf is a control flow. Note that the terms "agent" and "super-agent" are relative. In other words, super-agents can be assembled into super-super-agents, and so on. Moreover, super-super-agents and all their larger successors can be represented in the same way as super-agents, that is, by doubles, each containing a data flow and a control flow.

Some important types of super-agents are illustrated in Figs. 3.3 and 3.4.

Design as search

Think of designing an artifact as the process of obtaining its structure (the values of its designable variables), given its behavior. One technique for designing an artifact is this: construct (assemble) a space, each of whose points represents a different structural alternative; then, search this space for an alternative with the desired behavior. The better the construction of the space, the easier and more fruitful the search is likely to be.

There are both advantages and disadvantages to putting one's efforts into constructing and searching a space rather than using a more direct technique, such as constructing the desired artifact directly from its elemental parts. Among the advantages, a space helps in visualizing and understanding an entire range of artifacts. And search can be used to design artifacts that are unfamiliar, even counter-intuitive. Among the disadvantages, search-based approaches tend to be less efficient than more direct techniques. However, these direct techniques are usually the distillate of considerable experience and therefore, do not become available until such experience has been collected.

Many types of super-agents, especially those with null control flows, seem strange and counter-intuitive. Currently, there is little or no experience with designing them, and certainly no direct design techniques. Therefore, we will begin with a search-based approach to super-agent design, while recognizing that more direct and efficient approaches must be the ultimate goal.

Our definitions of data flows, control flows and super-agents provide the components from which to construct $S(k)$, a structural space of super-agents, as follows:

$$S(k) = S(Df)\ \mathbf{X}\ S(Cf)$$

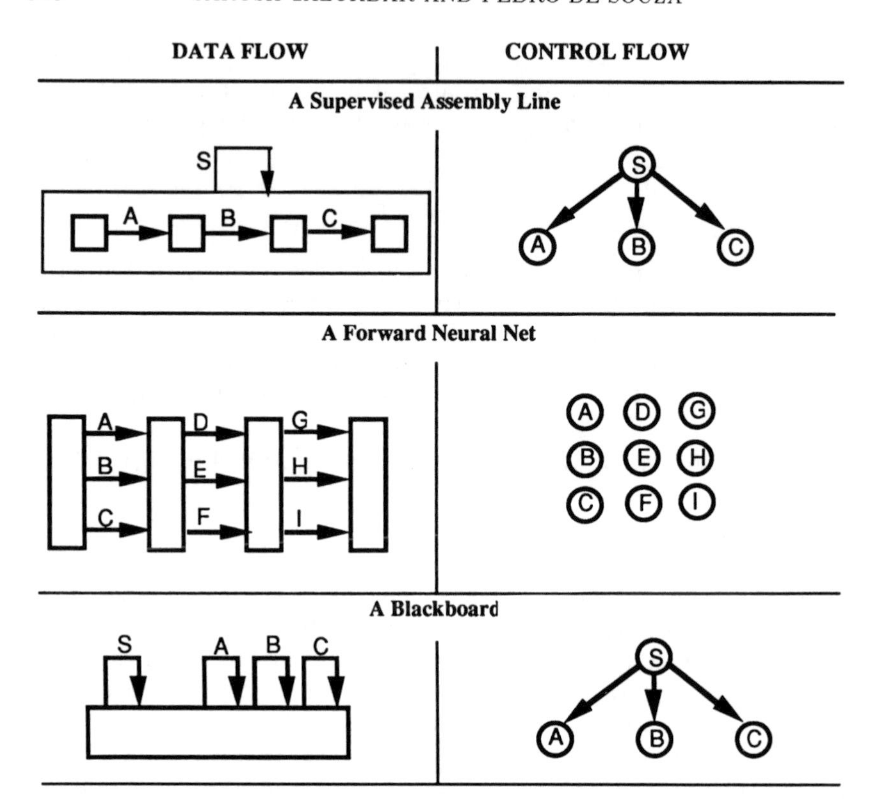

FIG. 3.3. *The data and control flows of: (1) a supervised assembly line, such as is used in mass production and most CAD packages; (2) a forward neural net with an input layer (neurons A, B and C), a hidden layer (neurons D, E and F), and an output layer (neurons G, H and I); and (3) a simple but powerful arrangement that has been rediscovered and renamed many times. When the agents are human, it is called a moderated group, when each agent is a single rule, it is called a production system, and when each agent is a largish program, it is called a blackboard. The agents modify the contents of the memory when they have access to it. The supervisor, S, decides the sequence in which the other agents gain this access.*

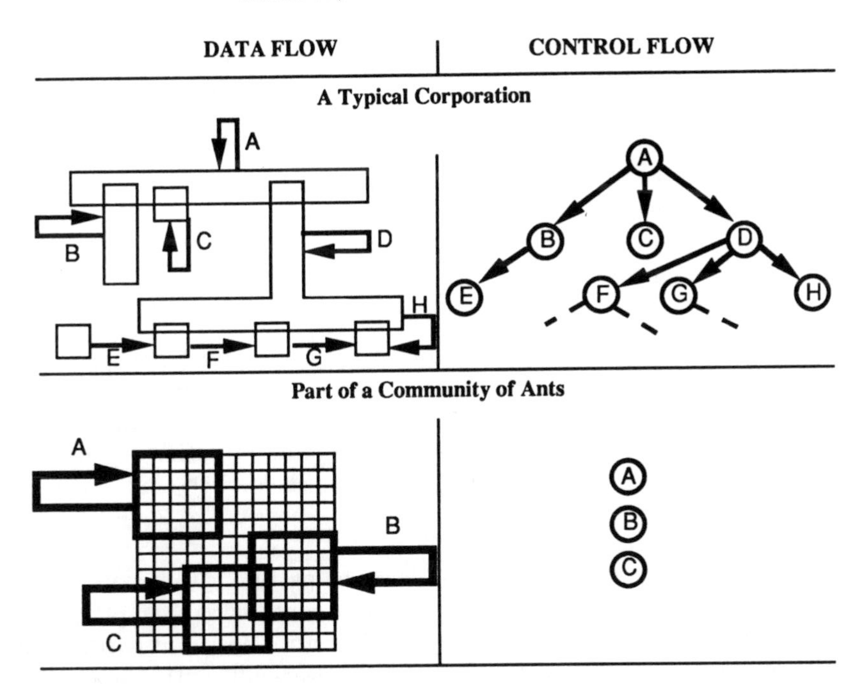

FIG. 3.4. *Portions of the data and control flows of a corporation and a community of ants. There are three main differences. First, corporations tend to use compound (multi-layered) hierarchies for their control flows, while ants, like other social insects—termites, certain bees and certain wasps—are essentially autonomous. Second, ants have comparatively modest reasoning and learning capabilities. And third, ants have a narrower view of the world. While a human may have access to much of what is happening in his corporation, an ant is probably aware of only its immediate surroundings. In other words, the input and output memories that represent an ant's view of the world in a data flow model, are coincident and cover only its immediate surroundings. Of course, these surroundings move with the ant and often intersect with those of its neighbors. The ant communicates with these neighbours through the products of its work, that is, by modifying its surroundings. Despite all these limitations, communities of ants are able to perform quite complex tasks, particularly, in the construction, maintenance and defense of their nests, and in the location, acquisition and storage of their food [5], [6]. Moreover, the abscence of supervisors makes an ant community very open.*

where $S(Df)$ is the space (set) of all data flows, $S(Cf)$ is the space (set) of all control flows, and **X** means Cartesian product. Now the design problem, (P2), can be reformulated as a search problem:

(P3)

given: T, a space of optimization tasks, and
A, a space of optimization agents,
search: $S(k)$ for a family of super-agents, K^*,
such that: the members of K^* are effective over T, and
the members of K^* are open to A.

4. A taxonomy of super-agents. $S(k)$ is a rather large space. In the following material we will partition it into four fuzzy regions, each containing a different kind of super-agent. The intent is to better understand the contents of $S(k)$ and simplify its search; instead of having to consider all of it, one can concentrate on the region with the desired kind of super-agent.

Four fuzzy regions

Let $S(Df)$, the space of all data flows, be partitioned into two fuzzy subsets: $U(Df)$, the subset of strongly cyclic data flows, and $\sim U(Df)$, its complement; let $S(Cf)$, the space of all control flows, be partitioned into two fuzzy subsets: $V(Cf)$, the subset of near-null control flows, and $\sim V(Cf)$, its complement. (Note: the specification of precise membership functions for these fuzzy sets is neither needed nor will be given here). Then $S(k)$ can be partitioned into the four fuzzy regions, CDNC (cyclic data, no control), CDSC (cyclic data, some control), ADNC (acyclic data, no control), and ADSC (acyclic data, some control), where:

$$\begin{aligned} \text{CDNC} &= U(Df)\ \mathbf{X}\ V(Cf) \\ \text{CDSC} &= U(Df)\ \mathbf{X}\ {\sim}V(Cf) \\ \text{ADNC} &= {\sim}U(Df)\ \mathbf{X}\ V(Cf) \\ \text{ADSC} &= {\sim}U(Df)\ \mathbf{X}\ {\sim}V(Cf) \end{aligned}$$

and **X** means Cartesian product. These partitions and a few of the families of super-agents they contain, are shown in Fig. 4.1.

Regional differences

There are qualitative differences among the regions of Fig. 4.1.

Super-agents in the two upper regions use supervisors. Consequently (c.f. Section 1), they tend to be more efficient in their use of agents and more familiar than super-agents from the two lower regions, but less open and more fragile.

Super-agents from the two regions on the right use acyclic data flows. This makes them well suited to repetitive and routine tasks, such as the mass production of automobiles, but ill suited to tasks that are uncertain, contain conflicts, call for iteration and feedback, or require interactions with customers (c.f. Section 1).

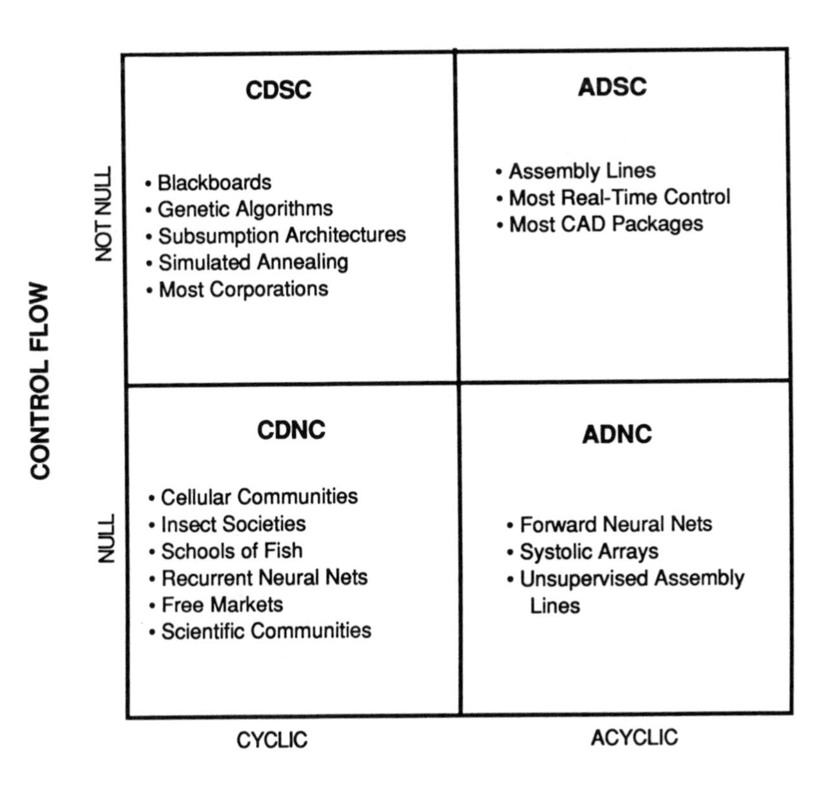

FIG. 4.1. *A taxonomy of super-agents obtained by partitioning $S(k)$, the space of all super-agents, into four fuzzy regions. All the regions contain natural and computer-based super-agents but the distribution is far from uniform. Specifically, there are few examples of computer-based super-agents in CDNC. In contrast, many natural super-agents have found it advantageous to reside there.*

If the tasks to be performed are routine, then a super-agent from one of the two regions on the right will, in all likelihood, suffice. For non-routine tasks, however, one should probably look elsewhere.

5. Asynchronous teams. Consider CDNC, one of the two regions appropriate for non-routine tasks. Its super-agents tend to be more open than those in other regions. A wide variety of natural super-agents, ranging from cellular colonies to scientific communities, have chosen to reside in it. Computer-based agents fall somewhere between cells and scientists in intelligence. Surely CDNC contains some good structures for them. But few of practical significance have been discovered. Indeed, we can think of no synthetic super-agents that lie in CDNC, aside from recurrent neural nets, and they remain to be made practical. Perhaps the reason is one of CDNC's relative disadvantages: its inefficient use of agents. In the past, this disadvantage was critical. But developments in computer networks are making it much less so. These developments make large amounts of computing power available at such low cost that large numbers of agents can be deployed and allowed to work in parallel. The productivity of any one of these agents is no longer of consequence. Some agents can be unproductive or even counter productive as long as the collection of agents, as a whole, is effective in performing its assigned tasks. How is this overall effectiveness to be achieved? One way is through the use of certain asynchronous modes of cooperation.

Asynchronous cooperation

Agents are said to cooperate asynchronously if they exchange data in such a way that no agent is delayed by another. That is, no agent has to wait for data that are to be produced or released by any other agent. Consequently, asynchronous cooperation allows all the agents to work in parallel all the time.

A-teams

Let Z be the subset of CDNC that contains super-agents capable of asynchronous cooperation. In other words, a super-agent k is in Z if k has a strongly cyclic data-flow, if k has a null or almost null control-flow (most if not all of its agents are autonomous), and if k's agents can cooperate asynchronously. The members of Z are called asynchronously cooperating teams or A-Teams, for short.

Advantages

The advantages of autonomous agents that cooperate asynchronously are considerable: extremely open super-agents that can mount massively parallel efforts without the need for centralized schedulers or any fear of deadlocks. But how is all this computational effort to be put to good use? Autonomous agents, each choosing what to do and when, if ever, to communicate, might seem like a recipe for anarchy. Actually, there are simple techniques for making asynchronous cooperation exceedingly effective. One just needs the right mixes of autonomous agents. To understand why, we must examine some issues in dynamics.

6. Super-agent dynamics

State flows

How can one visualize the dynamics (time-dependent behavior) of an A-Team, or for that matter, any other type of system? One way is to represent each state the system can occupy by a point in a space (called a state space). Then the trajectory traced by the point representing the current state, as this state varies with time, provides a visualization of the dynamics of the system.

Sometimes one can better visualize what is happening if the current state is represented, not by a single point, but by a population of points. In such cases, the state space is the union of all possible populations, and the dynamics are represented by an artifact, called a state flow, which is the trace of the current population as it varies with time. Note that a trajectory is a special case of a state flow; indeed, a trajectory is a state flow for a population of one point.

The state of a super-agent is characterized by the contents of its memories. Suppose that these contents are stored in the form of data-objects. Suppose also that a state space is assigned to each memory. Every object that can be stored in the memory is represented by a point in this space. Let Yj be the state space and $Xj(t)$ the state at time t of the j-th memory. That is, let $Xj(t)$ be the subset of Yj that represents the population of objects actually in the j-th memory at time t. Thus, the trace of $Xj(t)$, as t varies, is the state flow for the j-th memory.

Let Pj^* be the target or goal state of the j-th memory. The state flow, $Pj(t)$, for this memory is effective if $Pj(t)$ converges suitably quickly to Pj^* from any starting population, $Pj(0)$. The entire super-agent is effective if all its state flows are effective.

Making state flows effective

Many natural and synthetic methods have evolved for producing effective state flows. The following material lists some of them.

Descent Algorithms

Descent algorithms (their mirror images are called hill climbing algorithms) are used to solve optimization problems of the form:

(Q1):

$$\begin{aligned} &\text{Minimize } f(X) \\ &\text{subject to: } H(X) = 0 \\ &\qquad\qquad\quad G(X) < 0 \end{aligned}$$

where f is an objective function; X is an n-dimensional vector of real decision variables (the elements of X may be continuous or discrete); and H and G are function vectors embodying the constraints to be met by X.

The state space for (Q1) is Rn, the space of n-dimensional real vectors. Every possible value of X is denoted by a point in this space.

(Q1) can have multiple solutions. The best of these is called the global minimum. The rest are called local minima. Descent algorithms do not distinguish between local and global minima. In other words, the target state, X^*, is any solution, regardless of whether it is local or global. Convergence to this state is sought by iteratively changing X^i, the incumbent estimate to X^*. One can think of these changes as being made to decrease some merit function, $m(X^i)$, designed to reflect the value of the objective and the constraint violations, if any, of the point X^i.

Thus, the state flow of a descent algorithm is a trajectory that results in the monotonic decrease of some merit function. Descent algorithms vary only in their choice of merit function and in the techniques they use to reduce its value. They do not distinguish between local and global minima and invariably, head for the nearest minimum, regardless of its quality.

Simulated Annealing

Like descent algorithms, simulated annealing algorithms are iterative and apply to problem (Q1). Unlike descent algorithms, they allow for some randomness in the search for a solution. A descent algorithm will make a change to X^j, the incumbent estimate, only if the change produces a decrease in the chosen merit function. A simulated annealing algorithm, however, allows some changes that increase the value of the merit function. The decision to accept or reject such changes is made by simulating the toss of a coin. Heads: the change is accepted, tails: it is rejected. The coin is biased so that the probability of "heads" decreases as the number of iterations increases. Thus, the state flow of a simulated annealing algorithm is a trajectory that gradually changes from random directions to those dictated by a descent algorithm. The effect is that the chances of finding a global minimum are increased over those of a pure descent algorithm.

Genetic Algorithms

Like descent and simulated annealing algorithms, genetic algorithms are iterative and apply to problem (Q1). Unlike descent and simulated annealing, a genetic algorithm works on a population of estimates rather than a single estimate. As such, a genetic algorithm must deal with matters of population control. Is the population to be allowed to grow indefinitely? If not, how is it to be culled? A genetic algorithm deals with these matters by dividing each of its iterations into two stages. In the first, the current generation (population of estimates) is shrunk by destroying its weakest members (those with the poorest values of some preselected merit function). In the second stage, a new generation is created from the survivors of the first stage by algorithm-specific operators.

Thus, the state flow of a genetic algorithm is oscillatory; the population of estimates contracts and expands periodically as it is acted on by alternating processes of destruction and creation. The goal is to move each new generation closer to the target: the global minimum of (Q1). How fast this happens, if it happens at all, is determined by both the processes of destruction and creation. However, the designers of genetic algorithms have, so far, focused on the former, attempting to capture in mathematical operators the processes involved in cellular reproduction. Computational processes for destruction have received far less attention and as a result, are far less skillful.

Lamella Bone Growth

The long bones in mammals are continuously modified by two types of cells: Osteoclasts and Osteoblasts. The former converge on faces that have, during the last few weeks, either been under low stress or in tension and proceed to dissolve the bone there. The latter converge on faces that have been under high compressive stress and proceed to add bone there.

The net effect is to optimize the shape and mass of the bone for its past stresses [3].

Suppose that the bone and its neighborhood are divided into little cubes by a fine, three-dimensional mesh. Then, the set of these cubes constitutes a state space in which the current state of the bone is represented by a population, each of whose members is a cube filled with bone. This population changes continuously with time as new members of the population are created by Osteoblasts (as they add bone exactly where it was most needed), and old members are destroyed by the Osteoclasts (as they remove bone from where it was least used). The result is a state flow that is in constant pursuit of the optimum shape and mass for the current stress. The pursuit is maintained by creative and destructive process of equal intelligence. The contrast with descent, simulated annealing and genetic algorithms is interesting. Descent and simulated annealing devote none of their intelligence to destruction. Existing genetic algorithms devote some of their intelligence to destruction, but far less than goes into creation. In lamella bone growth, however, creation and destruction are exactly symmetrical. We suspect that this symmetry minimizes the total intelligence needed to solve many, if not most, optimization problems.

Weaver Ants

The Weaver ant builds its nests out of living leaves. "The construction of the nest requires the cooperation of many workers—some to hold the leaves in place while others spin the silk to bind them. When workers first attempt to fold a leaf they spread over its surface and pull up on the edge wherever they can get a grip. One part is turned more easily than others, and the initial success draws other workers who add their effort, abandoning the rest of the leaf margin" [10]. They hook their claws together to form long chains with the ant at the head holding on to a part of a leaf, the one at the tail, to a branch or other fixed object, and those in the middle, suspended in air. Through such chains they are able to exert forces sufficient to fold the leaves over.

Consider the problem of finding the end points for these chains. It would seem that the ants solve this problem by beginning with a breadth-first search. Almost immediately, however, they set to developing a consensus, that is, to pruning the alternatives until only a few of the very best remain for the concentrated effort of the entire work-force. The result is a state flow characterized by a single rapid expansion followed by a single gradual contraction.

Formations

Formations are the geometric patterns that are maintained by certain collections of moving things, such as flights of aircraft, flocks of birds and schools of fish. In optimization terms, the problem is to minimize deviations from a given pattern, or sequence of patterns, while avoiding collisions. In a flight of aircraft, the pattern is usually simple; a vee or diamond, for instance. One member of the flight acts as its leader and is responsible for choosing the overall direction. The other members maintain the pattern by periodically checking the positions of one or more of their neighbors and making corrections when their separations drift from their assigned values. These positional corrections are made asynchronously (each member works on essentially his or her own schedule), and intermittently (each member has other duties to perform, such as monitoring the cockpit's instruments).

Flocks of birds and schools of fish seem to use similar procedures except that they have no leaders and produce patterns very much more complex than those employed by aircraft. These complex patterns can be simulated by simple, strictly local interactions: each member reacts as if under the influence of a force that repels it from neighbors that are very close, attracts it to neighbors that are a little further away, and vanishes for members that are still farther away.

Observations

- There is no single form for an effective state flow. The examples above provide only a small sample of the types of state flows that can be effective.
- The contents of any memory are the end results of the actions of the agents connected to that memory. In other words, the mix of agents connected to a memory determines its state flows.
- How is one to assemble mixes of agents that will produce effective state flows? This question can be decomposed into a number of design decisions including:
  - the relative weights given to greedy strategies that concentrate on short term gains (descent algorithms, for instance), random strategies (those in simulated annealing, for instance), and long term views (such as human agents can provide).
  - the size of the population: one point, as in a descent algorithm, or many points, as in a genetic algorithm.
  - the balance between creation and destruction. The possibilities include: no destruction, as in descent algorithms, destruction of relatively limited intelligence, as in traditional genetic algorithms, and symmetrical creation and destruction, as in lamella bone growth.
  - the nature of the cohesive forces, if any, that bind agents together, as in formations, or cause them to seek a consensus, as in the nest building of Weaver ants.

7. Organizations as graph grammars. Consider the set of words: he, is, tired. These words can be combined to produce a variety of sentences, some that are legal in English, such as "he is tired," and some that are not, such as "tired he he tired."

Loosely speaking, a language is a subset of all the possible sentences that can be assembled from some collection of words, and a grammar is a set of rules for constructing the sentences in this subset, starting from some initial form or seed.

The notions of grammars and languages extend to artifacts other than words and sentences. For instance, think of the set of all possible houses. Now think of the subset of houses that Frank Lloyd Wright might have designed, if given infinite time. This subset constitutes an architectural language. A grammar for this language would be a set of rules for generating its members. (A number of such architectural grammars have actually been developed.)

A grammar provides a compact means for describing and constructing the members of a family of artifacts. But its development can take a good deal of effort. And success is predicated on an intimate understanding of the artifacts in the family and considerable experience with their design.

We have not acquired quite enough experience and understanding of any family of super-agents, to develop complete grammars for them. But we have devised precursors to grammars for A-Teams that solve problem (P3). Some of these precursors are described in the succeeding material. We feel that their refinement into complete grammars is only a matter of time, as is the development of grammars for other useful families of super-agents.

Recall that we defined an organization to be a set of rules for constructing a family of super-agents, and chose to represent each super-agent by two graphs (one for its data flow, the other, for its control flow). Thus, an organization, as viewed here, is a graph grammar.

Recall that in an A-Team the agents are autonomous (the control flow is null). Therefore, only the data flow of an A-Team needs to be designed. We will perform this design in two stages. First we will compose a data flow template (a data flow with "slots" where arbitrary numbers of agents can be inserted) , as in Fig. 7.1. Second we will choose mixes of agents to put in these slots.

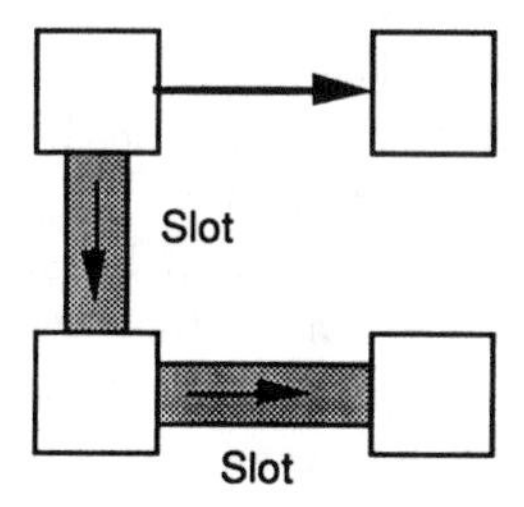

FIG. 7.1. *A data-flow-template for a family of super-agents. Members of the family are obtained by inserting agents into the slots. In principle, there is no bound on the number of agents that can be placed in a slot.*

Recall that the data flow of an A-Team must be strongly cyclic, that is, it must contain significant feedback loops. The purpose of the data-flow-template is to establish the topology of these loops. To obtain this template, we start with a seed consisting of a single memory for complete solutions to the task(s) to be performed, and a slot for agents that read from, and write to, this memory, as in Fig. 7.2. If experiment or analysis reveals this seed to be inadequate, then more loops will be added, as in Example 7, below.

Recall that the behavior of a super-agent is represented by the state flows of its memories and each of these flows is determined by the mix of agents connected to the memory in which it occurs. The purpose of the second stage is to adjust these mixes so the state flows are made effective. This is done by adding and deleting agents from the slots in the data-flow-template.

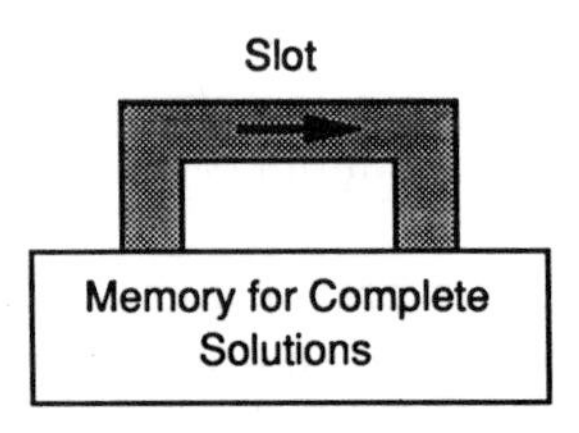

FIG. 7.2. *A seed for data-flow-templates. More elaborate templates are obtained by adding memories for partial solutions or other types of data, slots and agents.*

Example 7.1. The Traveling Salesman Problem. A traveling salesman problem is a single objective, combinatorial optimization problem of the form:

TSP:

given: the locations of N cities,

find: a minimum length tour (a closed path that passes through every city just once).

The TSP is a prototypical combinatorial problem; it shows up in a multitude of engineering applications, some that are immediately recognizable, such as sequences for soldering and testing circuit boards, others that are less obvious but can be modeled as TSPs, such as flow-shop scheduling.

Algorithms for solving TSPs and other optimization problems, can be divided into two categories: rigorous and heuristic. The former, produce optimal solutions if given enough time, the latter provide no guarantees of optimality, and usually fall far short of it, no matter how much time they are allowed.

Like many important problems, the TSP is NP-hard. Loosely speaking, this means that the rigorous algorithm which can solve all TSPs in less than exponential time (an amount that increases exponentially with the number of cities), remains to be discovered and may not exist. In other words, rigorous algorithms cannot be relied on when the number of cities becomes large, say 500 or more: they require too much time. However, there are many heuristics which are much faster. But none of them guarantee optimality. Can heuristic and rigorous algorithms be organized into A-Teams that find better, faster solutions than any of the algorithms could alone? Can such A-Teams be constructed rather than obtained by a much less efficient search through the set of all A-Teams? in other words, can we solve the following specialization of (P3):

(P4)

given: T', a space of TSPs,
A', a space of existing algorithms (heuristic and rigorous) for TSPs,
construct: K^*, a family of super-agents,
such that: the members of K^* produce state flows that converge more quickly, and to better tours than the individual members of A', and the members of K^* are A-Teams

Note that (P4) drops the requirement for the members of K^* to be open because it has become redundant: all A-Teams are open. Also, the replacement of the imperative "search" by "construct" is optimistic; it represents what we would like to be able to do rather than what we can do.

The first step in solving (P4) is to find an appropriate data-flow-template. We start with a seed: a memory of complete tours with a slot for agents that modify these tours. Next we insert agents in this slot, and if necessary, add other feedback loops. We do not have a direct procedure for doing this, nor even a systematic procedure. Instead, we grope around and fumble with the possibilities till we find a template that works well. The best so far is shown in Fig. 7.3. It has proven itself in tests on many difficult problems. Typical results for one of these problems are given in Figs. 7.4 and 7.5.

Note that:

- the team is very open. All TSP algorithms can be grouped into two classes: construction algorithms that convert partial tours (tours of $n < N$ cities) into complete tours, and modification algorithms that reconfigure complete tours. As such, all TSP algorithms fit directly into one or another of the slots in the data-flow-template, provided only that they be equipped with routines to read from, and write to, the appropriate memories.
- the team is very effective. It is able to find much better tours than its constituent TSP algorithms. It is also quicker. Even when the entire team must share a single computer, it converges faster than any of its constituents. This speed advantage increases linearly with the number of computers made available to the team.
- besides TSP algorithms, the team contains a number of destroyers

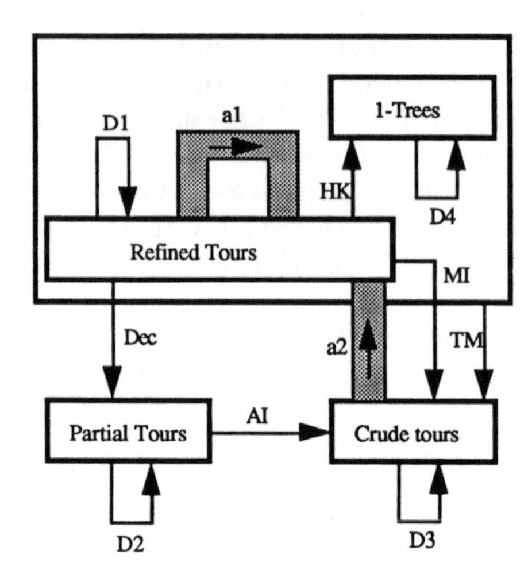

FIG. 7.3. *A data-flow-template for TSPs. Arbitrary numbers of agents embodying tour-modification algorithms can be inserted in the slots: a1 and a2. The remaining part of the structure provides feedback loops for these agents. Two of the agents in these loops are well known: AI (arbitrary insertion) and HK (held-Karp). The rest were specially designed for the template. The destroyers, D1 through D4 are for population control. Each uses a different heuristic. D1, for instance, selects elements for destruction on the basis of merit (tour length) and chance (the flip of a biased coin, as in simulated annealing, but with a linear rather than Boltzman distribution). Further details can be found in [20].*

(agents that delete tours from the memories with which they are associated). The net effect of the TSP algorithms and destroyers is to "herd" the population of complete tours towards optimality, as indicated in Fig. 7.5. The movements are gradual, like those of lamella bone growth rather than violently oscillatory, like the state flows of genetic algorithms.

Example 7.2. Robots-on-Demand. The robots-on-demand problem is a multi-criterion, mixed integer problem:

ROD:

given: a bin of robot parts (motors, joints and links), a robot task expressed as a set of kinematic and dynamic constraints, and a set of conflicting objectives (minimize robot weight and deflection, maximize robot dexterity)

find: those robot designs which are in the Pareto frontier with respect to the constraints and objectives.

By "robot design" we mean a string of twelve variables. Three are continuous and locate the base of the robot. The other nine are discrete and identify the motors, joints and links that constitute the robot's segments.

a1	a2	Average Error (Miles)	Time to 200*
–	MI	200	12.5
CLK	MI	150	7.5
CLK	MI+OrOpt	60	4.8
CLK+LK	MI+OrOpt	30	1.2

* average time in hours for the entire team, running on a single DECstation 5000/200, to come within 200 miles of the optimum-tour-length (27686 miles).

FIG. 7.4. *Typical results for A-Teams obtained by adding agents to the slots (a1 and a2) of the template in Fig. 7.3. The problem being solved has 532 cities and an optimum tour of 27686 miles. Two of the agents used, OrOpt and LK (Lin-Kernighan) are well known. CLK is a simplified version of LK. MI is an algorithm we have developed. In teams, these algorithms always generate better results than when they are working alone [20]. Notice that both the quality and speed of results improve as more algorithms are added to the slots, even though all the algorithms have to share a single computer. When the algorithms are given their own computers, the speed-up is much more dramatic [20].*

The intent in solving ROD is to obtain a robot design that is specialized for the task the robot is to perform. Once this task has been completed, the robot is disassembled so its parts can be reused in a new robot for the next task. Of course, when there are conflicting objectives, such as minimizing weight and deflection, there is no single optimum robot design. Rather, there are two sets of designs worth considering: F and P, where F is the set of designs that are feasible (meet the constraints), and P is the subset of F that contains designs with the very best possible tradeoffs among the objectives. P is called the Pareto frontier. Once P has been found, one of its members must be selected for actual implementation. This choice is subjective and will not be considered further here. There is no well established set of heuristics for solving ROD. In order to obtain A-Teams for solving RODs, one must not only design a data-flow-template and select the mixes of agents for its slots, but also, develop the agents themselves.

It happens that the seed (a single memory for complete robot designs with one slot for agents that modify these designs, as in Fig. 7.2) works fairly well, so we have not bothered to look for a better template. At this writing, the seed's slot has been stocked with some 60 agents that create new robot designs by modifying old ones, and about 10 that destroy designs. All the agents are very simple. Each creator acts on one of the twelve variables that constitute a design and takes into account only one criterion (one of the objectives or constraints). Its actions usually produce improvements in this criterion but, as often as not, are accompanied by a large degradation in the other criteria. For instance, one of the agents attempts to reduce kinematic constraint violations by replacing the link in the robot's third segment with the next longer link from the bin of parts.

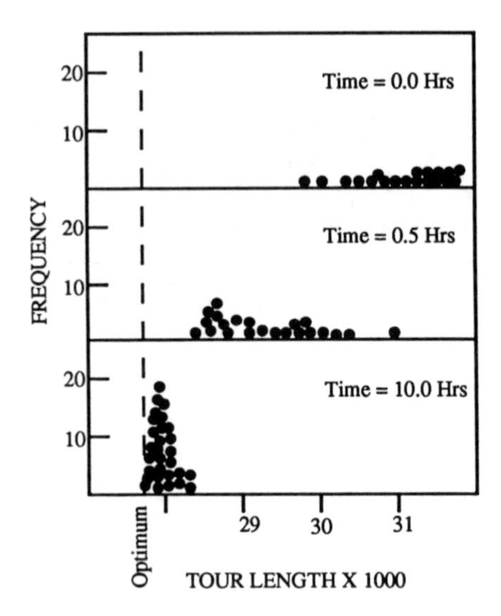

FIG. 7.5. *Three snapshots of a typical state flow for the refined-tours-memory (Fig. 7.3). Through the simultaneous effects of creation and destruction, the population of tours in this memory is herded from its initial state to the optimum.*

Invariably, this replacement causes a large degradation in deflection and dynamic constraint violations.

The little intelligence that is in each creator is concentrated in its input controller. This controller selects the next design for the agent to modify by performing a qualitative match of the design's needs and the agent's capabilities. In essence, the design is selected if it needs the improvement offered by agent and can afford the price (the accompanying degradation in other criteria).

Each destroyer also considers only a single criterion and its intelligence is also concentrated in its input controller. The selection process involves two steps. First the candidate design is tested with respect to the destroyer's criterion. If the design fails, the destroyer proceeds to the second step: flipping a coin that is biased so "tails" is more likely for designs that have failed badly. If the coin comes up "tails," the design is eliminated. Most of the destroyers use obvious criteria, such as kinematic-constraint-violations. However, one of them is a little more clever: it uses distance from the Pareto set as its criterion. That is, the distance from the candidate design to the subset of the best designs that currently exist.

All the 70 or so available agents are "plugged" into the data flow template. The resulting A-Team generates far better designs than any other process we know about. (A typical state flow is shown in Fig. 7.6).

Note that the agents are very narrow in their views. Each concerns itself with only one of the many coupled criteria in the problem. Yet

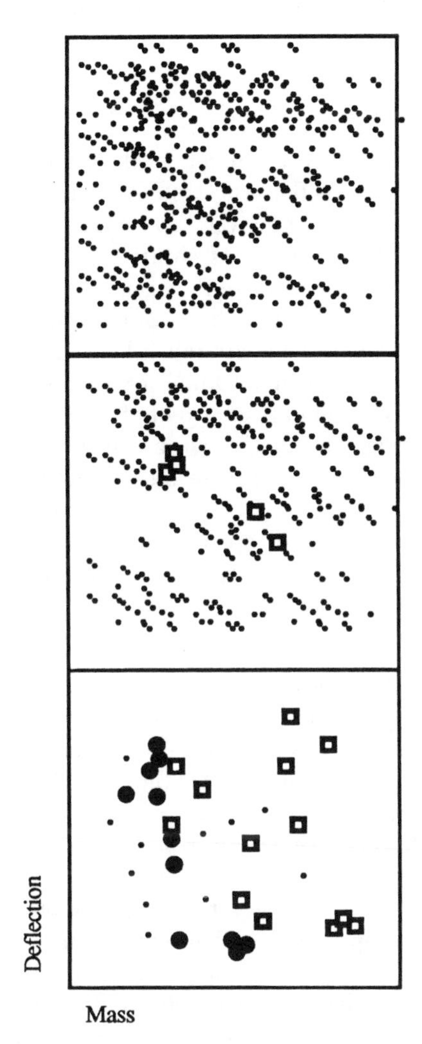

FIG. 7.6. *A two-dimensional section of a state flow for the ROD problem. Small dots denote infeasible designs, squares denote feasible designs and large dots denote designs in the Pareto frontier. A randomly selected initial population of robot designs (top) is herded by creators and destroyers through an intermediate state where some feasible designs have appeared (middle) to a final state where a number of members of the Pareto frontier have been found (bottom).*

together, they are able to find good (Pareto optimal) solutions. The lesson is that a set of simple agents with decoupled skills can solve a problem with multiple and profoundly coupled criteria, if each agent's selection controller is carefully designed and perhaps more important, if some of the agents are devoted to destruction and are as ingenious in performing this duty as the agents devoted to creation.

More about A-teams

What makes an A-Team so effective? The answer has three parts. First, all the agents in an A-Team can work in parallel without delaying one another. Second, an A-Team is strongly cyclic enabling iteration and feedback to be used to their greatest advantage. Third, an A-Team capitalizes on the diversity of its agents. Markov models of the dynamics of A-Teams show that the more diverse the agents, the greater the probability of finding good answers [20]. In other words: two heads are better than one and two very different heads are much better than one. For instance, in solving optimization problems, combining agents that use genetic algorithms (which tend to have low resolution but good coverage of the solution space) with agents that use quasi-Newton algorithms (which tend to have high resolution but low coverage) increases the probability of finding good solutions.

What makes an A-Team so open? Since the agents are autonomous, there is no managerial super-structure to modify when an agent is added or deleted. Consequently, the effort involved in adding or deleting agents is minimal.

8. Summary. A super-agent is an aggregation of lesser agents, just as a college is an aggregation of people and a university is an aggregation of colleges. An organization is a set of rules for constructing a super-agent. The structure of a super-agent can be thought of in terms of two flows: one of data, the other of control. Each of these flows can be represented by a directed graph. The behavior of a super-agent can be represented by a third flow: of state.

The Cartesian product of the sets of all data and control flows defines a structural space of super-agents. This space can be partitioned into four fuzzy regions. Three of these regions are fairly familiar; they contain families of natural and synthetic super-agents that make intuitive sense. The fourth region, CDNC, is notable for being much less familiar, for containing a number of natural families of super-agents that are both open and effective, and for containing virtually no synthetic families of super-agents.

We have demonstrated that at least one synthetic family of super-agents, called A-Teams, can be made to reside in CDNC with benefit. Members of this family use only autonomous agents that communicate asynchronously and work iteratively on populations of solutions. As such, they combine some of the best features of several existing super-agents, including cellular communities, insect societies, simulated annealing and genetic algorithms. The result is that A- Teams are open, parallel and distributable. They can also be made extremely effective in performing difficult optimization tasks, as has been demonstrated by an expanding list of practical applications and explained by the "two heads conjecture." (Markov models of the dynamics of A-Teams show that the more diverse the agents, the greater the probability of finding good answers. Colloquially: two heads are better than one and two very different heads are much better than one.)

Since optimization comes close to, if not actually providing, a domain-independent language for specifying engineering tasks, one might reasonably expect A-teams to be useful in a variety of engineering applications. One might also expect other families in CDNC to prove useful and worth the effort of further exploration.

The design of any super-agent can be decomposed into four coupled stages: 1) design the data flow topology, 2) design the control flow topology, 3) design a state space for each memory in the data flow (devise representations for the entities to be stored in the memories), and finally, 4) develop or acquire the agents to be used.

What have we learned about the design of super-agents? Perhaps the most important lesson is that complex tasks do not require complex agents. With the right flows of data and control and the right mix of agents, very complex and difficult tasks can be performed by simple and unintelligent agents. Of course, we knew this to be true at the hardware level. Everyone is aware that simple transistors, in sufficient numbers and with the right arrangement, form a processor with formidable capabilities, as do simple neurons combined into a neural net. But it is much less obvious that simple procedures or algorithms can be made to behave in the same way. Indeed, much of software engineering has been driven by the contrary belief, specifically, that each problem to be solved must be paired with a single, monolithic solution procedure. If a suitable procedure is not available, then the problem must be decomposed into sub problems. Thus the prevalent paradigm is: decompose the overall problem into suitably small sub problems, solve each sub problem with a single procedure, and then, compose the results into a solution of the overall problem. The alternate paradigm that we have learned is: compose all the relevant procedures (randomized and deterministic, symbolic and sub-symbolic, etc.) into a super-procedure and apply it directly to the overall problem. The availability of cheap, distributed computing makes this alternate paradigm practical.

A second lesson is the importance of using populations of solutions and intelligent destruction in controlling these populations. Most existing solution processes work with a single incumbent solution. Those that use populations, such as genetic algorithms, emphasize creation over destruction. For instance, the typical genetic algorithm has a multitude of operators for creating new members of its population but only one relatively dumb operator for destroying old members. We believe that the deconstruction of inferior members to recover their reusable parts and the destruction of hopeless members to prevent their further consideration, are at least as important as the creation of new members. In other words, the intelligence in the overall process should be allocated symmetrically between creation and destruction.

Besides the balance between creation and destruction, the selection of the mix of agents involves a number of design decisions that include the balance between short- and long-term views, and the type of cohesive forces (social instincts), if any, to be instilled in each agent. We have, as yet, no systematic methods for making these decisions nor for making those involved in the other three design stages.

What are the prospects for systematic methods? We believe they are good. By representing the structure of a super-agent by a pair of graphs we make it possible to think of an organization (a set of rules for constructing super-agents) as a graph grammar. The technology of graph grammars is fairly mature. By applying it to the experience gained from building new super-agents and the insights obtained from studying existing super-agents, we believe that systematic construction methods will inevitably emerge.

REFERENCES

[1] B.L.PARTRIDGE, *Structure and function of fish schools*, Scientific American **24** (6) (1982), 100–109.

[2] S.TALUKDAR, V.C.RAMESH, R.QUADREL, R.CHRISTIE, *Multi-agent organizations for real-time operations*, Proceedings of the IEEE **80** (5) May (1992).

[3] S.MURTHY, *Synergy in cooperating agents: designing manipulators from task specifications*, (*Ph.D. dissertation*) Carnegie Mellon University, September 1992.

[4] H.M.FROST, *The laws of bone structure*, Charles C. Thomas, Springfield, Illinois 1964.

[5] M.J.W.EBERHARD, *The social biology of polistine wasps*, (*miscellaneous publications*) Museum of Zoology, Univ. of Michigan **140** (1969), 1–101.

[6] C.W.RETTENMEYER, *Behavioral studies of army ants*, Kansas Univ. Sci. Bulletin **44** (9) (1963), 281–465.

[7] E.O.WILSON, The Insect Societies, Belnap Press of Harvard University: Cambridge, MA 1971.

[8] H.KALMUS, *Vorversuche über die orientierung der biene im stock*, Zeitschrift für Vergleichende Physiologie**24** (2) (1937), 166–187.

[9] B.A.WEISS, T.C.SCHNEIRLA, *Inter-situational transfer in the ant formica schaufussi as tested in a two-phase single choice-point maze*, Behaviour, **28** (3–4) (1967), 269–279.

[10] J.H.SUDD, *How insects work in groups*, Discovery, London, June (1963), 15–19.

[11] M.LINDAUER, *Ein beitrag zur frage der arbeitsteilung im bienenstaat*, Zeitschrift für Vergleichende Physiologi **34** (4) (1952), 299–345.

[12] C.EMERY, *Le polymorphisme des fourmis et la castration alimentaire*, Compte Rendu des Séances du Troisième Congrès International de Zoologie, Leyde 1895, 395–410.

[13] E.O.WILSON, *Chemical communication among workers of the fire ant solenopsis saevissima*, Animal Behavior **10** (1–2) (1962), 134–164.

[14] P.P.GRASSÉ, *Nouvelle expériences sur le termite de Müller (macrotermes mülleri) et considérations sur la théorie de la stigmergie*, Insectes Sociaux, **14** (1) (1967), 73–102.

[15] S.N.TALUKDAR, P.S.DE SOUZA, *Scale efficient organizations*, Proceedings of the 1992 IEEE International Conference on Systems, Man and Cybernetics, Chicago, Illnois, Oct. 18-21, 1992.

[16] S.N.TALUKDAR, S.S.PYO, R.MEHROTRA, *Distributed processors for numerically intense problems*, Final Report for EPRI Project, RP 1764-3, March 1983.

[17] R.QUADREL, *Asynchronous design environment: architecture and behavior*, (*Ph.D. dissertation*) Carnegie Mellon University, 1991.

[18] C.L.CHEN, S.N.TALUKDAR, *Causal nets for fault diagnosis*, 4th International Conference on Expert Systems Application to Power Systems, Melbourne, Australia, Jan 4-8, 1993.

[19] S.N.TALUKDAR, V.C.RAMESH, *Cooperative methods for security-planning*, 4th International Conference on Expert Systems Application to Power Systems, Melbourne, Australia, Jan 4-8, 1993.

[20] P.S.DE SOUZA, *Asynchronous organizations for multi-algorithm problems*, (*Ph.D. thesis*) Carnegie Mellon University, April 1993.

APPLICATION OF REAL-TIME PHASOR MEASUREMENTS IN POWER SYSTEM CONTROL

JAMES S. THORP*, STEVEN M. ROVNYAK*, AND CHIH-WEN LIU*

Abstract. Utilities are attempting to utilize newly available synchronized phasor measurements to improve system protection and control. Conventional static state estimators can be enhanced with the addition of phasor measurements. In addition, it may be possible to detect and respond to potentially unstable transient events in the electric power system. One method would be to predict the outcomes for a discrete set of control options, in effect implementing a discrete event control scheme. Modern microprocessor technology actually permits one to integrate a reduced-order system model in a small fraction of a second. For example the New England 10 generator system takes 0.07 seconds to integrate all the generator angles one second into the future. This paper describes efficient solution techniques for various models: constant impedance, constant $P-Q$, and dynamic Ward-type equivalent; and measures their execution speeds.

1. Introduction. Recent developments in satellite based dissemination of high accuracy time reference signals, and in the field of computer based substation functions has provided power system engineers with a new tools for achieving better overall system performance. We begin with an introduction to phasors measurements, and then describes the techniques and uses of synchronizing phasor measurements for improved monitoring, protection, and control of a power network.

With new systems capable of making real-time phasor measurements, the real-time assessment of the stability of a transient event in the power system has become an important area of investigation [1]. By synchronizing the sampling of microprocessor based systems, phasor calculations can be placed on a common reference [2]. Commercially available systems based on GPS (Global Positioning System) satellite time transmissions can provide synchronization to 1 microsecond accuracy. The phasors obtained from a period or more of samples from all three phases provide a precise estimate of the positive sequence voltage phasor at a bus. The magnitudes and angles of these phasors make up the state of the power system and are used in state estimation and transient stability analysis. By communicating time-tagged phasor measurements to a central location, the dynamic state of the system can be tracked in real time. Utility experience indicates that communication bandwidths can handle 12 complete sets of phasor measurements per second [3], which corresponds to one set every 5 cycles.

Using these phasor measurements for real-time transient stability prediction can advance the fields of protection and control. Out-of-step relaying is an obvious area of application. If an evolving swing could be determined to be stable or unstable, then the appropriate blocking or tripping could be initiated. A control application might involve determining

* Department of Electrical Engineering, 223 Phillips Hall, Cornell University, Ithaca, NY 14853-5401.

whether the event would be stable under a variety of control options. In both cases the determination of stability or instability must be accomplished faster than real time in order for effective action to be taken. In other words it is necessary to predict the outcome before it actually occurs.

Planning studies of power system dynamic instability have resulted in the development of a number of techniques used in simulation studies to save computer time. That is, if the stability of a large number of cases is to be studied, it would be efficient to devise a procedure that would not require the numerical integration of the complete trajectory for each of the cases. In the past decade two techniques have become popular for this application. They are: the Lyapunov or energy function approaches [4] and the Extended Equal Area criterion [5]. It should be recognized that these techniques (and all the variations: Venikov's method [6], Haque & Rahim [7], BCU method [8]) were not developed with real-time application in mind. The original intent was to determine the stability of a large number of contingency cases in a planning mode. The growing interest in on-line applications where a similar study is carried out in the control center in response to changing system conditions still falls short of being a real-time application. Nevertheless, the possibility of using one of these techniques directly with real-time phasor measurements was an early expectation of the work.

Many transient stability assessment techniques while simple in off-line application are too complicated for real-time use. Real-time monitoring obviates the need for some of these techniques since the system itself is actually solving the differential equations. What is required is a computationally efficient way of processing the real-time measurements to determine whether an evolving event will ultimately be stable or unstable. The availability of powerful workstations makes new approaches to the problem possible.

2. Phasors. A sinusodial quantity and its phasor representation are defined as follows:

2.1. Sinusodial quantity phasor

$$x(t) = X_m \cos(\omega t + \varphi) \longleftrightarrow X = \frac{X_m}{\sqrt{2}} \epsilon^{j\varphi} \tag{2.1}$$

The phasor computed from N samples taken over a one-cycle (of the fundamental frequency) window represents the fundamental frequency component of a waveform observed over a finite window. With the sampled data x_k, the phasor is given by

$$X = \frac{1}{\sqrt{2}} \frac{2}{N} \sum_{k=1}^{N} x_k \epsilon^{-jk\frac{2\pi}{N}} \tag{2.2}$$

or, with the sampling angle, $\theta = \frac{2\pi}{N}$, by

$$X = \frac{1}{\sqrt{2}}\frac{2}{N}(X_c - jX_s) \tag{2.3}$$

where

$$\begin{aligned} X_c &= \sum_{k=1}^{N} x_k \cos\ k\theta \qquad \text{and} \\ X_s &= \sum_{k=1}^{N} x_k \sin\ k\theta \end{aligned} \tag{2.4}$$

In the presence of noise, the fundamental frequency component of the Discrete Fourier Transform (DFT) given by equations 2.2–2.4, is a least-squares estimate of the phasor. In line relaying applications, versions of equations 2.2–2.4 for windows that are less than a cycle have been developed. The real-time phasor measurement applications is not as time critical as relaying applications so that longer windows are possible. Both two and five cycle windows are commonly used.

The instant at which the first data sample is obtained defines the orientation of the phasor in the complex plane. Using a recursive form of the algorithm is useful for real-time measurements. The phasor $X^r (X^r = X_c^r + j\ {}_s^r)$ corresponding to the data set

$$\{x_k\}, \quad \text{where } k = r, r+1, \ldots, N+r-1$$

is related to the phasor X^{r+1} by equation 2.5

$$X^{r+1} = X^r + \frac{1}{\sqrt{2}}\frac{2}{N}(x_{N+r} - x_r)\epsilon^{-jr\theta} \tag{2.5}$$

or

$$X_c^{r+1} = X_c^r + \frac{1}{\sqrt{2}}\frac{2}{N}(x_{N+r} - x_r)\cos(r\theta) \tag{2.6}$$

$$X_s^{r+1} = X_s^r - \frac{1}{\sqrt{2}}\frac{2}{N}(x_{N+r} - x_r)\sin(r\theta) \tag{2.7}$$

where θ is the sampling angle. The phasors produced by equations 2.5–2.7 are rotations of the phasors produced by equations 2.2–2.4 in the sense that when the input signal is a constant sinusoid, the recursive phasor calculated according to equations 2.5–2.7 is a constant complex number. The recursive phasor calculation, as given by equations 2.5 are computationally efficient involving only one multiplication for each update of the real and imaginary part. The factor $(\sqrt{2}/N)$ in equation 2.5 is can be omitted in the real-time computation. In addition, all three phase quantities are used to compute the positive sequence voltage. The additional filtering provided by using sequence quantities is particularly useful when the power system is operating at off-nominal frequency. The phasor measurement technique can also be used as an accurate frequency measurement device. The frequency measurement technique is based on the rotation of the phasor produced by equation 2.5 when the frequency is not exactly the nominal value. Local frequency is an important parameter in many important relaying and control function implementations. It should be emphasized that the frequency calculation is accurate only if the phase angle of the positive sequence voltage is used.

2.2. Time synchronization. Phasors, representing the bus voltages in the power system, define the state of the power system. Phasors at remote location can be measured with a common reference if the instant at which the samples are synchronized. The exact precision required of the time synchronization is debatable. Fortunately, synchronization accuracies of the order of 1 μsec are now achievable because of some recent technological developments. One micro second corresponds to $0.022°$ for a 60 Hz power system, which more than meets our needs. The Global Positioning System (GPS) [19], provides a common-view time transmission, which when decoded by appropriate receiver clocks, will provide 1 pulse-per-second (1 pps) at any location in the world with an accuracy of about 1 μsec. Commercially available units provide the sampling pulses needed by the phasor measurement units from the GPS transmission. The sampling pulses which are at frequencies at a multiple of the fundamental power system frequency are phase-locked with the 1 pps signal.

3. State estimation. The set of the positive sequence voltage phasors at all the buses of a power system constitutes the state vector of a power system in quasi-steady state. Existing state estimation procedures are based upon obtaining measurements of line power flows, line currents, and bus voltage magnitudes from throughout the power system at as fast a scan rate as possible. The scan rates may vary between a few seconds, to some minutes. Given the relationship between power and voltage, the state estimation is a nonlinear process. In addition, the metering systems tend to have fairly long time constants, so that some averaging of the measurements is implicit in the process. The resulting state estimate is understood to be an approximation.

Synchronized phasor measurements offer an opportunity to improve state-estimation. Since the measurements are precisely synchronized, they can be identified by a precise time-tag. By using the time-tags a true synchronous snap-shot of the power system state could be established at the control center. With improved communication channels, it is possible to consider dynamic (as opposed to static) state estimation. It is possible to include real-time phasor measurement in existing static state estimators as long as the issue of the reference angle for the two systems is resolved. The reference angle for existing static state estimators is typically a real system bus while the real-time phasor measurements have absolute time as a reference. In the extreme if only real-time phasor measurement are used by an estimator, the state estimation equations become linear. Even if positive sequence currents are measured in addition to voltages in order to provide redundancy in measurements, the measured positive sequence quantities are linearly related to the quantities to be estimated. It has been shown in reference [20] that the structure of the linear estimator can exploit system sparsity with sparse LU decomposition and Gaussian elimination techniques. Real-time phasor measurements, and the computed frequency can also be used to develop dynamic state estimates of generating stations. The extended Kalman filter or nonlinear observers can be used [21].

4. Control. It is possible to improve post-disturbance power system performance using control schemes based on real-time phasor measurements. Traditional controllers which act in such situations are restricted by the limited (local) nature of the measurements available to the controller. Traditional control schemes are local in nature while control based on real-time phasor measurements can be made centralized. If synchronized phasor measurements from throughout the system are available, the quality of the control can be improved considerably. Such controllers for dynamic stability enhancement of AC/DC systems have been reported in [21] while the application of real-time phasor measurements in generator exciters and speed governing systems has been reported in [22]. In the following, the general principles behind such centralized controllers will be given with references to both AC/DC systems and generator exciters and speed governing systems.

Let the state of the power system be denoted by χ. Depending on the level of modeling and the components included the state could include: generator angles and speeds, field flux linkages, exciter state variables, and governor state variables. The control vector, u, would include: the exciter control signals, the governor control signals, and the control signals for the HVDC line(s). Given a concern with large disturbances, the state equation is nonlinear in general and of the form

$$\dot{\chi}(t) = f(\chi(t), u(t)) \tag{4.1}$$

An output $y(t)$ which is used to quantify system performance is associated with the control problem. The variables in $y(t)$ could include: incremental changes in voltage magnitudes at generator terminals, rotor angles and speeds, and field voltages. To be completely general $y(t)$ can also be taken to be a nonlinear function of the state as

$$y(t) = h(\chi(t)) \tag{4.2}$$

Equation 4.2 would allow nonlinear functions of the state variables such as the generator terminal voltages to be considered.

It is assumed that the system has been moved away from equilibrium by some disturbance. The control problem is to steer the system along some desired trajectory (defined in terms of the desired output $y_d(t)$) to a desired state. Even with real-time measurements of all states it is not possible to solve such a nonlinear control problem in real time. Some approximations and simplifications are necessary. The use of real-time phasor measurements make it possible treat the nonlinear system in an unusual way. Let χ_0 be the pre-fault equilibrium of the uncontrolled system, i.e.

$$0 = f(\chi_0, 0) \tag{4.3}$$

If we let $x(t)$ represent the incremental changes in the system state about the equilibrium

$$\chi(t) = \chi_0 + x(t) \tag{4.4}$$

then we can write

$$\dot{x} = A\,x + B\,u + [f(\chi_0 + x, u) - A\,x - B\,u](12) \tag{4.5}$$

$$y(t) = C\,x + [h(\chi_0 + x) - C\,x] \tag{4.6}$$

It should be recognized that Equations 4.5 and 4.6 are valid for any matrices A, B, and C, and are not approximations or Taylor expansions. On the other hand, it is clear that reasonable choices of the matrices are necessary. If we take the matrices to be the gradients of the appropriate functions (f or h) with respect to x or u (i.e. the Taylor expansion terms) evaluated at the stable equilibrium χ_0, then we can write a state equation for the incremental system

$$\dot{x} = A\,x + B\,u + r(t) \tag{4.7}$$

$$y(t) = C\,x + s(t) \tag{4.8}$$

where $r(t)$ and $s(t)$ are residuals given by

$$r(t) = f(\chi_0 + x, u) - A\,x - B\,u = \dot{\chi} - A\,x - B\,u \tag{4.9}$$

and

$$s(t) = h(\chi_0 + x) - C\,x \tag{4.10}$$

Given measurements of χ and $\dot{\chi}$, the residuals may be computed or measured in real time. The system described by equations 4.7 and 4.8 is a little unusual from a control system point of view. Rather than the nonlinear system (equations 4.1, 4.2) we have a stable (stable because χ_0 is stable) linear system with inputs, $r(t)$ and $s(t)$, which can be measured in real time. Note that at an instant t_1, for example, $r(t)$, $t_0 < t < t_1$ is known but the future values of $r(t)$ for $t > t_1$ are not known.

We take as a cost function a quadratic function of control $u(t)$ and differences between the actual $y(t)$ and the desired value $y_d(t)$, i.e.

$$(4.11) \quad J(u) = \frac{1}{2}[e^T(t_f)H\ e(t_f)] + \frac{1}{2}\int_{t_0}^{t_f} [e^T Q\ e + u^T R\ u]dt$$

with

$$(4.12) \quad e(t) = y(t) - y_d(t)$$

where Q and R are positive semi-definite and positive definite respectively. The minimization of the cost function is an attempt to make the system output follow a desired output while using little control effort. The first term in equation 4.11 represents a penalty on the output error at the terminal time t_f. The problem of minimizing $J(u)$ subject to equations 4.7 and 4.8 given knowledge of $r(t)$ and $s(t)$ in the interval $t_0 < t < t_f$ is given by the well known equations

$$(4.13) \quad u(t) = R^{-1}B^T[g(t) - K(t)x(t)]$$
$$(4.14) \quad \dot{K}(t) = -A^T K - K\ A + K\ B\ R^{-1}B^T K - C^T Q\ C$$
$$(4.15) \quad \dot{g}(t) = -[A^T - K(t)B\ R^{-1}B^T]g - C^T Q[y_d - s] + K(t)r$$

The difficulty with the measurement of $r(t)$ and $s(t)$ in real time is that equations 4.14 and 4.15 are solved in reverse time with terminal conditions at time t_f.

$$(4.16) \quad K(t_f) = C^T H\ C; \qquad g(t_f) = C^T H[y_d(t_f) - s(t_f)]$$

That is, in order to determine the control at time t_1 from equation 4.13, we need the state at t_1 and the value of $g(t_1)$ (from equation 4.15). Equation 4.15 is solved backward from t_f, however, so that the future values of $r(t)$ and $s(t)$ are needed to determine $g(t_1)$. In [1] attempts were made to predict $r(t)$ and $s(t)$ to overcome this difficulty. Another solution to the problem which has been shown by simulation [1–2] to be quite acceptable, is to use K_s the steady state solution to equation 4.14, the Algebraic Riccati Equation (ARE).

$$(4.17) \quad 0 = -A^T K_s - K_s A + K_s B\ R^{-1}B^T K_s - C^T Q\ C$$

Equation 4.17 can be thought of as equivalent to taking $t_f = \infty$ in equation 4.11. The steady state solution has an additional advantage in that K_s can be computed off line and stored. Equation 4.14 would also be solved off line but would require considerably more storage. With a constant K_s in equation 4.15, the solution for $g(t)$ is not a large computational burden. Further simplification can be obtained by using the DC gain of the system described by equation 4.15 (i.e. assuming that the time varying inputs to the system are very slow). This gives

$$(4.18) \quad g_s(t) = -[A^T - K_s B\ R^{-1}B^T]^{-1}\{C^T Q[y_d(t) - s(t)] - K_s r(t)\}$$

Equation 4.18 only involves multiplication by known matrices and is an approximate but surprisingly effective control.

The above procedure has been tried on controllers for HVDC converters and generator excitation systems on realistic power systems. [22] It has been shown that with reasonable rates of state vector data collection at a central site, the controller designed according to the procedure described above is able to limit excursions in the voltage magnitudes, and in generator rotor angles, following power system disturbances. These ideas await field trials, which may follow after considerable experience with the phasor measurement hardware has been gained.

4.1. Detection of instability. [25] One of the key functions of many relaying and control functions is the detection of incipient instability of the power system following the occurrence of a fault. The out-of-step relaying function is of particular interest if a power system is on the verge of instability. Many impedance relays on the power system may perceive the swings as the machines in a power system begin to oscillate with respect to each other. It would be wrong to permit the impedance relays to trip for a swing that is not going to lead to instability. If it is determined that a particular swing is going to lead to instability (out-of-step) for some machines, it would be appropriate to separate the power system along preferred boundaries, so that the remaining islands of the power system are afforded a good chance of recovering on their own. Thus the function of the out-of-step relays is two-fold: to block tripping of some relays if a condition is unlikely to lead to instability, and force tripping of some relays if the system is judged to be going towards an out-of-step condition.

For a two-machine power system the determination of stability or instability can be accomplished by use of the Equal Area Criterion. Although the two-machine system would appear to be extremely simple, it turns out that many practical power systems approach this configuration under certain operating conditions. It has been shown that when a power system behaves as a two-machine system, that it is possible to correctly predict the stability or instability in about one-quarter of the electro-mechanical oscillation, so that appropriate control action can be taken to affect the outcome of the oscillation. Once the outcome of the oscillation is determined, it is a simple matter to decide the appropriate out-of-step relaying action for the distance relays on the system.

5. Direct numerical integration. Given that the purpose is to predict stability in real time, we ask if it is possible to simply integrate the system equations fast enough. This technique has been shown to be useful for determining first-swing stability by using step sizes as large as 0.2 seconds [9]. With a larger step size, however, accuracy diverges quickly as the integration length increases. The present work shows that integration can be accomplished in real time with a more reasonable step size, for example 0.025 seconds. Since lumping generators together within coherent groups greatly reduces the system dimension we begin with smaller, reduced-order systems. In this paper, we show how the stability of such systems can be computed in real time, fast enough to select a control option that is likely to mitigate the effects of a disturbance.

5.1. Computation speed required for control. In attempting to answer the question of how quickly a prediction must be calculated in order to be useful for control, we simulated hypothetical control options on the New England 39 bus system. The purpose was to determine whether controls initiated with 1/4 second delay, for example, after fault clearing time could significantly improve post-fault behavior. For comparison, decision trees for transient stability prediction [10] give a prediction immediately after observing 10 cycles of post-fault swing behavior (1/4 second = 15 cycles).

Hypothetical control options were tested on the New England system with loading increased by 25%, so that the system was marginally stable in any event. We found that a combination of load increases (hypothetical braking elements) and load decreases (hypothetical load shedding) in response to a transient event greatly improved stability. For example, there were some transmission lines which could not be tripped without causing instability. If however load shedding occurs within 1/4 second of clearing time, faults up to 8 cycles can be sustained in some cases. While braking resistors and load shedding are not attractive control options, this is evidence that real-time control actions, initiated shortly after the fault is cleared, have the potential to greatly benefit electric power systems. This is our motivation for accomplishing real-time transient stability prediction.

5.2. Integration methods. There is a very simple answer to the problem of real-time stability assessment for reduced-order systems. Modern microprocessor technology enables us to actually integrate the system trajectory in real-time. Given a model for the reduced system, and an initial condition, a microcomputer takes just a few hundredths of a second to calculate the system trajectory for seconds into the future. For example, an Intel 80486 based PC takes 0.06 seconds to calculate the swing angles of a 4-machine system one second into the future. A SUN SPARC Station IPX performs the task in 0.01 seconds. Thus numerical integration of the reduced system can be fast as well as accurate.

To be specific, the calculations from real-time integration will be as accurate as the models on which they are based. Since different models may give different results [11], it is desirable to have some flexibility in the model that is used. For this reason, we looked at the calculation speeds for several different models: constant impedance load model, constant $P - Q$ load model [12], and the dynamic Ward-type equivalent model [13]. The constant impedance model is the easiest and fastest to solve. The constant $P - Q$ model requires more computation because it retains the network structure, so an efficient algorithm was developed for its solution. The dynamic Ward-type equivalent model, also called the generator axis load-flow model, attempts to bridge the gap between speed and accuracy by making piecewise constant current approximations. In the following sections, constant impedance models are tested on 2, 3, and 4 machine systems, the constant $P - Q$ model is timed for a 4 machine system, and the dynamic Ward-type equivalent and the constant impedance models are compared for the New England 39 bus system.

This paper only concerns itself with timing, with the expectation that the user will select an appropriate model. Which model would be best, and it must be a reduced-order model, is left for further research. Before presenting results, direct integration is further motivated by illuminating some shortcomings of the Extended Equal Area Criterion as well as the energy function techniques.

6. Investigation of EEAC. The theory of the Extended Equal-Area Criterion (EEAC) [5] is investigated with respect to the problem of real-time stability prediction, and the potential for errors is demonstrated. In a typical transient stability study or a typical EEAC application, in which the critical clearing time is to be found, there are three stages that a power system goes through: pre-fault, fault-on, and post-fault stages. In this study, however, the critical clearing time is not of immediate interest to us. We are instead concerned with the transient stability of the power system when transmission lines are dropped. In fact, the loss of line is still due to the fault clearing, but we assume that the protection systems for transmission lines are extremely fast and the fault is removed immediately at the fault inception. This assumption transforms the study into a switching operation where one or more transmission lines are lost, and only pre-fault or base case and post-fault or contingency stages are involved.

Since the classical Equal-Area Criterion can be used to determine the transient stability for switching operations or losses of line for the one-machine-infinite bus (OMIB) system, the EEAC, which is claimed to be acceptably accurate for the determination of the critical clearing time in transient stability studies of a multi machine power system, should also be able to handle the determination of transient stability resulting from switching operations of the multi machine power system. The system would be regarded as stable if the stability margin computed following the procedures in the EEAC is positive. The EEAC procedures for the typical transient stability study for each outage or disturbance are as follows:

1. Decompose the system into two clusters, namely the critical cluster and the remaining cluster.
2. Transform the two clusters into two equivalent machines using their center of angle reference frames.
3. Transform the two-machine system into the OMIB system.
4. Apply the equal-area criterion to the OMIB system to determine the critical clearing angle and the transient stability margin.

5. Apply truncated Taylor series to find the critical clearing time corresponding to the critical clearing angle in Step 4.

Since each disturbance in an n-machine system permits 2^{n-1} possible ways of decomposing the system into two clusters, the inventors of the EEAC suggest that the initial accelerations of the machines should be used to decompose the system in an efficient way. First, the list of machines is drawn in descending order of their initial accelerations. Then, starting at the top of the list, each machine is added to the critical cluster one-by-one. This reduces the number of the candidate critical clusters to $n-1$. For each candidate critical cluster, we compute its critical clearing time. Finally, the actual critical cluster is the one having the smallest critical clearing time.

As mentioned earlier, in a switching operation, the critical clearing time is not a concern, and the stability margin is the only criterion to determine system stability in the equal-area-criterion. Consequently, it is assumed that in the application of the EEAC to the switching operation the actual critical cluster is the one having the smallest stability margin.

Our sample power system model is the New England 39-bus system [14] which is composed of ten generators and forty-six line segments. Eliminating cases that lead to system islanding, there are thirty-five possible single-line losses or switching operations. The time-domain solutions for those thirty-five cases predict the system to be stable. These results are in agreement with the solutions from the EEAC. In all thirty-five cases, the stability margins are positive regardless of how we decompose the system. A problem arises, however, if the actual critical cluster has to be identified, since in some contingencies (16 out of 35 cases) the critical cluster that gives the minimum stability margin does not belong to the candidate critical clusters identified by the initial acceleration criterion.

The EEAC performs less well as the system load increases. To simulate a heavy loading situation, we multiply the load and generation at all the buses by two. In the heavy loading situation, it becomes more evident that the application of the initial acceleration criterion to the decomposition of the system does not work satisfactorily. For some contingencies, all or some of the candidate critical clusters using the initial acceleration yield the following:

1. There is no load flow solution to the base case *and/or* contingency situation in the OMIB equivalent system while the original system does have such solutions (for all of the candidate critical clusters of 19 out of 35 cases).
2. If δ_0 is the stable equilibrium point of the angle of the internal emf of the equivalent machine in the OMIB for the base case, and δ_d is similarly for the contingency, $\delta_d < \delta_o$ while it is expected otherwise (for some of the candidate critical clusters of 14 out of 35 cases).
3. Positive stability margin or transiently stable in the EEAC while the time-domain solution is unstable (for all of the candidate critical clusters of 22 out of 35 cases).

For those contingencies listed in 2.3, some critical clusters other than the ones obtained from the initial acceleration criteria, produce an instability prediction that agrees with the time-domain solution. Yet there are a few contingencies (12 out of 35 cases) that, no matter what the critical cluster is, the application of the EEAC always incorrectly predicts stability for the system. In all but one such cases, the EEAC tends to be optimistic.

7. Constant impedance: 2, 3, 4 machine systems. Constant im-pedance load-model power systems with 2, 3, and 4 generators were timed for their speed of numerical integration on various small computers (Section 7 has the constant impedance timing results for 10 machines). In the classical model, each generator is represented as a constant voltage source behind its transient reactance and the loads are constant impedance. The generator angles are governed by the real-power swing equations:

$$\begin{aligned} \dot{\delta}_i &= \omega_i \\ M_i \dot{\omega}_i &= P_{mi} - \sum_j E_i E_j Y_{ij} \cos(\delta_i - \delta_j - \theta_{ij}) \end{aligned} \tag{7.1}$$

where

δ_i, ω_i	rotor angles and velocities
M_i	inertia coefficients
P_{mi}	mechanical input powers
E_i	generator voltages
Y_{ij}, θ_{ij}	admittance magnitudes and angles

The integration technique uses the fourth-order Runge-Kutte method, with a step size of 0.025. There were approximately 40 steps per swing period because our models oscillate at a frequency of about one Hertz. The computer program was written in FORTRAN, and the programs were tested on IBM PC compatibles using Microsoft FORTRAN 4.1 with optimizing compiler. The program was also tested on a SUN SPARC Station IPX.

The two-machine model represents the equivalenced Florida-Georgia system. One machine represents the Florida Power and Light Company equivalent, and the other represents the Eastern Seaboard equivalent. The equivalent machines had an oscillation period just slightly longer than one second. The three machine model was taken from Anderson and Fouad [15]. We created a simple four-machine model because none were readily available. Since the integration program executes in fixed time regardless of the model parameters, it is valid to pick parameters in this way.

The technique of numerical integration has the flexibility to handle machine damping in the system model. While the test models did not originally contain damping terms, we experimented by adding nonzero coefficients D_i in the swing equations:

$$
\begin{array}{rcl}
\dot{\delta}_i & = & \omega_i \\
{}_i\dot{\omega}_i + D_i\omega_i & = & P_{mi} - P_{ei} \ , \quad i = 1, ..., n
\end{array}
\tag{7.2}
$$

Since most of the computation comes from the terms P_{ei}, the addition of machine damping only slightly lengthened the execution time. The timing results presented in this section represent models with machine damping terms.

The FORTRAN computer code was written with speed performance in mind. One modification was particularly helpful. Numerical integration involves multiplying the right hand side of the differential equation by the step size and accumulating the result. The fourth-order Runge-Kutte method does this four times for every time step. By scaling the coefficients which appear on the right hand side of the differential equation before entering the integration loop, we reduced the total number of computations.

The simulation programs were developed on a 386/16 IBM compatible computer with a math coprocessor. The four-machine simulation was repeated on a 486/33 IBM compatible, and also on a SUN SPARC Station IPX. Table 7.1 gives the timing results.

TABLE 7.1
Program times to integrate machine angles one second into the future.

Execution Time in Seconds – Const. Z			
Power System Model	386/16	486/33	SPARC/IPX
2-machine	0.05–0.06	–	–
3-machine	0.11–0.17	–	–
4-machine	0.22–0.28	0.05–0.06	0.01

The timing results were obtained by placing timing routine calls within the program. The first call was placed before the computation, and the second was placed at the end. Because the programs execute so quickly, it was difficult to obtain great precision. The 386/16 in particular gave slightly varying results. When trajectories were integrated 100 seconds into the future, the results were consistent with the lower ends of the ranges.

8. Constant $P-Q$ model: 4 machine system. In this section, we investigate the possibility of real-time integration for power systems with constant PQ load models. The transient dynamics are described by the classical swing equations 7.1 for the generators together with the power flow balance equations at the load buses. The specifics are given below.

Efficient techniques already exist for solving such differential-algebraic equations [16]. One of these integration schemes, called implicit trapezoidal integration, gains speed by taking advantage of special features found in power systems. A novel integration technique is introduced in this paper, which uses the principle of the decoupled power flow method. The method is designed to reduce CPU time for solving power system transient dynamics. This technique is called the implicitly decoupled PQ integration technique. When applied to a 4-machine sample power system, the simulations are found to execute faster than real-time on a workstation computer.

8.1. Transient dynamics model. We assume that the generators are modeled by constant voltage sources behind their transient reactances. For a lossy transmission system the power flowing through the transmission lines connected to the ith bus can be written as

$$\begin{aligned} P_i &= \textstyle\sum_j E_i E_j Y_{ij} \cos(\delta_i - \delta_j - \theta_{ij}) \\ Q_i &= \textstyle\sum_j E_i E_j Y_{ij} \sin(\delta_i - \delta_j - \theta_{ij}) \end{aligned} \tag{8.1}$$

We shall assume that the real and reactive load powers are constant for each bus i.

$$\begin{aligned} P &= P^0 \\ Q_{li} &= Q_{li}^0 \end{aligned} \tag{8.2}$$

Kirchhoff's laws imply that the net real and reactive power entering a bus is zero.

$$\begin{aligned} P_i + P_{li} &= 0 \\ Q_i + Q_{li} &= 0 \end{aligned} \tag{8.3}$$

These equations lead to a set of differential-algebraic equations (DAE's).

$$\begin{aligned} \dot{\delta}_i &= \omega_i \\ \dot{\omega}_i &= (-D_i\omega_i + P_{mi} - P_{ei})/M_i \\ P_{li}^0 &= -\textstyle\sum_j E_i E_j Y_{ij} \cos(\delta_i - \delta_j - \theta_{ij}) \\ Q_{li}^0 &= -\textstyle\sum_j E_i E_j Y_{ij} \sin(\delta_i - \delta_j - \theta_{ij}) \end{aligned} \tag{8.4}$$

The above system can be re-expressed in vector form:

$$\begin{aligned} \dot{x} &= f(x, y) \\ 0 &= g(x, y) \end{aligned} \tag{8.5}$$

where x is a vector of generator voltage angles and frequencies and y is a vector of load bus voltage angles and magnitudes.

8.2. Implicitly decoupled PQ integration technique. The proposed integration technique is outlined as follows. Discretizing Eqn. 2.7 by the implicit trapezoidal rule yields

$$\begin{aligned} \frac{x_n - x_{n-1}}{h} &= \frac{f(x_n, y_n) + f(x_{n-1}, y_{n-1})}{2} \\ 0 &= g(x_n, y_n) \end{aligned} \tag{8.6}$$

where $h = t_n - t_{n-1}$ is a step size. The resulting system of nonlinear equations for x_n and y_n at each time step is then solved by the Quasi-Newton method using x_{n-1} and y_{n-1} as initial points.

Divide Eqn. 8.6 into two groups. Group 1 corresponds to real power equations and group 2 corresponds to reactive power equations. Evaluate the Jacobian matrix, J, of Eqn. 8.6 at the initial point on the trajectory. Partition the Jacobian matrix, J, into 4 sub matrices: J_{11}, J_{12}, J_{21}, J_{22} . The dimensions of square sub matrices J_{11} and J_{22} are equal to the number of equations in groups 1 and 2 respectively.

Hold the sub matrices J_{11} and J_{22} constant for finding the next several points on the segment of trajectory using the Quasi-Newton method. If convergence does not occur, then re-evaluate J_{11} and J_{22} over again, and repeat the Quasi-Newton algorithm. The advantages of this technique are the numerical stability of implicit integration, and a large reduction of computational burden by no new evaluation and inversion of Jacobian matrices, J_{11} and J_{22} during some segments of trajectory.

8.3. Simulation results. The implicitly decoupled PQ integration technique is applied to a 4-machine, 3 PQ-load sample power system. The line and load flow data are given in the paper by El-Abiad and Nagappan [17]. The simulation programs were developed on a SUN SPARC II in FORTRAN. The implicitly decoupled PQ integration technique is approximately 20 times faster than the traditional method.

TABLE 8.1
Program times to integrate machine angles one second into future.

Execution Time in Seconds – Const. $P-Q$	
Type of Swing	SPARC II
Stable →	0.06–0.66
Unstable →	0.80–0.82

These timing results show the proposed integration technique executing faster than real-time on a SUN SPARC II workstation.

9. Dynamic Ward equivalent and constant impedance models. The 10 machine system IEEE 39-bus, 10 generator system [14] was used to test the integration speed for the constant impedance model and the

Dynamic Ward Equivalent Model described in [13,18]. The latter attempts to provide fast solution speed while accurately modeling the dynamics of load behavior during a transient event, and is briefly introduced below.

The issue of reducing the size of the system while preserving the validity of the transient stability analysis is solved by employing a dynamic Ward-type equivalencing technique. In this approach the buses with static and passive loads are eliminated while the dynamic load buses are treated in a structure preserving manner. Therefore, a static-type equivalencing method can be used, specifically a modified version of the Ward equivalent. The focus is on the analytical derivation of the method for handling constant PQ loads. Unlike the classical Ward equivalent which is exact at only the operating point where the reduction has been performed, this equivalencing method retains its accuracy over a wide excursion of the generator angles. This is achieved by making use of sensitivity analysis. Sensitivity factors are incorporated into a correction formula expressing the incremental changes of the equivalent current injections in terms of the changes of the machine angles at the retained buses. These buses include the internal nodes of the generators along with the dynamic load buses. When the operating point moves beyond the validity of the linearization, the sensitivity factors are updated about a new operating point, yielding a piece-wise linear approximation. The sensitivity factors can be calculated off-line for different operating points and stored for use on-line. The correction factors are employed in a load flow calculation, referred to as machine axis load flow. This load flow calculation is required for any transient stability analysis method either for on-line implementation or for planning applications.

Both the constant impedance model program, and the machine axis load-flow model program for the 10 machine systems were written in C. The constant impedance program integrated the swing equations. The generator axis model contains extra terms in the expression for P_{ei}. The explicit expression for the generator axis model can be obtained from [13,18]:

$$\begin{aligned} P_{ei} &= \textstyle\sum_j E_i E_j Y_{ij}^{eq} \cos(\delta_i - \delta_j - \theta_{ij}^{eq}) \\ &\quad + E_i I_{gi0}^{eq} \cos(\delta_i - \psi_{gi0}^{eq}) + \textstyle\sum_j E_i W_{ij} (\delta_j - \delta_{j0}) \cos(\delta_i - \varphi_{ij}) \end{aligned} \tag{9.1}$$

where

E_i	voltage magnitudes
δ_i	rotor angles
Y_{ij}^{eq}	equivalent admittance magnitudes
θ_{ij}^{eq}	equivalent admittance angles
I_{gi0}^{eq}	base equivalent current magnitude injections
ψ_{gi0}^{eq}	base equivalent current angle injections
W_{ij}	sensitivity matrix magnitudes
φ_{ij}	sensitivity matrix angles
δ_{ij}	base case generator angles

Collecting the constants together and renaming them, we see that P_{ei} has the following form:

$$
\begin{aligned}
P_{ei} &= \textstyle\sum_j \beta_{ij}\cos(\delta_i - \delta_j - \theta_{ij}) + \gamma_i \cos(\delta_i - \psi_i) \\
&\quad + \textstyle\sum_j \aleph_{ij}(\delta_j - \delta_{j0})\cos(\delta_i - \varphi_{ij})
\end{aligned} \tag{9.2}
$$

where β_{ij}, θ_{ij}, etc... are constants. Multiple versions of the parameters γ_i, ψ_i, $\aleph_{ij}$, φ_{ij} and δ_{j0} must be stored in memory to preserve accuracy as the angles vary from their original operating point. The generator angles are tested at each step of the integration to determine which set of stored parameters to use. In order to time the execution speed of this algorithm, we assigned random values to the new parameters and incorporated the new terms. A realistic test on the generator angles was also performed at every integration step. Both programs were timed on IBM PC compatibles running DOS, and on workstations running Unix. Only the workstations gave practical results for execution speed. These were the Sun SPARC Station IPX, and the Hewlett Packard 9000/715. The execution time was approximately twice that of the constant impedance model. The HP 9000/735 is expected to run 2-3 faster than the 9000/715.

TABLE 9.1
Program times to integrate 10 machine angles one second into future.

Execution Time in Seconds		
Computer Model	Constant Z	Dynamic Equivalent
386/16 PC Compatible	2.26	4.67
786/35 PC Compatible	0.55	1.10
Suns SPARC Station IPX	0.14	0.27
HP 9000/715	0.07	0.14

10. Conclusions. A number of field installations of the synchronized phasor measurement systems are being planned at this time. About half a dozen such systems have already been installed. At the same time, the GPS receiver technology is undergoing significant improvements and cost reductions. It seems entirely likely that the complete synchronized phasor measurement computer, along with the GPS clock receiver can be packaged in one unit. It is also expected that the cost of such a unit will be reasonable — comparable to that of a modern relay terminal. As more experience with these systems is gained, it can be expected that more applications for this technique will be found. Certainly, in the context of an educational institution, this field of research is very attractive to the students, as it brings together the most modern developments in technology, and modern electric power engineering; and thereby makes attractive a field of engineering which has often been accused of being too old-fashioned and stodgy. From this point of view, it has been a very rewarding activity to initiate and develop the field of synchronized phasor measurements in power systems.

When it comes to assessing stability in the reduced system, we have shown that numerical integration can be very fast if it is coded efficiently. Unlike the equal area criteria numerical integration is not restricted to two-machine systems. Furthermore, we have shown that numerical integration is very fast on up to ten machines. Two typical workstations that we tested can solve the 10 machine system model one second into the future in 0.07 and 0.14 seconds respectively. A soon to be acquired HP 9000/735 workstation is expected to run the program in approximately 0.03 seconds.

The constant $P-Q$, and the dynamic Ward equivalent models were also timed and found to show promising results. The implicit decoupled $P-Q$ integration technique runs 20 times faster than the standard structure preserving model, and executes faster than real-time. The dynamic Ward equivalent, or the machine axis load flow model, as it is also referred to, takes almost exactly twice as long as the constant impedance model. In doing so, the assumption was made that certain parameters were precomputed and stored on-line for use in the simulation. The practicality of this remains to be resolved.

Areas of further research include more investigation into the implementation of the machine axis load flow model, including issues about storing the necessary sensitivity matrices needed as the operating point varies. Another important requirement for either method is to have an accurate reduced-order system model available in real-time. Although some work has been done in this area, further research remains.

REFERENCES

[1] A.G.Phadke, *Synchronized phasor measurements in power systems*, IEEE Computer Applications in Power **6** (2) (1993), 10–15.

[2] A.G.Phadke, J.S.Thorp, *Improved control and protection of power systems through synchronized phasor measurements*, Control and Dynamic Systems **43**, Academic Press, New York 1991, 335–376.

[3] R.P.Schulz, L.S.VanSlyck, S.H.Horowitz, *Applications of fast phasor measurements on utility systems*, PICA Proc., Seattle, May 1989, 49–55.

[4] M.Ribbens-Pavella, F.J.Evans, *Direct methods for studying dynamics of large scale power systems—a survey*, Automatica **32** January (1985), 1–21.

[5] Y.Xue, Th.Van Cutsem, M.Ribbens-Pavella, *A simple direct method for fast transient stability assessment of large power systems*, IEEE Trans. on Power Systems **3** (2) May (1988), 400–412.

[6] V.A.Venikov, S.N.Asambaev, *Quick estimation of the stability of a process on the basis of its initial stage*, Izvestiya Akademii Nauk SSSR. Energetika *i* Transport (Translation available from Allerton Press) **24** (3) (1986), 23–29.

[7] M.H.Haque, A.H.M.A.Rahim, *Determination of first swing stability limit of multimachine power systems through taylor series expansions*, IEE Proceedings, Part C : Generation, Transmission and Distribution, **136** (6) (1989), 373–379.

[8] J.Tong, H.-D.Chiang, T.P.Coneen, *A sensitivity-based BCU method for fast derivation of stability limits in electric power systems*, IEEE PES Winter

Meeting, 92 WM 149-5 PWRS 1992.

[9] Y.Dong, H.R.Pota, *Fast transient stability assessment using large step-size numerical integration*, IEE Proceedings–*C* **138** (4) July (1991), 377–383.

[10] S.M.Rovnyak, S.E.Kretsinger, J.S.Thorp, D.E. Brown, *Decision trees for real-time transient stability prediction*, IEEE PES Summer Meeting, 93 SM 530-6 PWRS 1993.

[11] M.H.Kent et al., *Dynamic modeling of loads in stability studies*, IEEE Trans. on Power Apparatus and Systems, PAS-88, (5) May 1969, 756–763.

[12] A.R.Bergen, D.J.Hill, *A structure preserving model for power system stability analysis*, IEEE Trans. on Power Apparatus and Systems **PAS-100** Jan 1981, 25–35.

[13] T.L.Baldwin, L.Mili, A.R.Phadke, *Dynamic ward equivalents for transient stability analysis*, IEEE PES 1993 Winter Meeting, WM 244-4 PWRS 1993.

[14] M.A.Pai, Energy Function Analysis for Power System Stability, Kluwer, Boston, 1989.

[15] P.M.Anderson, A.A.Fouad, Power System Control and Stability, Iowa State University Press, Ames 1977.

[16] K.E.Brean, S.L.Campbell, L.R.Petzold, Numerical Solution of Initial-Value Problems in Differential Algebraic Equations, North-Holland 1989.

[17] El-Abiad, K.Nagappan, *Transient stability region of multi-machine power systems*, IEEE Tran. on Power Apparatus and Systems, **PAS-85** (2) Feb 1966, 169–178.

[18] T.L.Baldwin, L.Mili, A.G.Phadke, *Ward-type equivalents for transient stability analysis*, Proceedings of the IFAC International Symposium on Control of Power Plants and Power Systems, Munich, Germany, March 9–11, 1992.

[19] Global Positioning System, **I, II, III** (papers published in) Navigation, (reprinted by) The Institute of Navigation, Washington, D.C. 1980.

[20] A.G.Phadke, J.S.Thorp, Computer Relaying for Power Systems, Research Studies Press Ltd., John Wiley & Sons Inc. (second printing) April 1990.

[21] P.Pillay, A.G.Phadke, D.K.Lindner, J.S.Thorp, *State estimation for a synchronous machine: observer and kalman filter approach*, Princeton Conference 1987.

[22] A.G.Phadke, J.S.Thorp, *Improved power system protection and control through synchronized phasor measurements*, Proceedings of the 6th National Conference, Bombay, McGraw-Hill Publishing Company, New Delhi, India June 4–7, 1990, 339–346,

[23] A.G.Phadke, S.H.Horowitz, *Adaptive relaying*, IEEE Computer Applications in Power **3** (3) July 1990, 47–51.

[24] A.G.Phadke, S.H.Horowitz, A.K.McCabe, *Adaptive automatic reclosing*, CIGRE 1990, Paper No. 34-204, Proceedings of the CIGRE General Assembly, Paris, September 1990.

[25] J.S.Thorp, A.G.Phadke, S.H.Horowitz, M.M.Begovic, *Some applications of phasor measurements to adaptive protection*, Proceedings of Power Industry Computer Applications (PICA) 1987, 467–474.

ON THE DYNAMICS OF DIFFERENTIAL-ALGEBRAIC SYSTEMS SUCH AS THE BALANCED LARGE ELECTRIC POWER SYSTEM

VAITHIANATHAN VENKATASUBRAMANIAN*, HEINZ SCHÄTTLER†, AND JOHN ZABORSZKY†

Abstract. This paper summarizes some recent results on the state space structure in differential-algebraic systems of the form $\dot{x} = f(x, y)$ and $0 = g(x, y)$ which are the slow dynamics of singularly perturbed nonlinear systems of the form $\dot{x} = f(x, y)$ and $\epsilon\dot{y} = g(x, y)$. After restricting the analysis to the portion of the slow manifold with local stability in the fast variables, the boundary of the region of attraction is characterized under some Morse-Smale like assumptions. It is proved that these constrained singular systems do possess a nice dynamical structure both locally (in general) and globally (under nongeneric assumptions) and the structure is of direct practical significance. The dynamics of the balanced large power system is a primary example.

1. Introduction. The dynamics of the balanced large electric power system can be modeled by singularly perturbed differential equations of the form

$$(1.1)\qquad \Sigma_\epsilon : \quad \dot{x} = f(x, y, p) \quad , \quad f : \mathbb{R}^{n+m+q} \to \mathbb{R}^n, \quad f \text{ is } C^\infty$$

$$(1.2)\qquad \epsilon\dot{y} = g(x, y, p) \quad , \quad g : \mathbb{R}^{n+m+q} \to \mathbb{R}^m, \quad g \text{ is } C^\infty$$

$$x \in X \subset \mathbb{R}^n, \quad y \in Y \subset \mathbb{R}^m, \quad p \in P \subset \mathbb{R}^q$$

with the lumped RLC representation of the transmission system [21]. A model of the form Σ_ϵ was earlier proposed in [13] for the power system under the assumption that the capacitors are absent in the system and was applicable to certain types of load models. Such limitations were eliminated in [21] using a notion of fast time varying phasors where dynamical equations of the form (1.1)–(1.2) were constructed for the balanced general large electric power system. It should be emphasized that singularly perturbed models of the form Σ_ϵ arise from a precise formulation of the lumped RLC network dynamics in [21], and it does not justify casually defined singular perturbations of the load flow equations which are commonly used in the literature.

Typical dynamic state variables x are the time dependent values of the generator voltages and rotor phases, the dynamic states of the control devices and dynamic load models, and the dynamic phasor states of the capacitor banks and large reactors. The fast dynamic variables y are typically the phasor states associated with the capacitors and inductors of

* School of Electrical Engineering and Computer Science, Washington State University, Pullman, WA 99164-2752.

† Department of Systems Science and Mathematics, Washington University, St.Louis, MO 63130.

the lumped RLC transmission system. The fast state variables of the synchronous generator [13] and any fast dynamics of the dynamic load models need also be modeled in the set of y variables. The parameter space P is composed of system parameters (the system topography, i.e. what is energized, and equipment constants e.g. inductances), and operating parameters (such as loads, generation, voltage set-points etc.). System hard limits and saturation are not considered here.

In the precise formulation of (1.1)–(1.2), each capacitor and inductor in the lumped RLC representation of the transmission system define two dynamic states for describing their fast dynamics, which however result in a very high dimensional system Σ_ϵ that is computationally not attractive. Moreover since there exists a natural time-scale separation between the fast dynamic states of the transmission system (in y) and the traditional quasi stationary phasor states (in x), the singular perturbation theory can be used to simplify the analysis of Σ_ϵ [8].

Essentially the solutions of the system Σ_ϵ can be well-approximated by those of the corresponding slow system Σ_s defined by

$$(1.3) \qquad \Sigma_s : \quad \dot{x} = f(x,y,p) \quad , \quad f : \mathbb{R}^{n+m+q} \to \mathbb{R}^n, \quad f \text{ is } C^\infty$$

$$(1.4) \qquad \qquad 0 = g(x,y,p) \quad , \quad g : \mathbb{R}^{n+m+q} \to \mathbb{R}^m, \quad g \text{ is } C^\infty$$

provided certain stability requirements on the fast dynamics are satisfied [8]. Integral manifold theory has also been applied extensively in the power system literature (e.g. [13]) previously for the simplification of the power system models. Traditional power system dynamic analysis has been based on the differential-algebraic systems of the form Σ_s where roughly the network equations $g = 0$ in (1.4) can be shown to reduce to the conventional load flow equations or the power balance equations after some simplifications. But unlike the conventional quasi-stationary power system models, the full model Σ_ϵ includes in it the fast dynamics as well, therefore the stability of the slow system Σ_s within the full dynamics, namely *the component stability problem*, can be directly assessed from the overall system stability of Σ_ϵ [21]. Note that the lumped parameter RLC transmission system representation is still limited in its scope and needs to be replaced by distributed parameter - stray capacitor and inductance models for very fast phenomena.

Let us first define the constraint $g = 0$ in (1.4) (also known as the slow manifold) to be L,

$$(1.5) \qquad L := \{(x,y,p) \in \mathbb{R}^{n+m+q} : g(x,y,p) = 0\}$$

From the singular perturbation theory [8], it follows that any trajectory of the system Σ_ϵ can be well-approximated by that of the slow constrained system Σ_s provided the "fast system" is locally uniformly asymptotically stable along the trajectory, e.g., if the Jacobian $D_y g$ (i.e. $\frac{\partial g}{\partial y}$) has all its eigenvalues in the open left half complex plane $\mathbb{C}^-$ along the trajectory, i.e.,

1. if the trajectory stays away from the singular subset S where the Jacobian $D_y g$ has zero eigenvalues where S is defined as

   $$S := \{(x, y, p) \in L : \Delta(x, y, p) := \det(D_y g)(x, y, p) = 0\} \tag{1.6}$$

   and

2. if the trajectory also stays away from the set H where the Jacobian $D_y g$ has purely imaginary eigenvalues where H is defined as

   $$H := \{(x, y, p) \in L \setminus S : \sigma(D_y g)(x, y, p) \cap \mathcal{I} \neq \emptyset\} \tag{1.7}$$

   where $\mathcal{I}$ denotes the imaginary axis in the complex plane.

The set H named here *the Hurwitzian surface* and the singularity S then outline the state space boundary where the component stability (or the stability of the fast dynamics) is lost.

The presence of the singular set (called the impasse surface in [7,6]) in power system models has been noted before [10,23,7] and a detailed analysis of the constrained singular dynamics of the form Σ_s has been presented elsewhere [20,18]. When a trajectory reaches the singular set S, the solution of Σ_ϵ typically undergoes jumps, i.e. the trajectory moves very fast, almost instantaneously, to a different point on the constraint L. Such jump behavior has been studied in some classes of singularly perturbed systems [14,12,3]. However in general, the jump behavior can be extremely complex even in two dimensional ($n = 2$) systems near special singular points (canards, [1,26]) and at the present time, it seems to us that there exists no general theory for analyzing the system behavior near the singular points S for the large system Σ_ϵ (nonstandard singular perturbation, [26]).

The existence of the Hurwitzian set H identified here can be readily confirmed mathematically in simplistic power system models but their existence in detailed power system models and their physical implications need careful analysis. Note that since the Jacobian $D_y g$ is typically not symmetric in power system models, the existence of the Hurwitzian set H cannot be ruled out in general power system models. More on this later in Section 2.

Here it should be remembered that the model (1.1)–(1.2) itself is only an approximation for the large power system and is valid only under the assumptions that 1) the lumped parameter RLC representation of the transmission system is valid and 2) the three phase signals are balanced [21]. When the trajectories start moving very fast such as near the singular set and off the constraint L, these two assumptions become questionable. It can be shown that the inherent randomness in the reference phase of the modulated power system signals together with the breakdown of the assumptions imply that the power system dynamics becomes a true stochastic process near the singular set [21].

For the reasons mentioned above, the analysis in this paper will be restricted to studying the dynamics of the constrained system Σ_s, specifically, around the subset L^s of the slow manifold L which is component stable,

$$L^s := \{(x,y,p) \in L : \sigma(D_y g)(x,y,p) \subset \mathbb{C}^-\} \tag{1.8}$$

Here $\sigma(D_y g)$ denotes the eigenspectrum of the Jacobian $D_y g$. The component stability condition in (1.8) implies the local asymptotic stability of the fast dynamics in y near the constraint surface L^s for the singularly perturbed system Σ_ϵ. Any trajectory which lies entirely within L^s is indeed an excellent approximation for the singularly perturbed trajectory of Σ_ϵ [8]. The dynamics of differential-algebraic systems of the form Σ_s has been analyzed extensively in [20,18] and a Lyapunov theory for such constrained systems has been proposed in [7,6]. The results in [20,18] assumed the component stability in the state space since an explicit condition was unavailable for checking this assumption. With the development of such a condition in [21] (e.g. $\sigma(D_y g) \subset \mathbb{C}^-$ being a sufficient condition), it becomes possible here to directly extend the results in [20,18] now including this additional stability criterion. The extended results are presented in this paper.

The basic state space structure of the constrained system Σ_s is briefly summarized in Section 2, with the precise definitions to follow in Section 3. Stability regions are defined for the large power system in Section 4 and the boundary of the region of attraction is characterized in Section 4 under certain transversality conditions.

2. The structure of the state space. The state space for the constrained system Σ_s displays a hierarchical structure summarized in Table 1. The constraint surface L is divided into connected components by the singular set S [20]. The state space components (called causal regions in [6]) are further decomposed into component stable and unstable regions by the interior component stability boundaries H where the Jacobian $D_y g$ has purely imaginary eigenvalues. Simulation of simple power system models indicates that this set H is typically nonempty. The Hurwitzian surface H together with the singular surface S then divide the constraint L into connected regions where the fast dynamics is stable (component stable regions) and those where the fast dynamics is unstable (component unstable regions).

To be precise, the connected components of the set L^s are defined here *the component stable regions* and the connected components of the complement $(L \setminus L^s) \setminus (S \cup H)$ are defined *the component unstable regions*. The system operating point being a stable equilibrium point for Σ_ϵ must belong to a component stable region. For the power system, the implications of the component unstable regions need a careful scrutiny. Even though the RLC dynamics which essentially defines the fast y dynamics in

(1.2) should be an internally asymptotically stable system being a passive RLC network, since the coupling equations with the loads at the network terminals are nonlinear power balance equations in the power system, the dynamic equations (1.2) are nonlinear and hence it seems plausible that the network dynamics (1.2) could be locally unstable near some points of the network solutions depending on the load behavior. Actual load measurements on real large power systems [15,9,11] yield load models which indicate the presence of the singularity S hence these load models also indicate the possibility of the component unstable regions in the real power system. The possible presence of such component unstable regions was conjectured in [4]. Recent results in [21] prove their existence theoretically though their practical implications need further study.

Since the Jacobian $D_y g$ of the fast dynamics for the system Σ_ϵ has zero eigenvalues in the set S, the singular set S can also be viewed as a static bifurcation for the fast dynamics, typically resulting in a saddle node type fold bifurcation [5] of the slow manifold solutions. In other words, two solutions of the constraint set L typically meet and disappear at the singularity S [3]. Similarly the Hurwitzian set H may lead to a Hopf bifurcation [5] in the fast dynamics of Σ_ϵ provided certain transversality conditions are satisfied by the fast dynamics at a given Hurwitzian point in H. The Hopf bifurcation can in turn be either supercritical resulting in fast stable oscillations in the fast dynamics near the component unstable regions in the vicinity of the component stability boundary $H \cap \partial L^s$, or be subcritical corresponding to fast unstable oscillatory behavior away from the component unstable regions. Such bifurcation theoretic analysis of the interaction between the fast and slow dynamics near the sets S and H could be combined with the analysis of the slow dynamics Σ_s near these points (to be presented in Section 3) for analyzing the overall jump phenoemena in the large singularly perturbed system Σ_ϵ. Moreover for the power system, the validity of the models of the form Σ_ϵ for global analysis needs to be re-evaluated [21], and the effect of the dynamic load models in their physical implications should be clarified.

As mentioned earlier, it can be shown that the slow dynamics Σ_s indeed is an excellent approximation for the general power system analysis near the component stable region L^s. Therefore, in this paper, the analysis is restricted to the dynamics of the constrained system Σ_s within component stable regions and the system behavior will be treated as unpredictable near the component unstable regions [21]. Component stable regions in turn contain regions of attractions of the slow dynamics Σ_s and other regions where trajectories either converge to more complex limit points or may simply diverge.

TABLE 2.1
Structure of the state space

	Region	Sections of the boundary
State space components	1) Component Stable Regions System moves on predictable trajectories 2) Component Unstable Regions	1) Singular surface 2) Hurwitzian surface
contain *various regions*	1) Region of attraction 2) Region of convergence to stable limit cycles 3) Region of convergence to complex limit sets 4) Region of no convergence 5) Additional types	Boundary of the Region of attraction 1) Stable manifolds of five types of anchors: i) Unstable equilibria ii) Unstable limit cycles iii) Pseudo-equilibrium surface iv) Semi-singular surface v) Semi-Hurwitzian surface 2) Singular surface 3) Hurwitzian surface

3. State space definitions. The dynamics of the constrained system Σ_s will be analyzed in this section. A fixed parameter $p = p_0$ is considered for the state space analysis and the parameter p will not be shown in the subsequent equations for simplicity. Also the state variables (x, y) will be denoted together as z. The following sets essentially define the dynamical hierarchy of the state space:

$$L = \{(x,y) \in \mathbb{R}^{n+m} : g(x,y) = 0\} \tag{3.1}$$

$$L^s = \{(x,y) \in L : \sigma(D_y g)(x,y) \subset \mathbb{C}^-\} \tag{3.2}$$

$$M = \{(x,y) \in L : \operatorname{rank}(D_x g(x,y), D_y g(x,y)) = m\} \tag{3.3}$$

$$S = \{(x,y) \in L : \Delta(x,y) := \det D_y g(x,y) = 0\} \tag{3.4}$$

$$H = \{(x,y) \in L \setminus S : \sigma(D_y g)(x,y) \cap \mathcal{I} \neq \emptyset\} \tag{3.5}$$

L is the constraint surface determined by the algebraic constraint. The slow system or the reduced system Σ_s must stay on L and henceforth all topological definitions are relative to L. M is a regular subset of L which, by the implicit function theorem, is an n-dimensional manifold. L^s is the subset of the constraint surface L where the fast dynamics is locally asymptotically stable. Therefore secure system operation is restricted to the subset L^s of the constraint L. L^s is bounded by the singular set S (also called the impasse surface [6]) and the Hurwitzian surface H within L. The singular surface S can be characterized as the zero set of the determinant $\Delta(x,y)$ by definition (3.4) within L. Similarly it can be shown that a certain Hurwitzian determinant $\chi(x,y)$ defined as

$$\chi(x,y) := \det(H_{n-1}(D_y g))(x,y) \tag{3.6}$$

vanishes on the Hurwitzian surface $H \cap \partial L^s$ and here $H_{n-1}(A)$ denotes the classical Hurwitz matrix of order $n-1$ for the characteristic polynomial of the matrix A [25].

LEMMA 3.1.

$$H \cap \partial L^s = \{(x,y) \in \partial L^s \setminus S : \chi(x,y) = 0\} \tag{3.7}$$

Proof. In the boundary $H \cap \partial L^s$, the Jacobian $D_y g$ has purely imaginary eigenvalues with no zero eigenvalues, and all the remaining eigenvalues have negative real parts by the definition of H. Therefore the result follows by the Hurwitz determinant property from the classical system theory [25]. □

Therefore the component stability boundary ∂L^s is entirely characterized by either $\Delta(x,y) = 0$ (the singular set S) or $\chi(x,y) = 0$ (the Hurwitzian set H). The set S is especially significant since the system dynamics defined by Σ_s is singular in S.

3.1. The singular set S [8,9]. Define the induced vector field of the constrained system Σ_s on the constraint surface L by

$$Z(x,y) := \begin{pmatrix} f(x,y) \\ -(D_y g(x,y))^{-1} D_x g(x,y) f(x,y) \end{pmatrix} \tag{3.8}$$

Clearly the vector field Z is singular at the singularity S. The set S can now be divided into the following three subsets which can be shown to have locally different solution behavior [20,18]:

$$\Psi = \{(x,y) \in S : \kappa(x,y) := \mathrm{adj}(D_y g(x,y)) D_x g(x,y) f(x,y) = 0\} \tag{3.9}$$

$$\Xi = \{(x,y) \in S \setminus \Psi : D_y \Delta(x,y) \kappa(x,y) = 0\} \tag{3.10}$$

$$R = S \setminus (\Psi \cup \Xi) \tag{3.11}$$

In the definition of κ, $\mathrm{adj}(D_y g)$ denotes the classical matrix adjoint of the Jacobian $D_y g$. The subset Ψ of S is called the *pseudo equilibrium surface*. This high dimensional subset displays the behavior of an equilibrium surface for the vector field Z (equation (3.8)) induced by the system Σ_s and indeed becomes an equilibrium surface for an "equivalent" smooth dynamical system defined by multiplying the vector field Z pointwise by the determinant $\Delta(x,y)$. Such a transformation was proposed in [14] for analyzing the local dynamics of the singular vector field Z. Using the classical matrix adjoint property, the resulting transformed vector field can be defined [20,18] as

$$
(3.12) \qquad \begin{aligned} Z^T(x,y) &:= \begin{pmatrix} f(x,y)\Delta(x,y) \\ -\text{adj}(D_y g)(x,y) D_x g(x,y) f(x,y) \end{pmatrix} \\ &= \begin{pmatrix} f(x,y)\Delta(x,y) \\ -\kappa(x,y) \end{pmatrix} \end{aligned}
$$

Since the transformation from Z to Z^T amounts to a pointwise rescaling of the vector field Z by the factor $\Delta(x,y)$, it follows that the transformation is just a singular time rescaling along the trajectories. Therefore, the orbits or the trajectories of the vector field Z (those of Σ_s), as integral curves, are identical to those of the those of the transformed vector field Z^T, excepting for a reversal of the orientation when $\Delta(x,y) < 0$. Hence the topological properties of the singular dynamics Z such as the features of the stability boundary can be proved in the transformed smooth system for Z^T, and the results can be directly interpreted to the original singular system Σ_s. More details on the singular transformation and its properties can be seen in [20,18].

From the definition of Ψ and Z^T, it follows that the pseudo equilibrium surface Ψ (3.9) is a true equilibrium surface for the transformed vector field Z^T. It is useful to distinguish the following subsets of Ψ:

$$
(3.13) \quad N_\psi = \{(x,y) \in \Psi \cap M : \text{rank}(D_y g) = m-1 \text{ and the Jacobian of } Z^T \text{ has exactly two eigenvalues with nonzero real parts}\}
$$

$$
(3.14) \quad B_\psi = \Psi \setminus N_\psi
$$

Typically, the Jacobian of Z^T (restricted to M) has at least $(n-2)$ zero eigenvalues at a point in Ψ. Hence N_ψ corresponds to the subset of Ψ where the most regular type behavior takes place. It can be shown that near every point in N_ψ the integral curves of the system locally behave like for a linear system, and the points in the set N_ψ will be called the *nice* pseudo equilibrium points. The local dynamics near the points in the set B_ψ is more complex, and we call these points the *bad* pseudo equilibrium points. Eventually certain assumptions will be imposed to make sure that the complement B_ψ is a truly lower dimensional subset.

The set Ξ corresponds to points where Z^T is tangent to the singularity. Henceforth points in Ξ will be called *semi-singular*. Let

$$
(3.15) \quad N_\xi = \{(x,y) \in \Xi : \text{rank} \begin{pmatrix} D_x g & D_y g \\ D_x\Delta & D_y\Delta \end{pmatrix} = m+1, \text{ and } D_y\{(D_y\Delta)\kappa\}\kappa \neq 0\}
$$

$$
(3.16) \quad B_\xi = \Xi \setminus N_\xi
$$

The points in the set N_ξ are called the *nice* semi-singular points, and the points in the set B_ξ the *bad* semi-singular points. The condition on the second derivative in the definition of N_ξ ensures that trajectories do not cross over as they touch the singularity.

The set R is the complement of Ψ and Ξ in S and it consists of the singular points where exactly two trajectories of Z originate or converge transversal to R at infinite speeds. The points in R are called *the transverse singular points*. By the definition of R (3.11), at every point in R, the term $(D_y\Delta)\kappa$ is nonzero. The set of transverse singular points where two trajectories are converging corresponds to the case when $(D_y\Delta)\kappa$ is negative, and the latter when $(D_y\Delta)\kappa$ is positive [17]. Therefore the set of transverse singular points R can be divided into *transverse sinks* R_{si} $((D_y\Delta)\kappa < 0)$ and *transverse sources* R_{so} $((D_y\Delta)\kappa > 0)$ respectively.

3.2. The Hurwitzian set H**.** The analysis in this section is restricted to the component stability boundary $H \cap \partial L^s$. Similar results follow for the other segments of H. The induced vector field Z is smooth on the Hurwitzian surface H from equation (1.7). Let us divide the set H into two parts $\yen$ where the trajectories are transversal to H and Υ where Z^T is tangential to H:

$$\begin{aligned} (3.17) \qquad \yen &:= \{(x,y) \in H \cap \partial L^s : v(x,y) \\ &:= ((D_x\chi, D_y\chi)(x,y)Z^T(x,y) \neq 0\} \\ (3.18) \qquad \Upsilon &:= (H \cap \partial L^s) \setminus \yen \end{aligned}$$

The points in the set $\yen$ are called *the transversal Hurwitzian points* and their properties are summarized next:

LEMMA 3.2. *If non-empty, then* $\yen$ *is an* $(n-1)$*-dimensional embedded submanifold of* M *and* Z^T *is transversal to* $\yen$.

Proof. At every point in $\yen$, the local neighborhood of $\yen$ in L can be described by $g(x,y) = 0$ and $\chi(x,y) = 0$ with additional inequality conditions imposed. Since the constraint set L is invariant under the transformed vector field Z^T, it follows that $(D_xg, D_yg)Z^T = 0$. Therefore

$$(3.19) \qquad \begin{pmatrix} D_xg & D_yg \\ D_x\chi & D_y\chi \end{pmatrix} Z^T = \begin{pmatrix} 0 \\ v(x,y) \end{pmatrix}$$

Let $z_0 \in \yen$. As $z_0 \notin S$ (by the definition of H), it follows that $z_0 \in M$ (Lemma 2.3, [18]), therefore rank $(D_xg, D_yg) = m$. Also from the definition of $\yen$, $v(z_0) \neq 0$ and so $(D_x\chi, D_y\chi)$ does not lie in the row span of (D_xg, D_yg) at $z = z_0$. But then the Jacobian

$$(3.20) \qquad \begin{pmatrix} D_xg & D_yg \\ D_x\chi & D_y\chi \end{pmatrix}$$

has maximal rank ($= m + 1$) everywhere on $\yen$. By the implicit function theorem and Lemma 3.1, the set $\yen$ is therefore an $(n-1)$-dimensional manifold. Equation (3.19) also shows that Z^T is transversal to $\yen$. □

Moreover since any point in the Hurwitzian surface cannot be singular, then Lemma 3.2 proves the following: Given any transversal Hurwitzian point $z_0 \in \yen$, there exists a single smooth trajectory of Z which passes through z_0 and this trajectory is transversal to $\yen$ at z_0. Note that the term $v(z_0)$ in equation (3.19) corresponds to the first time derivative of the function $\chi(x, y)$ at $z = z_0$ evaluated along this trajectory.

From the equation (3.19), it can be seen that the vector field Z^T is tangential to the set Υ. The points in the set Υ are called *the semi-Hurwitzian points* and the set Υ can be further divided into a nice subset N_v which displays the most regular structure in Υ and the reminder B_v.

$$N_v := \{(x, y) \in \Upsilon : (D_x v, D_y v)Z^T \neq 0\} \tag{3.21}$$

$$B_v := \Upsilon \setminus N_v \tag{3.22}$$

The points in the set N_v are called the *nice* semi-Hurwitzian points and those in B_v the *bad* semi-Hurwitzian points. The properties of the set N_v are summarized next:

LEMMA 3.3. *If non-empty, N_v is an $(n-2)$-dimensional embedded submanifold of M and Z^T has a nonzero normal component at every point in N_v.*

Proof. Note that for any point in N_v, the local neighborhood of N_v in L can be characterized by the equalities $g(x, y) = 0$, $\chi(x, y) = 0$ and $v(x, y) = 0$ along with some additional inequality conditions. Therefore the result can again be proved by the implicit function theorem and the proof is an extension of the argument in Lemma 3.2. Essentially it can be shown that in N_v,

$$\begin{pmatrix} D_x g & D_y g \\ D_x \chi & D_y \chi \\ D_x v & D_y v \end{pmatrix} Z^T = \begin{pmatrix} 0 \\ 0 \\ \varsigma \end{pmatrix} \tag{3.23}$$

where $\varsigma = (D_x v, D_y v)Z^T \neq 0$ in N_v by definition. Therefore it follows that the Jacobian

$$J_v := \begin{pmatrix} D_x g & D_y g \\ D_x \chi & D_y \chi \\ D_x v & D_y v \end{pmatrix} \tag{3.24}$$

is of rank $(m + 2)$ proving that the set N_v is of dimension $(n-2)$ and that J_v is a normal vector field for N_v. Again by (3.23), then the normal component of Z^T is nonzero. □

Therefore given any $z_0 \in N_\upsilon$, there exists a single smooth trajectory of Z which passes through z_0 and is tangential to N_υ. Moreover this trajectory upon reaching N_υ must immediately leave N_υ by Lemma 3.3 and the second time derivative of $\chi(x, y)$ at $z = z_0$ evaluated along this trajectory is given by ς in the equation (3.23). Since $\varsigma \neq 0$ at z_0, then the sign of the function $\chi(x, y)$ along the trajectory does not change at z_0 even though $\chi(x, y) = 0$ at $z = z_0$.

As introduced before, the constraint set L can be divided into connected regions by the singular surface S and the Hurwitzian surface H. The connected components of the component stable set $L^s \backslash (S \cup H)$ are called the component stable regions and the connected components of $(L \backslash L^s) \backslash (S \cup H)$ are called the component unstable regions. The previous Lemma 3.3 proves that given any $z_0 \in N_\upsilon$, the unique trajectory of Z^T which passes through z_0 tangential to N_υ does so by staying in the same component stable or component unstable region because the function $\chi(x, y)$ does not change sign. In section 4, under certain transversality conditions, it will be shown that these points anchor the stability boundary of the region of attraction along with the singular points, pseudo equilibrium points, semi-singular points, unstable equilibrium points and unstable periodic orbits as shown in Table 2.1.

3.3. Singular dynamics hierarchy theorem. The properties of the various subdivisions of the state space introduced in the previous two sections are now summarized as a theorem. The proof of the theorem for the singularity aspects can be seen in [20,18] and the results for the Hurwitzian set have been already proved in Section 3.2 in the form of Lemmas 3.2 and 3.3.

THEOREM 3.4. *(Singular dynamics hierarchy theorem) If the subsets are not empty, then*

1. *$L \setminus S$ is an n-manifold. The solutions of the system Σ_s are the integral curves of the vector field Z on $L \setminus S$.*
2. *R is an $(n-1)$-dimensional manifold. At each point in R, two trajectories approach or diverge transversally at infinite speeds.*
3. *N_ψ is an $(n-2)$-dimensional manifold. Ignoring orientations, the solution curves are locally equivalent to those of a linear system near the origin.*
4. *N_ξ is an $(n-2)$-dimensional manifold. Trajectories are tangent to the singularity at N_ξ, but do not cross over.*
5. *$¥$ is an $(n-1)$-dimensional manifold. At each point in $¥$, a unique smooth trajectory passes through transversally.*
6. *N_υ is an $(n-2)$-dimensional manifold. Trajectories are tangential to the Hurwitzian set at N_υ, but do not cross over.*

4. Stability regions. In this section, the boundary of region of attraction of a stable equilibrium point will be characterized under some

Morse-Smale like assumptions. For the system Σ_s, equilibrium points and periodic orbits are defined as usual, with the proviso that they do not intersect the singular set S or the Hurwitzian set H. When an equilibrium of Σ_s is at the singularity, the resulting bifurcation has been analyzed in [20,18,16] as *the singularity induced bifurcation* for the constrained system Σ_s. Similarly when an equilibrium of the system Σ_s is at the Hurwitzian surface H, it can be shown that the local stability at the equilibrium will undergo a change generically and the analysis of the resulting bifurcation called *the Hurwitzian induced bifurcation* will be presented elsewhere. Stable equilibrium points define the system operating points for general physical systems of the form Σ_s such as the large power system. Since the definition of equilibrium points here is restricted to L^s, it is indeed sufficient to check the local stability in the constrained system Σ_s for guaranteeing the local stability in the full system Σ_ϵ.

Generally for systems of the form Σ_ϵ, given a stable equilibrium point z_s, the set of all trajectories which converge to the point z_s should be defined as the region of attraction for z_s. However in this paper, the definition of the region of attraction is restricted to those trajectories which converge to z_s and stay entirely within the component stable region C^s (i.e. the connected component of L^s) which contains z_s. This definition is motivated by the fact that for the power system, the system dynamics becomes unpredictable away from L^s [21]. The *region of attraction* of a stable equilibrium z_s is defined as

$$A = \{z \in C^s : \Phi_t(z) \in C^s \ \forall t \geq 0,\ \Phi_t(z) \to z_s \text{ as } t \to \infty\} \tag{4.1}$$

where $\Phi_t(\cdot)$ denotes the flow of the induced vector field Z.

It should be pointed out that for general singularly perturbed systems of the form Σ_ϵ, the definition of the region of attraction (4.1) only provides a conservative estimate of the full region of attraction. The possibility that a trajectory of the system Σ_ϵ reaching the stable equilibrium z_s via a number of jumps [14,12] from different component regions is ruled out in this definition. Further research is indicated on these questions for general systems of the form Σ_ϵ. However for the power system, this definition seems to be the only practical definition because of fundamental modeling issues [21].

Similarly define the stable and unstable manifolds for equilibria and periodic orbits in the usual way, but with the restriction to the component stable region C^s as in the case of the region of attraction. The stable and unstable manifolds for a pseudo equilibrium in Ψ, a semi-singular point in Ξ and a semi-Hurwitzian point in Υ is similar to those for the equilibrium, but the convergence may occur in finite time. As an example, the definitions for a nice semi-Hurwitzian point $z_0 \in \Upsilon$ are shown below:

$$\begin{aligned} W^s(z_0) \ = \ & \{z \in C^s : \exists t_0 > 0 \text{ such that } \Phi_t(z) \in C^s \text{ for } 0 \leq t < t_0 \\ & \text{and } \Phi_t(z) \to z_0 \text{ as } t \to t_0\} \end{aligned} \tag{4.2}$$

$$W^u(z_0) = \{z \in C^s : \exists t_0 < 0 \text{ such that } \Phi_t(z) \in C^s \text{ for } 0 \geq t > t_0 \tag{4.3}$$
$$\text{and } \Phi_t(z) \to z_0 \text{ as } t \to t_0\}.$$

Note that ignoring the speed, the trajectories of the vector fields Z and Z^T are identical (topologically equivalent) in C^s (Lemma 2.1, [18]). Since the trajectories (integral curves to be more precise) of Z coincide with the trajectories of Z^T within the component stable retion of C^s, the stability regions defined above for the flow Φ_t of Z can be equivalently redefined in terms of the flow Φ_t^T of the transformed vector field Z^T, by restricting the trajectories of Z^T to the component stable region C^s as shown below for the region of attraction A.

$$A = \{z \in C^s : \Phi_t^T(z) \in C^s \; \forall t \geq 0, \; \Phi_t^T(z) \to z_s \text{ as } t \to \infty\} \tag{4.4}$$

Hence the boundary of the region of attraction ∂A (with respect to L) can be characterized equivalently for the transformed vector field Z^T.

By analyzing the stable manifolds as defined here for the transformed system Z^T, it can be shown that these definitions indeed result in smooth manifolds of trajectories for the points such as the nice pseudo equilibrium points N_ψ, nice semi-singular points N_ξ and the nice semi-Hurwitzian points N_υ. The detailed analysis of the local dynamics near N_ψ and N_ξ is available in [18,16]. The semi-Hurwitzian points, a new mathematical concept introduced in this paper, can be treated similar to the semi-singular points considered in [20,18,16] and dynamical properties such as λ-lemmas can also be developed for the nice semi-Hurwitzian points. Here these details are omitted due to space limitations but can be seen in [19]. The main result of this paper, the characterization of the boundary ∂A, will be presented next after making some Morse-Smale like assumptions.

4.1. Assumptions. Assumptions (A): It will be assumed that the following assumptions are satisfied in the closure of the region of attraction $\overline{A}$:

(A0) The nice sets N_ψ, N_ξ and N_υ are dense in Ψ, Ξ and Υ, respectively. Furthermore, at every point $\psi \in \Psi$, $\xi \in \Xi$ or $z_0 \in \Upsilon$, the sets B_ψ, B_ξ and B_υ have dimension at most $(n-3)$.

(A1) (a) Equilibria z_0 and periodic orbits γ_0 in the boundary of the region of attraction, ∂A, are hyperbolic.
(b) Except for sets of dimension at most $(n-3)$, all pseudo saddles in ∂A are transverse.

(A2) Stable manifolds from the set $\{W^s(z_0), W^s(\gamma_0), W^s(N_C), W^s(B_C)\}$ intersect transversally with unstable manifolds from the set $\{W^u(z_0), W^u(\gamma_0), W^u(N_C), W^u(B_C)\}$. Here N_C stands for a connected component of either N_ψ, N_ξ or N_υ and B_C stands for a connected component of B_ψ, B_ξ or B_υ.

(A3) All trajectories converge to an equilibrium point, a periodic orbit, a pseudo equilibrium point, a semi-singular point or a semi-Hurwitzian point.

It can be shown that generally assumptions (A0)–(A2) will be satisfied for the system Σ_s [16,18]. However Assumption (A3) is not generic, but will be satisfied if there exists a dissipating energy function as claimed next. The proof is a direct extension of the proof in [20,18] which now includes the Hurwitzian component stability boundary as well.

LEMMA 4.1. *If there exists a Lyapunov-function* $V = V(x, y)$ *such that*

1. $(D_x f, D_y f)Z \leq 0$ *in* A*;*

2. V *is proper in the closure of* A*;*

3. the ω*-limit set in* $\overline{A}$ *consists only of equilibria and periodic orbits; then assumption (A3) is satisfied. All trajectories in* ∂A *converge to equilibria, periodic orbits, pseudo equilibria, semi-singular points or semi-Hurwitzian points.*

4.2. Stability boundary theorem. The boundary of the region of attraction ∂A is shown to consist of ten different segments when the Assumption (A) are satisfied.

THEOREM 4.2. *(Stability Boundary Theorem) [19] Under assumptions (A) the stability boundary is composed of*

1. Stable manifolds of unstable equilibria $z_0 \in \partial A$*,*

$$\bigcup_{z_0 \in \partial A} \{z \in C^s : \Phi_t^T(z) \to z_0 \text{ as } t \to \infty \text{ and } \Phi_t^T(z) \in C^s \ \forall\, t > 0\} \tag{4.5}$$

2. Stable manifolds of unstable periodic orbits $\gamma_0 \subset \partial A$*,*

$$\bigcup_{\gamma_0 \subset \partial A} \{z \in C^s : \Phi_t^T(z) \to \gamma_0 \text{ as } t \to \infty \text{ and } \Phi_t^T(z) \in C^s \ \forall\, t > 0\} \tag{4.6}$$

3. Stable manifolds of transverse pseudo saddles $N_{tr.sa} \cap \partial A$*,*

$$\bigcup_{\psi \in (N_{tr.sa} \cap \partial A)} \{z \in C^s : \Phi_t^T(z) \to \psi \text{ as } t \to \infty \text{ and } \Phi_t^T(z) \in C^s \ \forall\, t > 0\} \tag{4.7}$$

4. Stable manifolds of semi-saddles $N_{se.sa} \cap \partial A$*,*

$$\bigcup_{\xi \in (N_{se.sa} \cap \partial A)} \{z \in C^s : \exists\, t_0 > 0 \text{ such that } \Phi_t^T(z) \to \xi \text{ as } t \to t_0 \text{ and} \Phi_t^T(z) \in C^s \ \forall\, 0 < t < t_0\} \tag{4.8}$$

5. Stable manifolds of other pseudo equilibria $(\Psi \setminus N_{se.sa}) \cap \partial A$*,*

$$\bigcup_{\psi \in ((\Psi \setminus N_{tr.sa}) \cap \partial A)} \{z \in \partial A : \Phi_t^T(z) \to \psi \text{ as } t \to \infty \text{ and } \Phi_t^T(z) \in C^s \ \forall\, t > 0\} \tag{4.9}$$

6. Stable manifolds of other semi-singular points $(\Xi \setminus N_{se.sa}) \cap \partial A$,

$$\bigcup_{\xi \in ((\Xi \setminus N_{se.sa}) \cap \partial A)} \{z \in \partial A : \exists\, t_0 > 0 \text{ such that } \Phi_t^T(z) \to \xi \text{ as } t \to t_0 \text{ and} \Phi_t^T(z) \in C^s \ \forall\ 0 < t < t_0\} \tag{4.10}$$

7. Transversal singular boundary pieces $R \cap \partial \overline{A}$,

8. Stable manifolds of semi-Hurwitzian saddles $N_{se.Hu.sa} \cap \partial A$,

$$\bigcup_{z_0 \in (N_{se.Hu.sa} \cap \partial A)} \{z \in C^s : \exists\, t_0 > 0 \text{ such that } \Phi_t^T(z) \to z_0 \text{ as } t \to t_0 \text{ and } \Phi_t^T(z) \in C^s \ \forall\ 0 < t < t_0\} \tag{4.11}$$

9. Stable manifolds of other semi-Hurwitzian points $(\Upsilon \setminus N_{se.Hu.sa}) \cap \partial A$,

$$\bigcup_{z_0 \in ((\Upsilon \setminus N_{se.Hu.sa}) \cap \partial A)} \{z \in \partial A : \exists\, t_0 > 0 \text{ such that } \Phi_t^T(z) \to z_0 \text{ as } t \to t_0 \text{ and } \Phi_t^T(z) \in C^s \ \forall\ 0 < t < t_0\} \tag{4.12}$$

10. Transversal Hurwitzian boundary pieces $\yen \cap \partial \overline{A}$,
In other words,

$$\begin{aligned} \partial A \;=\; & \cup_{z_0 \in \partial A} W^s(z_0) \cup_{\gamma_0 \in \partial A} W^s(\gamma_0) \cup W^s\,(N_{tr.sa} \cap \partial A) \\ & \cup W^s\,(N_{se.sa} \cap \partial A) \cup (W^s\,(\Psi \setminus N_{tr.sa}) \cap \partial A) \\ & \cup (W^s\,(\Xi \setminus N_{se.sa}) \cap \partial A) \cup (R \cap \partial \overline{A}) \cup (\yen \cap \partial \overline{A}) \\ & \cup W^s\,(N_{se.Hu.sa} \cap \partial A) \cup (W^s\,(\Upsilon \setminus N_{se.Hu.sa}) \cap \partial A) \end{aligned} \tag{4.13}$$

The proof of the theorem is quite technical based on the standard λ-lemma type proofs and the invariance properties. The proofs connected with the singularity S and its subsets Ψ and Ξ can be seen in [18]. The additional lemmas which extend the result in [18] to the transversal Hurwitzian boundary and the semi-Hurwitzian boundary pieces are similar to the arguments for the transversal singular points and the semi-singular points respectively and can be seen in [19]. Necessary and sufficient conditions can be derived for checking the presence of the variety of anchor points of the stable manifolds similar to such results in [18] and the quasi-stability boundary $\partial \overline{A}$ can also be characterized under similar assumptions. Rigorous energy function estimates of the region of attraction can be developed from Theorem 4.2 similar to the results in [18].

Even though the equations in Theorem 4.2 seem complicated, the structure of the stability boundary is essentially simple as shown in Table 1. It is hoped that the detailed analysis of the constrained system dynamics Σ_s here will provide some of the basic tools necessary for the nonstandard global analysis of large singularly perturbed systems of the form Σ_ϵ.

5. Conclusions. This paper summarizes the state space structure of a class of constrained singular systems and it is shown that the resulting dynamics is rich in structural details. The global structure if any in large singularly perturbed systems which include the jump phenomena needs to be explored. The parameter space structure can also be pursued and the implications of the singular surface and the Hurwitzian surface in the bifurcation analysis need investigation. The practical implications of the component instability and possible jump behavior in the network equations should be carefully analyzed for the large power system. A better understanding of the load behavior in the large sense is essential. It seems that rigorous energy function techniques for the transient stability analysis can be developed based on the stability boundary characterization. It is hoped that the detailed analysis of the constrained slow dynamics Σ_s summarized here will aid in the development of an overall theory of the power system dynamics which covers global transient behavior such as the jump phenomena outside the constraint surface.

REFERENCES

[1] E. Benoit, *Canards et enlancements*, Publications I.H.E.S., Paris 1991.

[2] H.D. Chiang, M. Hirsch and F. Wu, *Stability regions of nonlinear autonomous dynamical systems*, IEEE Transactions on Automatic Control, **33** (1) January (1988), pp. 16–27.

[3] L.O. Chua and A.C. Deng, *Impasse points—part II: analytical aspects*, International Journal of Circuit Theory (1989).

[4] C.L. DeMarco and T.J. Overbye, *Integrating voltage security measures: connections between steady state, small disturbance stability, and nonlinear dynamic criteria*, Proceedings of the International Workshop on Bulk Power System Voltage Phenomena—II: Voltage Stability and Security, Maryland, August 1991, pp. 169–182.

[5] J. Guckenheimer and P. Holmes, *Nonlinear oscillations, dynamical systems and bifurcations of vector fields*, Springer-Verlag, New York, 1983.

[6] D.J. Hill and I.M.Y. Mareels, *Stability theory for differential/algebraic systems with applications to power systems*, IEEE Transactions on Circuits and Systems, Nov 1990, pp. 1416–1423.

[7] I.A. Hiskens and D.J. Hill, *Energy function, transient stability and voltage behavior in power systems with nonlinear loads*, IEEE Transactions on Power Systems, **4** (4) October (1989), pp. 1525–1533.

[8] F. Hoppensteadt, *Properties of solutions of ordinary differential equations with small parameters*, *Singular Perturbations in Systems and Control*, (edited by P.V. Kokotovic and H.K. Khalil,) IEEE Press, 1986, pp. 79–94.

[9] S.A. Kalinowky and M.N. Forte, *Steady state load-voltage characteristic field tests at area substations and fluorescent lighting component characteristics*, IEEE Transactions on Power Apparatus and Systems, **PAS-100** (6) June (1991), pp. 3087–3094.

[10] H.G. Kwatny, A.K. Pasrija and L.Y. Bahar, *Static bifurcation in power networks: loss of steady state stability and voltage collapse*, IEEE Trans. CSAS, **CAS-33** (10) October (1986), pp. 981–991.

[11] T. Ohyama, A. Watanabe, K. Nishimura and S. Tsuruta, *Voltage dependance of composite loads in power systems*, IEEE Transactions on Power Systems, **PAS-104** (11) November (1985), pp. 3064–3073.

[12] S. Sastry and C. Dasoer, *Jump behavior of circuits and systems*, IEEE Trans. CSAS, **CAS-28** (12) Dec 1981, pp. 1109–1124.

[13] P.W. Sauer and M.A. Pai, *Modeling and simulation of multimachine power dynamics*, in *Advances in Control and Dynamic Systems*, **43** (edited by C.T. Leondes,) Springer-Verlag, Academic Press 1991.

[14] F. Takens, *Constrained equations: a study of implicit differential equations and their discontinuous solutions*, in *Lecture Notes in Mathematics*, **525**, Springer-Verlag, Berlin and New York, 1976, pp. 143–234.

[15] E. Vaahedi, M.A. Fl-Kady, J.A. Libaque-Easine and V.F. Carvalho, *Load models for large-scale stability studies from end-user compensation*, IEEE Transactions on Power Systems, **PWRS-2** (4) November (1987), pp. 864–871.

[16] V. Venkatasubramanian, *A taxonomy of the dynamics of large differential-algebraic systems such as the power system*, Doctoral dissertation, Department of Systems Science and Mathematics, Washington University, School of Engineering and Applied Science, Saint Louis, Missouri, 63130.

[17] V. Venkatasubramanian, *On a singular transformation for analyzing the global dynamics of a class of singular DAEs*, Proc. of the Symposium on Implicit and Nonlinear Systems, Dallas, Texas, 1992.

[18] V. Venkatasubramanian, H. Schättler and J. Zaborszky, *A stability theory of large differential algebraic systems—a taxonomy*, Report SSM 9201-Part I, Department of Systems Science and Mathematics, Washington University, School of Engineering and Applied Science, Saint Louis, Missouri 63130.

[19] V. Venkatasubramanian, H. Schättler and J. Zaborszky, *A stability theory of large differential algebraic systems—a taxonomy*, Report SSM 9201-Part II, Department of Systems Science and Mathematics, Washington University, School of Engineering and Applied Science, Saint Louis, Missouri, 63130.

[20] V. Venkatasubramanian, H. Schättler and J. Zaborszky, *A taxonomy of the dynamics of the large electric power system*, Proceedings of the International Workshop on Bulk Power System Voltage Phenomena—II: Voltage Stability and Security, Maryland, August 1991, pp. 9–52.

[21] V. Venkatasubramanian, H. Schättler and J. Zaborszky, *Fast time varying phasor analysis in the balanced three phase large electric power system* (submitted for publication).

[22] V. Venkatasubramanian, H. Schättler and J. Zaborszky, *Voltage dynamics: study of a generator with voltage control, transmission and matched MW load*, IEEE Transactions on Automatic Control, November 1992.

[23] J. Zaborszky, *Some basic issues in voltage stability and viability*, Proceedings of Bulk-Power Voltage Phenomena—Voltage Stability and Security, Potosi, MO September, 1988, pp. 1.17–1.60.

[24] J. Zaborszky, G. Huang, B. Zheng and T.C. Leung, *On the phase portraits of a class of large nonlinear dynamic systems such as the power system*, IEEE Transactions on Automatic Control, **33** (1) January (1988), pp. 4–15.

[25] L.A. Zadeh and C.A. Desoer, *Linear system theory: the state space approach*, McGraw-Hill, New York, 1963.

[26] A.K. Zvonkin and M.A. Shubin, *Non-standard analysis and singular perturbations of ordinary differential equations*, Russian Mathematical Surveys, **39**(2) (1984), pp. 69–131.

ROBUST STABILIZATION OF CONTROLS IN POWER SYSTEMS

V. VITTAL*, M.H. KHAMMASH* , AND C.D. PAWLOSKI*

Abstract. This paper presents a novel approach to analyze stability robustness of controls in power systems over a range of operating conditions. The controls considered include exciters, power system stabilizers, and governors. The analytical basis for the approach is presented. The robustness framework for power systems is formulated. Numerical results for a sample test system are obtained and compared with those obtained by conventional techniques.

1. Introduction. In recent years, electric power systems have become more complex. Available transmission and generation is highly utilized, with large amounts of power interchanged among companies and geographical regions. To a large extent adequate system dynamic performance now depends on proper performance of controls such as excitation systems, power system stabilizers (PSS), static var compensators, and special HVDC controls. Proper design and robust performance of these controls is essential for the reliable operation of the interconnected power system.

The North American interconnection has seen an increase in economy transfers as well as security transfers. This has lead to several instances [1,2] of low frequency oscillations being observed in practice that involve entire areas of the interconnection. This phenomenon is called the inter-area oscillations, and is receiving increased attention by the utility industry. Controls once again are the main tools used for the mitigation of the inter-area oscillation. Hence, a careful design and robust performance of controls is imperative for reliable and secure system performance.

To a large extent, the current industry practice for the analysis and design of controls consists of conventional linear analysis coupled with detailed nonlinear simulation of the designed control settings [3,4].

The basic design philosophy consists of linearizing the system around the *worst-case* loading condition and using conventional linear analysis tools for analysis and design of the control settings. In the linearization process the effects of the various nonlinearities are eliminated and only localization of the original nonlinear system around the operating point considered. The design tools consist mainly of conventional time domain and frequency domain analysis together with modern state space techniques using eigenvalues and eigenvectors.

The designed controls are then tested for robustness over a wide range of operating conditions using detailed time domain simulation of the nonlinear system. This predicates an accurate model of the system and consideration of a comprehensive range of operating conditions. This procedure

* Electrical Engineering and Computer Engineering Department, Iowa State University, Ames, IA 50011, U.S.A.

is practical and has served the purpose. It, however, lacks a systematic approach and does not guarantee robustness.

The past ten years or so have witnessed an expolosion of controls research mainly directed at understanding the robustness properties of control systems. The purpose of these efforts is to obtain closed loop systems which are stable and which meet performance objectives despite the presence of plant uncertainty and parameter variations. In 1978 V.L. Kharitonov addressed the problem of parameter uncertainty [5,6] and triggered several efforts directed at extending Kharitonov's results to more general settings.

In the H^{∞} framework, Doyle [7] introduced the structural singular value (SSV) approach alternatively referred to as the μ-function. This approach handles multiple perturbations which are norm bounded linear time-invariant. A similar concept has also been developed in [8]. In [9] Dahleh and Ohta give necessary and sufficient conditions for stability robustness in the presence of unstructured nonlinear possibly time-varying perturbations. Recent results [10,11,12] provide a complete solution to the robustness analysis problem in the presence of linear time-varying and/or nonlinear structured perturbations. Both the stability robustness problem and the performance robustness problem are considered. The necessary and sufficient conditions obtained for robustness are both simple and computationally feasible, making this approach very attractive especially for problems with a large number of perturbation blocks as is the case in large scale interconnected systems including power systems.

In the approach presented in this paper we use the procedure developed in [12]. The analytical details of the approach are presented in Section II. We formulate the robustness framework for the power system. Stability robustness of control settings for power system stabilizers, exciters, and governors is then analyzed over a range of operating conditions. The formulation has capability to include any additional control features and is not limited to the control features considered. The formulation is presented in Section III. One of the salient features of the robustness approach approach presented is the ability to provide systematic and computable necessary and sufficient conditions for stability robustness of large scale systems. The great reduction in computation is attributed to the simplicity of the derived conditions for robustness which consists of computing the spectral radius of a certain nonnegative matrix obtained from the dynamical equations governing the system. In Section IV a sample test system [13] specifically designed to analyze the effect of controls on the interarea mode is used to test the procedure. The results obtained are compared with those obtained using the conventional eigenvalue techniques.

2. Analytical background. The standard setup for a general robustness problem appears in Fig. 2.1. In the figure, G_0 is a nominal linear time-invariant plant. G_0 may be continuous-time or discrete-time. C is

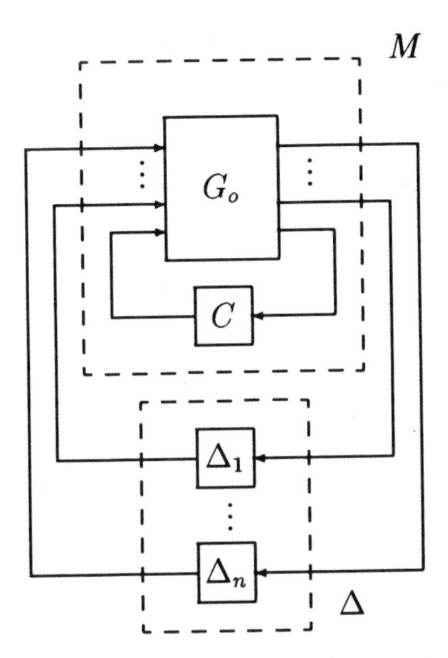

FIG. 2.1. *System with structured uncertainty*

a linear controller stabilizing G_0. For the analysis problem, C is assumed given and fixed. The uncertainty is modelled with perturbation blocks $\Delta_1, ..., \Delta_n$. Each perturbation Δ_i belongs to the following class of admissible perturbations:

$$\Delta = \left\{ \Delta : \Delta \quad \text{is causal, and} \quad ||\Delta|| := \sup_{u \neq 0} \frac{||\Delta u||_\infty}{||u||_\infty} \leq 1 \right\}. \tag{2.1}$$

where the norm used is the L^∞ norm (or ℓ^∞ norm for discrete-time systems). The perturbations may therefore be nonlinear or time-varying. The n perturbation blocks can be lumped into one perturbation block with a diagonal structure. Hence we can view the class of admissible perturbations as the class of all $\Delta \in D(n)$ where

$$D(n) := \{\text{diag}(\Delta_1, ..., \Delta_n) : \Delta_i \in \Delta\} \tag{2.2}$$

Similarly, G_0 and C can be lumped into one system M. M is therefore, linear, time-invariant, casual, and stable. Any weighting on any of the perturbations can be lumped into M. It follows that the system in Fig. 2.1 can be transformed into that in Fig. 2.2.

Consider the system in Fig. 2.2. The system is said to achieve robust stability if it is L^∞-stable for all admissible perturbations, i.e., for all

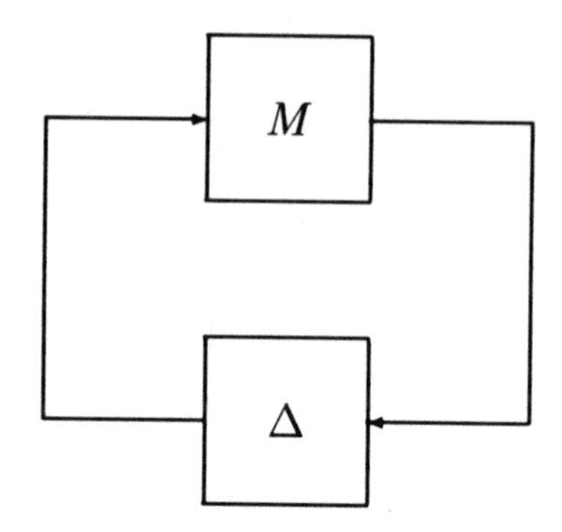

FIG. 2.2. *Stability robustness problem.*

$\Delta \in D(n)$. We next provide necessary and sufficient conditions for robust stability. These conditions will be stated in terms of M. From the figure, M has n inputs and n outputs corresponding to the inputs and outputs of the perturbations. Consequently, M can be represented by the system matrix.

$$M = \begin{pmatrix} M_{11} & & M_{1n} \\ \vdots & \cdots & \vdots \\ M_{n1} & & M_{nn} \end{pmatrix}$$

In particular, M_{ij} is given by the transfer function between the ouput of Δ_j and the input to Δ_j, with all the perturbations, $\Delta_1, \ldots, \Delta_n$ set to zero.

Whenever M is finite-dimensional, it admits a state space realization of the form:

$$\dot{x} = Ax + Bu \tag{2.3}$$

$$y = Cx + Du \tag{2.4}$$

where

$$C = \begin{pmatrix} C_1 \\ \vdots \\ C_n \end{pmatrix}, \quad B = (B_1 \ldots B_n) \quad D = \begin{pmatrix} D_{11} & \ldots & D_{1n} \\ \vdots & & \vdots \\ D_{n1} & \ldots & D_{nn} \end{pmatrix}$$

u is an n vector input to M which corresponds to the perturbations' output, y is an n vector output corresponding to the perturbations' inputs. Each M_{ij} has induced norm which we refer to as the A norm. It can be computed arbitrarily accurately since

$$||M_{ij}||_A = |D_{ij}| + \sum_{k=0}^{\infty} |C_i A^k B_j| \tag{2.5a}$$

in the discrete time case, and

$$\|M_{ij}\|_A = |D_{ij}| + \int_0^\infty |C_i e^{At} B_j| dt \tag{2.5b}$$

in the continuous-time case. We can therefore define the following matrix:

$$\hat{M} = \begin{bmatrix} \|M_{11}\|_A & \dots & \|M_{1n}\|_A \\ \vdots & & \vdots \\ \|M_{n1}\|_A & \dots & \|M_{nn}\|_A \end{bmatrix} \tag{2.6}$$

As the next theorem shows, it turns out that $\hat{M}$ plays a fundamental role in the robustness of the given system. We now state the main theorem establishing nonconservative conditions for robustness:

THEOREM 2.1. *The system in Fig. 2.2 achieves robust stability if and only if any one of the following conditions holds:*

1. $\rho(\hat{M}) < 1$, *where* $\rho(.)$ *denotes the spectral radius.*

2. $\inf_{R \in \mathcal{R}} \|R^{-1} M R\|_A < 1$, *where* $\mathcal{R} : \{\text{diag}(r_1, ..., r_n) : r_i > 0\}$.

This theorem, due to Khammash and Pearson in the discrete-time case [10,11] and to Khammash [12] in the continuous-time case, reduces the robustness analysis problem to that of computing the spectral radius of a nonnegative matrix. The theory for nonnegative matrices (see e.g., [14]) provides a power algorithms for fast computation of the spectral radius of a nonnegative matrix. As a result, the robustness condition can be computed exactly and efficiently and thus is especially suited for systems with a large number of uncertainty blocks.

3. Robustness framework for power systems. The robustness approach presented in Section 2 is now formulated for power systems. The stability robustness of a group of control settings (gains and time constants) for power system stabilizers, exciters, and governors is verified over a range of loading conditions. A nominal loading condition is considered and the system is linearized at the steady state solution.

Perturbations are then considered at each load bus for both the real power and reactive power portions. This introduces perturbations in the diagonal entries of the load bus elements in the network admittance matrix, since the loads are modelled as constant impedances. In addition, perturbations are introduced over the elements of the system A,B,C, and D matrices. The lower bound and upper bound on the ranges of these perturbations are determined by running power flows at the extreme load conditions.

We will now describe the development of the M matrix, discussed in Section II, using only the perturbations on the load bus entries of the admittance matrix. A similar approach is followed to include the perturbations on the A,B,C, and D matrices.

Perturbations on the diagonal entries of the load buses in the admittance matrix affect the relationship between the injected currents in the network and the terminal voltages. In the linearized equations in the network reference frame, this is governed by

$$\begin{bmatrix} I_{Q,D_\Delta} \\ \hline 0 \end{bmatrix} = \left[\begin{array}{c|c} Y_{nn} & Y_{nm} \\ \hline Y_{mn} & Y_{mm} \end{array}\right] \begin{bmatrix} V_{Q,D_\Delta} \\ \hline V_{\ell Q,D_\Delta} \end{bmatrix} \tag{3.1}$$

where

I_{Q,D_Δ} = vector of injected currents at the generator terminal buses.
V_{Q,D_Δ} = vector of generator terminal bus voltages.
$V_{\ell Q,D_\Delta}$ = vector of load bus voltages.

Equation(3.1) can be simplified to obtain

$$\begin{aligned} I_{Q,D_\Delta} &= [Y_{nn} - Y_{nm}Y_{mm}^{-1}Y_{mn}]V_{Q,D_\Delta} \\ &= \hat{Y}_{nn}V_{Q,D_\Delta} \end{aligned} \tag{3.2}$$

Based on the preliminaries presented in Section 2, the M matrix is derived by determining the relationship that exists between the outputs of the perturbations and the inputs to the perturbations taking into account the dynamics of the linearized system with the controllers.

In arriving at the linearized differential equations for the system we consider the following representation for the synchronous machines, exciters, power system stabilizers, and governors. The synchronous machines are represented by the two-axis model [15, Chap. 4] given by

$$\begin{aligned} \tau'_{do_i}\dot{E}'_{qi} &= E_{FD_i} - E'_{q_i} + (x_{d_i} - x'_{d_i})I_{d_i} \\ \tau'_{qo_i}\dot{E}'_{d_i} &= -E'_{d_i} - (x_{q_i} - x'_{q_i})I_{q_i} \\ M_i\dot{\omega}_i &= P_{m_i} - D_i\omega_i - [E'_{d_i}I_{d_i} + E'_{q_i}I_{q_i} - (x'_{q_i} - x'_{d_i})I_{d_i}I_{q_i}] \\ \dot{\delta}_i &= \omega_i - 1 \qquad i = 1, 2, ..., n \end{aligned} \tag{3.3}$$

A fast static exciter model given in [13] is used. The power system stabilizer is represented by a ETMSP [16] Type 1 model, and the governor-turbine by a ETMSP [16] Type 8 model.

The linearized system equations are then interfaced with the network equations to obtain the overall state space representation of the system. In interfacing the network equations to the system dynamic equations appropriate reference frame transformations [15, Chap. 9] have to be performed.

The process is summarized in Fig. 3.1, where

$$T_0 = \begin{bmatrix} \cos\delta_{10} & -\sin\delta_{10} & 0 & 0 & \cdots & 0 & 0 \\ \sin\delta_{10} & \cos\delta_{10} & 0 & 0 & \cdots & 0 & 0 \\ 0 & 0 & \cos\delta_{20} & -\sin\delta_{20} & \cdots & \vdots & \vdots \\ 0 & 0 & \sin\delta_{20} & \cos\delta_{20} & \cdots & \vdots & \vdots \\ \vdots & \vdots & \vdots & \vdots & & \vdots & \vdots \\ \vdots & \vdots & \vdots & \vdots & & \vdots & \vdots \\ 0 & 0 & 0 & 0 & \cdots & \cos\delta_{no} & -\sin\delta_{no} \\ 0 & 0 & 0 & 0 & \cdots & \sin\delta_{no} & \cos\delta_{no} \end{bmatrix}$$

$$V_{qdo} = \begin{bmatrix} -V_{d10} & 0 & \cdots & 0 \\ V_{q10} & 0 & \cdots & 0 \\ 0 & -V_{d20} & \cdots & 0 \\ 0 & V_{q20} & \cdots & 0 \\ \vdots & \vdots & & \vdots \\ \vdots & \vdots & & \vdots \\ 0 & 0 & \cdots & -V_{dno} \\ 0 & 0 & \cdots & V_{qno} \end{bmatrix} \quad I_{qdo} = \begin{bmatrix} -I_{d1o} & 0 & \cdots & 0 \\ I_{q1o} & 0 & \cdots & 0 \\ 0 & -I_{d2o} & \cdots & 0 \\ 0 & I_{q20} & \cdots & 0 \\ \vdots & \vdots & & \vdots \\ \vdots & \vdots & & \vdots \\ 0 & 0 & \cdots & -I_{dno} \\ 0 & 0 & \cdots & I_{qno} \end{bmatrix}$$

$$Z_{SM_\Delta} = \begin{bmatrix} r_1 & -x'_{d1} & 0 & 0 & \cdots & 0 & 0 \\ x'_{q1} & r_1 & 0 & 0 & \cdots & 0 & 0 \\ 0 & 0 & r'_2 & -x'_{d2} & \cdots & 0 & 0 \\ 0 & 0 & x'_{q2} & r_2 & \cdots & 0 & 0 \\ \vdots & \vdots & \vdots & \vdots & & \vdots & \vdots \\ 0 & 0 & 0 & 0 & \cdots & r_n & -x'_{dn} \\ 0 & 0 & 0 & 0 & \cdots & x'_{qn} & r_n \end{bmatrix}$$

The variables with the subscript o represent the steady state conditions for the variables of the nominal system.

In obtaining Fig. 3.1, the absolute values of the perturbations are normalized to 1. This is done by selecting the nominal value to be the midpoint of the upper and lower range. The perturbation is then normalized to 1 and weighting factor given by the difference between the nominal value and the upper or lower bound is pulled into the system matrix. In addition, the case of perturbations on the load results in complex perturbations associated with admittance entry corresponding to the load buses. The theory presented in Section II allows only real perturbations. As a result, we separate the real and imaginary parts of equation (8) as follows:

$$\begin{gathered}[I_{Q\Delta} + jI_{D\Delta}] = [\hat{G}_{nn} + j\hat{B}_{nn}][V_{Q\Delta} + jV_{D\Delta}] \\ \begin{bmatrix} I_{Q\Delta} \\ I_{D\Delta} \end{bmatrix} = \begin{bmatrix} \hat{G}_{nn} & -\hat{B}_{nn} \\ \hat{B}_{nn} & \hat{G}_{nn} \end{bmatrix} \begin{bmatrix} V_{Q\Delta} \\ V_{D\Delta} \end{bmatrix}\end{gathered} \tag{3.4}$$

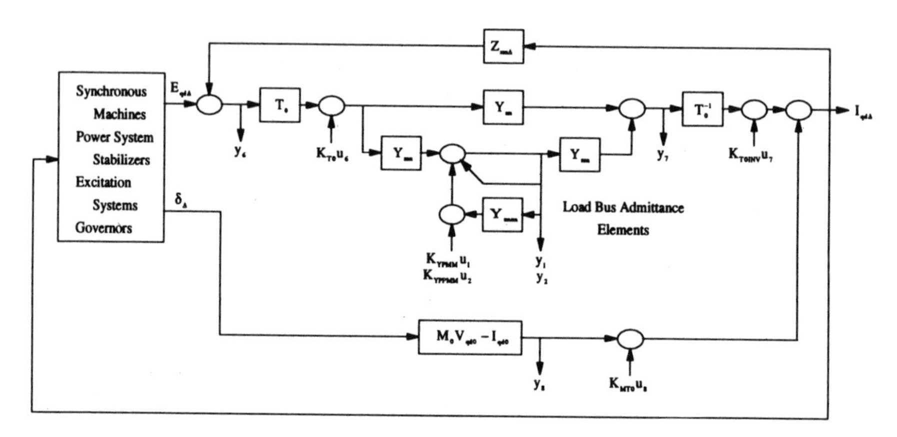

FIG. 3.1. *Block diagram of robust control formulation in power systems for load variation.*

From (3.4) one observes that the input and output relationship of the network results in a repetition of the conductance and admittance elements. If these elements are perturbed, then the perturbations have to be repeated to capture the change in conditions.

The range on the entries corresponding to the load admittance is determined by conducting power flows at the extreme loading conditions. The nominal case is selected at the mid point of the range. Then

$$|G_{\ell_{ii}}| \leq \gamma_{G_{\ell}i}, |B_{\ell_{ii}}| \leq \gamma_{B\ell_i}, \quad \text{where} \quad Y_{\ell_{ii}} = G_{\ell_{ii}} + jB_{\ell_{ii}}.$$

The perturbations are normalized to 1, and the weighting factor pulled into the system matrix. This is represented in Fig. 3.1 by K_{YPMM} where

$K_{YPMM} =$

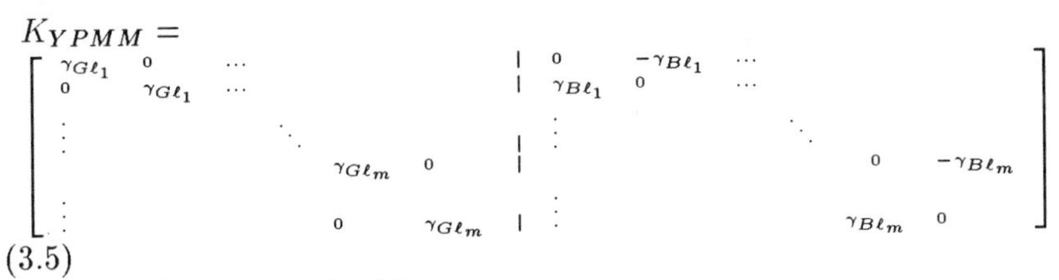

(3.5)

assuming there are m load buses.

Similarly in Fig. 3.1, K_{YPPMM} are the gains due to the perturbations on the shunt capacitor elements. K_{T_0} and K_{TOINV} are the gains due to perturbations on T_0 and T_0^{-1} respectively. In addition, due to the change in operating conditions perturbation gains are also obtained on the synchronous machine blocks and the exciter blocks.

The overall system representation is then given by

$$\begin{aligned} \dot{x} &= A_{sys}x + B_1 u_1 \\ y_1 &= C_{1Eqd\Delta}E_{qd\Delta} + C_{1\delta\Delta}\delta_\Delta + D_{11}u_1 \end{aligned} \tag{3.6}$$

where

$$\begin{aligned} C_{1Eqd\Delta} &= (Y_{mm}^{-1}Y_{mn}T_0)(U - Z_{SM\Delta}\hat{M}_0 M_0) \\ C_{1\delta\Delta} &= -(Y_{mm}^{-1}Y_{mn}T_0)(Z_{SM\Delta}\hat{M}_0\tilde{M}_0) \\ D_{11} &= -(U + Y_{mm}^{-1}Y_{mn}T_0 Z_{SM\Delta}\hat{M}_0 T_0^{-1}Y_{nm})Y_{mm}^{-1}K_{YPMM} \end{aligned}$$

and

$$\begin{aligned} &U = \text{Identity matrix} \\ &M_0 = T_o^{-1}\hat{Y}_{nn}T_0 \\ &\hat{M}_0 = (U + M_0 Z_{SM\Delta})^{-1} \\ &\tilde{M}_0 = M_0 V_{dq_0} - I_{qd_0} \\ &X = [X_{GEN'}X_{PSS'}X_{EXC'}X_{GOV}]^T \\ &E_{qd\Delta} = [E'_{q1\Delta}E'_{d1\Delta}\cdots E'_{qn\Delta}E'_{dn\Delta}]^T \\ &\delta_\Delta = [\delta_{1\Delta}, \delta_{2\Delta}, \ldots\delta_{n\Delta}]^T \end{aligned}$$

A_{sys} and B_1 can be derived knowing the linearized representation of the synchronous machine, exciters, PSS, and governors [17].

The representation (3.6) is then used to obtain the norm matrix shown in (2.6) using (2.5b). In the derivation of the norm matrix, it is important to note that the inclusion of the n-machine angles as states, the matrix A_{sys} for the nominal conditions has a zero eigenvalue which is unobservable but controllable. The minimal realization of the system is obtained before the norm matrix is evaluated. This ensures that the system matrix has all eigenvalues with negative real parts.

4. Numerical results. The robustness procedure developed in this paper is tested on a sample system developed in [13]. This system exhibits the interarea mode phenomenon, and has the complexity to verify the efficacy of the procedure developed. Figure 4.1 gives a one line diagram of the system. We consider two different network topologies:

a) Two tie-lines in service and Area 1 is exporting 400 MW to Area 2. We then analyze the robust performance of the controls for a range of variations of Load 1 and Load 2. In our analysis we consider a 200 MW variation in each load; Load 1 1127-1327 MW and Load 2 1767-1967 MW.

b) One tie-line in service and Area 1 is exporting 400 MW to Area 2. Load 1 is varied over the range 1247-1267 MW and Load 2 1927-1967 MW.

For the range of variations of loads considered we analyze the system for the following configurations.

CASE 1: Four fast exciters with four power system stabilizers (PSS).

CASE 2: Four fast exciters, with only one PSS, in the area exporting power on GEN2.

CASE 3: Four fast exciters, with only one PSS, in the area exporting power on GEN12.

CASE 4: Four fast exciters with no PSS.

The settings for the exciters are given in [13]. The PSS settings were set as follows: $K_s = 15$, $T_5 = 10$, $T_1 = 0.047$, $T_2 = 0.021$, $T_3 = 3.0$, $T_4 = 5.4$. For each case the M matrix was calculated including perturbations in the loads and the system A, B, C, and D matrices. In evaluating the M matrix, the integration of the exponential terms was done up to 50 secs. Beyond this value, the contribution of the integrand was negligible.

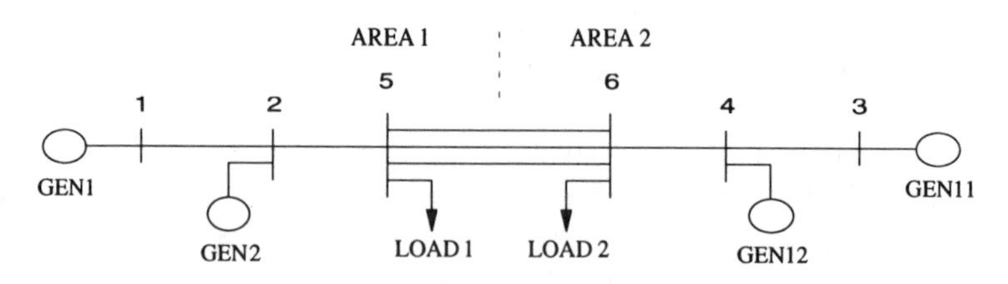

FIG. 4.1. *Two area system.*

Table 4.1 shows the spectral radius results for the two tie-line cases. In order to cover the 200 MW load variation range, we consider grids of 20 MW variation and perform the analysis over the entire range. For each case the nominal condition is specified and range of variation on Load 1 and Load 2 is given. Note that we consider the entire range of Load 2 for each range of Load 1 in order to cover all possible loading conditions. The results of Table 4.1 clearly show that the system is robustly stable for all the cases as the spectral radii in all cases are < 1. Furthermore, Case 1 is most robust and Case 2 is the least robust. Case 3 and Case 4 have spectral radii which have the same order of magnitude. These robustness results are now compared with those obtained by time simulation. To test the robustness, the nonlinear system was simulated at the extreme condition for each load setting. The initial conditions are calculated at the extreme condition by running a power flow. A three phase fault was then placed on bus # 3 and cleared in 5 ms. This is small disturbance which results in the system moving away from the equilibrium, enabling us to study its small signal stability. The stability is analyzed in terms of the behavior of the electrical power output of GEN12. A sample set of plots for the extreme conditions of Load Setting # 50 where Load 1 is 1327 MW and Load 2 is 1967 MW is shown in Fig. 4.2 where the different cases are compared. We note that all the cases are stable as predicted by the spectral radius test. In addition Fig. 4.2 shows that Case 1 which corresponds to the small spectral radius settles faster than Case 4 which has the larger spectral radius. The relationship between the settling and the relative value of the spectral radius has also been observed in other cases as well.

Table 4.2 shows the spectral radius results for the one tie-line case. The results indicate that Case 1 is the most robust, followed by Case 4 and Case 3. In addition, Case 2 is found to be unstable. The nonlinear time simulation was repeated for Setting # 8 Load 1 1267 MW and Load 2 1967 MW. These results are shown in Fig. 4.3. We again note that Fig. 4.3 confirms that Case 1 is the most robust of all cases. It shows that Case 4 is more robust than Case 3 as it settles down to lower value at about 12 secs than Case 3 even though it has a higher initial peak. The spectral radius predicts this as shown in Table 4.2 where the spectral radius for Case 4 is 0.5408 and Case 3 is 0.7400. Finally, Fig. 4.3 clearly indicates that Case 2 is unstable as predicted by the spectral radius of 3.1676. The spectral radius for the case remained well above 1 no matter how small a load range was chosen.

TABLE 4.1
Two Tie-Line Spectral Radius Results.

Load Setting Number	Load 1 Range (MW)	Load 2 Range (MW)	Nominal Load		Spectral Radius			
			Load 1 (MW)	Load 2 (MW)	Case 1	Case 2	Case 3	Case 4
1	1127-1147	1767-1807	1137	1787	0.5591	0.7878	0.6959	0.6209
2		1807-1847		1827	0.5492	0.7921	0.6768	0.6111
3		1847-1887		1867	0.5252	0.7582	0.6516	0.5870
4		1887-1927		1907	0.6142	0.9004	0.7583	0.6917
5		1927-1967		1947	0.5095	0.7427	0.6322	0.5755
6	1147-1167	1767-1807	1157	1787	0.5977	0.8540	0.7394	0.6660
7		1807-1847		1827	0.5511	0.7962	0.6785	0.6162
8		1847-1887		1867	0.5373	0.7741	0.6619	0.6037
9		1887-1927		1907	0.6156	0.9029	0.7588	0.6969
10		1927-1967		1947	0.5648	0.8203	0.7033	0.6432
11	1167-1187	1767-1807	1177	1787	0.5838	0.8223	0.7233	0.6531
12		1807-1847		1827	0.5642	0.8128	0.6933	0.6347
13		1847-1887		1867	0.5568	0.8084	0.6855	0.6288
14		1887-1927		1907	0.5594	0.8161	0.6880	0.6358
15		1927-1967		1947	0.5903	0.8632	0.7300	0.6740
16	1187-1207	1767-1807	1197	1787	0.5749	0.8059	0.7130	0.6469
17		1807-1847		1827	0.5437	0.7735	0.6705	0.6150
18		1847-1887		1867	0.5406	0.7837	0.6669	0.6143
19		1887-1927		1907	0.5422	0.7924	0.6692	0.6211
20		1927-1967		1947	0.5493	0.8044	0.6740	0.6298
21	1207-1227	1767-1807	1217	1787	0.5927	0.8341	0.7323	0.6704
22		1807-1847		1827	0.5459	0.7845	0.6687	0.6210
23		1847-1887		1867	0.5612	0.8168	0.6891	0.6410
24		1887-1927		1907	0.5525	0.8078	0.6805	0.6368
25		1927-1967		1947	0.5623	0.8287	0.6897	0.6513
26	1227-1247	1767-1807	1237	1787	0.5918	0.8290	0.7300	0.6742
27		1807-1847		1827	0.5422	0.7721	0.6660	0.6209
28		1847-1887		1867	0.5572	0.8095	0.6854	0.6429
29		1887-1927		1907	0.5232	0.7575	0.6478	0.6074
30		1927-1967		1947	0.5563	0.8222	0.6790	0.6472
31	1247-1267	1767-1807	1257	1787	0.6300	0.8978	0.7713	0.7220
32		1807-1847		1827	0.5589	0.8015	0.6844	0.6440
33		1847-1887		1867	0.5672	0.8252	0.6958	0.6585
34		1887-1927		1907	0.5422	0.7885	0.6688	0.6343
35		1927-1967		1947	0.5457	0.8041	0.6714	0.6422
36	1267-1287	1767-1807	1277	1787	0.6574	0.9407	0.7992	0.7580
37		1807-1847		1827	0.5989	0.8678	0.7272	0.6947
38		1847-1887		1867	0.5690	0.8302	0.6970	0.6662
39		1887-1927		1907	0.5333	0.7761	0.6577	0.6285
40		1927-1967		1947	0.5649	0.8390	0.6879	0.6680
41	1287-1307	1767-1807	1297	1787	0.5726	0.8177	0.6976	0.6642
42		1807-1847		1827	0.6261	0.9138	0.7563	0.7324
43		1847-1887		1867	0.6011	0.8823	0.7310	0.7083
44		1887-1927		1907	0.5515	0.8082	0.6772	0.6557
45		1927-1967		1947	0.5547	0.8222	0.6807	0.6643
46	1307-1327	1767-1807	1317	1787	0.6117	0.8704	0.7452	0.7160
47		1807-1847		1827	0.5668	0.8175	0.6899	0.6679
48		1847-1887		1867	0.6060	0.8920	0.7352	0.7212
49		1887-1927		1907	0.5885	0.8709	0.7157	0.7031
50		1927-1967		1947	0.5657	0.8477	0.6914	0.6855

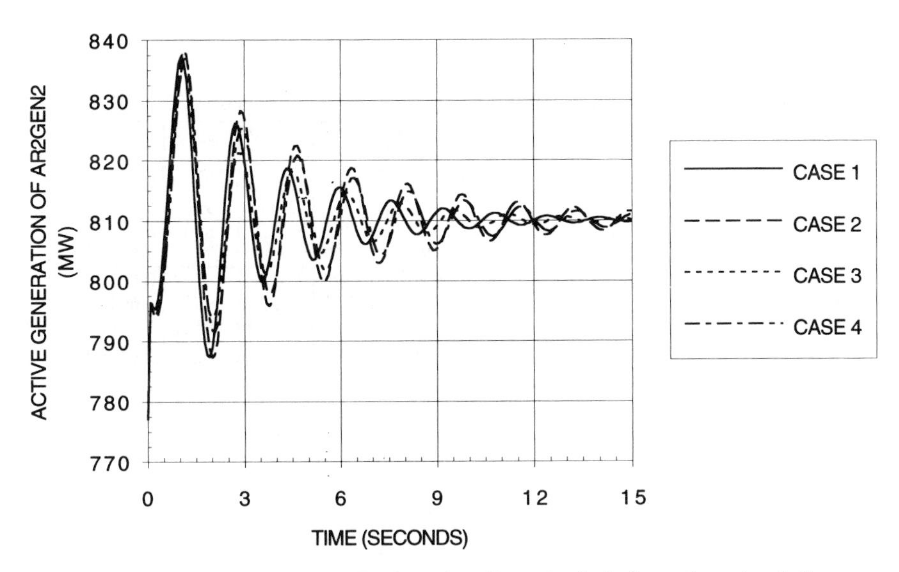

FIG. 4.2. *Two tie-line case. Load setting #50. Analysis from time simulation.*

TABLE 4.2
One Tie-Line Spectral Radius Results.

Load Setting Number	Load 1 Range (MW)	Load 2 Range (MW)	Nominal Load		Spectral Radius			
			Load 1 (MW)	Load 2 (MW)	Case 1	Case 2	Case 3	Case 4
1	1247-1257	1927-1937	1252	1932	0.3901	4.0283	0.9699	0.7020
2		1937-1947		1942	0.3852	3.7448	0.9562	0.6799
3		1947-1957		1952	0.3390	3.1827	0.8167	0.6029
4		1957-1967		1962	0.3077	2.7905	0.7316	0.5397
5	1257-1267	1927-1937	1262	1932	0.3198	3.8226	0.8017	0.6032
6		1937-1947		1942	0.3717	4.3228	0.9583	0.6915
7		1947-1957		1952	0.3394	3.6609	0.8547	0.6080
8		1957-1967		1962	0.3002	3.1676	0.7400	0.5408

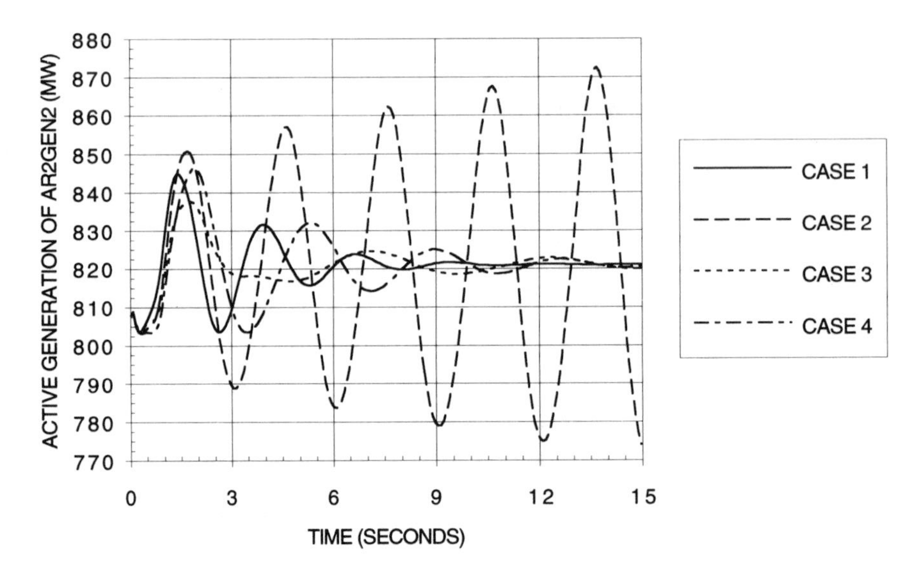

FIG. 4.3. *One tie-line case. Load setting #8. Results from time simulation.*

5. Conclusions. A novel technique is developed in this paper to analyze the robust stability of control settings(gains and time constants). The basic principles governing the technique are presented and is adapted to power systems.

The control features considered include exciters, power system stabilizers, and governors. The stability robustness of these controls is then analyzed for a range of loading conditions.

Preliminary results of the procedure applied to a sample test system clearly indicate the efficacy of the procedure in predicting stability robustness. The formulation of the procedure can be extended to include any modeling feature. The procedure can also be used to aid in robust design by applying it repetitively.

REFERENCES

[1] CRESSAP, R.L., AND J.R. HAUER, *Emergence of a New Swing Mode in the Western Power System*, IEEE Transactions on Power Apparatus and Systems, PAS-100 (1981) pp. 2037–2043.

[2] KUNDUR, P., M. KLEIN, G.J. ROGERS, AND M. ZWYNO. *Application of Power System Stablizers for Enhancement of Overall Systems Stability*, IEEE Transaction on Power Systems PS-4 (May 1989), pp. 614–622.

[3] ROGERS, G.J., AND P. KUNDUR, *Small Signal Stability of Power Systems*, Special Publication: IEEE Power Engineering Socity, 90TH0292-3-PWR: 5-16.

[4] MARTINS, N., AND L.T.G. LIMA, *Eigenvalue and Frequency Domain Analysis of Small-Signal Electromechanical Stability Problems*, Special Publication: IEEE Power Engineering Society, 90TH0292-3-PWR, pp. 17–33.

[5] KHARITONOV, V.L., *Asymptotic Stability of an Equilibrium Position of a Family of Systems of Linear Differential Equations*, Differntial'nye Uraveniya 14, no. 11 (1978), pp. 1483–1485.

[6] KHARITONOV, V.L., *On a Generalization of Stability Criterion*, Akademii mauk Kahskoi SSR, Fiziko-matematicheskaia 1, (1978), pp. 53–57.

[7] DOYLE, J.C., *Analysis of Feedback Systems with Structured Uncertainty*, IEE Proceedings 129, PtD, no. 6 (November 1982), pp. 242–250.

[8] SAFONOV, M.G., *Stability Margins of Diagonally Perturbed Multivariable Feedback Systems*, IEE Proceedings 129, PtD, no. 6 (November 1982), pp. 251–256.

[9] DAHLEH, M.A. AND Y. OHTA, *A Necessary and Sufficient Condition for Robust BIBO Stability*, Systems & Control Letters 11 (1988), pp. 271–275.

[10] KHAMMASH, M. AND J.B. PEARSON, *Performance Robustness of Discrete-Time Systems with Structured Uncertainty*, IEEE Transactions on Automatic Control AC-36, no. 4 (1991), pp. 398–412.

[11] KHAMMASH, M., AND J.B. PEARSON, *Analysis and Design for Robust Performance with Structured Uncertainty*, To appear in Systems and Control Letters, Vol. 20, Issue 1.

[12] KHAMMASH, M., *Necessary and Sufficient Conditions for the Robustness of Time-Varying Systems with Applications to Sampled-Data Systems*, Submitted to the IEEE Transactions on Automatic Control, To appear in January 1993.

[13] KLEIN, M., G.J. ROGERS, AND P. KUNDUR, *A Fundamental Study of Interarea Oscillations in Power Systems*, Paper no. 91WM015-8-PWRS, IEEE Winter Power Meeting (Feb. 1991).

[14] BERMAN, A, AND R. BLEMMONS, *Nonnegative Matrices in the Mathematical Sciences*, Academic Press, New York, 1979.

[15] ANDERSON, P.M., AND A.A. FOUAD, *Power System Stability and Control*, Ames, Iowa: Iowa State University Press, 1977.

[16] KUNDUR, P., G.J. ROGERS, AND D.Y. WONG, *Extended Transient-Midterm Stability Program Package: Version 2.0, Users Manuals* EPRI EL 6648 (December 1989).

[17] KUNDUR, P., G.J. ROGERS, AND D.Y. WONG, *The Small Signal Stability Program Package 1*, EPRI Final Report, EL-5798 (May 1988).